Advanced Chemistry

in

Creation

by Dr. Jay L. Wile

Advanced Chemistry in Creation

Manufactured in the United States of America
Third Printing 2003

Published By
Apologia Educational Ministries, Inc.
Anderson, IN

Printed by
The C.J. Krehbiel Company
Cincinnati, OH

Cover Photo: Electrostatic Fields around benzene

Need Help?

Apologia Educational Ministries Curriculum Support

If you have any questions while using Apologia curriculum, feel free to contact us in any of the following ways:

By Mail: Curriculum Help
Apologia Educational Ministries, Inc.
1106 Meridian Plaza, Suite 220
Anderson, IN 46016

By E-mail: help@highschoolscience.com

On the Web: http://www.highschoolscience.com

By FAX: (765) 608 - 3290

By Phone: (765) 608 - 3280

STUDENT NOTES

Advanced Chemistry in Creation

You are about to embark upon an amazing journey! In this text, you will take a deep look at the fascinating subject of chemistry. You will learn about the matter that makes up God's creation and how it changes. Although the course will be hard work, you will learn some truly amazing things. Hopefully, these things will help you develop an even deeper appreciation for the wonderful creation that God has given us!

I hope that you enjoy taking this course as much as I have enjoyed writing it.

Pedagogy of the Text

This text contains 16 modules. Each module should take you about 2 weeks to complete, as long as you devote 45 minutes to one hour of every school day to studying chemistry. At this pace, you will complete the course in 32 weeks. Since most people have school years which are longer than 32 weeks, there is some built-in "flex time." You should not rush through a module just to make sure that you complete it in 2 weeks. Set that as a goal, but be flexible. Some of the modules might come harder to you than others. On those modules, take more time on the subject matter.

To help you guide your study, there are several student exercises which you should complete.

- The "on your own" problems should be solved as you read the text. The act of working out these problems will cement in your mind the concepts you are trying to learn. Complete solutions to these problems appear at the end of the module. Once you have solved an "on your own" problem, turn to the back of the module and check your work. If you did not get the correct answer, study the solution to learn why.

- The review questions are conceptual in nature and should be answered after you have completed the entire module. They will help you recall the important concepts from the reading.

- The practice problems should also be solved after the module has been completed, allowing you to review the important quantitative skills from the module.

Your teacher/parent has the solutions to the review questions and practice problems.

Any information that you must memorize is centered in the text and put in boldface type. In addition, all definitions presented in the text need to be memorized. Words that appear in bold-face type (centered or not) in the text are important terms that you should know. Finally, if any student exercise requires the use of a formula or skill, you must have that memorized for the test.

Experiments

The experiments in this course are designed to be done as you are reading the text. I recommend that you keep a notebook of these experiments. This notebook serves two purposes. First, as you write about the experiment in the notebook, you will be forced to think through all of the concepts that were explored in the experiment. This will help you cement them into your mind. Second, certain colleges might actually ask for some evidence that you did, indeed, have a laboratory component to your chemistry course. The notebook will not only provide such evidence but will also show the college administrator the quality of your chemistry instruction. I recommend that you perform your experiments in the following way:

- When you get to the experiment during the reading, read through the experiment in its entirety. This will allow you to gain a quick understanding of what you are to do.

- Once you have read the experiment, start a new page in your laboratory notebook. The first page should be used to write down all of the data taken during the experiment and perform any calculation explained in the experiment.

- When you have finished the experiment, write a brief report in your notebook, right after the page where the data and calculations were written. The report should be a brief discussion of what was done and what was learned. You should not write a step-by-step procedure. Instead, write a brief summary that will allow someone who has never read the text to understand what you did and what you learned.

- **PLEASE OBSERVE COMMON SENSE SAFETY PRECAUTIONS. The experiments are no more dangerous than most normal, household activity. Remember, however, that the vast majority of accidents do happen in the home. Chemicals should never be ingested; hot beakers and flames should be regarded with care; and OSHA recommends that all chemistry experiments be performed while wearing some sort of eye protection such as safety glasses or goggles.**

Advanced Chemistry in Creation

Table of Contents

Module 5: Solutions and Colloids

Module 6: Solutions and Equilibrium

Module 7: Acid/Base Equilibria

Module 8: More on Equilibrium

Module 9: Electrochemistry - Part 1

Module 10: Electrochemistry - Part 2

Module 11: Chemical Kinetics

Module 12: An Introduction To Organic Chemistry

Module 13: Functional Groups in Organic Chemistry

Module 14: Nuclear Chemistry

Module 15: Review - Part 1

Module 16: Review - Part 2

Module #1: Units, Chemical Equations, and Stoichiometry Revisited

Introduction

There are probably no concepts more important in chemistry than the three listed in the title of this module. In your first-year chemistry course, I am sure that you learned quite a lot about each of these concepts. You certainly did not learn everything, however. Whether we are talking about units, chemicals equations, or stoichiometry, there is simply too much information to possibly learn in just one year. As a result, we will take another look at each of these concepts in this first module. This will help you "warm up" to the task of recalling all of the things you learned in your first year chemistry course, and it will help to learn each of these valuable concepts at a much deeper level.

Units Revisited

Almost regardless of the chemistry course, units are always covered first, because a great deal of chemistry is based on properly analyzing units. In your first year course, you were taught how to solve problems such as the one in the following example:

EXAMPLE 1.1

A sample of iron has a mass of 254.1 mg. How many kg is that?

In this problem, we are asked to convert from milligrams to kilograms. We cannot do this directly, because we have no relationship between mg and kg. However, we do know that a milligram is the same thing as 0.001 grams and that a kilogram is the same thing as 1,000 grams. Thus, we can convert mg into g, and then convert g into kilograms. To save space, we can do that all on one line:

$$\frac{254.1\ \cancel{\text{mg}}}{1} \times \frac{0.001\ \cancel{\text{g}}}{1\ \cancel{\text{mg}}} \times \frac{1\ \text{kg}}{1{,}000\ \cancel{\text{g}}} = 0.0002541\ \text{kg} = 2.541 \times 10^{-4}\ \text{kg}$$

The sample of iron has a mass of <u>2.541×10^{-4} kg</u>.

Did this example help dust the cobwebs out of your mind when it comes to units? This should all be review for you. I converted the units using the factor-label method. Because this is a conversion, I had to have the same number of significant figures as I had in the beginning, and even though it was not necessary, I reported the answer in scientific notation. If you are having trouble remembering these techniques, then go back to your first-year chemistry book and review them.

There are a couple of additional things I want you to learn about units now. I am not going to show you any new techniques; I am just going to show you new ways of applying the techniques that you should already know. Do you remember the concept of molarity? Molarity is a concentration unit that you learned in first year chemistry. It is expressed in moles per liter. There are other ways of expressing concentration, however. For example, you can express concentration in grams per milliliter instead. Well, if you have concentration in one unit, you should be able to convert it to another unit, right? Study the next example.

EXAMPLE 1.2

The concentration of HCl in a solution of muriatic acid is about 0.35 g per mL. What is the molarity of the HCl in muriatic acid?

Now remember, molarity is a concentration unit, just like grams per milliliter, so all we need to do is make a conversion. Hopefully, you remember how to convert grams into moles. You find the mass of HCl from the periodic chart and realize that the mass tells you the number of grams it takes to make a mole of HCl. The mass of HCl is 36.5 amu, so this tells us that it takes 36.5 grams of HCl to make one mole. We also know that a mL is the same as 0.001 L. Now that we know both of the relationships between what we have and what we know, we can set up the conversion:

$$\frac{0.35\text{ g}}{1\text{ mL}} \times \frac{1\text{ mole}}{36.5\text{ g}} \times \frac{\text{mL}}{0.001\text{ L}} = 9.6\,\frac{\text{mole}}{\text{L}} = 9.6\text{ M}$$

Although there is nothing new here, you probably haven't seen a conversion done in this way. Despite the fact that the unit is a derived unit (g/mL), you can still do conversions on it. I could have just converted grams to moles and gotten the unit moles/mL. I could have just converted mL to L and gotten g/L. In this case I did both. When working with derived units, remember that you can convert any unit that makes up the derived unit. Thus, 0.35 g/mL is the same thing as 9.6 M.

Okay, we are almost done reviewing units. There is just one more thing that you need to remember. Sometimes, units have exponents in them. You were probably taught how to deal with this fact in your first year chemistry course, but we need to review it so that you *really* know how to deal with it.

EXAMPLE 1.3

One commonly-used unit for volume is the cubic meter. After all, length is measured in meters, and volume is length times width times height. The more familiar unit, however, is cubic centimeters (cc) which is often used in medicine. If a doctor administers 512 cc of medicine to a patient, how many cubic meters is that?

Once again, this is a simple conversion. If, however, you do not think as you go through it, you can mess yourself up. We need to convert cubic centimeters to cubic meters. Now remember, a cubic centimeter is just a cm^3 and a cubic meter is just a m^3. Now we have no relationship between these units, but we do know that 1cm = 0.01 m. That's all we need to know, as long as we think about it. Right now, I have the following relationship:

$$1 \text{ cm} = 0.01 \text{ m}$$

This is an equation. I am allowed to do something to one side of the equation as long as I do the exact same thing to the other side of the equation. Okay, then, let's cube both sides of the equation:

$$(1 \text{ cm})^3 = (0.01 \text{ m})^3$$

$$1 \text{ cm}^3 = 0.000001 \text{ m}^3$$

Now look what we have. We have a relationship between cm^3 and m^3, exactly what we need to convert between!

$$\frac{512 \cancel{\text{cm}^3}}{1} \times \frac{0.000001 \text{ m}^3}{1 \cancel{\text{cm}^3}} = 5.12 \times 10^{-4} \text{ m}^3$$

So 512 cc's is the same as <u>5.12×10^{-4} m^3</u>.

Now when most students do a conversion like the example without thinking, they simply use the relationship between cm and m to do the conversion. That, of course does not work, because the cm^3 unit does not cancel out, and you certainly don't get the m^3 unit in the end:

$$\frac{512 \text{ cm}^3}{1} \times \frac{0.01 \text{ m}}{1 \cancel{\text{cm}}} = 5.12 \text{ m} \cdot \text{cm}^2$$

Do you see what happened? The cm unit canceled one of the cm out of cm^3, but that still left cm^2. Also, since m is the unit that survives from the conversion relationship, you get the weird unit of $m \cdot cm^2$! When you are working with units that have exponents in them, you need to be very careful about how you convert them. Make sure you are using a conversion relationship that will definitely give you the unit you want in the end.

ON YOUR OWN

1.1 The speed limit on many highways in the United States is 65 miles per hour. What is the speed limit in meters per second? (There are 1609 meters in a mile.)

1.2 The instantaneous rate of disappearance of NaOH in a chemical reaction is 1.02 $\frac{\text{moles}}{\text{L} \cdot \text{s}}$. What is the rate in $\frac{\text{grams}}{\text{mL} \cdot \text{hr}}$?

1.3 The size of a house is 1600 square feet. What is the square yardage of the house?

1.4 In physics, acceleration is expressed in the unit m/s^2. If a car accelerates at $1.1\ m/s^2$, what is the acceleration in km/hr^2?

A New Look at Chemical Equations

In your first year chemistry course, you learned how to balance chemical equations and use them in stoichiometry. When you were balancing equations, you probably learned that chemical equations can be treated, in many ways, the same as you treat algebraic equations. For example, as you learned to balance chemical equations, you were taught that you can multiply both sides of a chemical equation by any number, just as if it were an algebraic equation. If you have the following chemical equation:

$$N_2\ (g) + 3H_2\ (g) \rightarrow 2NH_3\ (g) \qquad (1)$$

You can multiply this equation by 3 to get:

$$3N_2\ (g) + 9H_2\ (g) \rightarrow 6NH_3\ (g) \qquad (2)$$

Both of these are legitimate, balanced chemical equations. Why in the world would you want to do such a thing? Well, you'll find out in a moment.

Before you find out about that, however, I need to point out something that you probably didn't realize when you learned how to multiply equations by numbers. You probably didn't realize that the same property which allows you to multiply chemical equations by numbers also allows you to *add chemical equations together*. For example, when you burn carbon in a limited amount of oxygen, the following chemical reaction occurs:

$$2C\ (s) + O_2\ (g) \rightarrow 2CO\ (g) \qquad (3)$$

If, however, more oxygen becomes available, the carbon monoxide formed in Equation (3) can further be burned to make carbon dioxide, according to this chemical equation:

$$2CO\ (g) + O_2\ (g) \rightarrow 2CO_2\ (g) \qquad (4)$$

What is the overall chemical reaction, then? The carbon first burns to make carbon monoxide, but then the carbon monoxide burns to form carbon dioxide. What is the overall reaction? Well, we can *add Equations (3) and (4) to get that overall reaction*:

$$\begin{array}{l} \quad 2C\ (s) + O_2\ (g) \rightarrow 2CO\ (g) \\ \underline{+\ 2CO\ (g) + O_2\ (g) \rightarrow 2CO_2\ (g)} \\ \quad 2C\ (s) + 2CO\ (g) + O_2\ (g) + O_2\ (g) \rightarrow 2CO\ (g) + 2CO_2\ (g) \end{array} \qquad (5)$$

Is Equation (5) the overall reaction, then? Not quite. Just like in algebraic equations, you can *cancel equivalent terms* from each side of the equation. Notice, for example, there is a "2CO (g)" on both sides of the equation. Thus, we can cancel them:

$$2C\ (s) + \cancel{2CO\ (g)} + O_2\ (g) + O_2\ (g) \rightarrow \cancel{2CO\ (g)} + 2CO_2\ (g) \qquad (6)$$

This leaves us with:

$$2C\ (s) + O_2\ (g) + O_2\ (g) \rightarrow 2CO_2\ (g) \qquad (7)$$

Is that the overall equation, then? Once again, not quite. Notice that there are 2 "O_2 (g)" terms on the left-hand side of the equation. Just like in algebraic equations, you can group like terms in chemical equations. This leaves us with:

$$2C\ (s) + 2O_2\ (g) \rightarrow 2CO_2\ (g) \qquad (8)$$

In principle, we are done, because there are no longer any equivalent terms to group or cancel. However, most chemists like to keep the stoichiometric coefficients as simple as possible. Since all of the coefficients in Equation (8) are divisible by two, we will also divide both sides of the equation by 2 to get:

$$C\ (s) + O_2\ (g) \rightarrow CO_2\ (g) \qquad (9)$$

Finally, we see that Equations (3) and (4) add up to this very simple equation.

You should see, then, that sometimes complicated chemical processes can be simplified by adding chemical equations. Study the following example which illustrates this process again.

EXAMPLE 1.4

The destruction of ozone by chlorine atoms from chlorofluorohydrocarbons (CFCs) proceeds as follows: First, a chlorine atom reacts with ozone:

$$\mathbf{Cl + O_3 \rightarrow ClO + O_2}$$

The compound ClO, however, is not stable, and it immediately reacts with free oxygen atoms in the air:

$$ClO + O \rightarrow Cl + O_2$$

What is the overall chemical reaction for the destruction of ozone?

To find the overall reaction, we can add these two reactions up:

$$\begin{array}{l} Cl + O_3 \rightarrow ClO + O_2 \\ \underline{+\, ClO + O \rightarrow Cl + O_2} \\ Cl + O_3 + ClO + O \rightarrow ClO + O_2 + Cl + O_2 \end{array}$$

Notice there are two ClO terms on each side of the equation. They can therefore be canceled. The same goes for the Cl terms on each side.

$$\cancel{Cl} + O_3 + \cancel{ClO} + O \rightarrow \cancel{ClO} + O_2 + \cancel{Cl} + O_2$$

This gives us the following equation:

$$O_3 + O \rightarrow O_2 + O_2$$

To finish up, we can group the two O_2 terms on the right side of the equation:

$$\underline{O_3 + O \rightarrow 2O_2}$$

So once again, the final equation is much simpler than the previous equations which describe the process.

Now I want you to notice something about the example you just studied. Notice that while Cl is a reactant in the first equation, it is not a reactant in the overall equation. Why? Well, although Cl is used up in the first equation, it is produced in the second equation. As a result, the amount of Cl present never changes. Thus, when you look at the overall reaction, it looks like Cl just isn't a part of the reaction. Now you should remember from your first year course that we give a name to substances that take part in chemical reactions but do not get used up or produced. We call them **catalysts**. In this case, Cl is acting like a catalyst. When someone shows you the way ozone is destroyed on a step-by-step basis, you see Cl being both used and then later produced. However, when you view the overall reaction, you lose all information regarding Cl.

You should also remember from your first year chemistry course that when a chemical reaction is presented in a step-by-step manner, we call that a **reaction mechanism**. In chemical terminology, then, the first two equations I presented in the example are the reaction mechanism for the overall reaction that we determined by the end of the example. Which way is best to show the reaction? That depends on what you are looking for. The reaction mechanism is a detailed description of the reaction process, but it is so detailed that you might miss the overall effect of the chemical reaction. The overall equation gives you a good view of what happens, but you lose detail. Make sure that you can add chemical equations together by performing the following "on your own" problem.

ON YOUR OWN

1.5 When nitrogen dioxide gas is mixed with carbon monoxide gas, two reactions occur. In the first reaction, two NO_2 molecules react as follows:

$$2NO_2\ (g) \rightarrow NO_3\ (g) + NO\ (g)$$

The nitrogen trioxide formed in that reaction then reacts with the carbon monoxide present as follows:

$$NO_3\ (g) + CO\ (g) \rightarrow NO_2\ (g) + CO_2\ (g)$$

What is the overall chemical reaction that occurs?

A New Look At Hess's Law

Okay, so now we can add chemical equations together. What's the big deal? Well, do you remember Hess's Law? You learned it in your first year course as a way to calculate the **change in enthalpy** of a chemical reaction. Remember, the change in enthalpy (abbreviated as ΔH) tells us how much energy is released into the surroundings (when ΔH is negative) or absorbed by the reaction (when ΔH is positive). You should have learned Hess's Law as:

$$\Delta H = \Sigma \Delta H_f\,(\text{products}) - \Sigma \Delta H_f\,(\text{reactants}) \qquad (10)$$

Where ΔH is the change of enthalpy of the reaction, and ΔH_f is the enthalpy of formation of the reactants and the products. When using Equation (10), you take the ΔH_f for each product (you look them up in a table), multiply by the stoichiometric coefficient of the product, and subtract from that the sum of the ΔH_f's of each reactant times its stoichiometric coefficient. When you are done, you have the ΔH of the chemical reaction.

Well, it turns out that this formulation of Hess's Law is really only a special application of Hess's Law. Remember, Hess's Law is stated as follows:

Hess's Law - Energy is a state function and is therefore independent of path.

In your first year course, you should have learned that Equation (10) is a consequence of this statement. However, the statement itself tells us something much more dramatic. Hess's Law tells us that the ΔH of a chemical reaction is independent of *how* the reaction takes place. In other words, suppose I looked at the overall chemical equation for the destruction of ozone:

$$O_3 + O \rightarrow 2O_2 \quad (11)$$

Hess's Law states *regardless of how this reaction occurs, the ΔH is the same.*

What do I mean by this? Well, as you saw in the previous section, chlorine that comes from CFCs is a catalyst in the destruction of ozone. There are, however, many other catalysts from ozone destruction, including hydrogen atoms. Here's how a hydrogen atom catalyzes the decomposition of ozone:

$$H + O_3 \rightarrow OH + O_2 \quad (12)$$
$$OH + O \rightarrow H + O_2 \quad (13)$$

You can verify for yourself that the overall chemical reaction produced in these two steps is precisely the same as Equation (11), which is the overall chemical equation produced by the chlorine-catalyzed decomposition sequence shown in the previous section.

Hess's Law states that whether ozone is destroyed with a chlorine catalyst or a hydrogen catalyst, the overall change in enthalpy will be the same, because the overall chemical reaction is the same. Thus, it does not matter *how* the reaction occurs. The ΔH will always be the same, as long as the overall chemical reaction is the same.

How is this possible? Think about it. Each chemical equation has its own ΔH. Thus, Equation (12) and Equation (13) each have their own ΔH. Similarly, each step of the reaction mechanism presented in Example 1.4 also has its own ΔH. The energetics work out so that when you add the ΔH's of each individual step in a reaction mechanism (just like you added the equations), the total ΔH will always be the same as long as the final reaction is the same!

If all of that is just a little over your head, study the next example to see how it works.

EXAMPLE 1.5

When solid iron and chlorine gas are reacted together, the product is solid $FeCl_3$. This reaction (most likely) proceeds in two steps:

$$\textbf{Fe (s) + Cl}_2\textbf{ (g)} \rightarrow \textbf{FeCl}_2\textbf{ (s)} \qquad \Delta H = -341.8 \text{ kJ}$$

$$\textbf{FeCl}_2\textbf{ (s)} + \frac{1}{2}\textbf{Cl}_2\textbf{ (g)} \rightarrow \textbf{FeCl}_3\textbf{ (s)} \qquad \Delta H = -57.7 \text{ kJ}$$

What is the overall reaction and its ΔH?

To get the overall reaction, we add these two reactions together:

$$\begin{array}{l} Fe\ (s) + Cl_2\ (g) \rightarrow FeCl_2\ (s) \\ + FeCl_2\ (s) + \frac{1}{2}Cl_2\ (g) \rightarrow FeCl_3\ (s) \\ \hline Fe\ (s) + Cl_2\ (g) + FeCl_2\ (s) + \frac{1}{2}Cl_2\ (g) \rightarrow FeCl_2\ (s) + FeCl_3\ (g) \end{array}$$

Once we cancel the $FeCl_2$ (s) on each side of the equation and group the Cl_2 (g) terms on the left-hand side of the equation together, we get the overall reaction:

$$Fe\ (s) + \frac{3}{2}Cl_2\ (g) \rightarrow FeCl_3\ (s)$$

Now don't be concerned about the fraction in the equation. Although there is no such thing as 3/2 of a Cl_2 molecule, there is no real problem with the fraction in the equation. If we wanted to, we could multiply the equation by 2 to get rid of the fraction, but as we will see in a moment, that can lead to some complications with Hess's Law.

Well, now that we figured out the overall equation, how do we get the ΔH? Since we just added the two equations together to get the overall equation, we can just add the two ΔH's together to get the overall ΔH.

$$\Delta H = -341.8\ kJ + -57.7\ kJ = -399.5\ kJ$$

The overall ΔH, then, is -399.5 kJ.

So Hess's Law states that if we can add two equations together to get an overall chemical equation, we can add the ΔH's to get the overall ΔH. Now notice in the example that I said the reaction mechanism presented is *probably* the way that iron and chlorine gas react. Well, if we're not correct and the reaction mechanism is different, does that mean that the overall ΔH we just calculated is wrong? NO! That's the beauty of Hess's Law. Since enthalpy is a state function and is independent of path, it doesn't matter whether or not the reaction actually occurs that way! If it occurs that way, or if it occurs another way, the overall ΔH will be the same!

Think, for a moment, about the implications here. Whether or not we know *how* a chemical reaction works, if we can add other equations together to get the overall equation, we can calculate the ΔH! So, if we know the ΔH of a few equations and we can add up those equations to get an overall chemical equation, we can add the individual ΔH's to get the overall equation's ΔH.

Okay, you might be thinking, that doesn't sound so bad. Well, in principle, it's not. However, things can get really tricky when you start realizing that since we can multiply equations by a number as well as add them together, we can find ways to construct *a lot* of overall chemical equations. Study the following example to see what I mean.

EXAMPLE 1.6

Given the following information:

$$2Mn\ (s) + O_2\ (g) \rightarrow 2MnO\ (s) \qquad \Delta H = -770.4\ kJ$$
$$2MnO_2\ (s) \rightarrow O_2\ (g) + 2MnO\ (s) \qquad \Delta H = 269.7\ kJ$$

Calculate the ΔH for

$$Mn\ (s) + O_2\ (g) \rightarrow MnO_2\ (s)$$

Now that we know Hess's Law allows us to add up chemical equations and their corresponding ΔH's all we need to do is figure out how to get the first two equations to add up to the third. Now if you look at the first two equations, it is clear that we cannot just add them together. For example, our final equation has MnO_2 (s) as a product, but neither of the first two reactions have MnO_2 (s) as a product. If we just add the two equations together, then, there will not be an MnO_2 (s) on the products side. Are we out of luck? Not at all.

You should remember from your first year chemistry course that *any* chemical equation can be written backwards, with the products as reactants and reactants as products. The second equation written backwards is as follows:

$$O_2\ (g) + 2MnO\ (s) \rightarrow 2MnO_2\ (s)$$

We now have MnO_2 (s) as a product. Wait a minute, though, if we write the equation backwards, what happens to ΔH? Well, the ΔH given above indicates that when the reaction runs as written, the reaction absorbs 269.7 kJ of energy. Well, if we write the equation backwards, then we are talking about the *reverse* of the reaction. If the reaction as written *absorbs* 269.7 kJ of energy, then the reverse reaction will *release* 269.7 kJ of energy. How do we denote that? When a reaction releases energy, its ΔH is negative. Thus, for the reaction as written above, $\Delta H = -269.7$ kJ.

Okay, so now we have two reactions that we can add up to get the third, right? Not quite. After all, if we add this equation to the very first one we were given, then we would have $2MnO_2$ (s) as a product. What we are looking for is an equation that has MnO_2 (s) as a product *without* the 2. Are we out of luck, then? Not at all. Remember, we can multiply chemical equations by numbers, so let's multiply the equation above by one-half:

$$\tfrac{1}{2}O_2\ (g) + MnO\ (s) \rightarrow MnO_2\ (s)$$

Now we have MnO_2 (s) as a product, just as it is in the equation we want to end up with. What about the ΔH, however? Is the ΔH the same? No. Remember, the ΔH tells us about how much energy is released or absorbed in a reaction. If I multiply the equation by 1/2, then only half as much energy will be released. Thus, I need to multiply ΔH by 1/2 as well, making the new ΔH = -134.9 kJ.

Are we ready to add the equations, then? Not quite. Look at the first equation. It has 2Mn (s) as a reactant. The equation we want, however, has only one Mn (s) as a reactant. Thus, we need to multiply that equation (and its ΔH) by one half:

$$Mn\ (s) + \tfrac{1}{2}O_2\ (g) \rightarrow MnO\ (s) \qquad \Delta H = -385.2\ kJ$$

Now we can finally add the equations together, along with their ΔH's:

$$Mn\ (s) + \tfrac{1}{2}O_2\ (g) \rightarrow MnO\ (s) \qquad \Delta H = -385.2\ kJ$$
$$+ \tfrac{1}{2}O_2\ (g) + MnO\ (s) \rightarrow MnO_2\ (s) \qquad \Delta H = -134.9\ kJ$$

$$Mn\ (s) + \tfrac{1}{2}O_2\ (g) + \tfrac{1}{2}O_2\ (g) + \cancel{MnO\ (s)} \rightarrow \cancel{MnO\ (s)} + MnO_2\ (s) \qquad \Delta H = -520.1\ kJ$$

Once we cancel the MnO (s) on each side of the equation and group the O_2 (g) terms on the reactants side of the equation, we get:

$$Mn\ (s) + O_2\ (g) \rightarrow MnO_2\ (s)$$

This is exactly the equation for which we wanted to calculate the ΔH. Thus, ΔH = -520.1 kJ.

Do you see what happened in the example? Hess's Law says that we can add *any* equations together and, when we do that, the ΔH of the overall reaction will be the sum of the ΔH's for the individual reactions that we added together. So, in order to calculate the ΔH of the reaction, I just kept manipulating the equations until they could add together to make the chemical equation for which I wanted to calculate the ΔH. If I reversed the chemical equation, I changed the sign on the ΔH. If I multiplied the equation by a number, then I multiplied the ΔH by the same number. In the end, when the equations could add up to the equation in which I was interested, the ΔH's would add to the ΔH of the reaction in which I was interested.

Let's see how that happens in another example. As you study the example, keep in mind the following statements:

When you reverse a chemical equation, the ΔH of the reaction changes sign.

When you multiply a chemical equation by a number, you must multiply the ΔH by that same number.

EXAMPLE 1.7

Calculate the ΔH of the following reaction:

$$Zn\ (s) + S\ (s) + 2O_2\ (g) \rightarrow ZnSO_4\ (s)$$

from the following information:

$$ZnS\ (s) \rightarrow Zn\ (s) + S\ (s) \qquad \Delta H = 206.0\ kJ$$
$$\tfrac{1}{2}ZnS\ (s) + O_2\ (g) \rightarrow \tfrac{1}{2}ZnSO_4\ (s) \qquad \Delta H = -388.4\ kJ$$

To solve this problem, we just need to manipulate these two equations until they add to the one in which we are interested. To start out, the equation that we are interested in has Zn (s) and S (s) as reactants. Of the two equations we are given, only the first one has these two elements in it. Unfortunately, they are on the wrong side of the equation. We need them as reactants, not products. So, we need to reverse the chemical equation. When we do that, we need to change the sign on the ΔH:

$$Zn\ (s) + S\ (s) \rightarrow ZnS\ (s) \qquad \Delta H = -206.0\ kJ$$

In the second equation we are given, the $ZnSO_4$ (s) is on the products side, as it is in our equation of interest. In addition, the O_2 (g) is on the reactants side, as it is in our equation of interest. Thus, we need not reverse the reaction here. However, notice that we need one $ZnSO_4$ (s) as a product, but the equation has only 1/2 of that substance. Therefore, we need to multiply the second equation by 2. When we do that, we need to multiply the ΔH by 2 as well.

$$ZnS\ (s) + 2O_2\ (g) \rightarrow ZnSO_4\ (s) \qquad \Delta H = -776.8\ kJ$$

Now we have 2 equations that have everything we need all in the right places. Thus, we just need to add them up. If they add to the equation of interest, then the sum of their ΔH's will add to the ΔH of the equation of interest.

$$Zn\ (s) + S\ (s) \rightarrow ZnS\ (s) \qquad \Delta H = -206.0\ kJ$$
$$+\ ZnS\ (s) + 2O_2\ (g) \rightarrow ZnSO_4\ (s) \qquad \Delta H = -776.8\ kJ$$

$$Zn\ (s) + S\ (s) + \cancel{ZnS\ (s)} + 2O_2\ (s) \rightarrow \cancel{ZnS\ (s)} + ZnSO_4\ (s) \qquad \Delta H = -982.8\ kJ$$

Once we cancel the ZnS (s) from both sides of the equation, we have the chemical equation we originally wanted, so we also have the ΔH:

$$Zn\ (s) + S\ (s) + 2O_2\ (s) \rightarrow ZnSO_4\ (s)$$

Which means <u>ΔH = -982.8 kJ</u>.

So when we use Hess's Law, it sometimes works like a puzzle. If you have a few equations with known ΔH's, then you can manipulate them to get the equation in which your are interested. Once you do that, you can get the ΔH of that equation. If you noticed how I went about attacking the problem, you will see that I look at each equation individually and decide first whether or not it has the substances I want on the correct side of the equation. If not, I reversed the equation and changed the sign of the ΔH. Next, I looked at the stoichiometric coefficients. If the equation did not have the proper stoichiometric coefficient, then I multiplied the equation as well as the ΔH.

Now I want you to notice something about the example problem. Think back to the classifications that you learned for chemical reactions in your first year chemistry course. How would you classify the reaction whose ΔH you calculated? It's a formation reaction, because it forms $ZnSO_4$ from its constituent elements. Now if you remember, when you used Hess's Law in your first year chemistry course, the ΔH's you looked up in tables were ΔH's of formation. Did you ever wonder where those numbers came from? Well, some came directly from experiment, because it is possible to run some formation reactions in a calorimeter. Other formation reactions, however, do not lend themselves to measurements of the ΔH. To measure the ΔH of those formation reactions, chemists measure the ΔH's of related chemical reactions and then perform calculations like the ones in Examples 1.6 and 1.7.

ON YOUR OWN

1.6 Calculate the ΔH of formation for Co_3O_4 (s) given the following information:

$$CoO\ (s) \rightarrow Co\ (s) + \tfrac{1}{2}O_2\ (g) \qquad \Delta H = 237.9\ kJ$$
$$6CoO\ (s) + O_2\ (g) \rightarrow 2Co_3O_4\ (s) \qquad \Delta H = -355.0\ kJ$$

1.7 Calculate the ΔH for the combustion of carbon:

$$C\ (s) + O_2\ (g) \rightarrow CO_2\ (g)$$

Given that

$$C\ (s) + \tfrac{1}{2}O_2\ (g) \rightarrow CO\ (g) \qquad \Delta H = -111\ kJ$$
$$2CO_2\ (g) \rightarrow 2CO\ (g) + O_2\ (g) \qquad \Delta H = 566\ kJ$$

Stoichiometry and Limiting Reagents

The subject of stoichiometry is probably the most fundamental aspect of chemistry. Although you might think you learned all there is to know about stoichiometry, there is still plenty more to learn. Before we hit that, however, I want to go through one quick example of stoichiometry in an attempt to further dust the cobwebs from your mind.

EXAMPLE 1.8

When hydrogen and oxygen are ignited, a violent reaction occurs, releasing a huge amount of energy, usually in the form of an explosion. This reaction, which produces water, is responsible for the 1937 explosion that destroyed the famous dirigible known as the Hindenburg. If 10.0 grams of oxygen are ignited in an excess of hydrogen, how many grams of water will be formed?

In this problem, we are given the number of grams of oxygen and are also told that oxygen is the limiting reagent because there is an excess of hydrogen. Thus, the amount of products produced will be dependent only on the amount of oxygen in the reaction. Well, we are told (in words) what the reaction is, so we start there:

$$H_2\ (g) + O_2\ (g) \rightarrow H_2O\ (g)$$

The equation is not balanced, so we have to do that first:

$$2H_2\ (g) + O_2\ (g) \rightarrow 2H_2O\ (g)$$

Now we know how oxygen relates to water in this reaction. For every one mole of oxygen gas, 2 moles of water are formed. Of course, we do not have the number of moles of oxygen, we have grams. Thus, to be able to use the chemical equation at all, we must first convert to moles:

$$\frac{10.0\cancel{g\,O_2}}{1} \times \frac{1\text{ mole } O_2}{32.0\ \cancel{g\,O_2}} = 0.313\text{ moles } O_2$$

Now that we have moles of oxygen, we can use the information in the chemical equation's stoichiometric coefficients to convert from moles of oxygen to moles of water:

$$\frac{0.313\ \cancel{\text{moles } O_2}}{1} \times \frac{2\text{ moles } H_2O}{1\ \cancel{\text{mole } O_2}} = 0.626\text{ moles } H_2O$$

That tells us the answer in moles, but the nasty guy who wrote the problem wants the answer in grams, so we have one more conversion to go:

$$\frac{0.626\ \cancel{\text{moles } H_2O}}{1} \times \frac{18.0\text{ g } H_2O}{1\ \cancel{\text{mole } H_2O}} = 11.3\text{ g } H_2O$$

The reaction produces <u>11.3 g H_2O</u>.

Does that example ring a few bells? It should! You did those kinds of exercises over and over again in your first year chemistry course. There were a couple of things you probably didn't cover, however. Notice that in the example, you were told what the limiting reagent was because you were told that the other reagent (hydrogen) was "in excess." Suppose you weren't told that. Suppose you were just told how much of each reagent was in the experiment. Is there any way that you can determine what the limiting reagent is on your own? Study the following example to find out.

EXAMPLE 1.9

PCl_5 is an important ingredient in the making of certain insecticides. It can be made by the following reaction:

$$P_4\,(s) + 10Cl_2\,(g) \rightarrow 4PCl_5$$

How many grams of PCl_5 can be made from 10.0 g of P_4 and 30.0 g of Cl_2?

In this problem, we are not told what reagent is in excess and what one is the limiting reagent. We will have to determine that for ourselves. How can we do this? Well, the chemical equation says that for every 1 mole of P_4, there must be 10 moles of Cl_2. Let's convert to moles to see if that is really the relationship between moles of P_4 and moles of Cl_2.

$$\frac{10.0\,\text{g}\,\cancel{P_4}}{1} \times \frac{1\text{ mole }P_4}{124\text{ g}\,\cancel{P_4}} = 0.0806\text{ moles }P_4$$

$$\frac{30.0\,\text{g}\,\cancel{Cl_2}}{1} \times \frac{1\text{ mole }Cl_2}{71.0\,\text{g}\,\cancel{Cl_2}} = 0.423\text{ moles }Cl_2$$

Looking at those two numbers, we see that there are many, many more moles of Cl_2 than of P_4. Does this mean that P_4 is the limiting reagent? No! We have to look at the chemical equation to do that. The chemical equation tells us that for every one mole of P_4, there must be 10 moles of Cl_2. Well, since we have 0.0806 moles of P_4, there need to be 0.806 moles of Cl_2. Are there that many moles of Cl_2? No, there are only 0.423 moles of Cl_2. What does that tell us? Well, it tells us that there isn't enough Cl_2 to take care of the P_4 that we have. This means that Cl_2 will run out before P_4 does, making Cl_2 the limiting reagent!

Now that we know Cl_2 is the limiting reagent, we know that all products are limited by the amount of Cl_2. Thus, the amount of P_4 is irrelevant, and we can continue on in our calculation using just the number of moles of Cl_2. We can use the information in the chemical equation's stoichiometric coefficients to convert from moles of Cl_2 to moles of PCl_5:

$$\frac{0.423\,\cancel{\text{moles }Cl_2}}{1} \times \frac{4\text{ moles }PCl_5}{10\,\cancel{\text{moles }Cl_2}} = 0.169\text{ moles}PCl_5$$

That tells us the answer in moles, but we need the answer in grams, so we have one more conversion to go:

$$\frac{0.169\ \cancel{moles\,PCl_5}}{1} \times \frac{208.5\ g\ PCl_2}{1\ \cancel{mole\,PCl_5}} = 35.2\ g\ PCl_2$$

The reaction produces 35.2 g PCl_5.

Do you see what I did to determine what the limiting reagent was? I looked at the chemical equation and determined how many moles of one reactant it would take to completely use up the other reactant. I then calculated the number of moles of each reactant and looked for that relationship. In this case, there needed to be 10 times as many moles of Cl_2 as PCl_5. There weren't. This told me that Cl_2 would get used up first and was therefore the limiting reagent. What if there had been more than 10 times as many moles of Cl_2 as there were moles of P_4? Well, then I would know that after all of the P_4 was used up, there would still be Cl_2. Thus, P_4 would be used up first and would therefore be the limiting reagent.

Now before I go on to show you two more examples, I want to point out a couple of things. Notice first that even though there were 3 times as many grams of Cl_2 as there were of P_4, Cl_2 was still the limiting reagent. Why? Remember that grams *have nothing to do with a chemical equation.* The number of grams of a reactant doesn't tell you anything about how much product will be made. In order to know anything about how a reactant relates to another reactant or to a product, you must compare *moles*.

Okay, then, let's compare moles. There are many, many more moles of Cl_2 than there are of P_4 in the example. Why, then, was Cl_2 the limiting reagent. Well, even though we are talking moles now, we still need to look at the chemical equation to see *how many* moles of each reactant are needed. Looking at the chemical equation, 10 times as many moles of Cl_2 were needed as compared to P_4. Since there was not anything close to 10 times as many moles of Cl_2, then Cl_2 was the limiting reagent. It is *very* important for you to realize that you cannot use number of grams or even number of moles by itself to determine limiting reagent. You *must* take a look at the chemical equation to determine *how many moles* of each reactant is needed. Only then can you determine the limiting reagent. See how this happens in the next two examples.

EXAMPLE 1.10

Calcium acetate [$Ca(CH_3CO_2)_2$] is a chemical important in the process of dyeing fabrics. It is made with the following reaction:

$$Ca(OH)_2 + 2CH_3CO_2H \rightarrow Ca(CH_3CO_2)_2 + 2H_2O$$

How many grams of calcium acetate can be made from 10.0 grams of each reactant?

We need to determine the limiting reagent before we can really answer the question. Looking at the chemical equation, there need to be 2 moles of CH_3CO_2H for every one mole of $Ca(OH)_2$. Let's convert to moles to see if that is really the relationship between the number of moles of each reactant.

$$\frac{10.0\text{ g}\cancel{CH_3CO_2H}}{1} \times \frac{1\text{ mole } CH_3CO_2H}{60.0\text{g}\cancel{CH_3CO_2H}} = 0.167\text{ moles } CH_3CO_2H$$

$$\frac{10.0\text{g}\cancel{Ca(OH)_2}}{1} \times \frac{1\text{ mole } Ca(OH)_2}{74.1\text{g}\cancel{Ca(OH)_2}} = 0.135\text{ moles } Ca(OH)_2$$

Are there 2 moles of CH_3CO_2H for every one mole of $Ca(OH)_2$? Well, since there are 0.135 moles of $Ca(OH)_2$, there would need to be 0.270 moles of CH_3CO_2H. There aren't quite that many, so CH_3CO_2H will run out first.

Now that we know CH_3CO_2H is the limiting reagent, we know that all products are limited by the amount of CH_3CO_2H. Thus, the amount of $Ca(OH)_2$ is irrelevant, and we can continue on in our calculation using just the number of moles of CH_3CO_2H.

$$\frac{0.167\ \cancel{\text{moles } CH_3CO_2H}}{1} \times \frac{1\text{ mole } Ca(CH_3CO_2)_2}{2\ \cancel{\text{moles } CH_3CO_2H}} = 0.0835\text{ moles} Ca(CH_3CO_2)_2$$

$$\frac{0.0835\ \cancel{\text{moles} Ca(CH_3CO_2)_2}}{1} \times \frac{158.1\text{g } Ca(CH_3CO_2)_2}{1\ \cancel{\text{mole} Ca(CH_3CO_2)_2}} = 13.2\text{ g } Ca(CH_3CO_2)_2$$

The reaction produces <u>13.2 g $Ca(CH_3CO_2)_2$</u>.

Rutile (TiO_2) is now used as a pigment in white paints. It replaces the toxic pigment known as white lead, whose chemical formula is $Pb_3(OH)_2(CO_3)_2$. Rutile is made from a natural ore ($FeTiO_3$) via the following reaction:

$$\mathbf{2FeTiO_3 + 4HCl + Cl_2 \rightarrow 2FeCl_3 + 2TiO_2 + 2H_2O}$$

How many grams of Rutile will be made from 50.0 g of $FeTiO_3$, 100.0 g HCl, and an excess of Cl_2?

We need to determine the limiting reagent before we go any further. We know that Cl_2 is not the limiting reagent because we are told there is an excess of Cl_2. Thus, the only real question is whether $FeTiO_3$ or HCl is the limiting reagent. Looking at the chemical equation, there need to be 2 moles of $FeTiO_3$ for every 4 moles of HCl. Let's convert to moles to see if that is really the relationship between the number of moles of each reactant.

$$\frac{50.0\ \text{g}\ \text{FeTiO}_3}{1} \times \frac{1\ \text{mole}\ \text{FeTiO}_3}{151.7\ \text{g}\ \text{FeTiO}_3} = 0.330\ \text{moles}\ \text{FeTiO}_3$$

$$\frac{100.0\ \text{g}\ \text{HCl}}{1} \times \frac{1\ \text{mole}\ \text{HCl}}{36.5\ \text{g}\ \text{HCl}} = 2.74\ \text{moles HCl}$$

Are there 2 moles of $FeTiO_3$ for every 4 moles of HCl? Well, since there are 0.334 moles of $FeTiO_3$, there would need to be twice as many moles of HCl, or 0.668 moles. There are a LOT more than that present, which means that after all of the $FeTiO_3$ is used up, there will still be a lot of HCl left over. Thus, $FeTiO_3$ is the limiting reagent.

Now that we know $FeTiO_3$ is the limiting reagent, we know that all products are limited by the amount of $FeTiO_3$. Thus, the amount of HCl (and Cl_2) is irrelevant, and we can continue on in our calculation using just the number of moles of $FeTiO_3$.

$$\frac{0.330\ \text{moles}\ \text{FeTiO}_3}{1} \times \frac{2\ \text{moles}\ \text{TiO}_2}{2\ \text{moles}\ \text{FeTiO}_3} = 0.330\ \text{moles}\ \text{TiO}_2$$

$$\frac{0.330\ \text{moles}\ \text{TiO}_2}{1} \times \frac{79.9\ \text{g}\ \text{TiO}_2}{1\ \text{mole}\ \text{TiO}_2} = 26.4\ \text{g}\ \text{TiO}_2$$

The reaction produces 26.4 g TiO_2.

In order to determine the limiting reagent, then, we convert all reactants to moles and try to determine which will run out first by comparing the amount of moles we have to the mole relationship in the chemical equation. Whichever reactant runs out first is the limiting reagent, and we use it to determine everything else. Try this on your own.

ON YOUR OWN

1.8 One of the ores which we mine for silver is $KAg(CN)_2$. In order to get silver from this, it must be processed in the following reaction:

$$2KAg(CN)_2\ (aq) + Zn\ (s) \rightarrow 2Ag\ (s) + Zn(CN)_2\ (aq) + 2KCN\ (aq)$$

How many grams of silver will be produced when 100.0 g of $KAg(CN)_2$ is reacted with 75.0 g Zn?

1.9 Many of the barbecue grills that people have in their backyard use the combustion of propane (C_3H_8) for heat. How many grams of water would be produced by the combustion of 10.0 g of propane with 75.0 g of oxygen gas? You have to remember the definition of a combustion reaction in order to get the chemical equation you need to solve this problem!

Stoichiometry, Percent Yield, and Multiple Reactions

Now that you've dusted the cobwebs out of your mind when it comes to stoichiometry, there is one more skill I want you to learn in this module. Stoichiometry is a powerful tool in chemistry, and it is used over and over again, especially in situations where chemists are synthesizing compounds. Since you have the basic tools of stoichiometry in your mind now, it is important to make you aware of at least two things that complicate stoichiometry in the "real world." The first concept is **percent yield** and the second is the fact that most chemical processes in the real world use more than just one reaction to make the product of interest.

The first concept is the easiest of the two to learn, and it is really just a consequence of experimental error. When you use stoichiometry to calculate how much product will be produced in a chemical reaction, the number that you get is really an upper limit. In other words, no matter how careful an experimenter you are, when you actually perform the reaction in a lab, you will never, never get as much product as stoichiometry indicates that you should. This isn't because stoichiometry is wrong, but it is because experimental error will *always* result in lost product. Even the most talented experimenter in the world will not make as much product as he or she *should* make, because of unavoidable experimental errors.

Because you can never make as much product as stoichiometry indicates that you should, chemists use the concept **percent yield**.

Percent yield - The actual amount of product you make in a chemical reaction, divided by the amount stoichiometry indicates you should have made, times 100

If you look at the definition, you can see what percent yield tells us. Suppose stoichiometry says that I should make 10.0 grams of a substance in a chemical reaction. When I do the reaction and measure the mass of the product, I see that I actually made 9.0 grams. The percent yield would be:

$$\frac{9.0\text{ g}}{10.0\text{ g}} \times 100\% = 90\%$$

This means that I made 90% of what I should have made. Percent yield, therefore, is high when there are few experimental errors and is low when there are many. In a moment, you will see some problems with percent yield.

The second concept is a little harder to master. When chemists synthesize compounds, the process is usually quite long and complicated, involving several different reactions. Chemists still have to employ stoichiometry to determine how much starting materials to use, but they must apply stoichiometry to several reactions. How can that be done? Study the following example to find out.

EXAMPLE 1.11

Sulfuric acid (H_2SO_4) is one of the Unites States' most important exported commodities. The following three equations illustrate one of the ways it is prepared commercially from iron sulfide:

$$4FeS_2\,(s) + 11O_2\,(g) \rightarrow 2Fe_2O_3\,(s) + 8SO_2\,(g)$$

$$2SO_2\,(g) + O_2\,(g) \rightarrow 2SO_3\,(g)$$

$$SO_3\,(g) + H_2O\,(l) \rightarrow H_2SO_4\,(l)$$

If a chemist starts with 1.13 kilograms of FeS_2 and an excess of oxygen and water, how much H_2SO_4 should be produced? If the chemist actually produces 1.51 kg of H_2SO_4, what is the chemist's percent yield?

How do we go about solving this? We know that FeS_2 is the limiting reagent, but how do we calculate the amount of H_2SO_4 made? Well, we could calculate how much SO_2 is made in the first reaction, and then use it to calculate how much SO_3 is made in the second reaction. Then we could use that result in the third reaction to finally find out what we want to know. Sounds like a lot of work, doesn't it? Well, it is a lot of work and, quite frankly, it is not necessary. After all, look at the product in which we are interested: H_2SO_4. It has hydrogen atoms, sulfur atoms, and oxygen atoms in it. Where do the hydrogen atoms come from? They come from the water that is a reactant in the third reaction. Where do the oxygen atoms come from? They come from the water and oxygen used as reactants in all three equations. Where do the sulfur atoms come from? They come from the FeS_2.

Now think about it a minute. The hydrogen and oxygen atoms come from reactants which we are told are in excess. The sulfur atoms come from the limiting reagent: FeS_2. So, the sulfur atoms in sulfuric acid all come from the limiting reagent FeS_2. What if I have one mole of FeS_2. Given an excess of water and oxygen, how much H2SO4 can come from 1 mole of FeS_2? Well, 1 mole of FeS_2 will give us two moles of sulfur atoms, right? Given plenty of hydrogen and oxygen atoms (from water and oxygen), then 2 moles of sulfur atoms will give us 2 moles of sulfuric acid. In the end, then, these three chemical equations work together to tell us that when water and oxygen are in excess, each mole of FeS_2 will give us 2 moles of H_2SO_4. In other words:

$$1 \text{ mole } FeS_2 = 2 \text{ moles } H_2SO_4$$

This relationship is all we need to do the stoichiometry:

$$\frac{1.13 \times 10^3 \text{ g } \cancel{FeS_2}}{1} \times \frac{1 \text{ mole } FeS_2}{120.0 \text{ g } \cancel{FeS_2}} = 9.42 \text{ moles } FeS_2$$

$$\frac{9.42 \cancel{\text{moles } FeS_2}}{1} \times \frac{2 \text{ moles } H_2SO_4}{1 \cancel{\text{moles } FeS_2}} = 18.8 \text{ moles } H_2SO_4$$

$$\frac{18.8 \cancel{\text{moles } H_2SO_4}}{1} \times \frac{98.1 \text{ g } H_2SO_4}{1 \cancel{\text{mole } H_2SO_4}} = 1.84 \times 10^3 \text{ g } H_2SO_4$$

So these three reactions should produce 1.84 kg H_2SO_4. Since the chemist produced only 1.51 kg of sulfuric acid, the percent yield is:

$$\frac{1.51 \cancel{\text{kg}}}{1.84 \cancel{\text{kg}}} \times 100\% = 82.1\%$$

The percent yield, then, is 82.1%.

Do you see what I did in the example? I tried to determine the source of each element in the product of interest. Once I did that, I saw what element came from the limiting reagent. Then I used the number of that element present in the limiting reagent and in the final product to develop a mole relationship between the limiting reagent and the product. After that, the stoichiometry was a breeze. Notice that I did not do anything with the stoichiometric coefficients that were in the chemical reactions. I was relating two chemicals *solely* by their elements. As a result, I did not need to worry about the coefficients. Remember, the coefficients in chemical equations relate the moles of one substance to the moles of another substance *within that chemical equation*. Since I was working across three chemical equations, the stoichiometric coefficients really held no meaning for this technique.

Now it's important to realize that this method will only work when the element you use comes *only* from one source, and that source is the limiting reagent. Notice that the first equation produces SO_2, and that SO_2 is then used in the next equation. Suppose if, in addition to the SO_2 from the first equation, another sulfur-containing compound (like Na_2S) was used as a reactant in the second equation. At that point, FeS_2 would not be the only source of sulfur in the production process, and we would not be able to use the technique we used. Also, if FeS_2 had not been the limiting reagent, we could not have done what we did. This technique, therefore, is quite powerful, but you need to know when you can and cannot use it. Try this technique yourself by solving the following "on your own" problem.

ON YOUR OWN

1.10 The following three equations represent a process by which iron (Fe) is extracted from iron ore (Fe_2O_3) in a blast furnace:

$$3Fe_2O_3 + CO \rightarrow 2Fe_3O_4 + CO_2$$

$$Fe_3O_4 + CO \rightarrow 3FeO + CO_2$$

$$FeO + CO \rightarrow Fe + CO_2$$

If a blast furnace starts with 1.00 kg of iron ore and ends up making 293 g of iron, what is the percent yield? Assume that CO is in excess.

Did you struggle a little bit through this module? It's okay if you did. Part of the problem is due to the fact that it had been a while since you did stoichiometry and chemical equations. As time goes on, you will remember more and things will get easier. For right now, don't worry if you have to go back to your first-year chemistry book in order to bring all of the concepts back to mind.

ANSWERS TO THE ON YOUR OWN PROBLEMS

1.1 To solve this problem, we simply have to convert miles to meters and hours to seconds. It doesn't matter that they are both a part of our original unit. As long as we use the factor-label method, everything will work out:

$$\frac{65\ \cancel{\text{miles}}}{1\ \cancel{\text{hour}}} \times \frac{1609\ \text{m}}{1\ \cancel{\text{mile}}} \times \frac{1\ \cancel{\text{hour}}}{3600\ \text{s}} = 29\ \frac{\text{m}}{\text{s}}$$

Sixty-five miles per hour doesn't sound as fast when you say it as 29 m/s.

1.2 To get the unit of interest, we need to convert moles to grams using the mass of NaOH; we need to convert L to mL; and we need to convert seconds to hours. The mass of NaOH from the periodic chart is 39.0 amu, which means there are 39.0 grams in a mole of NaOH. The conversion from mL to L and seconds to hours you should know.

$$\frac{1.02\ \cancel{\text{moles}}}{1\ \cancel{\text{L}} \cdot \text{s}} \times \frac{39.0\ \text{g}}{1\ \cancel{\text{mole}}} \times \frac{3600\ \text{s}}{1\ \text{hour}} \times \frac{0.001\ \cancel{\text{L}}}{1\ \text{mL}} = 143\ \frac{\text{grams}}{\text{mL} \cdot \text{hr}}$$

The rate is 143 g/(mL·hr).

1.3 We can't use the conversion of 3 feet in 1 yard right away, because we want to convert from square feet into square yards. Thus, we need a relationship between those quantities. We can get such a relationship by squaring both sides of the relationship that we do have:

$$3\ \text{ft} = 1\ \text{yd}$$

$$9\ \text{ft}^2 = 1\ \text{yd}^2$$

Now we can do the conversion:

$$\frac{1600\ \cancel{\text{ft}^2}}{1} \times \frac{1\ \text{yd}^2}{9\ \cancel{\text{ft}^2}} = 180\ \text{yd}^2$$

The house has an area of 180 square yards. If your answer was 177.8, go back and review significant figures from your first year course.

1.4 This problem requires us to convert between m and km as well as s^2 and hr^2. The first is easy; the second requires us to square the relationship between hours and seconds:

$$1\ \text{hr} = 3600\ \text{s}$$

$$1\ \text{hr}^2 = 12960000\ \text{s}^2$$

Notice that I did not round significant figures here. This is because the relationship between hours and seconds is exact, so the precision is infinite. As a result, I keep all figures. Now we can do the conversion:

$$\frac{1.1\ \cancel{\text{m}}}{\cancel{\text{s}^2}} \times \frac{1\ \text{km}}{1{,}000\ \cancel{\text{m}}} \times \frac{12960000\ \cancel{\text{s}^2}}{1\ \text{hr}^2} = 1.4 \times 10^4\ \frac{\text{km}}{\text{hr}^2}$$

An acceleration of 1.1 m/s^2 is the same as <u>1.4×10^4 km/hr^2</u>.

1.5 To get an overall reaction, we just add up the two reactions given:

$$\begin{array}{l} 2NO_2\ (g) \rightarrow NO_3\ (g) + NO\ (g) \\ + NO_3\ (g) + CO\ (g) \rightarrow NO_2\ (g) + CO_2\ (g) \\ \hline \end{array}$$

$$2NO_2\ (g) + NO_3\ (g) + CO\ (g) \rightarrow NO_3\ (g) + NO\ (g) + NO_2\ (g) + CO_2\ (g)$$

Now we can cancel terms that appear on both sides of the equation:

$$\cancel{2}NO_2\ (g) + \cancel{NO_3\ (g)} + CO\ (g) \rightarrow \cancel{NO_3\ (g)} + NO\ (g) + \cancel{NO_2\ (g)} + CO_2\ (g)$$

Notice that there are two NO_2 (g) molecules on the reactants side and only one on the products side. Thus, I can use the one on the products side to cancel one of the NO_2 (g) on the reactants side, but I cannot cancel the other. That's why I marked out the "2" next to the NO_2 (g) on the reactants side and not the entire molecule. Since there are no like terms to group together, we are done:

$$\underline{NO_2\ (g) + CO\ (g) \rightarrow NO\ (g) + CO_2\ (g)}$$

1.6 Before we begin, we need to know what equation we are trying to calculate the ΔH of. A formation equation takes the elements in a molecule and reacts them together to form that molecule. We must use the elements in their natural form, however. Since oxygen is a homonuclear diatomic, its natural form is O_2. The formation reaction is:

$$3Co\ (s) + 2O_2\ (g) \rightarrow Co_3O_4\ (s)$$

Now we need to get the two equations above to add to this one. In order to get that to happen, the first equation will have to be reversed, because currently Co (s) and O_2 (g) are products, but we need them as reactants:

$$Co\ (s) + \tfrac{1}{2}O_2\ (g) \rightarrow CoO\ (s) \qquad \Delta H = -237.9\ \text{kJ}$$

This is the only equation from which we get Co (s). Thus, the stoichiometric coefficient must be 3, because we need a 3Co (s) term in our final equation. Thus, I need to multiply both the equation and the ΔH by 3:

$$3Co\ (s) + \tfrac{3}{2}O_2\ (g) \rightarrow 3CoO\ (s) \qquad \Delta H = -713.7\ kJ$$

Notice that this does not give us the right coefficient with the oxygen. That's okay, though, because we get oxygen from the next equation as well. That equation has Co_3O_4 (s) as a product, so it need not be reversed. However, it has two of them, and we only want one in our equation. Thus, we need to multiply the equation and the ΔH by one-half:

$$3CoO\ (s) + \tfrac{1}{2}O_2\ (g) \rightarrow Co_3O_4\ (s) \qquad \Delta H = -177.5\ kJ$$

Now we can add the equations and the ΔH's:

$$\begin{array}{ll} 3Co\ (s) + \tfrac{3}{2}O_2\ (g) \rightarrow 3CoO\ (s) & \Delta H = -713.7\ kJ \\ +\ 3CoO\ (s) + \tfrac{1}{2}O_2\ (g) \rightarrow Co_3O_4\ (s) & \Delta H = -177.5\ kJ \\ \hline 3Co\ (s) + \tfrac{3}{2}O_2\ (g) + \cancel{3CoO\ (s)} + \tfrac{1}{2}O_2\ (g) \rightarrow \cancel{3CoO\ (s)} + Co_3O_4\ (s) & \end{array}$$

Once we combine like terms, we get our original equation:

$$3Co\ (s) + 2O_2\ (g) \rightarrow Co_3O_4\ (s)$$

The ΔH is then just the sum of the two ΔH's listed with those reactions, or <u>-891.2 kJ</u>.

1.7 Calculate the ΔH for the combustion of carbon :

$$C\ (s) + O_2\ (g) \rightarrow CO_2\ (g)$$

Given that

$$C\ (s) + \tfrac{1}{2}O_2\ (g) \rightarrow CO\ (g) \qquad \Delta H = -111\ kJ$$

$$2CO_2\ (g) \rightarrow 2CO\ (g) + O_2\ (g) \qquad \Delta H = 566\ kJ$$

The first reaction is fine. It has C (s) and O_2 (g) as reactants, which is what we need. The coefficient with the C (s) is correct, but the coefficient with the O_2 (g) is not. That's okay, however, because we will get more O_2 from the next reaction. In that reaction, we must reverse it because that's the only way to get CO_2 (g) as a product. Also, we will have to divide the equation (and its ΔH) by 2 in order to get the correct coefficient next to the CO_2.

$$CO\ (g) + \tfrac{1}{2}O_2\ (g) \rightarrow CO_2\ (g) \qquad \Delta H = -283\ kJ$$

Now we can add the equations:

$$\begin{array}{ll} \text{CO (g)} + \tfrac{1}{2}\text{O}_2\text{ (g)} \rightarrow \text{CO}_2\text{ (g)} & \Delta H = -283 \text{ kJ} \\ + \text{ C (s)} + \tfrac{1}{2}\text{O}_2\text{ (g)} \rightarrow \text{CO (g)} & \Delta H = -111 \text{ kJ} \\ \hline \end{array}$$

$$\cancel{\text{CO (g)}} + \tfrac{1}{2}\text{O}_2\text{ (g)} + \text{C (s)} + \tfrac{1}{2}\text{O}_2\text{ (g)} \rightarrow \text{CO}_2\text{ (g)} + \cancel{\text{CO (g)}}$$

After canceling the CO (g) on both sides of the equation and combining the O_2 (g) terms on the reactants side of the equation, we are left with the equation of interest:

$$\text{C (s)} + \text{O}_2\text{ (g)} \rightarrow \text{CO}_2\text{ (g)}$$

So the ΔH is simply the sum of the ΔH's listed for the two equations after we manipulated them, or <u>-394 kJ.</u>

1.8 We need to determine the limiting reagent before we can really answer the question. Looking at the chemical equation, there need to be 2 moles of $KAg(CN)_2$ for every one mole of Zn. Let's convert to moles to see if that is really the relationship between the number of moles of each reactant.

$$\frac{100.0 \text{ g } \cancel{\text{KAg(CN)}_2}}{1} \times \frac{1 \text{ mole KAg(CN)}_2}{199.0 \text{ g } \cancel{\text{KAg(CN)}_2}} = 0.5025 \text{ moles KAg(CN)}_2$$

$$\frac{75.0 \text{ g } \cancel{\text{Zn}}}{1} \times \frac{1 \text{ mole Zn}}{65.4 \text{ g } \cancel{\text{Zn}}} = 1.15 \text{ moles Zn}$$

Are there 2 moles of $KAg(CN)_2$ for every one mole of Zn? Well, since there are 1.15 moles of Zn, there would need to be 2.30 moles of $KAg(CN)_2$. There aren't nearly that many, so $KAg(CN)_2$ will run out first.

Now that we know $KAg(CN)_2$ is the limiting reagent, we know that all products are limited by the amount of $KAg(CN)_2$. Thus, the amount of Zn is irrelevant, and we can continue on in our calculation using just the number of moles of $KAg(CN)_2$.

$$\frac{0.5025 \cancel{\text{ moles KAg(CN)}_2}}{1} \times \frac{2 \text{ moles Ag}}{2 \cancel{\text{ moles KAg(CN)}_2}} = 0.5025 \text{ moles Ag}$$

$$\frac{0.5025 \cancel{\text{ moles Ag}}}{1} \times \frac{107.9 \text{ g Ag}}{1 \cancel{\text{ mol Ag}}} = 54.22 \text{ g Ag}$$

The reaction produces 54.22 g of silver.

1.9 Before we can do anything, we need to know what equation we are working with. Combustion is defined as a reaction that adds oxygen and produces carbon dioxide and water. The combustion of propane, then, would be:

$$C_3H_8 + O_2 \rightarrow CO_2 + H_2O$$

This reaction is not balanced, so we need to do that next:

$$C_3H_8 + 5O_2 \rightarrow 3CO_2 + 4H_2O$$

Now that we have the equation, we can determine the limiting reagent. Looking at the chemical equation, there need to be 5 moles of O_2 for every one mole of C_3H_8. Let's convert to moles to see if that is really the relationship between the number of moles of each reactant.

$$\frac{10.0\ \text{g}\ \cancel{C_3H_8}}{1} \times \frac{1\ \text{mole}\ C_3H_8}{44.1\ \text{g}\ \cancel{C_3H_8}} = 0.227\ \text{moles}\ C_3H_8$$

$$\frac{75.0\ \text{g}\ \cancel{O_2}}{1} \times \frac{1\ \text{mole}\ O_2}{32.0\ \text{g}\ \cancel{O_2}} = 2.34\ \text{moles}\ O_2$$

Are there 5 moles of O_2 for every one mole of C_3H_8? Well, since there are 0.227 moles of C_3H_8, there would need to be 1.14 moles of O_2. There are more than that, so after all of the C_3H_8 molecules are used up, there will still be O_2. Thus, C_3H_8 will run out first.

Now that we know C_3H_8 is the limiting reagent, we know that all products are limited by the amount of C_3H_8. Thus, the amount of O_2 is irrelevant, and we can continue on in our calculation using just the number of moles of C_3H_8.

$$\frac{0.227\ \cancel{\text{moles}\ C_3H_8}}{1} \times \frac{4\ \text{moles}\ H_2O}{1\ \cancel{\text{mole}\ C_3H_8}} = 0.908\ \text{moles}\ H_2O$$

$$\frac{0.908\ \cancel{\text{moles}\ H_2O}}{1} \times \frac{18.0\ \text{g}\ H_2O}{1\ \cancel{\text{mol}\ H_2O}} = 16.3\ \text{g}\ H_2O$$

The reaction produces 16.3 g of water.

1.10 To do this problem, we look for an element that is in the product and also in a reactant of the first equation. In this case, it's easy. After all, the product is Fe. Thus, the element we need to look for must also be Fe. It is in the Fe_2O_3 which is a reactant in the first reaction. This is the only source of Fe, because the Fe-containing compounds that are reactants in the next two equation come from the previous equation. Thus, Fe_2O_3 is the only source of Fe. There are 2 Fe's in each Fe_2O_3 molecule, so one mole of Fe_2O_3 will make 2 moles of Fe.

$$1 \text{ mole } Fe_2O_3 = 2 \text{ moles Fe}$$

With that relationship, we can do stoichiometry.

$$\frac{1.00 \times 10^3 \text{ g } \cancel{Fe_2O_3}}{1} \times \frac{1 \text{ mole } Fe_2O_3}{159.6 \text{ g } \cancel{Fe_2O_3}} = 6.27 \text{ moles } Fe_2O_3$$

$$\frac{6.27 \cancel{\text{moles } Fe_2O_3}}{1} \times \frac{2 \text{ moles Fe}}{1 \cancel{\text{mole } Fe_2O_3}} = 12.5 \text{ moles Fe}$$

$$\frac{12.5 \text{ moles } \cancel{Fe}}{1} \times \frac{55.8 \text{ g Fe}}{1 \cancel{\text{mole Fe}}} = 698 \text{ g Fe}$$

So these three reactions should produce 698 g Fe. Since the furnace produced only 293 g of sulfuric acid, the percent yield is:

$$\frac{293 \text{ g}}{698 \text{ g}} \times 100\% = 42.0\%$$

The percent yield, then, is 42.0%.

REVIEW QUESTIONS

1. A chemist presents a possible reaction mechanism for a chemical reaction she has been studying. If she adds up every equation in the reaction mechanism, what should she get in the end?

2. State Hess's Law.

3. When you reverse a chemical equation, what must you do to its ΔH?

4. When you multiply a chemical equation by a number, what must you do to its ΔH?

5. Suppose you had a chemical equation with H_2O (l) on one side and H_2O (g) on the other, could you cancel those terms?

6. Suppose you had a chemical equation with $2H_2O$ (l) on one side and H_2O (l) on the other, could you cancel those terms?

7. Why is the limiting reagent such an important thing to know in stoichiometry?

8. Other chemistry books teach another way to determine the limiting reagent. In those books, you are told to do the stoichiometry assuming one reactant is the limiting reagent and then go back and do the stoichiometry *again*, assuming the other reactant is the limiting reagent. I think that's too much work. However, suppose you did a problem that way. You will get two answers for the amount of product. One answer will be bigger than the other. Which is the correct answer?

9. Suppose you did a limiting reagent problem as described above and got two answers that were equal. What would that tell you?

10. Two students do the same experiment. One gets a percent yield of 45% and the other a percent yield of 65%. Which student is the more careful experimenter?

PRACTICE PROBLEMS

1. The volume of a box is 0.034 m^3. What is the volume in cm^3?

2. Electrical fields are measured in Volts per meter. If an electrical field has a strength of 45 V/m, what is the strength in milliVolts per cm? (MilliVolt has the same relationship to Volt as milliliter does to liter.)

3. The density of a substance is 1.2 g/cm^3. What is the density in kg/m^3?

4. Given the following reaction:

$$2S\ (s) + 3O_2\ (g) \rightarrow 2SO_3\ (g) \qquad \Delta H = -792\ kJ$$

What is the ΔH of this reaction?

$$4SO_3\ (g) \rightarrow 4S\ (s) + 6O_2\ (g)$$

5. The following series of equations is a possible reaction mechanism for a certain reaction. What is the reaction?

$$2NO \rightarrow N_2O_2$$

$$N_2O_2 + H_2 \rightarrow N_2O + H_2O$$

$$N_2O + H_2 \rightarrow N_2 + H_2O$$

6. Calculate the ΔH for the following reaction:

$$2NO\ (g) + O_2\ (g) \rightarrow 2NO_2\ (g)$$

given the following data:

$$N_2\ (g) + O_2\ (g) \rightarrow 2NO\ (g) \qquad \Delta H = 180.6\ kJ$$

$$N_2\ (g) + 2O_2\ (g) \rightarrow 2NO_2\ (g) \qquad \Delta H = 66.4\ kJ$$

7. Calculate the enthalpy of formation of acetic acid, CH_3CO_2H (l), given the following data:

$$CH_3CO_2H\,(l) + 2O_2\,(g) \rightarrow 2CO_2\,(g) + 2H_2O\,(l) \qquad \Delta H = -871\text{ kJ}$$

$$C\,(s) + O_2\,(g) \rightarrow CO_2\,(g) \qquad \Delta H = -394\text{ kJ}$$

$$H_2\,(g) + \tfrac{1}{2}O_2\,(g) \rightarrow H_2O\,(l) \qquad \Delta H = -286\text{ kJ}$$

8. Ethylene glycol is a popular component of automobile antifreeze. In order to make ethylene glycol, chemists must first make ethylene oxide (C_2H_4O) from the following reaction:

$$2C_2H_4 + O_2 \rightarrow 2C_2H_4O$$

If a chemist starts with 11.0 g of C_2H_4 and 11.0 g of O_2, how much ethylene oxide should the chemist make?

9. In order to make acetylene (C_2H_2) for acetylene torches, chemists must first make calcium carbide (CaC_2) according to this reaction:

$$CaO + 3C \rightarrow CaC_2 + CO$$

A chemist reacts 1.00 kg of CaO with 1.00 kg of carbon. If the chemist ends up with 1.00 kg of calcium carbide, what is the percent yield?

10. Nitric acid, HNO_3, is manufactured industrially with the following sequence of reactions:

$$4NH_3 + 5O_2 \rightarrow 4NO + 6H_2O$$

$$2NO + O_2 \rightarrow 2NO_2$$

$$3NO_2 + H_2O \rightarrow 2HNO_3 + NO$$

In this process, the NO made in the last reaction is recycled back into the second reaction and used as a reactant. In the end, *all* of the NO produced is recycled back so that when the manufacturing process is complete, no NO remains. If a chemist starts with 35.0 grams of NH_3 and an excess of water and oxygen, how much nitric acid can be made?

THE PERIODIC CHART OF ELEMENTS

1A	2A	1B	2B	3B	4B	5B	6B	7B	8B	9B	10B	3A	4A	5A	6A	7A	8A
1 **H** 1.01																	2 **He** 4.0
3 **Li** 6.94	4 **Be** 9.00											5 **B** 10.8	6 **C** 12.0	7 **N** 14.0	8 **O** 16.0	9 **F** 19.0	10 **Ne** 20.2
11 **Na** 22.0	12 **Mg** 24.3											13 **Al** 27.0	14 **Si** 28.1	15 **P** 31.0	16 **S** 32.1	17 **Cl** 35.5	18 **Ar** 39.9
19 **K** 39.1	20 **Ca** 40.1	21 **Sc** 45.0	22 **Ti** 47.9	23 **V** 50.9	24 **Cr** 52.0	25 **Mn** 54.9	26 **Fe** 55.8	27 **Co** 58.9	28 **Ni** 58.7	29 **Cu** 63.5	30 **Zn** 65.4	31 **Ga** 69.7	32 **Ge** 72.6	33 **As** 74.9	34 **Se** 79.0	35 **Br** 79.9	36 **Kr** 83.8
37 **Rb** 85.5	38 **Sr** 87.6	39 **Y** 88.9	40 **Zr** 91.2	41 **Nb** 92.9	42 **Mo** 95.9	43 **Tc** 97.0	44 **Ru** 101.1	45 **Rh** 102.9	46 **Pd** 106.4	47 **Ag** 107.9	48 **Cd** 112.4	49 **In** 114.8	50 **Sn** 118.7	51 **Sb** 121.8	52 **Te** 127.6	53 **I** 126.9	54 **Xe** 131.3
55 **Cs** 132.9	56 **Ba** 137.3	57 **La** 138.9	72 **Hf** 178.5	73 **Ta** 180.9	74 **W** 183.9	75 **Re** 186.2	76 **Os** 190.2	77 **Ir** 192.2	78 **Pt** 195.1	79 **Au** 197.0	80 **Hg** 200.6	81 **Tl** 204.4	82 **Pb** 207.2	83 **Bi** 209.0	84 **Po** (209)	85 **At** (210)	86 **Rn** (222)
87 **Fr** 223.0	88 **Ra** 226.0	89 **Ac** 227.0	104 (261)	105 (262)	106 (263)	107 (264)	108 (265)	109 (266)									

58 **Ce** 140.1	59 **Pr** 140.9	60 **Nd** 144.2	61 **Pm** 145.0	62 **Sm** 150.4	63 **Eu** 152.0	64 **Gd** 157.3	65 **Tb** 158.9	66 **Dy** 162.5	67 **Ho** 164.9	68 **Er** 167.3	69 **Tm** 168.9	70 **Yb** 173.0	71 **Lu** 175.0
90 **Th** 232.0	91 **Pa** 234.0	92 **U** 238.0	93 **Np** 237.0	94 **Pu** (244)	95 **Am** (243)	96 **Cm** (247)	97 **Bk** (247)	98 **Cf** (251)	99 **Es** (252)	100 **Fm** (257)	101 **Md** (258)	102 **No** (259)	103 **Lr** (260)

Module #2: The Atom Revisited

Introduction

In the previous module, we spent time adding on to some of the things you already learned. We will be doing that for a few more modules, including this one. In this module, we will take another look at the structure of the atom. In your previous chemistry course, you learned about the Bohr model of the atom as well as the quantum mechanical model of the atom. Now we are going to add to your knowledge of both of those models.

You should remember that the Bohr model is not considered an accurate model of the atom any more. Instead, the quantum mechanical model of the atom is generally considered the best description of the structure and function of the atom. Well, if that's the case, why do we spend time learning more about the Bohr model? If it's not really a good model, why bother?

There are two answers to this question. The first answer is that the Bohr model is *much* easier to learn. To really learn the quantum mechanical model of the atom, you need to have 2-3 years of *post-calculus* mathematics. Obviously, then, we cannot delve deeply into this highly mathematical model. The Bohr model, on the other hand, refers only to algebra. This makes it much easier for high school students to learn. The second answer is a bit more appealing than the first. Despite the fact that the Bohr model is incorrect, it still illustrates some of the fundamental chemistry and physics of the atom. Thus, it is worth learning so as to explore these fundamental concepts.

The Atom - What a Bohr!

As you learned in your previous chemistry course, the Bohr model pictures the atom as follows:

FIGURE 2.1
The Bohr Model of the Atom

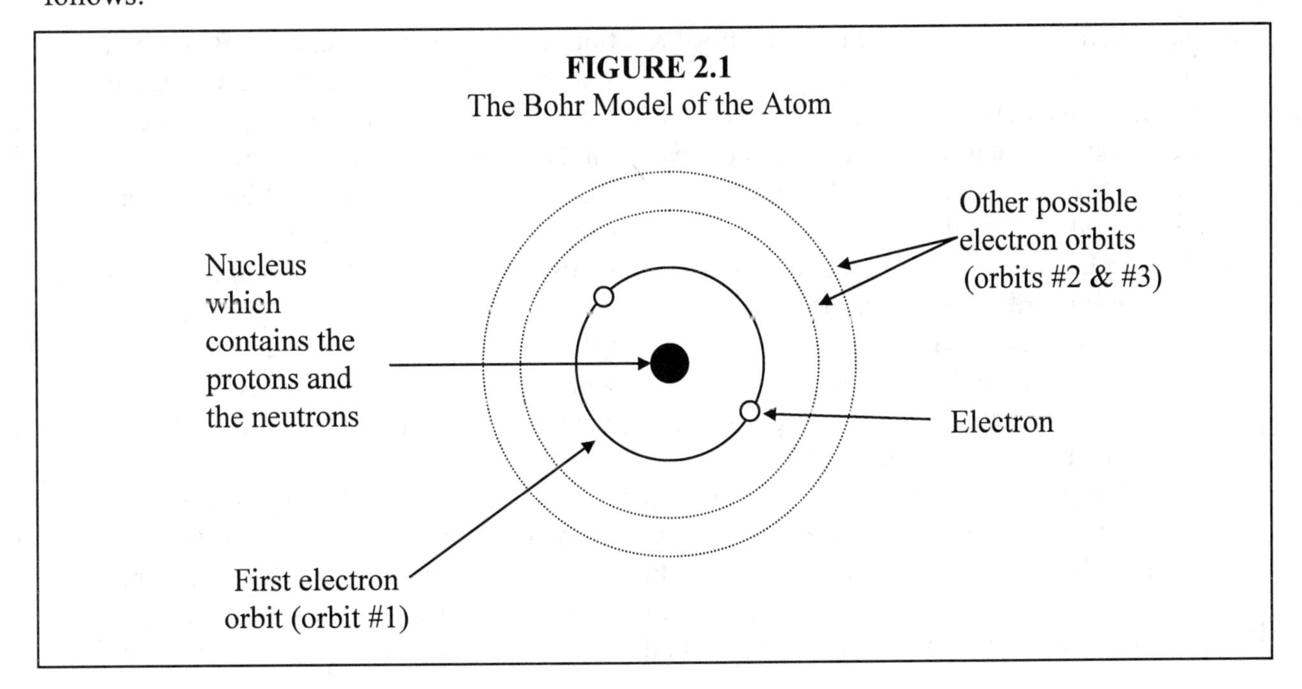

In this model, protons and neutrons cluster together in the center of the atom, called the **nucleus**. Electrons orbit the nucleus in circles. The important thing to realize about the Bohr model is that there are many, many orbits that electrons can occupy. In the helium atom pictured in the figure, for example, both electrons are in the orbit drawn with the solid line. Notice, however, the orbits drawn with dotted lines. These are also possible orbits for electrons. There are many, many more possible orbits as well. I just chose to show you two of them. If an electron absorbs energy (from heat or light, for example), then the electron can jump to another orbit that is farther away from the nucleus. If the electron then wants to drop down to an orbit that is closer to the nucleus, it must get rid of energy. Typically, it does this by emitting light. The following experiment will help guide our discussion on these points.

EXPERIMENT 2.1

The Colors of Chemistry

Supplies

From the laboratory equipment set:
- Two test tubes
- One test tube cap
- Measuring scoop
- Plastic dropper
- Cobalt chloride
- Sodium ferrocyanide
- Ferric ammonium sulfate
- Phenolphthalein solution
- Sodium carbonate
- Paper towel (**Do not** use a regular towel.)

In this experiment, you will see how chemical reactions make interesting colors. Before you do this experiment, however, you need to understand that some of the chemicals you will use in this course are **extremely toxic**. Thus, you do not want to ingest them in any way. Now, of course, I wouldn't expect you to want to eat the chemicals, but the problem is that there are many ways you can ingest things without realizing it. First of all, you can get the chemicals on your hands and then handle food. In that way, the chemicals can get into your system. As a result, **always wash your hands thoroughly when you are done with an experiment**. Also, **do not clean up in the kitchen sink**. Use a bathroom sink to do all of your cleaning and, once you are all done, clean the sink with a cleaner. Finally, notice that I told you to use a paper towel, not a cloth one. That's because you will throw the paper towel away in the end, and there will not be a chance of chemicals on the towel being transferred somewhere else in the laundry. With that out of the way, you can now do the experiment.

Fill two test tubes 1/4 of the way with water. Take the measuring scoop (the plastic spoon-like thing) and use it to place two measures of cobalt chloride into one of the test tubes with water. When I refer to a "measure," I am talking about the small end of the measuring scoop. To add a "measure" of solid to the test tube, use the small end of the chemical scoop to scoop up the solid. When the little rectangle at the end of the scoop is filled with solid, you have a "measure" of solid. Do not be real exacting about this. It is a qualitative measurement, not a quantitative one. Cap the test tube and shake it to dissolve the solid. Now rinse off the scoop and test tube cap and dry them with a paper towel.

Add two measures of sodium ferrocyanide to the other test tube that has water in it. Cap it and shake to dissolve the solid. Now, take the cap off of the test tube and pour its contents into the test tube that has the cobalt chloride solution. Cap the test tube with the new solution in it and shake. What happens?

Rinse out both test tubes, and fill two of them 1/4 of the way full of water again. Rinse and dry the cap and the scoop. Now add two measures of sodium ferrocyanide to one of the test tubes, and then cap and shake the test tube. Rinse and dry the scoop and cap, and then add two measures of ferric ammonium sulfate to the other test tube with water in it. Cap and shake the test tube. Uncap the test tube and pour the solution into the test tube with the sodium ferrocyanide solution in it. Cap and shake. What happens? **Be careful with this solution. It will stain your clothes!**

Rinse out both test tubes, and fill one of them 1/2 of the way full of water again. Rinse and dry the cap and the scoop. Add four measures of sodium carbonate to the water, cap the test tube, and shake. Uncap the test tube and use the plastic dropper to add 10 drops of phenolphthalein solution to the sodium carbonate solution. Cap and shake the test tube. What happens?

Where did the colors come from? How can two colorless solutions mix to form such nice colors? The answers to these questions can be found be studying the Bohr model of the atom, the quantum mechanical model of the atom, and the fundamentals of chemical bonding. This discussion will take the balance of this module and all of the next module, however, so don't expect to learn the answer right away!

A Detailed Look at the Bohr Model

To truly understand the results of Experiment 2.1, you need to see the mathematics behind the Bohr model. To make the mathematics simple, let's concentrate on the *simplest* atom in Creation, the hydrogen atom. According to the periodic chart, hydrogen has only one proton and one electron. If we ignore all isotopes of hydrogen except ^{1}H, then we do not even have to worry about neutrons. Thus, the simplest atom in Creation has just one proton in its nucleus and has just one electron orbiting that nucleus.

What keeps the electron orbiting the nucleus? Why doesn't the electron just go off on its own? Well, electrons are negatively charged and protons are positively charged. The mutual attraction between these opposite charges keeps the electron orbiting the proton. Physics tells us that in order for the electron to continue to orbit the proton, the attraction between the positive and negative charges must equal the force required to keep the electron moving in a circle:

$$\frac{m \cdot v^2}{r} = \frac{k \cdot q_1 \cdot q_2}{r^2} \qquad (1)$$

If you've already taken physics, this ought to look really familiar to you. If not, don't worry about it. The term on the left side of the equation represents the force required to keep an object of mass "m" traveling in a circle of radius "r" at a speed of "v." We call this force the **centripetal force**, and it is required whenever an object moves in a circle. The term on the right is the attractive force between two charges, "q_1" and "q_2", separated from each other with a distance of "r." The "k" in the equation is the Coulomb constant, which is equal to $8.99\text{x}10^9 \ \frac{\text{N} \cdot \text{m}^2}{\text{C}^2}$.

Now if Equation (1) mystifies you a bit, don't worry. If you think about it, the electron orbits the proton in a circle of radius "r." The term on the left side of the equation simply calculates the force necessary to keep the electron orbiting like that. Since the electron orbits the proton in a circle, then the proton and electron will *always* be a distance of "r" away from each other. The term on the right just calculates the strength of the attractive force between the electron and the proton. The total equation, then, just states that the force necessary for keeping the electron orbiting in a circle must be equal to the attraction between the two charges. In other words, the electrical attraction is what keeps the electron orbiting the nucleus in a circle.

Now to make the equation specific to our situation, we need to fill in the particulars of what we are studying. For example, we are studying an electron moving in a circle, so we will replace the m in the equation with m_e, to explicitly denote that this is an electron. Also, since we are dealing with the attraction between an electron and a proton, we can actually put in the charge of an electron and a proton. Let's keep the equation a bit general, though, by introducing some notation. Remember that the electrical charge on an electron is equal and opposite that of a proton. It is also the smallest amount of electrical charge in Creation, so we call the electrical charge of a proton and electron the "fundamental unit of charge." It's value is $1.6\text{x}10^{-19}$ C, but it is usually abbreviated as "e." That's what we'll put in for the electrical charge of the electron. Remember, however, that an electron in an atom is not attracted to *just one* proton in the nucleus, instead, it is attracted to the *entire* nucleus. Thus, we need to put in the charge of the entire nucleus. Well, in atomic chemistry and physics, the atomic number of an atom is called "Z." If an atom has an atomic number of "Z," then that means that there are "Z" protons in it. Since each proton has a charge of "e," the charge of the nucleus is "Z·e." This turns Equation (1) into:

$$\frac{m_e \cdot v^2}{r} = \frac{k \cdot e \cdot Z \cdot e}{r^2} \tag{2}$$

which simplifies to:

$$\frac{m_e \cdot v^2}{r} = \frac{k \cdot Z \cdot e^2}{r^2} \tag{3}$$

Now once again, think about what this equation means. The term on the right is the force of the attraction between the electron and the nucleus. The term on the left is the force necessary to keep an electron orbiting the nucleus at a distance of "r" and a speed of "v." Thus, the equation simply says that the attractive force between the electron and the nucleus must be strong enough to keep the electron orbiting in a circle.

That's all well and good, but so what? How does this help us understand the nature of the atom? So far, it doesn't. This kind of physics was well-known long before Bohr developed his theory of the atom, but Bohr added something. He added an assumption. You see, according to Equation (3), an electron can orbit the nucleus at *any* distance. As long as the electrical force was equal to the centripetal force necessary, then the value of "r" could be anything. Bohr looked at this situation and decided to add a constraint to the value of "r." His assumption was that electrons could not orbit the nucleus from any distance they wanted. Instead, he assumed that there were only certain orbits that an electron could occupy. In Figure 2.1, then, the electron could occupy orbit #1 or orbit #2, but it could not be anywhere in between. Scientists call this assumption a **quantum assumption**.

It is important to realize that Bohr had no explanation for this assumption. He couldn't point to any real reason that an electron could be in certain orbits but nowhere in between. He just decided to make such an assumption and see what happened in the math. To do this, he proposed the following equation:

$$m_e \cdot v \cdot r = n \cdot \hbar \quad (4)$$

where "m_e" is the mass of the electron, "v" is the electron's speed, "r" is the radius of the electron's orbit, and "n" is any integer (1, 2, 3, 4...). The "$\hbar$" in the equation is a special constant in atomic physics and chemistry. In your first-year chemistry course, you should have been introduced to **Planck's constant**, "h," which is equal to 6.63×10^{-34} J·s. The constant "$\hbar$" (called "h-bar") is simply Planck's constant divided by $2 \cdot \pi$.

$$\hbar = \frac{h}{2 \cdot \pi} \quad (5)$$

It turns out that because of the circular nature of the electron's orbit, dividing Planck's constant by $2 \cdot \pi$ just makes the math less messy in the end. That's the only significance to this constant.

How in the world does Equation (4) restrict the electron to be in only certain orbits? Well, that's where the "n" comes in. Since "n" can only have integer values (1, 2, 3, 4...), then the equation is only valid for certain values of "r." Thus, Equation (4) is simply a mathematical way of stating Bohr's quantum assumption that the electron can be only in certain orbits and nowhere else. To incorporate this assumption into the model of the atom, we simply have to solve Equation (4) for "r" and plug the result into Equation (3). First, then, we need to re-arrange Equation (4) to solve for "r":

$$r = \frac{n \cdot \hbar}{m_e \cdot v} \quad (6)$$

Then we plug that value of "r" into Equation (3) and simplify:

$$\frac{\frac{m_e \cdot v^2}{\frac{n \cdot \hbar}{m_e \cdot v}}} = \frac{k \cdot Z \cdot e^2}{\left(\frac{n \cdot \hbar}{m_e \cdot v}\right)^2}$$

$$m_e \cdot v^2 = \frac{k \cdot Z \cdot e^2}{\left(\frac{n \ \hbar}{m_e \cdot v}\right)}$$

$$m_e \cdot v^2 = \frac{k \cdot Z \cdot e^2 \cdot m_e \cdot v}{n \cdot \hbar}$$

$$v = \frac{k \cdot Z \cdot e^2}{n \cdot \hbar} \qquad (7)$$

Notice that aside from "n," there are no real variables on the right side of the equation. The letters "k," and "$\hbar$," and "e" stand for physical constants, while "Z" is determined by the atom that we are investigating. Equation (7), then, tells us the speed of the electron in any orbit (n) of the Bohr atom. Now it turns out that even though speed is a very important physical quantity to measure, chemists usually like to talk about the energy of an electron instead. Well, physics tells us that energy and speed are related as follows:

$$E = \tfrac{1}{2} \cdot m \cdot v^2 \qquad (8)$$

Plugging Equation (7) into Equation (8), then, will give us the energy of an electron in any orbit of the Bohr atom:

$$E = \frac{1}{2} \cdot m_e \cdot \left(\frac{k \cdot Z \cdot e^2}{n \cdot \hbar}\right)^2$$

$$E = \frac{m_e \cdot k^2 \cdot Z^2 \cdot e^4}{2 \cdot \hbar^2} \cdot \left(\frac{1}{n}\right)^2 \qquad (9)$$

Notice that I segregated the "1/n" term out. I did that to emphasize the fact that the "n" is the only real variable. All of the other terms in the equation are either physical constants or, in the case of "Z," determined by the atom of interest.

So now we know the energy of an electron in any orbit of the Bohr atom, right? Well, not quite. I now need to do something that drives students nuts. I want to introduce a negative sign. The electron is not moving freely in the Bohr orbit. It is bound to move in a circle by the proton. When an object is bound, we generally refer to its energy as negative. Thus, we will put a negative sign into Equation (9).

$$E = -\frac{m_e \cdot k^2 \cdot Z^2 \cdot e^4}{2 \cdot \hbar^2} \cdot \left(\frac{1}{n}\right)^2 \qquad (10)$$

Although you might think that I am pulling this negative sign out of the air, I am not. It is a consequence of the fact that the electron is not free to move on its own. If you don't like my explanation, then just accept it and move on.

Equation (10), then, gives us the energy of an electron in any Bohr orbit. In order to simplify this equation, we will actually put in the numbers for all of the physical constants in the equation. The mass of the electron (m_e) is 9.11×10^{-31} kg, the Coulomb constant (k) is 8.99×10^9 $N \cdot m^2 \cdot C^{-2}$, the fundamental charge unit is 1.60×10^{-19} C, and $\hbar$ is 1.05×10^{-34} J·s. When we do the math, then, Equation (10) turns into:

$$E = -(2.18 \times 10^{-18}\ J) \cdot Z^2 \cdot \left(\frac{1}{n}\right)^2 \qquad (11)$$

The number in Equation (11) actually has a name. It is called the "Rydberg constant" in honor of Swedish scientist J. R. Rydberg, who came up with the number *before* Bohr came up with his theory. You will learn how that happened in the next section. In the end, then, the energy of an electron in any Bohr orbit can be calculated using the following equation:

$$\boxed{E = -R_h \cdot Z^2 \cdot \left(\frac{1}{n}\right)^2} \qquad (12)$$

Where "R_h" is the Rydberg constant, 2.18 x 10^{-18} J, Z is the atomic number of the nucleus, and "n" is the number corresponding to the Bohr orbit which the electron occupies.

Equation (12) has a box around it because it is important. You will therefore have to memorize it. In addition to knowing the equation, you will also need to know *when you can* use it and *when you can not*. As I derived this formula, I made a really big assumption. You might not see it, but it is there. In Equation (1), I assumed that there was only one electron in the atom. How did I make that assumption? Well, Equation (1) attempts to balance all of the forces in the problem. I assumed there were two: the attractive force between the electron and the nucleus and the force required to hold it in its orbit. Had there been another electron in the atom, there would have been a third force: the electrical repulsion between the two electrons. That would have completely changed Equation (1), and it would have ultimately made the equation too hard to solve. In the end, then, although Equation (12) is important, its use is limited. It can be used only in situations where there is one electron.

Does this mean that Equation (12) can only be used for hydrogen atoms? Well, not exactly. It can be used for He^+ ions as well. After all, He^+ is the ion that results when helium loses an electron. Since helium has 2 electrons, He^+ has 1 electron. In the same way, Equation (12) works on Li^{2+} ions as well, since Li^{2+} ions also have only one electron. Even so, the use of

Equation (12) is rather limited. Nevertheless, without it, we would not know nearly as much about atoms as we know today. Thus, the use of Equation (12) is important.

EXAMPLE 2.1

What is the energy of an electron in the third Bohr orbit of a He^+ ion?

Since helium has two electrons, an He^+ ion has only one electron, so Equation (12) works. The atomic number of helium is 2, so Z = 2. Since the electron is in the third Bohr orbit, n = 3.

$$E = -R_h \cdot Z^2 \cdot \left(\frac{1}{n}\right)^2$$

$$E = -(2.18 \times 10^{-18}\ J) \cdot 2^2 \cdot \left(\frac{1}{3}\right)^2 = -9.69 \times 10^{-19}\ J$$

So the electron has an energy of $\underline{-9.69 \times 10^{-19}\ J}$. Notice that the answer has three significant figures. This is because the Rydberg constant has three significant figures, while "Z" and "n" are perfect integers. As a result, they have infinite precision. Thus, the number of significant figures is determined only by the Rydberg constant. Now don't worry about the energy being negative. That just means the electron is bound to the nucleus and not free to travel on its own. Also, don't worry that this number means nothing to you. You will see in the next section how this equation can be used to calculate something quite meaningful.

ON YOUR OWN

2.1 For which of the following atoms or ions can we use Equation (12)?

H, Li^+, Be, Li^{2+}, H^+, Be^{3+}, He

2.2 An electron in a hydrogen atom has an energy of -5.45×10^{-19} J. Which Bohr orbit does it occupy?

The Bohr Model and Atomic Spectra

Now we are finally to the point where we can understand a bit of what happened in Experiment 2.1. Electrons in the Bohr model have a certain amount of energy when they sit in a given Bohr orbit. Based on Equation (12), we can say that when "n" is small, the energy is large and negative. On the other hand, when "n" is large, the energy is small and negative. Well, the smaller the negative number the larger the value, so in the end, orbits with large values of "n" are considered **high energy orbits** while orbits with low values of "n" are considered **low energy orbits**. Well, suppose an electron is sitting in a low energy orbit. If it can absorb some energy, it will jump up to a high energy orbit. This energy can be absorbed from the heat of the surroundings, or it can be absorbed by capturing light. In nature, however, everything wishes to end up in its lowest energy state, which is called the **ground state**. Thus, the electron will not stay in the high energy orbit to which it jumped. Instead, it will want to go back to the lowest energy orbit that it can find. The only way it can do this is to release energy. How does an electron release energy? Usually, it does so by emitting light.

So, in order to jump up to a higher energy orbit, electrons need to absorb energy. Sometimes they absorb the energy from heat, sometimes from light, sometimes from other sources such as electricity. In order to jump back down into a lower energy orbit, the electron must release energy, usually in the form of light. How much energy does an electron need to absorb or release? Well, that depends on what orbit it is in and what orbit it is going to. For example, suppose an electron is in the third Bohr orbit and wants to jump back down to the first Bohr orbit. Well, according to Equation (12), the electron has a certain amount of energy in the third Bohr orbit, and it needs to have a certain amount of energy to be in the first Bohr orbit. The *difference* between these two energies is what the electron must release. Mathematically, we would say:

$$\Delta E = E_{initial} - E_{final} = -(2.18 \times 10^{-18}\ J) \cdot Z^2 \cdot \left(\frac{1}{n_{initial}}\right)^2 - \left[-(2.18 \times 10^{-18}\ J) \cdot Z^2 \cdot \left(\frac{1}{n_{final}}\right)^2\right]$$

This simplifies to:

$$\Delta E = (2.18 \times 10^{-18}\ J) \cdot Z^2 \cdot \left[\left(\frac{1}{n_{final}}\right)^2 - \left(\frac{1}{n_{initial}}\right)^2\right] \qquad (13)$$

In the end, then, Equation (12) leads us to another equation which allows us to calculate the energy that an electron must either absorb or release in order to change orbits in the Bohr model of the atom.

So what? Well, think about it. If an electron must absorb or emit light when it moves from one orbit to another, then Equation (13) allows us to calculate the energy of that light. As you should have learned in your previous chemistry course, the energy of light can be used to calculate the wavelength of the light, and the wavelength of light tells you its *color*! So, by using some relatively "simple" mathematics, we have derived a formula that lets you determine the color of light emitted or absorbed by an electron as it moves from one orbit to another in an atom! In order to help you do this yourself, here is a figure that tells you what wavelengths of light correspond to what color:

FIGURE 2.2

The Visible Spectrum (Wavelengths are in nanometers)

Red	Orange	Yellow	Green	Blue	Indigo	Violet
λ=700-655	λ=655-615	λ=615-570	λ=570-505	λ=505-460	λ=460-420	λ=420-390

Now remember, it is possible for light to have virtually *any* wavelength, so a great deal of the light in Creation is not in the visible spectrum. Nevertheless, if the energy of light corresponds to a wavelength between 390 nm and 700 nm, then the light is visible and its color is given by the figure above. Now I'll pull all of this together with an example.

EXAMPLE 2.2

What color of light is emitted when the electron of a hydrogen atom moves from the fourth Bohr orbit to the second Bohr orbit?

In this example, Z = 1 because we are dealing with hydrogen. The electron starts out in the fourth Bohr orbit, so $n_{initial} = 4$. It ends up in the second Bohr orbit, so $n_{final} = 2$.

$$\Delta E = (2.18 \times 10^{-18}\text{ J}) \cdot Z^2 \cdot \left[\left(\frac{1}{n_{final}}\right)^2 - \left(\frac{1}{n_{initial}}\right)^2\right]$$

$$\Delta E = (2.18 \times 10^{-18}\text{ J}) \cdot 1^2 \cdot \left[\left(\frac{1}{2}\right)^2 - \left(\frac{1}{4}\right)^2\right] = 4.09 \times 10^{-19}\text{ J}$$

This tells us, then, that the electron must lose 4.09×10^{-19} J to make the transition, so that is the energy of the light that it emits. From your first-year class in chemistry, you should have learned how to go from the energy of light to its wavelength. It starts by using Planck's constant (6.63×10^{-34} J·s) to go from energy to frequency.

$$E = h \cdot f$$

$$4.09 \times 10^{-19}\text{ J} = (6.63 \times 10^{-34}\text{ J} \cdot \text{s}) \cdot f$$

$$f = 6.17 \times 10^{14}\ \frac{1}{\text{s}}$$

Then, we use the speed of light to go from frequency to wavelength.

$$f = \frac{v}{\lambda}$$

$$6.17 \times 10^{14} \frac{1}{s} = \frac{3.00 \times 10^{8} \frac{m}{s}}{\lambda}$$

$$\lambda = 4.86 \times 10^{-7} \text{ m} = 486 \text{ nm}$$

The electron must emit light with a wavelength of 486 nm, which is blue light, according to Figure 2.2.

Now that you have seen how to use Equation (13) to calculate something really useful, it is important to realize a few things. First, when an electron has energy to absorb, it can jump up to any orbit, provided there is enough energy at its disposal. When the electron goes down to a lower energy orbit, it need not jump directly down to the lowest orbit available. Instead, it can make several jumps on its way down. Suppose, for example, that an electron jumps to the n=5 orbit. In order to jump back down to the n=1 orbit, it can do so in one jump, or it can go from n=5 to n=3 and then from n=3 to n=1. In that case, it would release light twice. First, it would release light that has the same energy as the difference between the energy in the fifth Bohr orbit and the third Bohr orbit, and then it would emit light that has energy equal to the difference in energy between the third Bohr orbit and the first Bohr orbit. Alternatively, it could jump down one orbit at a time, releasing light of four different energies. Because there are so many paths that an electron can take when jumping from a high energy orbit back to its ground state, there are many different energies of light that electrons will emit once they have absorbed energy. As a result, when atoms are exposed to a large amount of energy, they emit many different wavelengths of light. The sum total of all of those different wavelengths is called the **atomic spectrum** of that atom.

The next thing that you need to realize is that most of the light electrons emit is not visible. Remember, there are many different kinds of light, and only a small portion of it is visible. Thus, the visible spectrum is only a tiny fraction of an atom's atomic spectrum.

Finally, although this gives us some idea of how to explain the results of Experiment 2.1, it doesn't give us the entire explanation. After all, in Experiment 2.1, we were not dealing with individual atoms. We were dealing with molecules. Molecules have their own electronic structures, and we will get into that in the next module. Furthermore, those molecules were in solution, and that makes the situation even more complicated. For right now, however, we finally have a basic explanation for why colors appear in chemistry. Colors show up in chemistry because electrons emit or absorb light as a means of moving from one orbit to another.

Think, for example, about neon lights. Neon lights are made of tubes full of neon gas. Electricity is used to excite the neon atoms in the tube and, as a result, the electrons in the neon atoms move up into higher energy orbits. When they move back to lower energy orbits, they emit light. One of those electron transitions results in the emission of yellow-orange light that we call "neon" light. Originally, that was the only color in neon lights. However, as we learned more about chemistry, other gases which emit other colors when excited by electricity were put in the tubes, and even though those gases are not neon, we still refer to signs that use such tubes of gas as "neon" signs. Fluorescent lights operate in the same way, but they contain a mixture of gases that emit so many different colors when excited that, in the end, we see the light that is emitted as white light.

Now although you might think that in order for a substance to have color, it must emit light, a substance can actually have color by absorbing light as well. For example, a black substance appears black to us because it absorbs all light that hits it. Thus, when we see it, we see the absence of light, and our mind interprets that as the color black. In the same way, suppose a substance were to absorb all red, orange, yellow, and green light that hit it, but it did not absorb the blue, indigo and violet light that hit it. How would our brain interpret that? Well, our eyes would still see the blue, indigo, and violet portions of the visible light that hits the substance, but not the red, orange, yellow, and green light. As a result, we would see a purple-blue color. It turns out that the vast majority of color we see in Creation is the result of such a process. A substance absorbs certain colors and, by the absence of those colors, our eyes and brain construct a color image based on what light is not absorbed.

ON YOUR OWN

($h = 6.63 \times 10^{-34}$ J·s, speed of light = 3.00×10^{8} m/s)

2.3 An electron in a He^{+} ion starts out in the first Bohr orbit and jumps to the fourth Bohr orbit. How much energy must it absorb?

2.4 An electron jumps from the n=3 orbit of a hydrogen atom to the n=2 orbit. What is the wavelength of the light emitted? Is it visible? If it is visible, what color is it?

2.5 An electron jumps from the fifth Bohr orbit to a lower orbit in a Li^{2+} ion. If the light emitted has a frequency of 6.67×10^{14} Hz, what orbit did the electron end up in?

Before we leave this section, I want to point out a couple of things. First of all, long before Bohr came along, scientists knew that atoms emit light when they absorb energy. In fact, this phenomenon puzzled scientists quite a bit. They understood that if an electron absorbed energy it must eventually release that energy again, and it made sense to them that one way in which an electron could release energy was in the form of light. What puzzled them, however, was that only light of *certain* wavelengths were emitted. For example, when an electrical discharge was sent through a tube full of hydrogen gas, only 5 wavelengths of visible light were emitted: 397 nm (violet) ,411 nm (violet), 434 nm (indigo), 486 nm (blue), 656 nm (red). Scientists at the time could not understand why just those wavelengths were emitted. They assumed that an electron should be able to emit any wavelength of light. Why, then, did hydrogen atoms, when excited by an electrical discharge, only emit those wavelengths of visible light?

Lots of scientists went to great pains to try and explain this mysterious data. Indeed, Swedish scientist J. R. Rydberg actually spent so much time puzzling over the data that he was able to "cook up" an equation to fit the data. Long before Bohr came along, he developed the following equation for the energy of visible light emitted from an excited hydrogen atom:

$$E = (2.18 \times 10^{-18}\ \text{J}) \cdot \left[\left(\frac{1}{2}\right)^2 - \left(\frac{1}{n}\right)^2\right] \text{ where n = 3,4,5,6 and 7} \tag{14}$$

Look at what that is. It is Bohr's equation, with $n_{final} = 2$ and $n_{initial} = 3, 4, 5, 6,$ and 7. Now Rydberg did not derive this formula. He had no idea what it meant. He just looked at the data so long and so hard that he finally dreamed up an equation that would fit it!

What Bohr's model did was explain the data by making the assumption that only certain orbits were possible. That way, there would be only certain energies involved, and that would explain the existence of only certain wavelengths of emitted light. Once he worked out the math, he showed that Rydberg's equation was just a special case of his general equation. It was the case in which the electron was starting in the third through seventh orbit and was jumping down to the second orbit. Thus, even though Rydberg's equation was not all that useful, we still honor him by naming the constant after him. After all, he was actually able to stumble onto part of Bohr's equation before Bohr did! Thanks to Bohr's quantum assumption, however, Bohr was able to derive the full equation.

Now you must realize that *to this day* we do not understand *why* there are only certain possible orbits in an atom. As a matter of fact, that very assumption goes against common sense. After all, why shouldn't an electron be wherever it "wants" to be, provided it has the correct energy? We have no idea. We do know this, however. By making this wild assumption, Bohr was able to explain a wealth of scientific data that could not have been explained before. Thus, *despite the fact that the quantum assumption which Bohr made goes against common sense, we believe in it because it explains the data.* Even though we know that there are some mistakes in the Bohr model, our current model of the atom (the quantum mechanical model) makes a quantum assumption very similar to what Bohr's model makes. Thus, science today holds strong to the quantum assumption, despite the fact that it makes no sense. Science holds to the quantum assumption *only* because it explains the data.

Now I hope that you can make the obvious connection between faith in Christ and faith in the quantum assumption. The only reason to believe in the quantum assumption is that it explains the data. Scientists did not believe in the Bohr model because it made sense. Scientists do not currently believe in the quantum mechanical model because it makes sense. In fact, both models *go against* sense. They believe in it because it explains the data. It is the same with faith in Christ. It doesn't matter whether Christ as Lord and Savior makes any sense to you. There are certainly things in the Christian faith (like the Trinity) that are really *hard* to understand and probably do not make sense to you. Just like our current models of the atom, however, that's irrelevant. The only question (from a scientific point of view) is, "Does it explain the data?" After studying Creation, the scientist must admit that an all-powerful Designer does "explain the data." After studying the scientific evidence for the validity of the Bible (see Josh McDowell's *Evidence That Demands a Verdict* or Jay Wile's *Reasonable Faith: The Scientific Case for Christianity*), you will also find that the Christian faith "explains the data." Thus, the next time someone tells you that faith in Christ is "not scientific," you can tell them that it is just as scientific as current atomic science!

The other thing I wanted to point out to you is rather simple. Because the atomic number (Z) is in Equation (13), the light emitted and absorbed by any element will be unique because each element has its own atomic number. As we will learn later, each molecule also has its own electronic structure and, as a result, the light that it can absorb or emit is unique to that molecule. As a result, if we excite an element or compound (typically by heat or electricity) and look at the pattern of wavelengths (visible and not visible) of light that it emits, we can unambiguously determine that element or compound. This technique is called **emission spectroscopy**, and it is used to examine the elemental makeup of stars and to analyze chemicals. Alternatively, we could shine light through a compound or element and see what wavelengths of light are absorbed. This is called **absorption spectroscopy** and is typically used more frequently, because it does not risk destroying the compound or element being studied.

The Size of an Atom

Although you are aware that atoms are small, it is time to get a good handle on exactly how small they are. You see, the Bohr model not only allows us to calculate the energy of an electron in any given orbit, but it also allows us to calculate the radius of the orbits that the electrons occupy. As I have said before, we now know that the Bohr model is not completely accurate, but for all one-electron atoms, it does a pretty good job, so it is still rather instructive, especially because the mathematics associated with it are so "simple."

Now remember that the major advance Bohr made in his model was his quantum assumption. This assumption can be mathematically stated with Equation (5), which we then re-arranged to Equation (6):

$$r = \frac{n \cdot \hbar}{m_e \cdot v} \qquad (6)$$

Well, in our derivation of Equation (12), we came up with an expression for the velocity of an electron in the nth Bohr orbit:

$$v = \frac{k \cdot Z \cdot e^2}{n \cdot \hbar} \qquad (7)$$

If we put that expression for "v" into Equation (6), look what happens:

$$r = \frac{n \cdot \hbar}{m_e \cdot \left(\frac{k \cdot Z \cdot e^2}{n \cdot \hbar}\right)} = \left(\frac{\hbar^2}{m_e \cdot k \cdot e^2}\right) \cdot \frac{n^2}{Z} \quad (15)$$

If we take that equation and replace the physical constants in parentheses with numbers, we get:

$$r = (5.29 \times 10^{-11}\ \text{m}) \cdot \frac{n^2}{Z} \quad (16)$$

Since atoms are so small, we usually replace the meters unit with the Angstrom unit, which is equal to 10^{-10} meters. Thus, the usual way you see the equation is:

$$r = (0.529\ \text{Å}) \cdot \frac{n^2}{Z} \quad (16)$$

where "Å" is the symbol for the Angstrom unit.

Using this equation for the first Bohr orbit of a hydrogen atom, n=1 and Z=1. Thus, the radius of the first Bohr orbit in the hydrogen atom, according to Equation (16), is 0.529 Å, or 5.29×10^{-11} meters. Since there is only one electron in a hydrogen atom, in its ground state, the first Bohr orbit is the only orbit that is occupied. Thus, this tells us that the hydrogen atom itself has a radius of 0.529 Å, in its ground state. Now think about that number for a moment. Take out a metric ruler and look at the markings that indicate millimeters. They are pretty close together, aren't they? Well, based on the size we just calculated, approximately 19,000,000 hydrogen atoms in their ground state could fit in between two of those millimeter dashes! Hopefully, you now have some idea how small atoms really are!

ON YOUR OWN

2.6 What it the radius of a Li^{2+} ion in its ground state?

2.7 An electron in a hydrogen atom has an energy of -2.42×10^{-19} J. How far from the nucleus is it orbiting?

Moving From the Bohr Model to the Quantum Mechanical Model

As I have said many times before in this module, the Bohr model of the atom is flawed and, as a result, scientists do not accept it any more as the standard view of the atom. Today, atomic scientists are committed to the quantum mechanical model of the atom. In fact, there are many similarities between the quantum mechanical model and the Bohr model. As its name implies, the quantum mechanical model of the atom still makes use of the quantum assumption. Instead of talking about orbits, however, the quantum assumption in the quantum mechanical model says that there are only specific energies that an electron in an atom can have. An electron is not free to take on any energy. Instead, it is constrained to have only certain, distinct energies.

Instead of circular orbits like those of the Bohr model, the quantum mechanical model of the atom says that electrons, depending on their energy, can orbit the nucleus in "clouds" that sometimes have quite interesting shapes. The shapes of these clouds (called **orbitals**), as well as their size, are governed by certain numbers called **quantum numbers**. In your first-year chemistry course, you learned the shapes of some of these orbitals when you learned about electron configurations. In this course, however, we will go back and take a more detailed look at electron orbitals so that you can get a better idea of where they come from.

The quantum mechanical model of the atom is governed by a very complicated equation called the **Schrodinger equation**. This equation takes two years of post calculus mathematics to understand and an additional year of math beyond that to really learn how it is solved. Obviously, then, I am not going to try and make you learn it! However, I will tell you that when you apply the Schrodinger equation to the situation of an electron orbiting a nucleus, you get some reasonably "simple" results. The solution of the Schrodinger equation tells us that electrons orbiting a nucleus can have only certain, distinct energies and positions within the atom. Those energies and positions are determine by a series of four numbers called **quantum numbers**. Once you determine those four quantum numbers for an electron, you know how much energy it has and roughly where it is in the atom.

The first quantum number, called the **principal quantum number**, is abbreviated as "n." This is no accident. The principal quantum number turns out to be the same integer that Bohr used in his model to describe the orbit that the electron is in. Much like the Bohr model, this principal quantum number indicates how far the electron is from the nucleus. Just as is the case with the Bohr model, the larger the value for "n," the farther away the electron is (on average) from the nucleus. In addition, the value for "n" affects energy greatly. When two electrons in the same shape orbital have two different values for "n," the electron with the larger value for "n" has *substantially more energy* than the other electron. Once again, as is the case for the Bohr atom, the principal quantum number can have any integer value other than zero.

Principal Quantum Number: n = 1, 2, 3, 4...

The second quantum number that results from solving the Schrodinger equation is called the **azimuthal** (az uh myoo' thuhl) **quantum number** and it is usually abbreviated as "ℓ." This quantum number tells you what shape orbital the electron is using to orbit the nucleus. It also affects the energy, but not nearly as significantly as does the principal quantum number. Now the azimuthal quantum number can have any integer value, including zero, but the range of integer values is restricted by the value of n. For example, when "n" equals one, the only possible value of "ℓ" is zero. When "n" equals 2, "ℓ" can be either zero or one. In general, then, "ℓ" can have any integer value from zero to one less than the principal quantum number.

Azimuthal Quantum Number: ℓ = 0, 1, ... (n-1)

Now at this point that may make no sense to you, but don't worry. I'll explain why this is the case in a minute. For right now, I need to tell you a little more about "ℓ."

As I said before, the azimuthal quantum number tells you what shape orbital the electron occupies. For some strange reason, chemists associate letters with these orbital shapes. Figure 2.3 illustrates the basic shapes of electron orbitals for the first 6 values of the azimuthal quantum number, along with the letter associated with that shape:

FIGURE 2.3
Electron Orbitals

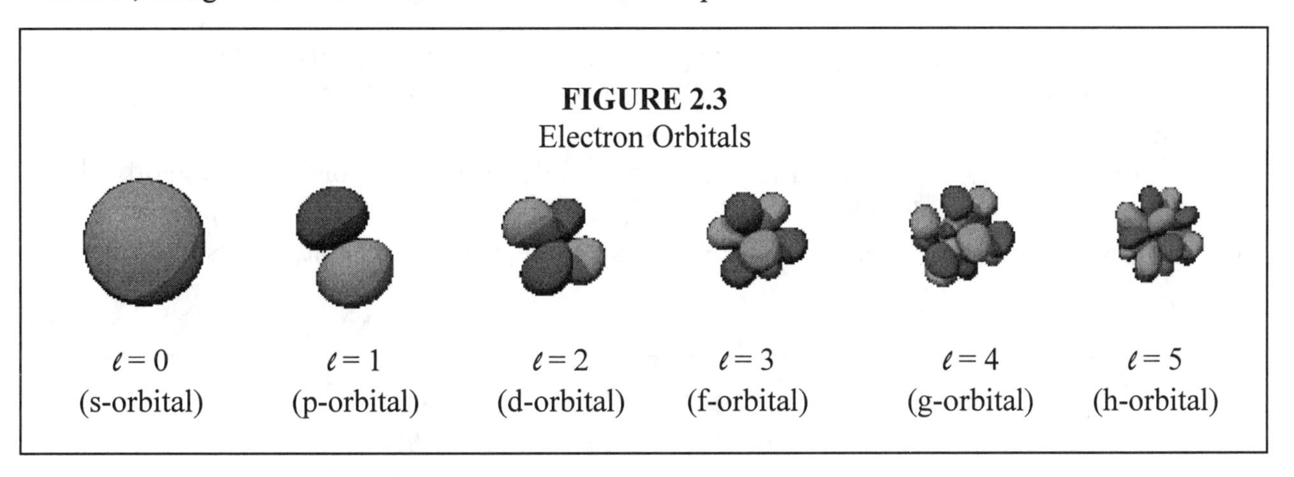

You learned the shape and letters associated with the first few orbitals (s, p, and d) in your previous chemistry course. In this course, you see that these orbitals are determined by the azimuthal quantum number, which comes from solving the Schrodinger equation in the quantum mechanical model of the atom.

Now why does the value of "ℓ" depend on the value of "n?" Well, think about what the value of "n" tells us. It tells us the average distance that the electron is from the nucleus. As the value for "n" gets larger, the larger that distance becomes. Well, when n=1, the electron is rather close to the nucleus; thus, there is not much room. As a result, only one orbital can "fit" within that space. Thus, only the s-orbital (ℓ=0) is present when n=1. When n=2, however, the electron is farther away from the nucleus. As a result, there is more room, so more orbitals can "fit" within that space. As we will learn in a moment, there are actually three p-orbitals, so when n=2, there is room for an s-orbital and three p-orbitals, so "ℓ" can equal either 0 or 1. Likewise, when n=3, the electron is even farther out, so now there is room for an s-orbital, three p-orbitals, and five d-orbitals. As a result, when n=3, "ℓ" can equal 0, 1, or 2.

So the reason the value of "ℓ" depends on the value of "n" all relates to space. There are spatial limitations that slowly go away as "n" increases. Thus, for low values of "n," there are only a few possible orbitals. For large values of "n," however, many more orbitals are available. How do we know what values of "n" and "ℓ" to use for any given electron in any given atom? Don't worry about it. We'll get to that in a moment. There are still two more quantum numbers to learn!

The next quantum number is called the **magnetic quantum number**, and it is usually abbreviated as "m." This number affects certain details of an orbital's shape, as well as its orientation in space. Just as the value of "ℓ" depends on the value of "n," the value of "m" depends on the value of "ℓ." The magnetic quantum number can have all integer values from $-\ell$ to $+\ell$. Thus, when ℓ=0, the only possible value for "m" is zero. When ℓ=1, however, "m" can take on any of three values: -1, 0, or 1. Likewise, when ℓ=2, the value of "m" can be -2, -1, 0, 1, or 2.

Magnetic Quantum Number: m ranges from $-\ell$ to $+\ell$ in integer steps

Once again, don't get wrapped up in worrying over how you know what quantum numbers to use when. Just understand the rules for how "m" relates to "ℓ." The rest will come later.

As I said, the value for "m" affects the shape and the orientation of the electron orbital. When ℓ=0, the only possible value for "m" is 0. This tells us that there is only one possible s-orbital. If we think of the atom in three-dimensional space, then, the s-orbital looks like this:

FIGURE 2.4
The S-orbital (ℓ=0, m=0)

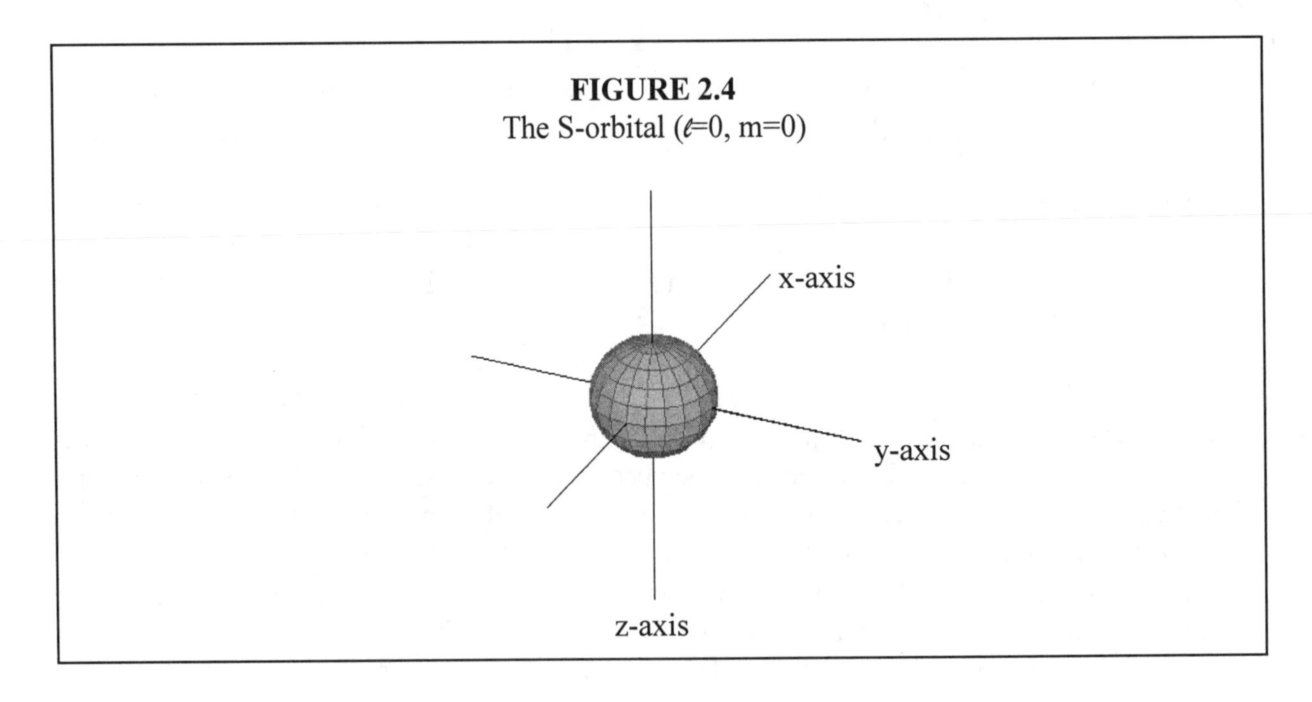

When ℓ=1, three values for "m" are possible, -1, 0, or 1. Since ℓ=1 means we are dealing with p-orbitals, then this tells us that there are 3 different kinds of p-orbitals. There is a p-orbital for which m=1, a p-orbital for which m=0, and a p-orbital for which m=-1. Those three orbitals are shown below:

FIGURE 2.5
The Three P-orbitals (ℓ=1, m=-1, 0, 1)

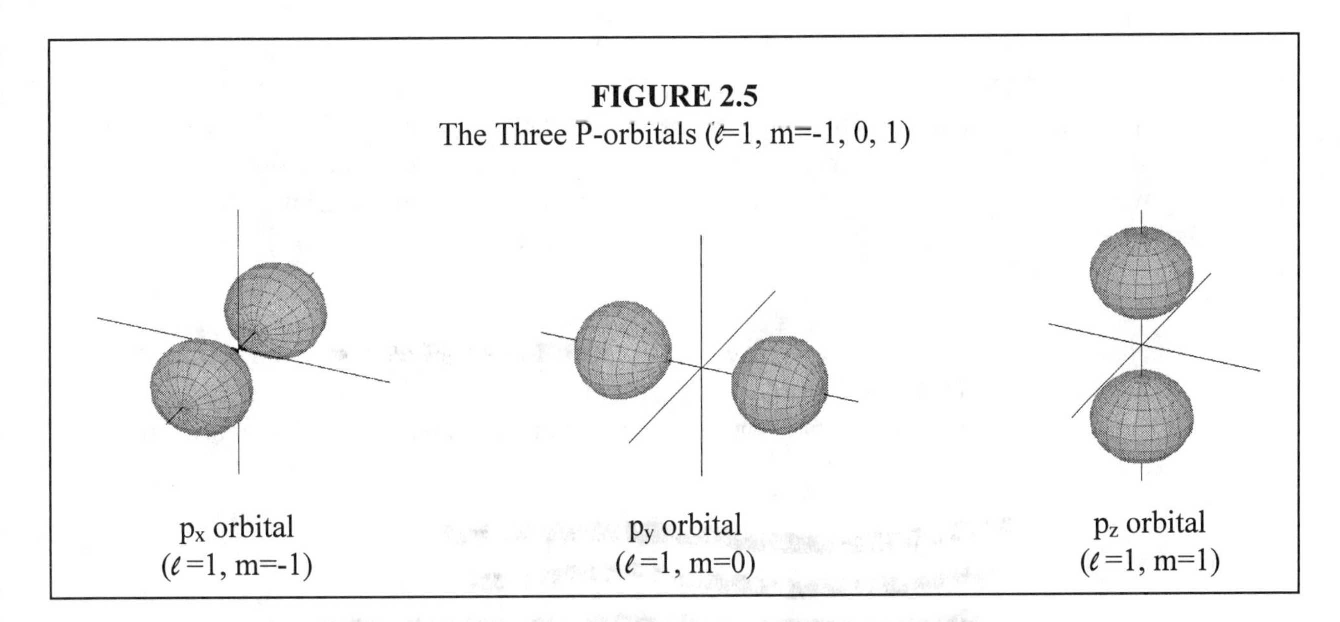

Notice that the basic shape of these three orbitals are the same. That's because, for the most part, orbital shape is determined by "ℓ." What's different between the three is the orientation. If we image the nucleus of the atom at the origin of our three-dimensional axis, you can see that all three of these orbitals can exist simultaneously around the atom. Now realize that the electron can be *anywhere* within the orbital, but it cannot be anywhere outside of the orbital. Thus, you can see that the electron cannot be anywhere right near the nucleus, but it can be anywhere within the two lobes of the orbital.

To carry the illustration just one step further, when ℓ=2, there are five possible values of "m": -2, -1, 0, 1, and 2. This means there are five different types of d-orbitals, as shown below:

FIGURE 2.6
The Five D-orbitals (ℓ=2, m = -2, -1, 0, 1, 2)

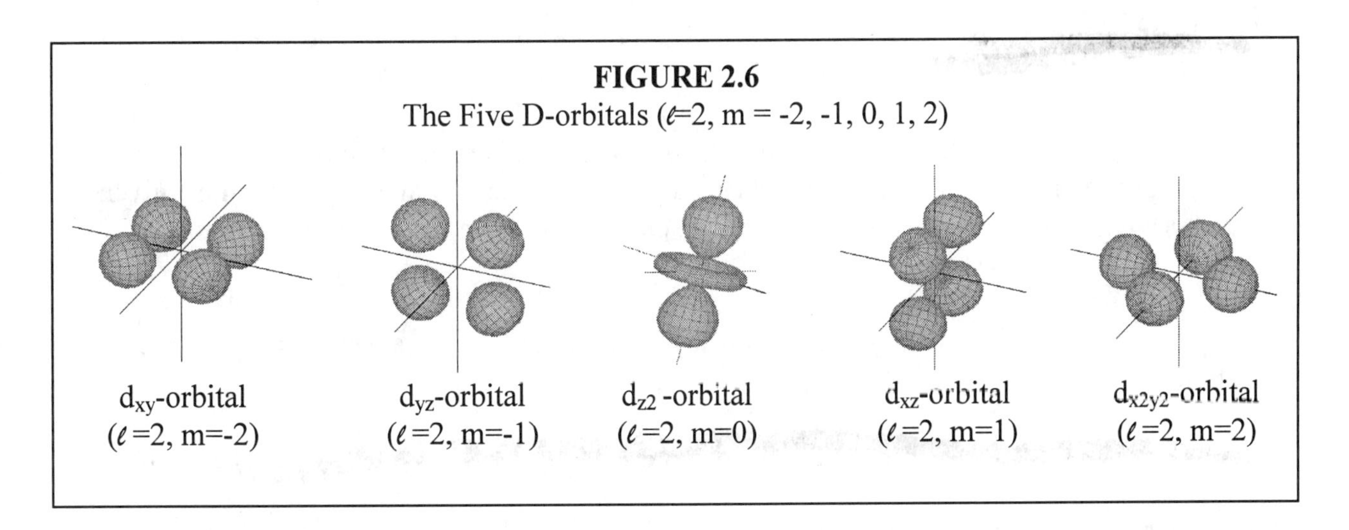

Notice that the value of "m" mostly affects the orientation of the orbital, except in the case where m=0. There, the value of m does affect the shape. If you are good at three-dimensional visualization, you can see that all of these orbitals can exist simultaneously with the nucleus at the origin of the three-dimensional axis. I could go on with ℓ=4, 5, and 6, but we will stop there. The point should be clear. Because of the relationship between "ℓ" and "m," the number of possible orbitals increases as the value of "ℓ" increases. In general, the value of "m" determines the orientation of the orbital in space, and it has an effect on the shape of the orbital in some cases. When you see drawings of orbitals such as those shown in these figures, remember what they mean. The electron can be found *anywhere* within the orbital, but nowhere outside of the orbital. These orbitals, then, represent regions around the nucleus in which the electrons can be found.

The last quantum number to learn is called the **spin quantum number**, and it is usually abbreviated with an "s." This is the simplest quantum number. It can have one of only 2 values, $+\frac{1}{2}$ or $-\frac{1}{2}$. These values are independent of any of the other quantum numbers. To review, then, there are four quantum numbers:

Principal Quantum Number: n = 1, 2, 3, 4...
Azimuthal Quantum Number: ℓ = 0, 1, ... (n-1)
Magnetic Quantum Number: m ranges from $-\ell$ to $+\ell$ in integer steps
Spin Quantum Number: $+\frac{1}{2}$ or $-\frac{1}{2}$

When you put together a list of four quantum numbers that meet these criteria, you have uniquely determined the average distance from the nucleus, general orbital shape, orbital orientation, and energy of an electron in an atom.

Now it's time to tell you *how* to determine what quantum numbers to use when you are analyzing an atom. To start with, you need to learn a few principles on how to use quantum numbers. The first one is a fundamental principle that comes with the quantum mechanical model, called the **Pauli exclusion principle.**

Pauli exclusion principle - No two electrons in the same atom can have the same 4 quantum numbers.

The reason for this principle is simple. The four quantum numbers listed above determine all of the properties of an electron in an atom. Thus, they *uniquely* identify an electron within a given atom. As a result, it is simply impossible for two electrons to have the same four quantum numbers within the same atom.

The next principle is one you have already learned before. It is a general driving force in Creation:

Everything in Creation strives to reach its lowest energy state, called the "ground state."

Thus, when you start trying to assign quantum numbers to an electron in an atom, you need to do so in such a way as to allow the electrons to have as little energy as possible. How do you do that? Well, remember that the quantum numbers "n" and "ℓ" are what determine the energy of an electron orbital. The value for "n" affects the energy of an orbital in a large way whereas the value for "ℓ" affects the energy a smaller way. For both cases, however, the larger the value of the quantum number, the higher the energy of the electron.

How in the world do you determine what values of "n" and "ℓ" to choose to keep an electron's energy as low as possible? Well, believe it or not, you have already learned how to do that! In your first year chemistry course, you learned how to do electron configurations. When you did an electron configuration, you were actually using the principle quantum number and the azimuthal quantum number, you just didn't know it. When an electron configuration has a "1s" orbital in it, for example, that means n=1 and ℓ=0. When an electron configuration has a 2p orbital in it, that means n=2 and ℓ=1! As you build an atom up, then, you choose the "n" and "ℓ" values in the same way that you choose orbitals when doing an electron configuration. You'll see how this is all done in just a moment.

The last principle you need to learn is called **Hund's rule**, and it deals with choosing the values for "m" and "s."

Hund's rule - For a given value of "ℓ," all orbitals must be singly filled with equivalent values of "s" (usually $+\frac{1}{2}$) before the orbitals are filled with a second electron.

You probably have no idea what this rule means. Don't worry about it. Study the following example and (hopefully) it will all come together for you.

EXAMPLE 2.3

Assign quantum numbers to all electrons in a helium atom.

A helium atom has 2 electrons. We cannot assign them both the same 4 quantum numbers, because that would violate the Pauli exclusion principle. Also, we need to assign the quantum numbers so that the electrons have their lowest possible energy state. First, we need to assign "n" and "ℓ." That's easy. We can already do the electron configuration of helium ($1s^2$). This means that both electrons are in the 1s orbital. Remember, all s-orbitals have ℓ=0, and the number in front of the orbital tells us the value of the principle quantum number. This tells me that n=1 and ℓ=0. Now, what about "m" and "s?" Well, the value of "m" ranges from -ℓ to + ℓ. Since ℓ=0, then this means m=0 as well. Now I just have to assign "s." How do I do that? Well, we generally assign $s=\frac{1}{2}$ first and $s=-\frac{1}{2}$ second. If I give $s=\frac{1}{2}$ to the first electron, then that makes its quantum number assignment complete. The second one can have the same values for n, ℓ, and m, and just have a different value for "s." Thus, I will give it $s=-\frac{1}{2}$. My quantum numbers, then, are:

First electron: n=1, ℓ=0, m=0, $s=\frac{1}{2}$
Second electron: n=1, ℓ=0, m=0, $s=-\frac{1}{2}$

Notice that neither of the electrons have the same 4 quantum numbers, and that I have kept the values of "n" and "ℓ" as low as possible, ensuring that the electrons are in their lowest energy state.

Assign the quantum numbers of all electrons in a nitrogen atom.

The best way to assign quantum numbers is to do the electron configuration for an atom. That tells you the values of "n" and "ℓ" to use so that the electrons stay in their ground state. The electron configuration for nitrogen is $1s^2 2s^2 2p^3$. This tells us that the first two electrons have n=1 and ℓ=0. We already know the other quantum numbers, because we just did helium above. Thus:

First electron: n=1, ℓ=0, m=0, s=$\frac{1}{2}$
Second electron: n=1, ℓ=0, m=0, s=-$\frac{1}{2}$

The next portion of the electron configuration tells us that two electrons go in the 2s orbital, which means two electrons have n=2 and ℓ=0. Remember, "ℓ" can range from 0 to n-1, so zero is certainly a legal value for "ℓ." This just leaves "m" and "s." Well, we are in the same boat we were in before. When ℓ=0, the only possible value for "m" is 0. Thus, to keep these two electrons from having the same four quantum numbers, we will give them both m=0 because we must, but we can give one of them s=$\frac{1}{2}$ and the other s=-$\frac{1}{2}$. In the end, then, the next two electrons have the following quantum numbers:

Third electron: n=2, ℓ=0, m=0, s=$\frac{1}{2}$
Fourth electron: n=2, ℓ=0, m=0, s=-$\frac{1}{2}$

There are three electrons left, and the electron configuration tells us that they go in the 2p orbitals. This is where Hund's rule comes into play. Because the value of "m" ranges from -ℓ to +ℓ, there are three possible values for "m" when ℓ=1. The possible values for "m" are -1, 0, or 1, and each of those values corresponds to its own p-orbital. How do we know which one to choose? Well, Hund's rule says that you place electrons in each orbital singly before you pair them up in the same orbital. Thus, if we put the first electron in the m=1 orbital (we always assign the positive numbers first), we can give it s=$\frac{1}{2}$. According to Hund's rule, however, we cannot put another electron into the m=1 orbital, because for a given value of "ℓ," we need to fill all orbitals singly before we go back and put a second electron in that orbital. Thus, we will give the next electron the m=0 orbital and the last electron the m=-1 orbital. That way, all three p-orbitals have only one electron in them, in accordance with Hund's rule. In each case, we will give each electron s=$\frac{1}{2}$, because we always assign that value of "s" until we have to assign the other value. The final electrons, then, have the following quantum numbers:

Fifth electron: n=2, ℓ=1, m=1, s=$\frac{1}{2}$
Sixth electron: n=2, ℓ=1, m=0, s=$\frac{1}{2}$
Seventh electron: n=2, ℓ=1, m=-1, s=$\frac{1}{2}$

Is your head swimming from that? If it is, don't worry. Quantum numbers are one of the most confusing things in chemistry when you first look at them. Once you get the knack, however, they are really quite easy to do. After all, it starts with just an electron configuration, which you became a master at in your last chemistry course. That gives you the "n" and "ℓ" values for all of the electrons in the atom. From that point, all you do is fill in the values of "m" and "s," being careful to follow Hund's rule.

Now don't get so wrapped up in all of the rules of quantum numbers that you forget the chemical point to it all. When you assign quantum numbers, you assign where an electron is in an atom. Look, for example, at the fifth electron in the last example problem above. It has an "n" value of 2, indicating that it is farther away from the nucleus than the closest electrons (the ones with n=1), but still relatively close to the nucleus. It also has an "ℓ" value of 1, indicating that it is in a p-orbital. The "m" value of 1 indicates that the *particular* p-orbital it occupies is the one oriented on the z-axis (see Figure 2.5), and the "s" value of $\frac{1}{2}$ indicates that it is the first one to go into that orbital. Thus, these quantum numbers tell you a *lot* about the electrons in an atom. Don't forget that!

You should also notice something about how quantum numbers relate to electron configurations. In your first year chemistry course, you learned that an individual orbital can hold only two electrons. You should now see why that is the case. An orbital is determined by the first three quantum numbers (n, ℓ, and m). The value for "n" tells you the average distance from the nucleus for the orbital, the "ℓ" value tells you the general shape, and the "m" value tells you the specific orbital for that general shape. This means that when you finally get down to assigning an electron to an orbital, you have used three of the four quantum numbers at your disposal. The last quantum number has only two values, so once you have chosen an orbital, you can only have 2 sets of unique quantum numbers. The Pauli exclusion principle, then, along with the nature of the spin quantum number, is what limits the number of electrons to two per orbital.

Also, in learning electron configurations, you learned that there were three p-orbitals, each of which has 2 electrons. In the end, then, a given set of p-orbitals can hold six electrons. You should now see that the reason there are 3 p-orbitals is that there are three possible values for "m" when "ℓ" has a value of 1, so there are three possible p-orbitals. In the same way, when ℓ=2 (d-orbitals), there are 5 possible values for "m" (-2,-1,0,1,2), so there are 5 d-orbitals. Since each orbital can hold 2 electrons, a given set of d-orbitals can hold 10 electrons. Thus, the relationships between quantum numbers explains *why* there is only 1 s-orbital while there are 3 p-orbitals and 5 d-orbitals.

You really do need to be clear on assigning quantum numbers. Also, you need to understand the relationships between quantum numbers and what that tells you. Study the following examples and solve the "on your own" problems to make sure this is coming clear in your mind.

EXAMPLE 2.4

Assign quantum numbers to the valence electrons in a bromine atom.

Remember that valence electrons are those electrons which are farthest away from the nucleus. Thus, we are going to be dealing with electrons with the largest value for "n." To start, we determine the electron configuration:

$$1s^2 2s^2 2p^6 3s^2 3p^6 4s^2 3d^{10} 4p^5$$

The valence electrons, then, are the ones in the 4s and 4p orbitals, since they have the largest values for "n." We will first work on the 4s orbital. For that orbital, n=4 and ℓ=0. When ℓ=0, m=0. To keep the electrons unique, then, they must have different values of "s." Their quantum numbers are:

$n=4, \ell=0, m=0, s=\frac{1}{2}$ and $n=4, \ell=0, m=0, s=-\frac{1}{2}$

That takes care of the 4s electrons. Now for the 4p electrons, of which there are five. These electrons have n=4 and ℓ=1. When ℓ=1, the value of "m" can be -1, 0, or 1. Hund's rule says we use each of those numbers once with $s=\frac{1}{2}$ before we go back and use them again. We always start with the most positive number and work our way down to the negative numbers; thus, the next three electrons have the following quantum numbers:

$n=4, \ell=1, m=1, s=\frac{1}{2}$ and $n=4, \ell=1, m=0, s=\frac{1}{2}$ and $n=4, \ell=1, m=-1, s=\frac{1}{2}$

We still have two electrons to go, so at this point, we must go back and put a second electron in two of the orbitals that already have one electron in them. The only way to do that and keep the 4 quantum numbers unique is to have $s=-\frac{1}{2}$. Thus, the final two electrons have the following quantum numbers:

$n=4, \ell=1, m=1, s=-\frac{1}{2}$ and $n=4, \ell=1, m=0, s=-\frac{1}{2}$

What is wrong with the following quantum number assignment?

$$n=3, \ell=2, m=3, s=\frac{1}{2}$$

This quantum number assignment is impossible because "m" can only range from $-\ell$ to $+\ell$. Since ℓ=2, that means "m" can only have values of -2, -1, 0, 1, or 2.

What is wrong with the following quantum number assignment for three electrons in the same atom?

$n=3, \ell=2, m=2, s=\frac{1}{2}$ **and** $n=3, \ell=1, m=2, s=-\frac{1}{2}$ **and** $n=3, \ell=2, m=2, s=\frac{1}{2}$

These assignments break the Pauli exclusion principle because the first and the third are identical. That's impossible.

The f-orbitals have ℓ=3. How many f-orbitals are there and how many electrons total can they hold?

When ℓ=3, "m" can have values of -3, -2, -1, 0, 1, 2, or 3. This makes 7 possible values which correspond to 7 different orbitals. Since each orbital can hold two electrons, the f-orbitals can hold 14 electrons total.

ON YOUR OWN

2.8 Give the quantum numbers for the valence electrons in a sulfur atom.

2.9 What is the problem with each set of quantum numbers given below?

a. n=2, ℓ=2, m=1, s= $\frac{1}{2}$ b. n=0, ℓ=0, m=0, s= $\frac{1}{2}$ c. n=2, ℓ=1, m=2, s= -$\frac{1}{2}$

2.10 When ℓ=4, the orbitals are called g-orbitals. How many g-orbitals are there and how many total electrons can they hold?

Before we leave this subject, there is one point of terminology that I want you to know. Remember that the only quantum numbers which affect the energy of an electron are "n" and "ℓ." That means that values of "m" and "s" do not affect the energy of an electron at all. Well, think about electrons in the 2p orbitals of an atom. There are three p-orbitals, each of which can contain two electrons. Thus, there can be up to six electrons in these three p-orbitals. All of those electrons will have the same energy, because they all have the same values for "n" and "ℓ." When you have many orbitals that all have electrons of the same energy, they are called **degenerate orbitals**.

Degenerate orbitals - Orbitals that contain electrons of the same energy

Since "n" and "ℓ" are the only quantum numbers that affect energy, any orbitals with the same values of "n: and "ℓ" will be degenerate orbitals. Thus, all five of the 3d orbitals are degenerate, as are all three of the 4p orbitals, for example.

Hopefully you now have a deeper understanding of atomic structure. Unfortunately, I have still not fully explained the results of Experiment 2.1. Don't worry, though. That will come in the next module when I discuss molecular orbitals. For right now, shore up your understanding of atomic orbitals because, as you might have already guessed, molecular orbitals will be awfully difficult to understand if you are not really confident with atomic orbitals!

ANSWERS TO THE ON YOUR OWN PROBLEMS

2.1 The Bohr model can only be used on atoms or ions with just one electron. Thus, the following atoms or ions will work: H, Li^{2+}, Be^{3+}.

2.2 In this problem, we know that Z=1 because we are dealing with a hydrogen atom. We also know the energy. Equation (12), therefore, can be used to solve for n:

$$E = -R_h \cdot Z^2 \cdot \left(\frac{1}{n}\right)^2$$

$$-5.45 \times 10^{-19}\ J = -(2.18 \times 10^{-18}\ J) \cdot 1^2 \cdot \left(\frac{1}{n}\right)^2$$

$$\left(\frac{1}{n}\right)^2 = \frac{-5.45 \times 10^{-19}\ J}{-2.18 \times 10^{-18}\ J} = 0.25$$

$$\frac{1}{n} = 0.5$$

$$n = 2$$

The electron is in the second Bohr orbit.

2.3 To jump from low energy orbits to high energy orbits, electrons must gain energy. The amount of energy is determined by the difference in energy between the two orbits, which is calculated using Equation (13). Since we are dealing with a helium ion here, Z=2.

$$\Delta E = (2.18 \times 10^{-18}\,J) \cdot Z^2 \cdot \left[\left(\frac{1}{n_{final}}\right)^2 - \left(\frac{1}{n_{initial}}\right)^2\right]$$

$$\Delta E = (2.18 \times 10^{-18}\,J) \cdot 2^2 \cdot \left[\left(\frac{1}{4}\right)^2 - \left(\frac{1}{1}\right)^2\right] = -8.18 \times 10^{-18}\ J$$

The negative sign simply means that the electron absorbs energy rather than emits it. Thus, the electron must absorb -8.18×10^{-18} J of energy.

2.4 In this problem, Z = 1 because we are dealing with hydrogen. The electron starts out in the third Bohr orbit, so $n_{initial} = 3$. It ends up in the second Bohr orbit, so $n_{final} = 2$.

$$\Delta E = (2.18 \times 10^{-18}\,J) \cdot Z^2 \cdot \left[\left(\frac{1}{n_{final}} \right)^2 - \left(\frac{1}{n_{initial}} \right)^2 \right]$$

$$\Delta E = (2.18 \times 10^{-18}\,J) \cdot 1^2 \cdot \left[\left(\frac{1}{2} \right)^2 - \left(\frac{1}{3} \right)^2 \right] = 3.03 \times 10^{-19}\ J$$

This tells us, then, that the electron must emit light with energy of 3.03×10^{-19} J to make the transition. From the energy, we can get the frequency:

$$E = h \cdot f$$

$$3.03 \times 10^{-19}\ J = (6.63 \times 10^{-34}\ J \cdot s) \cdot f$$

$$f = 4.57 \times 10^{14}\ \frac{1}{s}$$

Then, we use the speed of light to go from frequency to wavelength.

$$f = \frac{v}{\lambda}$$

$$4.57 \times 10^{14}\ \frac{1}{s} = \frac{3.00 \times 10^{8}\ \frac{m}{s}}{\lambda}$$

$$\lambda = 6.56 \times 10^{-7}\ m = 656\ nm$$

The electron must emit light with a wavelength of 656 nm, which is red light, according to Figure 2.2.

2.5 In this problem, $Z = 3$ because we are dealing with Li^{2+}. The first thing to realize is that we have the frequency of the light emitted. In order to find the final orbit (n_{final}), we are going to need energy. Thus, we first need to calculate the energy:

$$E = h \cdot f$$

$$E = (6.63 \times 10^{-34}\ J \cdot s) \cdot (6.67 \times 10^{14}\ Hz) = 4.42 \times 10^{-19}\ J$$

If you forgot about frequency since you last took chemistry, the unit "Hz" is the same as "1/s," and that's why the final unit is Joules.

Now that we know the energy of the light emitted, we can use Equation (13) to determine n_{final}:

$$\Delta E = (2.18 \times 10^{-18}\ J) \cdot Z^2 \cdot \left[\left(\frac{1}{n_{final}}\right)^2 - \left(\frac{1}{n_{initial}}\right)^2\right]$$

$$4.42 \times 10^{-19}\ J = (2.18 \times 10^{-18}\ J) \cdot 3^2 \cdot \left[\left(\frac{1}{n_{final}}\right)^2 - \left(\frac{1}{5}\right)^2\right]$$

$$\frac{4.42 \times 10^{-19}\ \cancel{J}}{(2.18 \times 10^{-18}\ \cancel{J}) \cdot 3^2} = \left(\frac{1}{n_{final}}\right)^2 - \frac{1}{25}$$

$$0.0225 + \frac{1}{25} = \left(\frac{1}{n_{final}}\right)^2$$

$$\frac{1}{n_{final}} = 0.250$$

$$n_{final} = 4$$

The electron lands in the fourth Bohr orbit.

2.6 This is a direct application of Equation (16). Since we are dealing with Li^{2+}, $Z = 3$, and since we are talking about the ground state (lowest possible energy), $n=1$.

$$r = (0.529\ \text{Å}) \cdot \frac{n^2}{Z}$$

$$r = (0.529\ \text{Å}) \cdot \frac{1^2}{3}$$

$$r = \underline{0.176\ \text{Å}}$$

So we see that a Li^{2+} ion is actually *smaller* than a hydrogen atom. Now don't confuse Li^{2+} with Li. A lithium atom has three electrons, so the second Bohr orbit is occupied. Since the radius depends on n^2, the radius of the second Bohr orbit is 4 times larger than the radius of the first

Bohr orbit, so a lithium *atom* is bigger than a hydrogen atom. A lithium 2+ *ion*, however, is smaller than a hydrogen atom.

2.7 We are dealing with hydrogen again here, so Z = 1. In order to get the radius, we need to know "n." We don't have that, however. We only have the electron's energy. Well, Equation (12) relates energy to "n":

$$E = -R_h \cdot Z^2 \cdot \left(\frac{1}{n}\right)^2$$

$$-2.42 \times 10^{-19} \text{ J} = -(2.18 \times 10^{-18} \text{ J}) \cdot 1^2 \cdot \left(\frac{1}{n}\right)^2$$

$$\left(\frac{1}{n}\right)^2 = \frac{-2.42 \times 10^{-19} \text{ J}}{-2.18 \times 10^{-18} \text{ J}} = 0.111$$

$$\frac{1}{n} = 0.333$$

$$n = 3$$

Now that we know "n," calculating the radius is a snap:

$$r = (0.529 \text{ Å}) \cdot \frac{n^2}{Z} = (0.529 \text{ Å}) \cdot \frac{3^2}{1} = \underline{4.76 \text{ Å}}$$

Notice how much bigger the hydrogen atom is when the electron jumps to a higher energy orbit. When the electron moves from n=1 to n=3, the size of the hydrogen atom increases by a factor of 9!

2.8 To start, we determine the electron configuration:

$$1s^2 2s^2 2p^6 3s^2 3p^4$$

The valence electrons, are the ones in the 3s and 3p orbitals, since they have the largest values for "n." We will first work on the 3s orbital. For that orbital, n=3 and ℓ=0. When ℓ=0, m=0. To keep the electrons unique, then, they must have different values of "s." Their quantum numbers are:

$\underline{n=3, \ell=0, m=0, s=\frac{1}{2}}$ and $\underline{n=3, \ell=0, m=0, s=-\frac{1}{2}}$

That takes care of the 3s electrons. Now for the 3p electrons, of which there are four. These electrons have n=3 and ℓ=1. When ℓ=1, the value of "m" can be -1, 0, or 1. Hund's rule says we use each of those numbers once with s=$\frac{1}{2}$ before we go back and use them again. We always start with the most positive number and work our way down to the negative numbers; thus, the next three electrons have the following quantum numbers:

n=3, ℓ=1, m=1, s=$\frac{1}{2}$ and n=3, ℓ=1, m=0, s=$\frac{1}{2}$ and n=3, ℓ=1, m=-1, s=$\frac{1}{2}$

We still have one electron to go, so at this point, we must go back and put a second electron in one of the orbitals that already has one electron in it. The only way to do that and keep the 4 quantum numbers unique is to have s=-$\frac{1}{2}$. Thus, the final electron has the following quantum numbers:

n=3, ℓ=1, m=1, s=-$\frac{1}{2}$

2.9 a. This set is impossible because "ℓ" can only go from 0 to n-1. When n=2, then, "ℓ" can only be 0 or 1.

b. This set does not work because "n" can never equal zero.

c. This set is wrong because "m" can only range from -ℓ to ℓ. Thus, when ℓ=1, the only possible values for "m" are -1, 0, or 1.

2.10 When ℓ=4, "m" can have values of -4, -3, -2, -1, 0, 1, 2, 3, or 4. This makes 9 possible values which correspond to 9 different orbitals. Since each orbital can hold two electrons, the g-orbitals can hold 18 electrons total.

REVIEW QUESTIONS

1. What assumption did Bohr make that led to the success of his theory?

2. What justification do we use to support Bohr's assumption?

3. In the Bohr model, an electron can be in orbit #1 or orbit #2, but it cannot be anywhere in between. If that's the case, how does the electron jump from orbit #1 to orbit #2?

4. Suppose you excite an atom and you see no light coming from it. Does that mean electrons are not moving up and down in orbitals?

5. Suppose an electron in the Bohr atom absorbs energy and jumps to the n=4 orbit. How many wavelengths of light can the electron possibly emit in getting back to the n=1 orbit?

6. What is the difference between atomic emission spectroscopy and atomic absorption spectroscopy?

7. List the rules for the values of the four quantum numbers.

8. Three electrons are examined. Their quantum numbers are:

#1: n=2, ℓ = 1, m=0, s = $\frac{1}{2}$ #2: n=1, ℓ=1, m=0, s = $\frac{1}{2}$ #3: n=2, ℓ=1, m=1, s = -$\frac{1}{2}$

Which two electrons are in degenerate orbitals?

9. Three electrons are examined. Their quantum numbers are:

#1: n=2, ℓ=1, m=0, s = $\frac{1}{2}$ #2: n=3, ℓ=1, m=0, s = $\frac{1}{2}$ #3: n=2, ℓ=1, m=1, s = -$\frac{1}{2}$

Which electron is farthest from the nucleus?

10. Three electrons are examined. Their quantum numbers are:

#1: n=2, ℓ=0, m=0, s = $\frac{1}{2}$ #2: n=3, ℓ=1, m=0, s = $\frac{1}{2}$ #3: n=3, ℓ=2, m=1, s = -$\frac{1}{2}$

Which electron is in a d-orbital?

PRACTICE PROBLEMS

1. What is the energy of an electron in the n=2 Bohr orbit of a He^{+} ion?

2 An electron in a hydrogen atom starts out in the first Bohr orbit and jumps to the fourth Bohr orbit. How much energy must it absorb?

3. An electron jumps from the n=3 orbit of a Li^{2+} ion to the n=1 orbit. What is the wavelength of the light emitted?

4. An electron jumps from the fifth Bohr orbit to a lower orbit in a hydrogen atom. If the light emitted has a frequency of 3.16×10^{15} Hz, what orbit did the electron end up in?

5. What is the radius of a He^{+} ion in its ground state?

6. An electron in a hydrogen atom has an energy of -6.06×10^{-20} J. How far from the nucleus is it orbiting?

7. An electron in a He^{+} ion is orbiting the nucleus at a distance of 0.265 Å. What is the electron's energy?

8. What are the quantum numbers for the valence electrons of oxygen?

9. What is wrong with the following quantum number assignments?

a. n=0, ℓ=0, m=0, s= $\frac{1}{2}$ b. n=2, ℓ=2, m=1, s= $\frac{1}{2}$ c. n=2, ℓ=1, m=-2, s= $\frac{1}{2}$

10. The g-orbitals have ℓ=4. How many g-orbitals are there and how many electrons do they hold?

Module #3: The Electronic Structure of Molecules

Introduction

In the previous module, I went through a (reasonably) detailed view of the quantum mechanical atom. You now know the general shapes of the orbitals, the relationship of the four quantum numbers to atomic structure, and how to assign those quantum numbers to electrons within an atom. Hopefully, you have come to a strong appreciation of the *complexity* of the atom. Even the smallest unit of matter has incredibly complex structure! How in the world can anyone view even a bit of what you learned in the last module and still deny the existence of a supernatural Creator? If even the "simplest" unit of matter is as exceedingly complex as what you learned in the last chapter, there is simply no way that the universe we see today could have occurred by chance!

The exceedingly complex atom is just the beginning of an even more complex world of matter made up mostly of molecules. As you already know, atoms form molecules by either giving up or taking valence electrons (in the case of ionic molecules) or sharing valence electrons (in the case of covalent and polar covalent molecules). Think for a moment about ionic molecules, in which one set of atoms (the metals) give up electrons to make positive ions while the other set of atoms (the non-metals) take on additional electrons to form negative ions. In such molecules, the ions stay together not because they share electrons. Instead, they stay together because of the mutual electrical attraction that exists between positive ions and negative ions. Thus, ionic molecules can be viewed as a collection of independent ions which simply stay together because they are electrically attracted to one another. With this view, it should be rather easy to see that the electrons in ionic molecules are stored in atomic orbitals with basically the same properties as those discussed in the previous module.

What happens, however, when polar covalent and covalent molecules are formed by sharing electrons? When electrons are shared between two atoms, they must spend time orbiting around two different nuclei: one for each atom that participates in the sharing. How does that happen? What kind of orbitals can allow for such a situation? That's what we are going to study next. In this module, then, we are focusing only on molecules which share electrons. Since ions have orbitals essentially equivalent to atomic orbitals, there is no reason to study them further. Polar covalent and covalent molecules, however, have electronic structures that are profoundly different than the electronic structure of atoms, so it is important to take a good, hard look at them!

How Atoms Share Electrons

When two atoms share electrons, those electrons must be able to spend time orbiting each atom. That way, each atom has the electrons for some fraction of time. Well, if you look at the orbital diagrams presented in the previous module, you will quickly realize that those orbitals will not allow that to happen. After all, atomic orbitals are designed so that the electrons orbit around *one* nucleus. Thus, in order for a polar covalent or covalent molecule to form, something must happen to the atomic orbitals of each atom that is a part of that molecule.

What happens is actually quite surprising, and it occurs in two steps. When an atom becomes part of a molecule, it first *mixes some or all of its atomic orbitals together* to make what we call **hybrid orbitals**.

Hybrid orbital - An electron orbital that forms when the atomic orbitals of an atom mix together to form a new kind of orbital

These hybrid orbitals, along with any left-over atomic orbitals, then overlap with the orbitals of the other atom with which electrons are shared. When the orbitals of two individual atoms overlap, they form a **molecular orbital**.

Molecular orbital - The energy and region in space in which an electron can be found orbiting the nuclei that share it

Molecular orbitals allow shared electrons to orbit the nuclei of both atoms that participate in the sharing.

All of that sounds rather confusing right now because I haven't given you any particulars. Don't worry. That's what I'm going to do now. I will first concentrate on the formation of hybrid orbitals in an atom, and then I will proceed to tell you how hybrid orbitals overlap to form molecular orbitals. Along the way, you will learn the actual reason behind the molecular geometries you learned in your previous chemistry course.

Hybrid Orbitals

As I already mentioned, the first step in forming a molecule is for one or more atoms to mix their electron orbitals together to make hybrid orbitals. Not surprisingly, this process is called **hybridization** (hi' brid uh zay' shun). The first thing to understand about hybridization is that it really only happens on the orbitals that contain the valence electrons. Now this should make sense to you. Remember from your previous chemistry course that the valence electrons are the ones used in chemical bonding. Thus, only the orbitals that contain valence electrons are the ones that will hybridize.

What does the process of hybridization entail? Well, it entails taking the valence electron orbitals and actually mixing them together to form orbitals which are completely different. For example, the majority of the atoms that we deal with (groups 1A - 8A) store their valence electrons in one s-orbital and three p-orbitals. When the valence orbitals of one of these atoms hybridize, those orbitals mix together and form completely different orbitals. Before I show you what these orbitals look like, I want you to think about *how many* hybrid orbitals will form when one s-orbital and three p-orbitals mix together. Well, these four orbitals have "room" enough to hold eight electrons. When they all mix together, there should still be "room" enough for eight electrons. Thus, these four atomic orbitals will mix together and form *four hybrid orbitals*. This is an important point:

When a certain number of atomic orbitals hybridize, they result in the same number of hybrid orbitals.

Thus, when an s-orbital and three p-orbitals (a total of 4 atomic orbitals) hybridize, the result is four hybrid orbitals.

Okay, so we know *how many* hybrid orbitals are formed when an s-orbital and three p-orbitals hybridize, but what do they *look like*? Well, we know that s-orbitals are spherical and p-orbitals have a dumbbell-like shape. Thus, the resulting orbital should look something like a cross between these two shapes, as illustrated in Figure 3.1.

FIGURE 3.1
One s-Orbital and Three p-Orbitals Make Four sp^3 Hybrid Orbitals

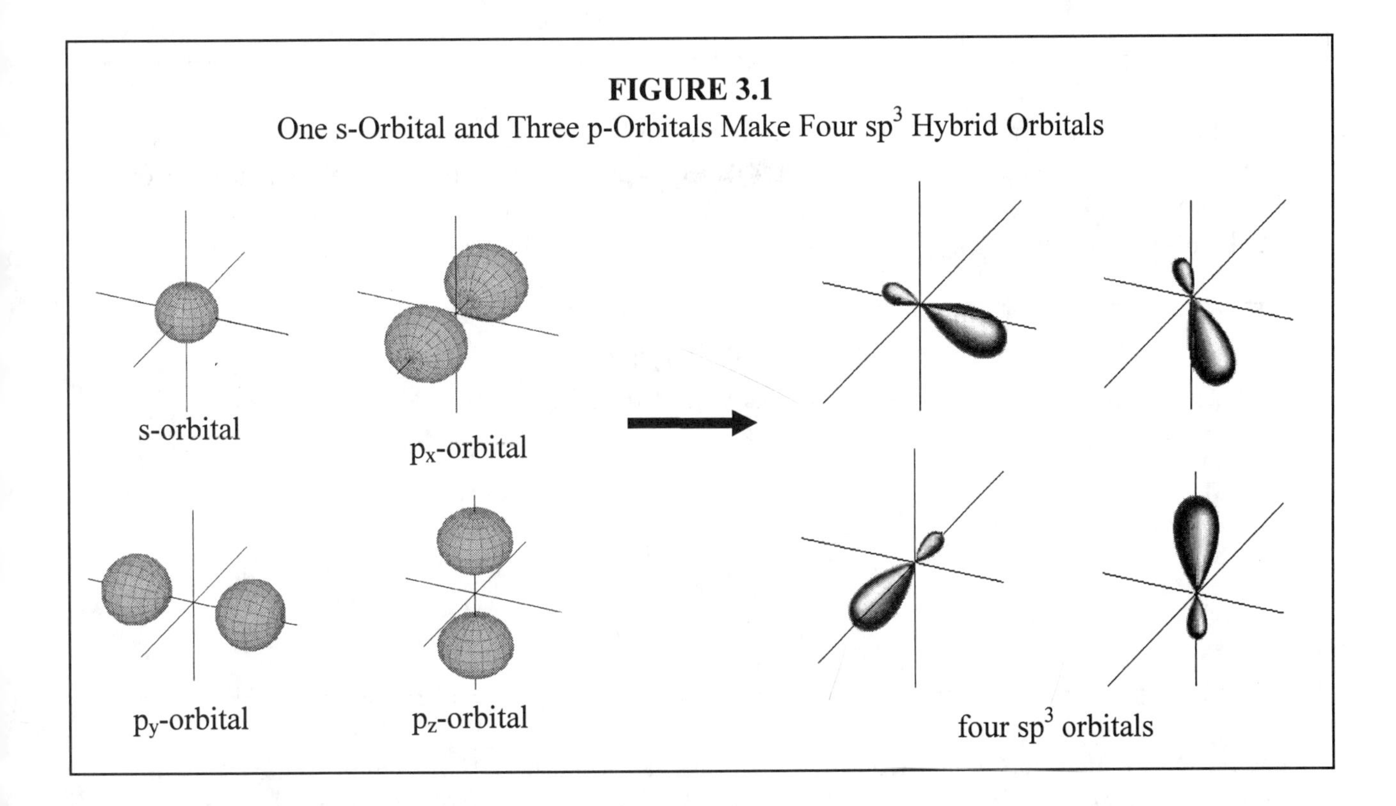

There are a couple of things you need to notice about the figure. First, notice the name. We call these orbitals "sp^3 orbitals" because they are formed by the mixing of one s-orbital and three p-orbitals. Second, notice the orientation of the orbitals. Remember that the three p-orbitals are each oriented along the three axes (x-axis, y-axis, and z-axis). The s-orbital, on the other hand, is spherical, pointing in all directions. The orientations of the resulting hybrid orbitals are a mixture of the orientations of the original atomic orbitals.

Now, of course, all four of these hybrid orbitals exist in the individual atoms whose atomic orbitals hybridized in the first place. Figure 3.2 illustrates this fact.

FIGURE 3.2
Four sp^3 Orbitals As They Exist in an Atom

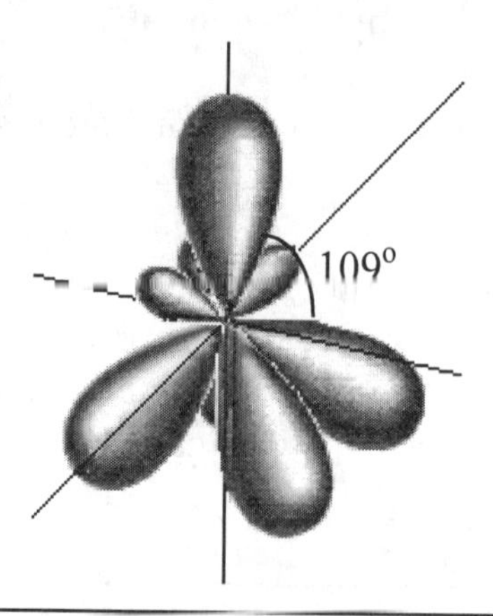

Does this look at all familiar? Do you remember the molecule geometries you learned last year? The main geometry you learned was the **tetrahedron**, in which each bond was 109° away from every other bond in the molecule. This is *why* the molecules you studied last year form a tetrahedron. For example, one of the classic cases of the tetrahedron is methane (CH_4):

FIGURE 3.3
The Molecular Geometry of Methane

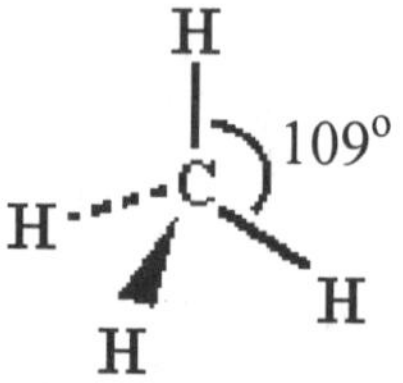

H
109°
H C
H
H

The reason methane has a tetrahedral geometry is that the carbon at the center of the molecule hybridizes its atomic orbitals to form sp^3 hybrid orbitals, which (by virtue of the orbitals in the hybridization) form a tetrahedron. Thus, the molecular geometries you learned last year are a result of *orbital hybridization.*

What about the other geometries you learned last year? You learned about **pyramidal** and **bent** geometries, but remember, those are just tetrahedrons without all of the legs. For example, ammonia (NH_3) has a pyramidal shape. That's because there are four groups of electrons around the central nitrogen atom, so the basic shape is still a tetrahedron. However, one of those four groups is a lone electron pair, not a chemical bond. Thus, ammonia forms the shape of a tetrahedron with one leg missing, which we called pyramidal. In the same way, water (H_2O) forms a bent shape, but once again, that's really a tetrahedron with two legs missing. So you see that even those two molecular shapes are caused by sp^3 orbital hybridization.

Do all atoms form sp^3 hybrids? The answer to that is no. There are many more basic shapes for molecules, and some of them are not based on a tetrahedron. For example, some molecules take on a triangular shape, which chemists usually call **trigonal** (trig' uh nul). The trigonal shape results when one of the p-orbitals *does not* participate in the hybridization. This occurs when an atom hybridizes one s-orbital and *two* p-orbitals. The result is three sp^2 hybrid orbitals:

FIGURE 3.4
Three sp^2 Orbitals in an Atom

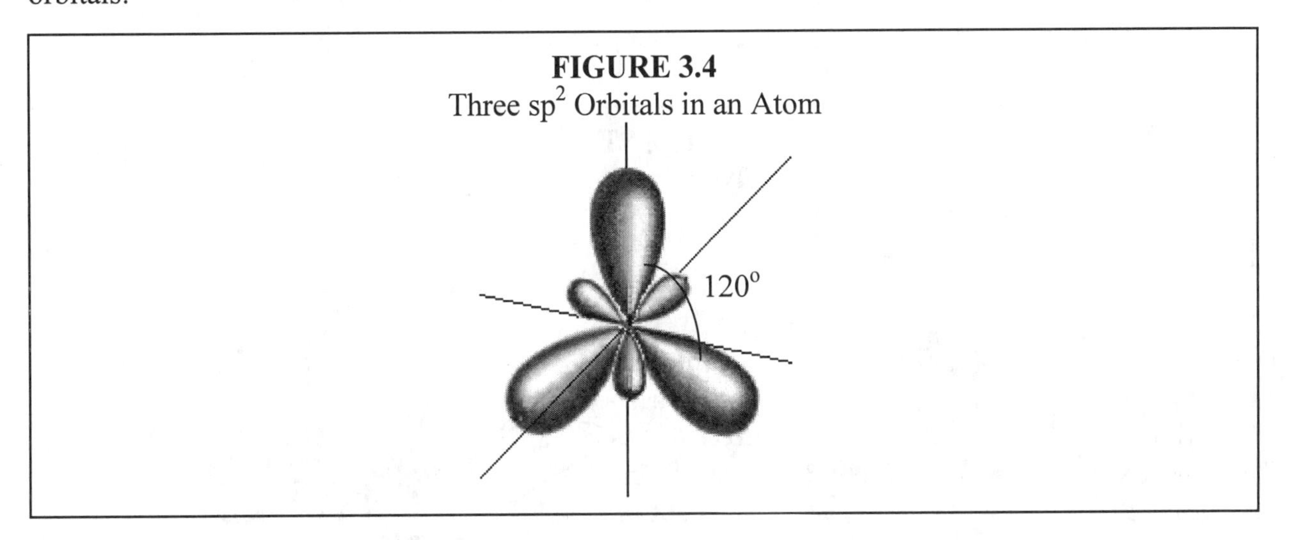

Since these orbitals all lie in the same plane 120° apart from each other, any molecule formed by an atom with this hybridization will appear triangular.

You should be asking yourself what happened to that other p-orbital. If it does not participate in the hybridization, what happens to it? Nothing! The p-orbital stays right where it is. Thus, the atom has three sp^2 hybrid orbitals and one p-orbital holding its valence electrons. Why would this happen? You'll find the answer to that question in the next section of this module.

If one p-orbital can keep from participating in hybridization, can two of them do so? The answer to that question is yes. Under certain conditions which I will discuss in the next section, an s-orbital will hybridize with only *one* p-orbital, making two sp hybrid orbitals.

FIGURE 3.5
Two sp Hybrid Orbitals in an Atom

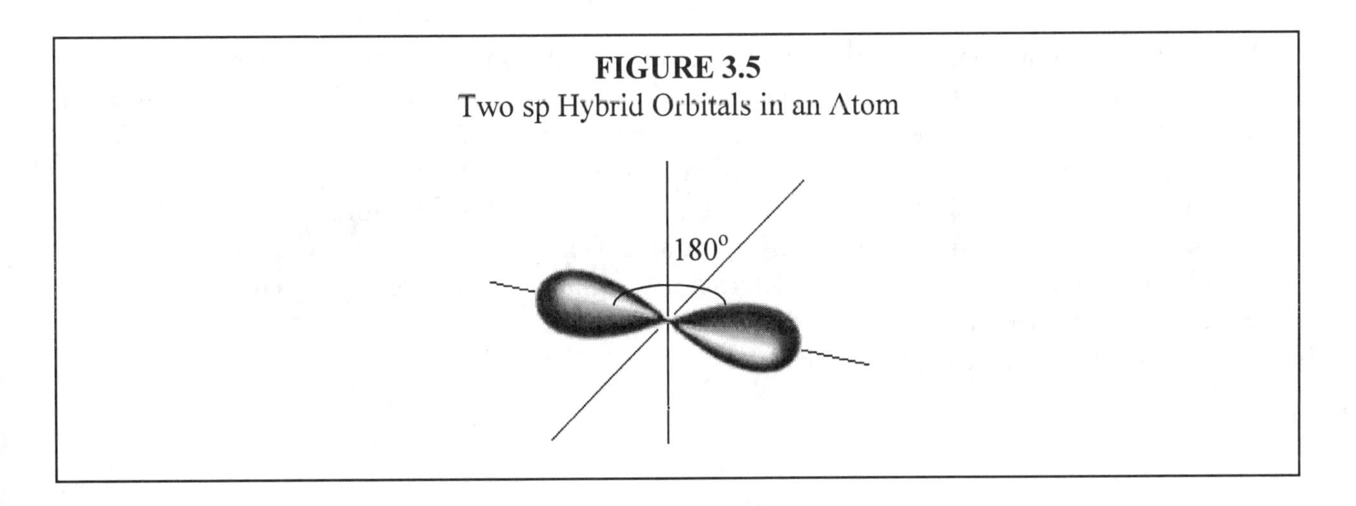

As the figure illustrates, these two orbitals are oriented 180° away from each other and end up forcing a **linear** geometry on any molecule formed by the atom.

That's it then. There are no more hybrid orbitals to worry about, right? Wrong! Remember that there are other orbitals in atoms as well. Sometimes, d-orbitals enter into the hybridization picture. These hybrid orbitals and the geometries that they form are illustrated in Figure 3.6.

FIGURE 3.6
Two Types of Hybridization Involving d-Orbitals

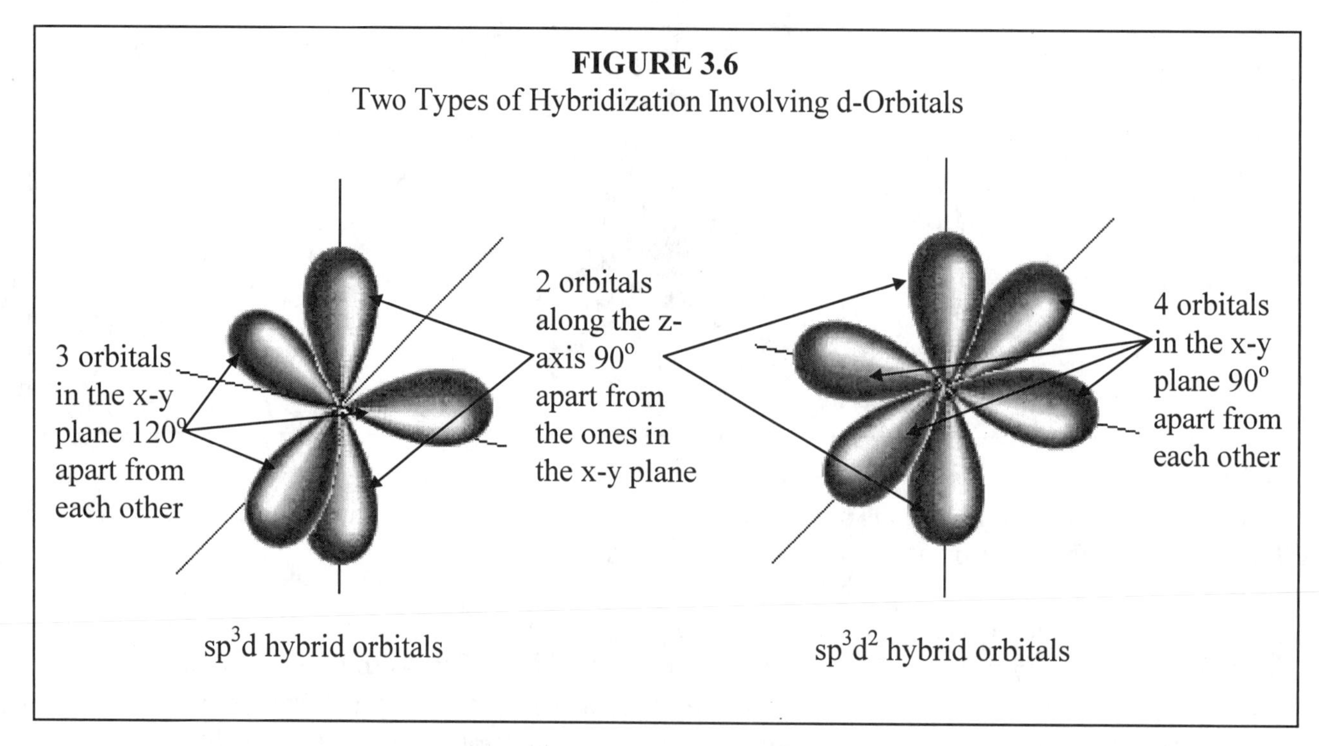

Notice the geometries that result from these two hybridizations. In sp^3d hybridization, one s-orbital, three p-orbitals, and one d-orbital mix to form 5 sp^3d hybrid orbitals in a **trigonal bipyramidal** (bi' puh ram' uh dul) geometry. In this geometry, three of the orbitals are 120° apart from each other, hence the "trigonal" part of the name. The other two orbitals are out of the plane formed by those orbitals, forming a 90° angle with them. Finally, sp^3d^2 hybridization results in the **octahedral** geometry, where six orbitals are all 90° apart from each other.

There are other possibilities of hybridization, but (thankfully) I will stop here. To summarize, when atomic orbitals mix, they form the same number of hybrid orbitals as there are atomic orbitals that participate in the mixing. The resulting hybrid orbitals have distinct shapes and orientations in space. These orientations are responsible for the geometry that we see in molecules. If you do not quite see the point here, don't worry about it. For right now, just understand the summary that exists in this paragraph. Get to the point that you can recognize the hybridization when shown the orbitals, determine how many orbitals exist in a given hybridization, and determine the geometry that results. The material in the next section should bring the chemistry involved into focus.

ON YOUR OWN

3.1 Suppose there were an atom that could form an sp^2d^2 hybridization. How many hybrid orbitals would result?

3.2 An atom hybridizes its atomic orbitals to form three hybrid orbitals. What is the hybridization and what geometry will result in a molecule formed with this atom?

Molecular Orbitals - Part 1

Once an atom has hybridized its valence orbitals, it is ready to bond with another atom. In the bonding process, the orbitals of the two atoms overlap, forming a **molecular orbital**, which spans both nuclei, allowing the shared electrons to orbit around both atoms. The simplest such overlap to show you is what happens in a methane molecule :

FIGURE 3.7
Orbital Overlap and Molecular Orbitals in Methane

H
H–C–H →
H

In your previous chemistry course, you learned that the Lewis structure of methane (left of the arrow) results in a tetrahedral geometry (right of the arrow)

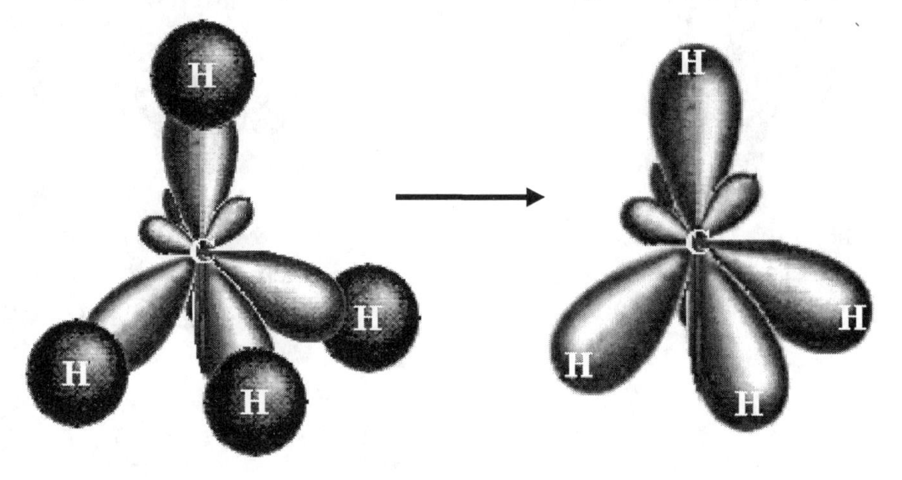

In this course, you see how methane gets its tetrahedral geometry. The bonds in the Lewis structure are the result of carbon's sp^3 orbitals overlapping with the hydrogens' s-orbitals (left of the arrow). Those overlapping orbitals mix to form molecular orbitals (right of the arrow) in a tetrahedron.

Do you see what happens in the formation of a methane molecule? The central carbon atom hybridizes its orbitals into four sp^3 orbitals. The four hydrogen atoms, whose electrons exist only in s-orbitals, do not hybridize their orbitals. Instead, each hydrogen atom overlaps its s-orbital with one of the sp^3 hybrid orbitals of the carbon atom. This results in four molecular orbitals, each stretched out far enough to allow the shared electrons to orbit both the central carbon atom and the hydrogen atom. Since the orbitals are arranged in a tetrahedron, the resulting geometry is tetrahedral.

So the first thing you should learn from the figure is that the lines you draw in a Lewis structure actually represent two electrons shared in a molecular orbital. The second thing you should see is that the geometry of a molecule is actually a result of the hybridization of the central atom's atomic orbitals. Since sp^3 orbitals form a tetrahedron, methane molecules are tetrahedral, because methane's molecular orbitals result from the overlap of sp^3 orbitals and s-orbitals. As you can see from the next figure, this same kind of thinking explains the shape of pyramidal and bent molecules:

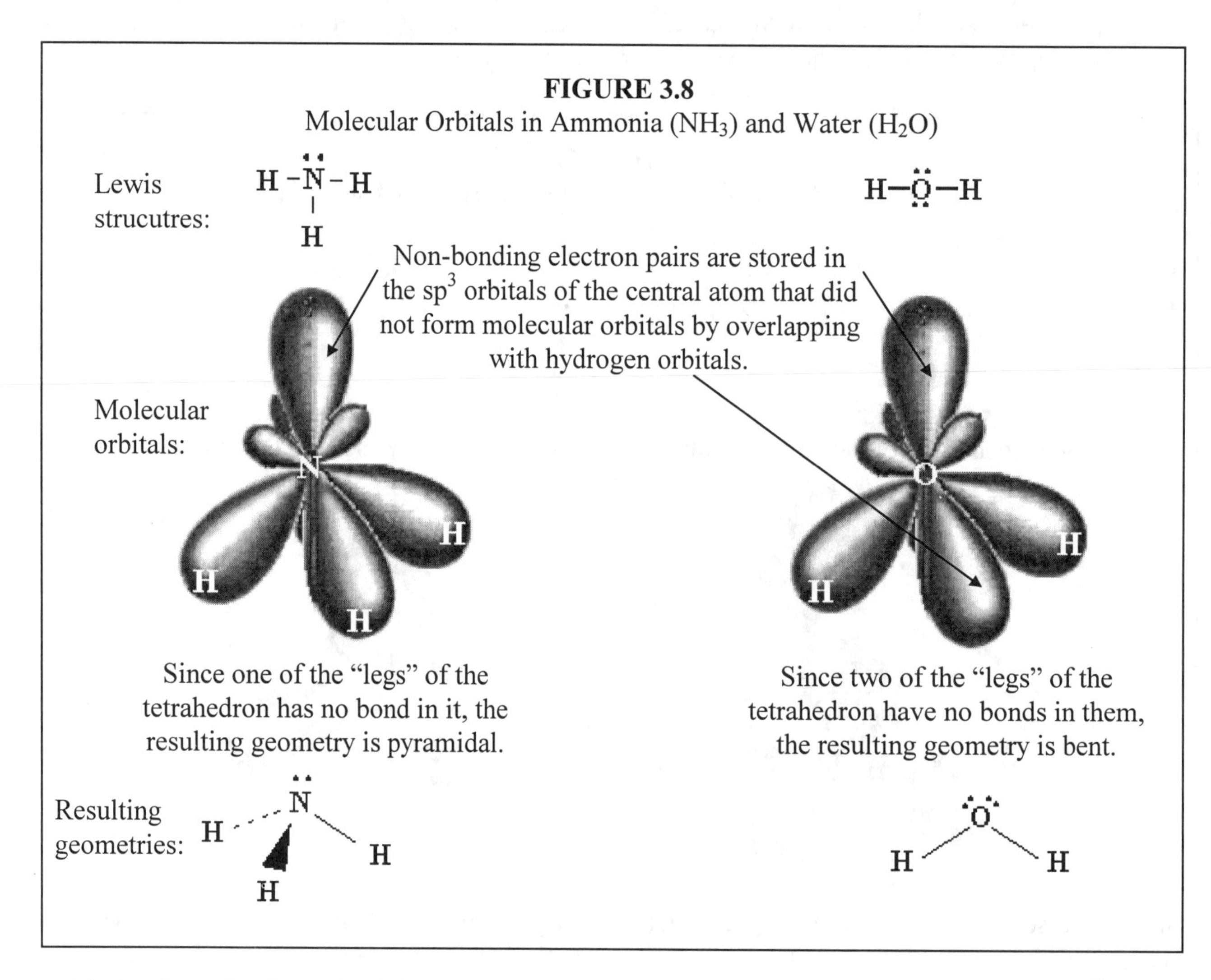

Notice from the figure that the unpaired electrons in both molecules are stored in atomic sp^3 orbitals so that those electrons orbit only the central atom. The shared electrons, on the other hand, are stored in molecular orbitals which allow them to orbit both of the atoms that share them.

At this point, you need to be introduced to a couple of definitions. We now know that covalent and polar covalent bonds are formed as the result of overlap between two orbitals. In all of the molecules that we studied so far, the overlap occurred between a hybrid orbital (sp^3) and an unhybridized s-orbital. Well, it turns out that there are two different ways that orbitals can overlap. This results in two distinctly different types of bonds. So far, we have studied how **sigma-bonds** are formed. Often, chemists use the Greek letter for "sigma" instead of writing out the word "sigma," so you will also see these bonds referred to as **σ-bonds.**

Sigma-bond - A bond in which the electron density is concentrated along the internuclear axis

What does this definition mean? Well, remember that the molecular orbital contains two electrons. These electrons are moving throughout the orbital. Thus, you cannot pin down exactly where they are. You can, however, point to any region within the orbital and determine the *probability* of finding the electron there. When you have a large probability of finding the electron, then chemists say that the **electron density** is high. If there is not much chance of finding the electron, we say that the electron density is low. Well, if I were to draw a line between the two atoms, I would call that the **internuclear axis**. Looking at Figure 3.8, then, if you were to draw an imaginary line between the nitrogen atom and one of the hydrogen atoms, you would see that the line runs right through the orbital. This tells you that the probability of finding the electron along that line is very high. Thus, the electron density is high along the internuclear axis.

The other kind of overlap results in what we call **pi-bonds**, which are often written as **π-bonds**.

Pi-bond - A bond in which the electron density is not concentrated along the internuclear axis

Study Figure 3.9 to see the difference between σ-bonds and π-bonds.

FIGURE 3.9

Sigma-Bonds and Pi-Bonds Between Two Oxygen Atoms

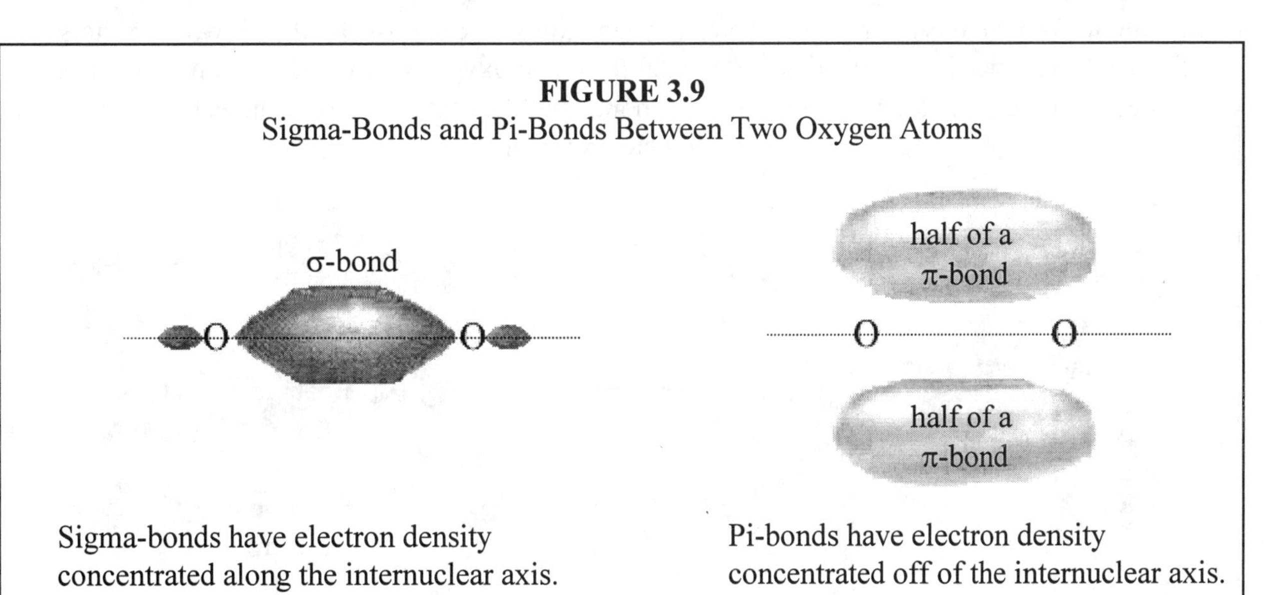

It is clear how sigma-bonds are formed, but how are π-bonds formed? Well, they are formed when p-orbitals overlap parallel to one another. This happens whenever a molecule has a *double or triple bond.*

Consider, for example, an oxygen molecule, O_2. From what you learned in your previous chemistry course, you should be able to draw the Lewis structure for this molecule:

:Ö=Ö:

As you have just seen, the bonds in a Lewis structure are actually the result of overlapping orbitals that mix to form molecular orbitals. Thus, the double bond in oxygen's Lewis structure must be the result of *two different* molecular orbitals. If you think about it, it is simply impossible to squeeze two molecular orbitals directly *between* the atoms! After all, if a sigma-bond forms between the two oxygen atoms, then the space in between the two atoms is full. Where does the second bond go, then? It goes *above and below* the internuclear axis as a pi-bond!

How does this happen? Well, in order to form any bond, atomic orbitals must overlap to form molecular orbitals. If you look at Figure 3.9, a pi-bond has equal electron density both above and below the internuclear axis. If such a molecular orbital is to form, it must be made from the overlap of atomic orbitals with equal electron density both above and below the atom. What kind of orbital has that characteristic? A p-orbital! Thus, a pi-bond is the result of p-orbitals which overlap. Where does the p-orbital come from? Examine Figure 3.10 to find out.

FIGURE 3.10

Molecular Orbitals and Bonds in an Oxygen Atom

In order to form an oxygen molecule, two oxygen atoms mix one s-orbital and two p-orbitals to form sp^2 hybridized orbitals. One sp^2 orbital from one oxygen atom will overlap with an sp^2 orbital from the other atom. This forms a σ-bond, and the other two sp^2 orbitals hold the non-bonding electron pairs .

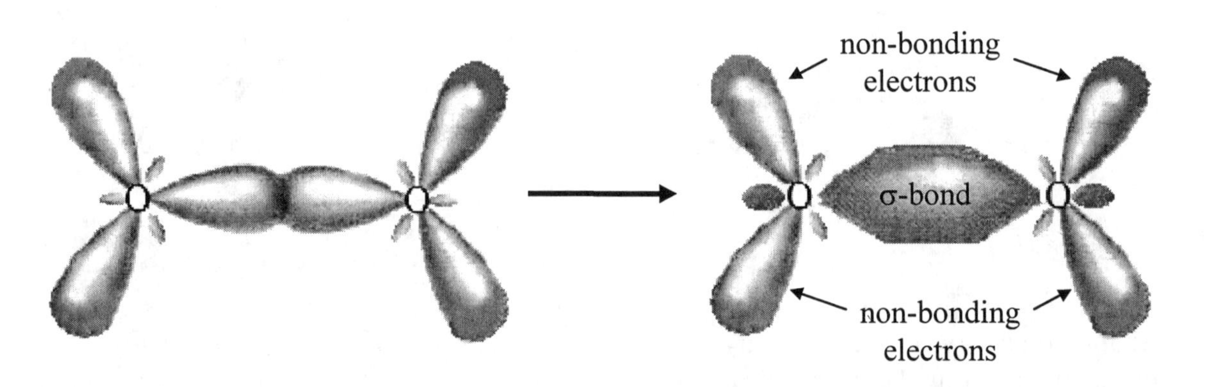

Since each oxygen atom uses only two p-orbitals in the hybridization, each atom has an extra p-orbital. Those line up and overlap, forming a π-bond:

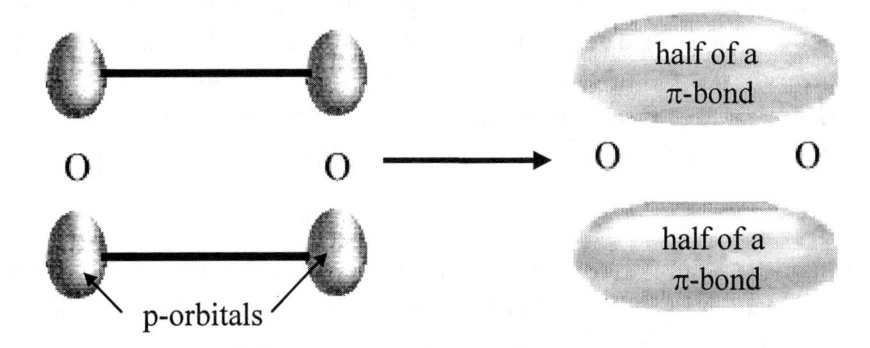

The overall picture, then, is as follows:

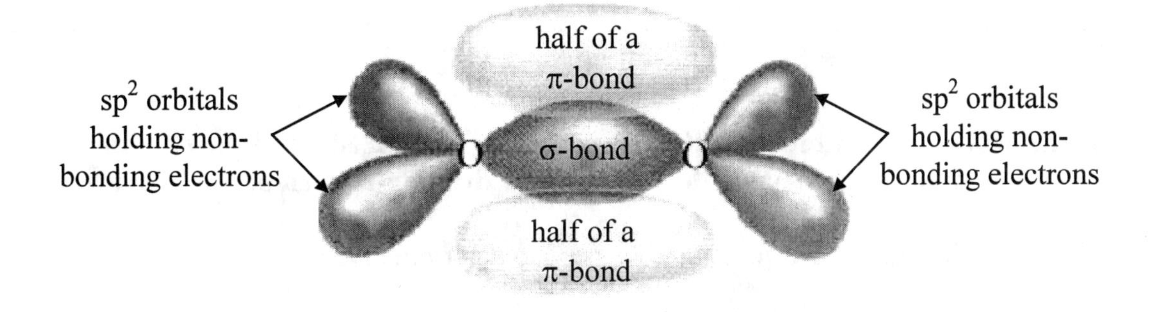

You should see, then, that π-bonds form when there are multiple bonds between atoms. In the case of a double bond, one of the two bonds is a sigma-bond which fills the space between the two atoms, while the other bond is a pi-bond above and below the internuclear axis. What happens in the case of a triple bond? Well, one of the three bonds is a sigma-bond, while the other two are π-bonds. If one of the π-bonds is above and below the nuclear axis, where is the other? Think three-dimensionally. There is still space *in front of and behind* the internuclear axis. Examine Figure 3.11 to see what I mean:

FIGURE 3.11
Molecular Orbitals in Acetylene (C_2H_2)

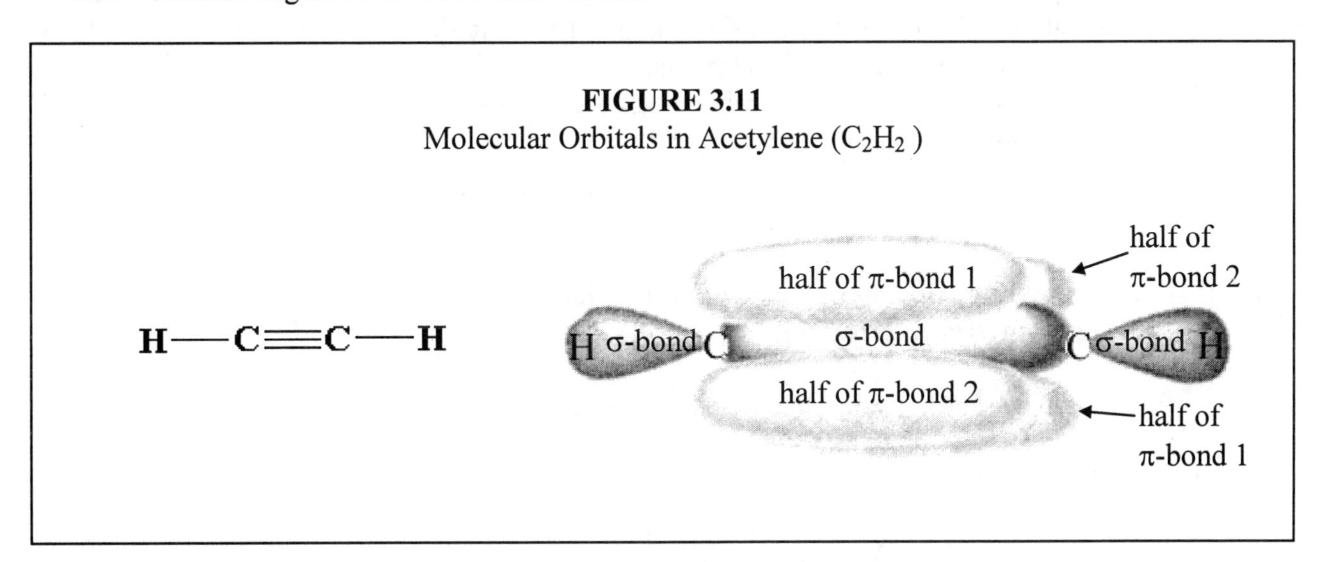

So now you know why sp^3 hybridization isn't always the hybridization that you find in molecules. If a molecule contains a double bond, it needs a π-bond. The only way that can happen is for p-orbitals to overlap. Thus, when the atoms that are double bonded hybridize, they leave a p-orbital out of the hybridization process, so that it can participate in the π-bond. As a result, the hybridization is sp^2, and the resulting geometry is trigonal. In the same way, atoms that form triple bonds need two p-orbitals each in order to make two π-bonds. As a result, only one p-orbital is used in hybridization and the resulting hybrid orbitals are sp, making a linear geometry.

In this module, I want you to be able to predict the orbital hybridization and the resulting geometry for any molecule I give you. In order to do this, you must first be able to determine what kinds of bonds exist in the molecule. For that, you will need to be able to draw Lewis structures! If you have forgotten how to draw Lewis structures, then go back and study it from your first-year course, as I will not review that process here.

EXAMPLE 3.1

What is the orbital hybridization of the central sulfur atom in SF_2 and the resulting geometry of the molecule? What is the orbital hybridization in each of the fluorine atoms?

To determine orbital hybridization, we need to determine the Lewis structure:

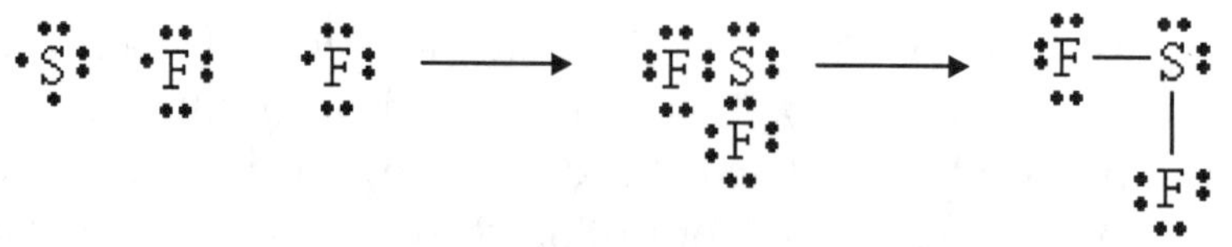

Now look at the bonds. There are no double bonds. This means that the atoms can hybridize their orbitals as sp^3. As you learned last year, the central atom determines the geometry. Thus, we need to look at the hybridization of the sulfur atom in order to determine the geometry. As I just said, it is sp^3. That means that the base geometry is tetrahedral. However, two of the sp^3 orbitals have no bonds; they have only non-bonding electron pairs. This means that two of the "legs" in the tetrahedron are missing, and the resulting geometry is therefore bent. Thus, <u>all atoms have sp^3 hybridization and the geometry of the molecule is bent</u>.

What is the orbital hybridization of each atom in a formaldehyde (CH_2O) molecule? What is the resulting geometry?

Once again, we start by doing the Lewis structure.

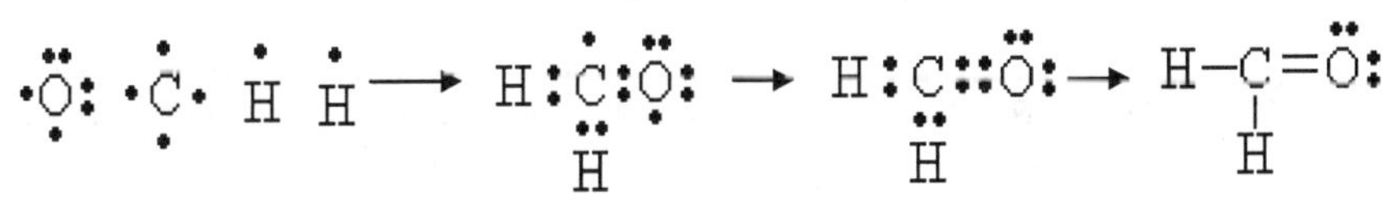

$\cdot\ddot{O}: \; \cdot\dot{C}\cdot \; \dot{H} \; \dot{H} \longrightarrow H:C:\ddot{O}: \rightarrow H:C::\ddot{O}: \rightarrow H-C=\ddot{O}:$ (with H bonded below C)

Now look at the bonds. The central carbon atom, which controls the geometry, has one double bond. This means that it needs a π-bond. The only way that can happen is if it holds a p-orbital out of the hybridization process. This means that <u>the carbon atom is sp^2 hybridized and the resulting geometry is trigonal</u>. The question also asks about the hybridization of the other atoms. Well, the oxygen also has a double bond, which means that <u>the oxygen atom is also sp^2 hybridized</u>. What about the hydrogen atoms? They only have single bonds, so their hybridization is sp^3, right? Wrong! They only have one pair of electrons, so they need only one orbital, not four. As a result, <u>the hydrogen atoms have no hybridization</u>. So we see again that hydrogen is an exception. In Lewis structures it is an exception because it wants only 2 electrons around it. In molecular orbital studies it is an exception because it does not hybridize.

What is the orbital hybridization of each atom in a carbon disulfide molecule? What is the resulting geometry?

First, you must determine the Lewis structure:

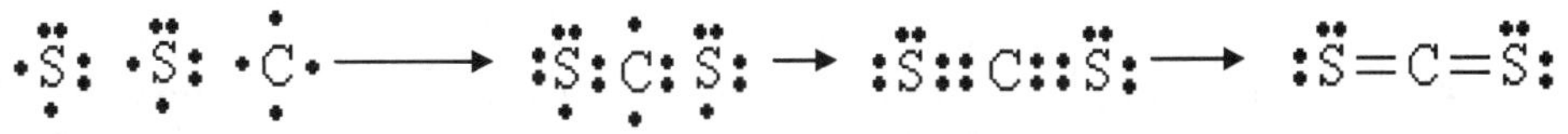

$\cdot\ddot{S}: \; \cdot\ddot{S}: \; \cdot\dot{C}\cdot \longrightarrow :\ddot{S}:\dot{C}:\ddot{S}: \rightarrow :\ddot{S}::C::\ddot{S}: \rightarrow :\ddot{S}=C=\ddot{S}:$

Look at the bonds around the central carbon atom, and be very careful. The central carbon atom has *two* double bonds. It needs a p-orbital for *each*, which means it must keep 2 p-orbitals out of the hybridization process. This means that <u>the carbon atom is sp-hybridized and the resulting molecular geometry is linear</u>. What about the sulfur atoms? Well, each sulfur atom has *only one* double bond, so they each must keep only one p-orbital out of the hybridization process. Therefore, the <u>sulfur atoms are sp^2 hybridized</u>.

In order to determine the orbital hybridization, then, the first thing you must do is draw the Lewis structure. Once you have that, you can look at each individual atom, taking a p-orbital away for every multiple bond that the atom has. The p-orbitals left over will hybridize with the s-orbital, and that will give you the orbital hybridization for the atom in question. If the atom is the central atom, you can use the hybridization to give you the geometry of the molecule. The only big exception to these rules is hydrogen, which never hybridizes its orbitals. See if you can do this on your own.

ON YOUR OWN

3.3 Given the following Lewis structure of acetic acid, determine the orbital hybridization of each atom.

```
         ••
    H    O:
    |    ||   ••
H—C—C—O—H
    |         ••
    H
```

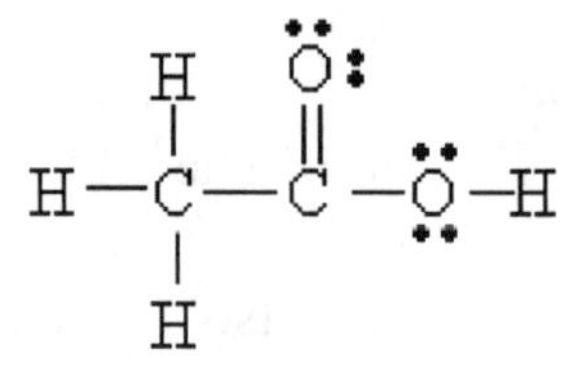

3.4 What is the orbital hybridization of each atom in a molecule of hydrogen cyanide (HCN)? What is the resulting geometry?

3.5 What is the orbital hybridization of each atom in a molecule of PCl_3? What is the resulting geometry?

3.6 What is the orbital hybridization of each atom in a molecule of H_2SiS? What is the resulting geometry?

Molecular Orbitals Part 2: The Rule-Breakers

I am not done discussing hybridization and geometry yet! That's because until now, we have left an entire class of molecules out of the discussion. When you first learned Lewis structures, you learned that with the exception of hydrogen atoms, all atoms try to get eight electrons around them in a Lewis structure. Although that is the general rule, there are more exceptions than just hydrogen. Beryllium, for example, is small enough that it does not have room for eight electrons in its valence shell. As a result, it strives for only four electrons. Because of this, it uses only one p-orbital in its hybridization. Thus, when beryllium has only single bonds, its orbital hybridization is sp. In the same way, boron can only accept six electrons in its valence shell. As a result, when boron has only single bonds, its orbital hybridization is sp^2.

The more important rule-breakers, however, are those atoms that strive for *more than eight* electrons in their valence shell. How is that possible? After all, the reason most atoms strive for eight valence electrons is that this number of electrons fills the valence s- and p-orbitals. Since there are always one s-orbital and three p-orbitals in any valence shell, then there should only be eight total electrons, right?

If we are talking about "normal" orbitals, that is right. Remember, however, that atoms will *hybridize* their orbitals before bonding. Suppose one or two d-orbitals participate in this hybridization? Two sections ago (Figure 3.6) I showed you what happens when one or two d-orbitals do this. In the case of one d-orbital participating in the hybridization process, the result is sp^3d hybridization. This gives us five hybrid orbitals, which leaves room for *ten electrons*. If two d-orbitals participate in the hybridization, then we have sp^3d^2 hybridization, and there are a total of six hybrid orbitals, giving the atom room for *twelve electrons* in its valence shell.

Because of these two possible hybridizations, some atoms can have ten or even twelve electrons around them in their Lewis structures. This may be a shock to you, since you are so used to striving to get eight electrons around every atom in a Lewis structure. Nevertheless, because d-orbitals can participate in the hybridization process, this is a real possibility for some atoms.

Is there any way of knowing what atoms can have ten or twelve electrons in their Lewis structure? Not really. There is, however, a quick way of telling what atoms cannot. In order for an atom to have more than eight electrons in its Lewis structure, it must have d-orbitals which can participate in the hybridization process. This eliminates all atoms in the first two rows of the periodic chart, because when n=2, ℓ has a maximum value of 1, which means the most complex orbital the atom has is a p-orbital. As soon as you move down to the third row of the periodic chart, then n=3 and ℓ can have a value of 2, which means those atoms can have d-orbitals. In principle, then, all atoms in rows three or greater of the periodic chart can get more than eight electron in their valence shell.

If we can't really tell which atoms break the rule of eight, then how in the world can we draw Lewis structures? Well, there are a couple of guidelines. First, you try to draw the Lewis structure without breaking the rule of eight. If you cannot, then you see if the central atom is in row three or greater of the periodic chart. If it is, then you should see if the Lewis structure can be drawn with ten or twelve electrons around that atom. If so, then that atom probably uses d-orbitals in its hybridization to get more than eight electrons in its valence shell.

Now I realize that this is a somewhat fuzzy guideline, but in real chemistry there are no hard and fast rules. Every general rule has exceptions, and the rule of eight is one that has many, many exceptions. Hydrogens always strive to get two electrons in their valence shell, beryllium atoms strive for four, and boron atoms strive for six. As a result, hydrogens never hybridize, berylliums have only sp hybridization, and borons have only sp^2 hybridization. In addition, any atom in rows three or greater of the periodic chart can use d-orbitals in its hybridization to get either ten electrons in its valence shell (with sp^3d hybridization) or twelve electrons (with sp^3d^2 hybridization). I'll go through plenty of examples in a moment, but before I do that, I must talk about the geometries that result in hybridizations that involve d-orbitals.

As shown in Figure 3.6, the base geometry for sp^3d hybridization is trigonal bipyramidal. That's not the only geometry that can result from this hybridization, however. Just as sp^3 hybridization can lead to pyramidal and bent geometries when some of the orbitals contain non-bonding electrons, sp^3d hybridization can lead to geometries other than trigonal bipyramidal. Examine Figure 3.12 to see what other geometries can result.

FIGURE 3.12

Molecular Geometries Other than Trigonal Bipyramidal Based on sp^3d Hybridization

Lewis structures:

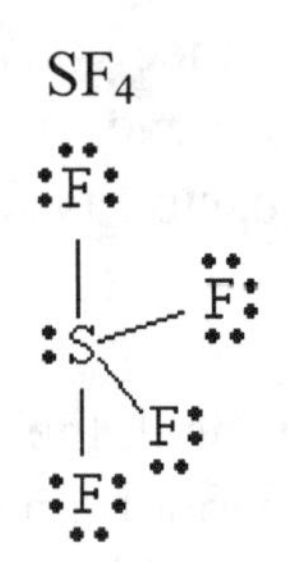

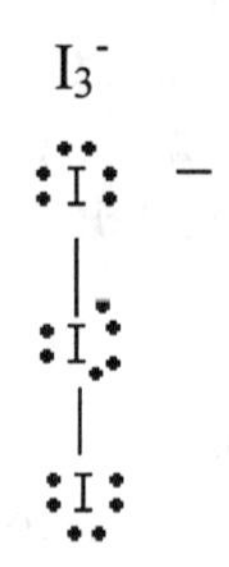

Molecular orbitals:

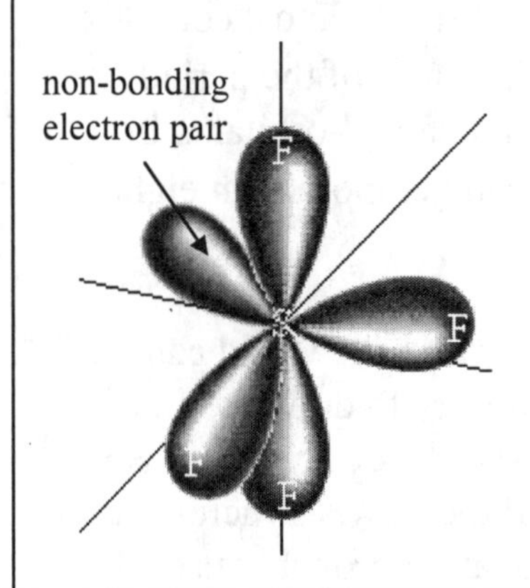

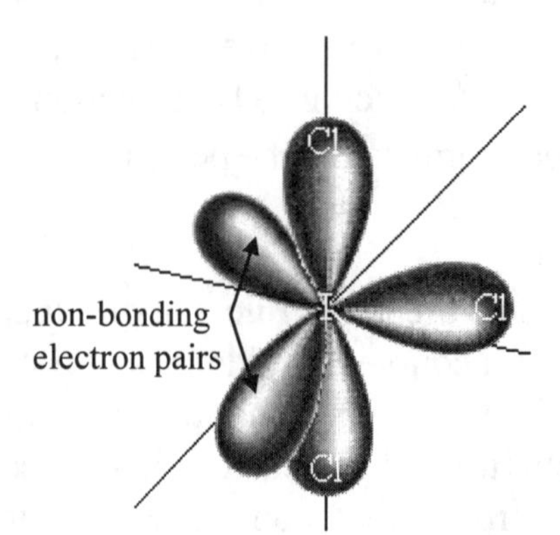

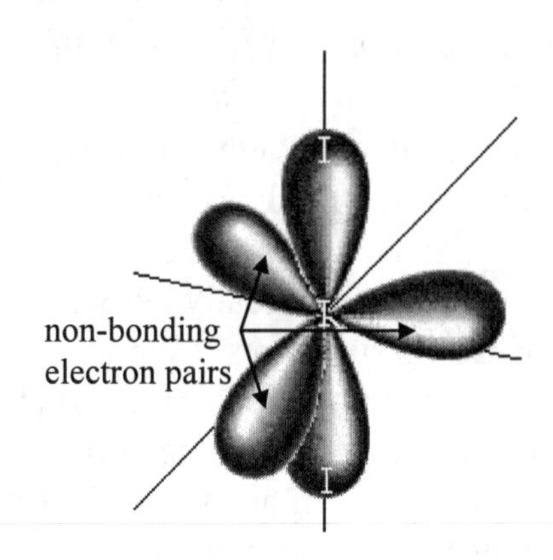

Resulting geometries:

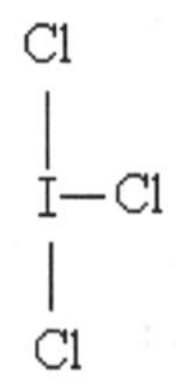

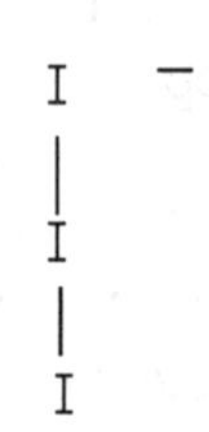

Since the central atom (S) has ten electrons around it, the hybridization is sp^3d and the base geometry is trigonal bipyramidal. One of the legs is missing, however, because of the non-bonding electron pair. The resulting geometry is **see-saw**.

Since the central atom (I) has ten electrons around it, the hybridization is sp^3d and the base geometry is trigonal bipyramidal. Two of the legs are missing, however, because of the two non-bonding electron pairs. The resulting geometry is **T-shaped**.

Since the central atom (I) has ten electrons around it, the hybridization is sp^3d and the base geometry is trigonal bipyramidal. Three of the legs are missing, however, because of the two non-bonding electron pairs. The resulting geometry is **linear**.

Much like molecules that involve sp^3 hybridization, these rule-breaking molecules have a base geometry that then gets altered by the number of non-bonding electrons pairs that surround the central atom. If the central atom has ten electrons in the form of 5 bonds around it, its hybridization is sp^3d and its geometry is called **trigonal bipyramidal.** If there are ten electrons around the central atom but one of them is a non-bonding electron pair, one of the "legs" of the trigonal bipyramidal shape is gone, and the result is a **see-saw** shape. If there are ten electrons around the central atom but two of them are non-bonding electron pairs, then two of the "legs" of the trigonal bipyramidal shape are gone, and the result is a **T-shape.** Finally, if there are ten electrons around the central atom but three of them are non-bonding electron pairs, then three of the "legs" of the trigonal bipyramidal shape are gone, and the result is a **linear** shape.

You should notice one thing from the figure. In each case, a non-bonding electron pair is placed in one of the orbitals that lie in the x-y plane. There is actually a very good reason for this. In your first year course, you should have been told that non-bonding electron pairs actually repel other electron pairs more than bonding electron pairs. As a result, molecules like to "give" non-bonding electron pairs "more room." There is "more room" in the x-y plane because those orbitals are 120 degrees apart, so that's why non-bonding electron pairs will always go into the molecular orbitals on the x-y plane.

Bear with me here. We still need to get through the geometries that result from sp^3d^2 hybridization. When a Lewis structure ends up with a central atom that has twelve electrons around it, then the orbital hybridization must be sp^3d^2 and the resulting base geometry is called **octahedral.** Once again, however, if there are non-bonding electron pairs, "legs" within the octahedral geometry are missing, and as a result the geometry changes. Examine Figure 3.13 to see how this works.

FIGURE 3.13

Molecular Geometries Other Than Octahedral Based on sp^3d^2 hybridization

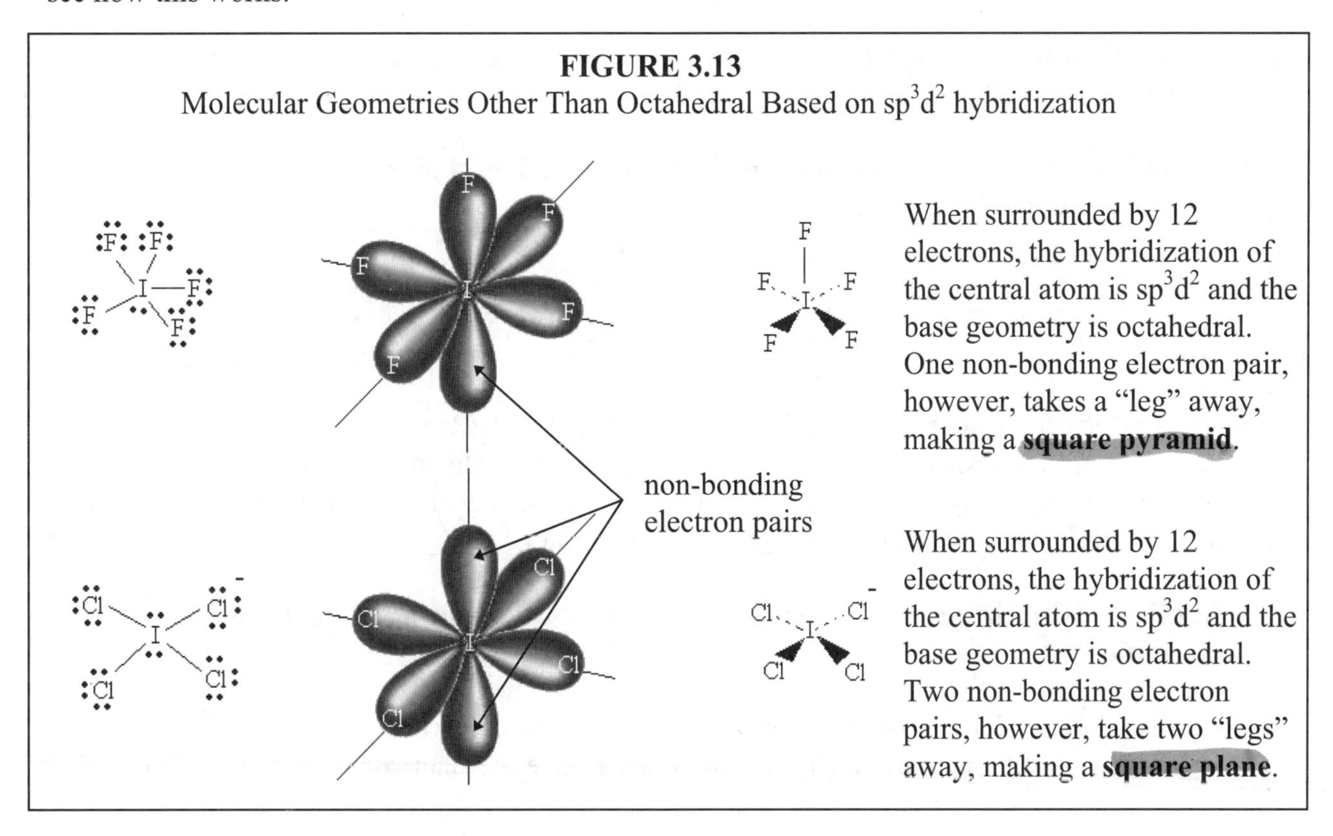

When surrounded by 12 electrons, the hybridization of the central atom is sp^3d^2 and the base geometry is octahedral. One non-bonding electron pair, however, takes a "leg" away, making a **square pyramid**.

When surrounded by 12 electrons, the hybridization of the central atom is sp^3d^2 and the base geometry is octahedral. Two non-bonding electron pairs, however, take two "legs" away, making a **square plane**.

When dealing with molecules that break the rule of eight, then, you still need to determine the Lewis structure and the resulting hybridization. Study how I solve the four example problems below to see the reasoning that goes on here.

EXAMPLE 3.2

What is the orbital hybridization of the central atom and the resulting geometry in beryllium dichloride?

First, we have to draw the Lewis structure, keeping in mind that beryllium is an exception, striving for only four electrons in its valence.

Be • •Cl: •Cl: ⟶ Be:Cl: (with :Cl: above Be) ⟶ Be—Cl: (with :Cl: bonded above Be)

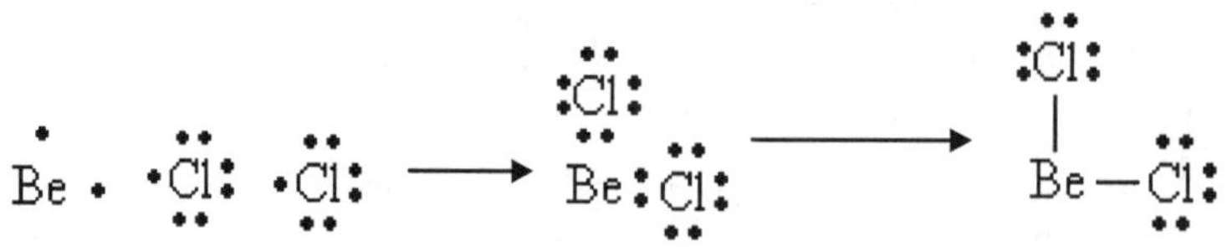

Since Be is the central atom, its hybridization is important. There are no double bonds, but beryllium is an exception. It only wants four electrons in its valence and therefore only has sp hybridization. This results in a linear geometry. Thus, even though the Lewis stricture does not look linear, the result of the sp hybridization is that the molecule takes on a linear geometry.

What is the hybridization of the central atom and the resulting geometry of a boron trifluoride molecule?

First, we have to draw the Lewis structure, keeping in mind that boron is an exception, striving for only six electrons in its valence.

•B• •F: •F: •F: ⟶ :F:B:F: (with :F: above B) ⟶ :F—B—F: (with :F: bonded above B)

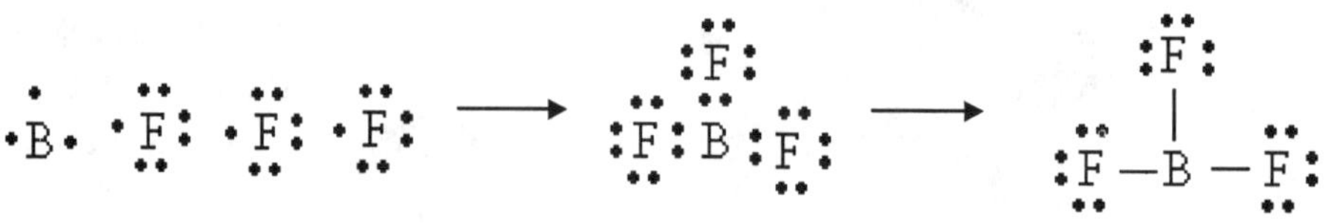

Since B is the central atom, its hybridization is important in determining the geometry. There are no double bonds, but boron is an exception. It only wants six electrons in its valence and therefore only has sp^2 hybridization. This results in a trigonal geometry. Thus, the central atom has sp^2 hybridization and the molecule takes on a trigonal geometry.

What is the hybridization of the central Cl atom and the resulting geometry of the ClF_4^+ polyatomic ion?

To answer the question, we need to draw the Lewis structure. How do you draw the Lewis structure of a polyatomic ion? Well, you work it the same way that you draw Lewis structures of non-ions, but you first add or remove electrons based on the charge. The charge of this polyatomic ion is 1+. As a result, we must *remove an electron* before we start the process. When removing electrons, it is best to do it from the central atom. The problem tells us that the central atom is the Cl atom, so that's where I will remove an electron:

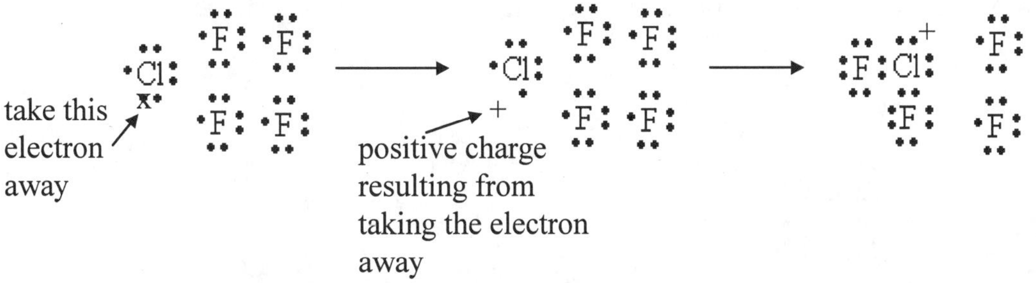

take this electron away

positive charge resulting from taking the electron away

Now what do I do? I was told that the Cl atom is the central atom, and there is simply no more room on it to put the other two fluorine atoms. Well, since Cl is in row 3 of the periodic chart, it can have d-orbital hybridization, allowing for ten or twelve electrons around it. So, if I split up one of the non-bonding electron pairs around the Cl into two single, unpaired electrons, there would be room for each fluorine atom to come in and bond:

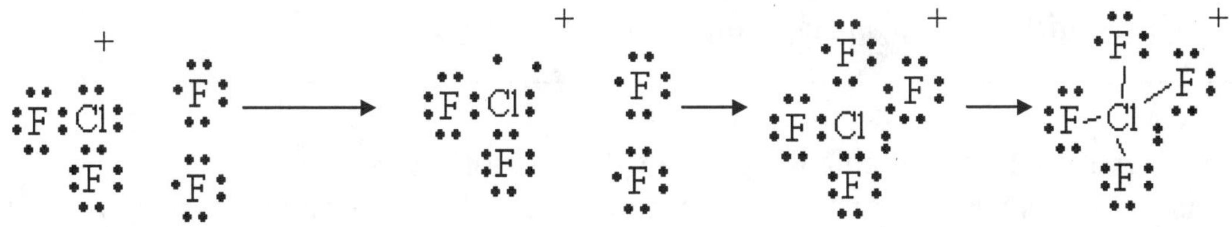

There's our Lewis structure. Cl is clearly breaking the rule of eight, but that's okay, since Cl does have access to d-orbitals. The total number of electrons around the Cl atom is 10, so there need to be 5 hybrid orbitals. This means that the hybridization is sp^3d, and the base geometry is trigonal bipyramidal. However, because one of the electron pairs is non-bonding, one of the "legs" is missing. Thus, the sp^3d hybridization leads to a see-saw geometry.

What is the orbital hybridization of the central phosphorous atom and the resulting geometry in the PCl_6^- polyatomic ion?

Well, we need to draw a Lewis structure again, but this time the ion is negative. What does that mean? It means that after we draw in all of the electrons from the phosphorous and chlorine atoms, we get to draw in one extra electron to account for the negative charge.

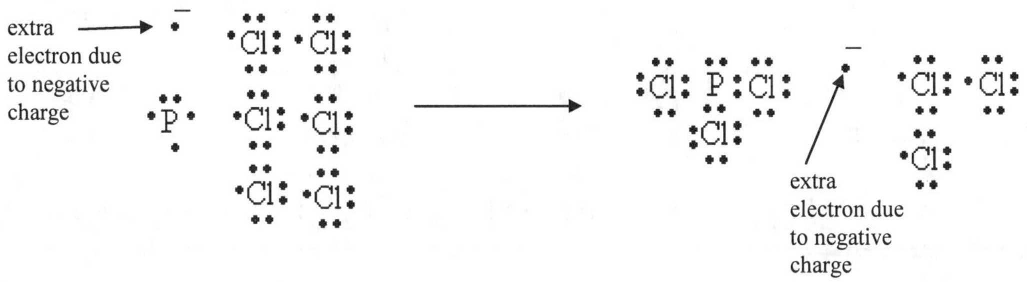

extra electron due to negative charge

extra electron due to negative charge

What do I do now? The phosphorous atom has to be the central atom, because the problem says so. However, I still have three chlorines and an extra electron that I need to stick somewhere. What do I do? Well, if I split up the non-bonding electron pair on the phosphorous, I will have two lone electrons to which I can attach two chlorine atoms. There will still be one chlorine, however. What do I do with that? Well, there is also an extra electron. I'll stick that on the phosphorous as well, giving me one more lone electron, to which I can attach the last Cl!

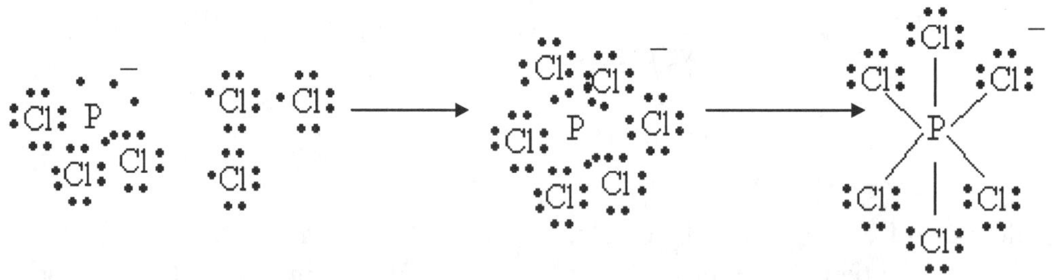

That's our Lewis structure. Notice that there are 12 electrons around the phosphorous, which demands six hybrid orbitals. The only way this can happen is with sp^3d^2 hybridization, so the base geometry is octahedral. There are no non-bonding electron pairs around the central atom, so the sp^3d^2 hybridization leads to an octahedral geometry.

In the end, then, determining the orbital hybridization and the resulting molecular geometry is not an easy thing. First, you draw a Lewis structure that might be tricky if you are dealing with a rule-breaker. Next, you determine the number of electrons around the central atom as well as the number of multiple bonds to determine the orbital hybridization. Finally, you use the base geometry given by the orbital hybridization and use the non-bonding electrons to determine what "legs" are missing off of the base geometry. This gives you the final geometry of the molecule. Whew! To help you out with this process, Table 3.1 lists the type of hybridization and the geometries to which it can lead.

TABLE 3.1
Hybridizations and Molecular Geometries

Orbital Hybridization	Base Geometry (no non-bonding electron pairs)	Geometry with one non-bonding electron pair	Geometry with two non-bonding electron pairs	Geometry with three non-bonding electron pairs
sp	linear	linear	**not possible**	**not possible**
sp^2	trigonal	bent	linear	**not possible**
sp^3	tetrahedral	pyramidal	bent	linear
sp^3d	trigonal bipyramidal	see-saw	T-shaped	linear
sp^3d^2	octahedral	square pyramid	square plane	**not considered**

Now don't go memorizing this table! That's not the point. All you need to memorize is the base geometry for each type of orbital hybridization. After you know that, determining the shape is merely the result of taking legs away for each lone pair and envisioning the result in your mind. Try this on your own to see if you can deal with this difficult process.

ON YOUR OWN

3.7 What is the hybridization of the central boron atom and the resulting geometry of BH_3?

3.8 What is the hybridization of the central iodine atom and the resulting geometry of IF_3?

3.9 What is the hybridization of the central tin atom and the resulting geometry of $SnCl_5^-$?

3.10 What is the hybridization of the central xenon atom and the resulting geometry of XeF_5^+?

Explanation of Experiment 2.1 Plus Another Experiment

In the previous module, I promised that you would get an explanation of Experiment 2.1 at the end of this module. Well, you have finally arrived. If you recall, the whole concept of color in chemistry is based on the idea that electrons absorb energy to move from a low-energy orbital to a high-energy orbital. When they move the other way, they release energy. If this energy is in the form of visible light, then color results.

If you see an object "glowing" with a particular color of light, then you know that the color comes from the electrons releasing light as they travel from high-energy orbitals to low-energy orbitals. When we see neon lights, flame, glow-in-the-dark objects, or objects that are fluorescent under black light, that's what's happening. Most of the colors we see in everyday life, however, are a result of the reverse process. When visible light strikes a substance, the electrons within that substance might absorb some of that light in order to jump from low-energy orbitals to high-energy ones. The light that is not absorbed then bounces off of the object and hits our eyes. When our eyes see this light, they miss the wavelengths that are not there and, as a result, a color is formed in our mind. Since white light is made up of red, orange, yellow, green, blue, indigo, and violet (ROY G. BIV), then any time a color is missing, the result is colored light. If, for example, an object absorbs red, orange, and yellow light, then the light that gets bounced off of the object and into your eyes will be missing these colors and will appear greenish-blue.

So what happened in Experiment 2.1? Well, when the chemicals you started out with appeared clear, that was because the electrons within the molecules did not absorb visible light. When a chemical reaction produced a new chemical, however, the electrons in those molecules did absorb visible light. The result, then, was a solution that had color. For example, when you mixed clear solutions (sodium ferrocyanide and ferric ammonium sulfate) to make the blue solution, you took two molecules that did not absorb visible light and reacted them to make a molecule whose electrons absorbed all visible light except the blue wavelengths. Thus, when you looked at the solution, the light that bounced off of the test tube was devoid of all colors except blue, and what you saw was a blue solution.

At this point, you should be asking yourself something. In Module #2, you learned that electrons in atomic orbitals absorb only *specific wavelengths* of light. This is a consequence of the quantum assumption, which indicates that electrons can have only certain specific amounts of energy. In order to get rid of all colors except blue, the blue solution that you made in Experiment 2.1 must absorb a *whole host* of wavelengths. How can that happen? Well, there are two important differences between the atomic situations we discussed in Module #2 and what went on in Experiment 2.1.

First of all, Experiment 2.1 dealt with molecules, not atoms. As you have hopefully seen as a result of going through this module, molecular orbitals are quite a bit more complex than atomic orbitals! As a result, the energetics of electrons in molecules is much more complex than the energetics of electrons in atoms. In fact, molecular orbitals are a lot more complex than what I discussed in this chapter! For example, two atoms in a molecule actually vibrate back and forth while they are bonded to one another. This distorts the molecular orbitals, constantly changing the energies that the electrons in those orbitals are allowed to have. As a result, electrons have a broader range of allowed energies in molecules than they do in atoms. Thus, rather than absorbing individual wavelengths of light, the electrons in molecules tend to absorb ranges of wavelengths.

The second difference is also very significant. Perform the following experiment to find out what this second difference is.

EXPERIMENT 3.1
The Effect of a Solvent on the Color of a Substance

Supplies:

From the laboratory equipment set:
- Cobalt chloride
- Chemical scoop
- Test tube
- Cap for test tube
- Q-tip

From your home:
- White paper (preferably without lines)
- Hair dryer

Fill the test tube halfway up with water and add three measures of cobalt chloride. Cap the test tube and shake vigorously to dissolve all of the cobalt chloride. Remove the cap and dip your Q-tip into the solution. Then make a design on the paper, using the cobalt chloride solution as "paint." Once you are done (you can use all of the solution if you want), plug in the hair dryer and use it to dry the paper. What happens? If you stop blowing on the paper with the hair dryer and allow it to set out, you should see it return to its original state. If you live in a very arid climate, that may not work. If humidity is above 50-60%, however, it should work quite well. Once the sheet has returned to its original state, you can make the change happen again with the hair dryer. You should be able to do that as many times as you like.

Okay, this was another color change lab, but what does it illustrate? It illustrates the fact that when we have two substances mixed together, they can interact to change the color of the mixture. In the experiment that you just performed, you were working with cobalt chloride, $CoCl_2$. This is an ionic compound; thus, in solution, it splits up into Co^{2+} and $2Cl^-$. Now remember, water dissolves substances because it is attracted to them. Well, when water molecules get close to the cobalt ion, they "see" sp^3d^2 hybridized orbitals with room in them. As a result, the water molecules will actually allow some of their electrons to share those orbitals! Since there are six orbitals, six water molecules will surround the ion, each allowing some of its electrons to sit in the cobalt's hybridized orbitals for a fraction of the time! The result looks something like this:

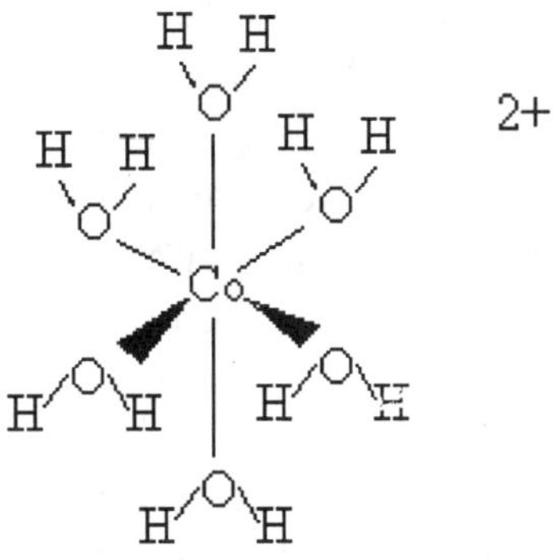

in which each water molecule occupies a sp^3d^2 orbital in an octahedral geometry. We call this a **complex ion**.

Complex ion - An ion that incorporates another substance (usually a solvent) into its orbitals

We denote such an ion as $[Co(H_2O)_6]^{2+}$. The square brackets make it clear that the entire group has a 2+ charge. Pretty complex, isn't it?

Well, this is what cobalt does when there is plenty of water around. The result is a complex ion that weakly absorbs the green and blue ends of the visible spectrum, giving the overall solution a pinkish color. The color is weak, however, so when you put the solution on paper, you probably saw no color. When water becomes scarce, the complex ion changes. As you get rid of water (by blowing on it with a hair dryer, for example), the complex changes into

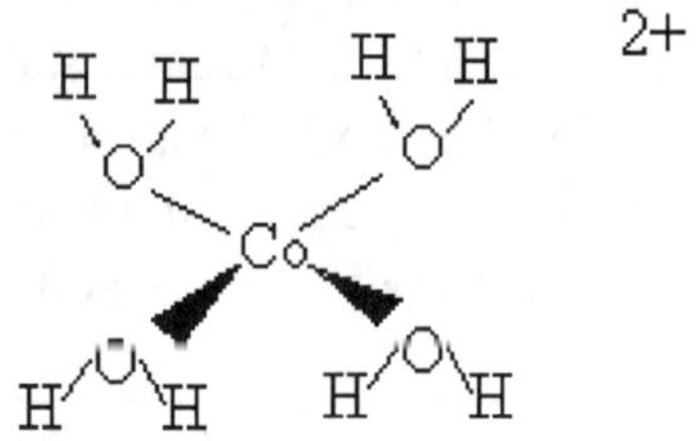

Now this complex ion, $[Co(H_2O)_4]^{2+}$, still uses sp^3d^2 orbitals, but since it lost two of the water molecules, two of the "legs" from the octahedral geometry are missing, and the result is a square planar geometry. This complex absorbs red, orange, yellow, and green light, resulting in a blue color.

In the experiment, then, you saw that the color of a substance not only depends on the molecule which makes up the substance, but it also depends on the type and amount of other molecules which surround the substance. In the presence of a lot of water, cobalt chloride is light pink. In the presence of very little water, cobalt chloride is blue.

So what have we learned? The basis for all of the color we see in Creation is, in fact, the energies that electrons can have within a substance. In the "simple" case of atoms, electrons have specific energies that are allowed, resulting in specific wavelengths of light that can be either absorbed or emitted. With molecules, the situation is more complicated. The complex nature of molecular orbitals results in a *range* of wavelengths that can be absorbed by the electrons in the molecule. In addition, interaction with surrounding molecules can further change the range of allowed energies for the electrons, which results in a change in the color of light that can be absorbed or emitted by electrons.

Now if all of this really confused you, don't worry too much about it. Chemistry is a broad field, and the subjects we studied in this module and the last one are contained in a field called *physical chemistry.* There are many excellent chemists who despise and do not understand physical chemistry. Thus, if all of this is a bit mystifying, that does not mean you aren't cut out to be a chemist. It probably means you should shy away from physical chemistry, however. The next module will also deal with physical chemistry concepts, but after that, we will move on to other branches of chemistry.

ANSWERS TO THE ON YOUR OWN QUESTIONS

3.1 This hybridization would result in 5 hybrid orbitals. The notation indicates that there is one s-orbital, two p-orbitals, and two d-orbitals. This makes a total of 5 atomic orbitals, which will form 5 hybrid orbitals.

3.2 The only way to form 3 hybrid orbitals is with three atomic orbitals. Thus, there must have been one s-orbital and two p-orbitals, so it is sp^2 hybridization. The resulting geometry will be trigonal.

3.3 First of all, since hydrogen is an exception to the rules, we know that all hydrogen atoms have no hybridization. Looking at the first carbon, we see only single bonds. This means that the first carbon has sp^3 hybridization. The second carbon, however, has a double bond, which requires a p-orbital. This leaves a p-orbital out of the hybridization process, resulting in the second carbon having sp^2 hybridization. The first oxygen also has a double bond, which means it, too, must save a p-orbital from the hybridization process. This results in the first oxygen having sp^2 hybridization. The second oxygen, however, has only single bonds, so the second oxygen has sp^3 hybridization.

3.4 The first thing we need to do here is draw the Lewis structure:

H •C• •N• ⟶ H:C:N• ⟶ H:C:::N ⟶ H—C≡N

First of all, the hydrogen does not hybridize its orbitals. Secondly, the carbon has a triple bond, for which it must reserve two p-orbitals. This means the carbon has sp hybridization. Since the carbon is the central atom, it controls the geometry, so the geometry is linear. Finally, the nitrogen also has a triple bond, so the nitrogen is also sp-hybridized.

3.5 The Lewis structure for PCl_3 is:

•P• •Cl: •Cl: •Cl: ⟶ :Cl:P:Cl: (:Cl: below P) ⟶ :Cl–P–Cl: (| :Cl: below P)

There are no multiple bonds anywhere, so all atoms are sp^3 hybridized. The phosphorous atom controls the geometry, since it is the central atom. Since the hybridization of phosphorous is sp^3, the base geometry is a tetrahedron. However, one of the orbitals in the hybridization is filled with a non-bonding pair. Thus, we have a tetrahedron with a "leg" missing. This means the geomctry is pyramidal.

3.6 First, we draw the Lewis structure:

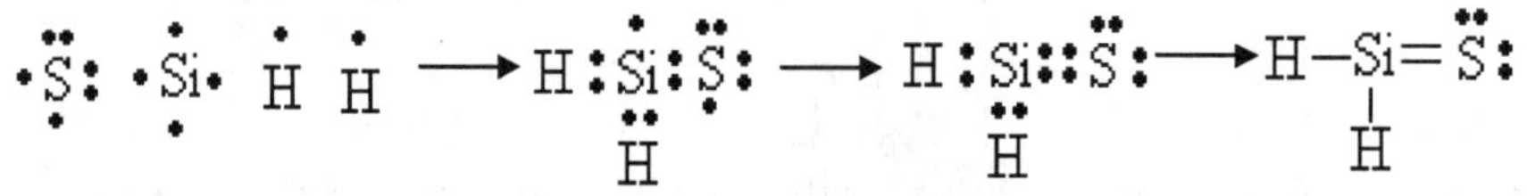

•S: •Si• H H → H:Si:S: → H:Si::S: → H–Si=S:
H H H

We know that all hydrogens have no hybridization. Silicon, however, has a double bond for which it must reserve a p-orbital. Thus, silicon has sp^2 hybridization. Since silicon is the central atom, it controls the geometry, making the geometry trigonal. Finally, the sulfur also has a double bond, so the sulfur is also sp^2 hybridized.

3.7 First, we have to draw the Lewis structure, keeping in mind that boron is an exception, striving for only six electrons in its valence.

•B• •H •H •H → H:B:H (H) → H–B–H (H)

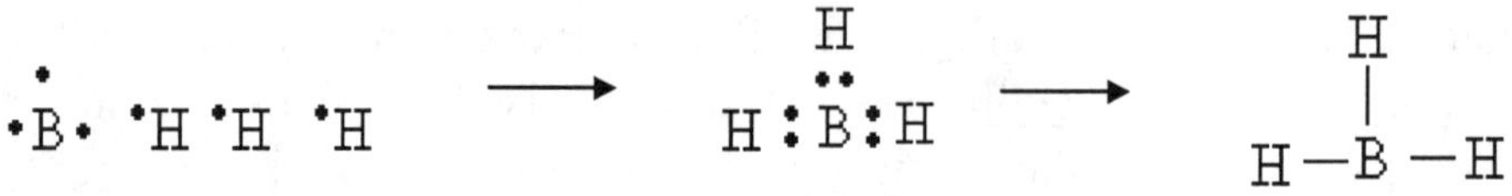

Since B is the central atom, its hybridization is important in determining the geometry. There are no double bonds, but boron is an exception. It only wants six electrons in its valence and therefore only has sp^2 hybridization. This results in a trigonal geometry. Thus, the central atom has sp^2 hybridization and the molecule takes on a trigonal geometry.

3.8 First, we draw the Lewis structure:

•I: •F: •F: •F: → :F:I: •F: •F:

Am I stuck now? No, because iodine has access to d-orbitals so it can have 10 or 12 electrons instead of just 8. I can split one of the electron pairs around the iodine into two single electrons, providing a place where each fluorine can attach:

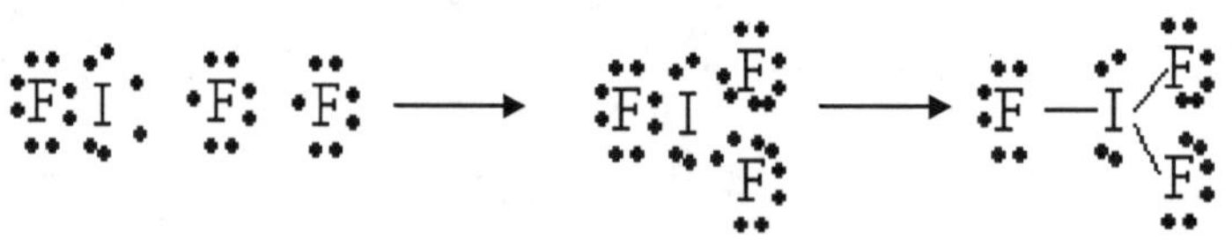

:F:I •F: •F: → :F:I (F) (F) → F–I(–F)(–F)

Now I know that the central iodine atom has ten electrons surrounding it. This calls for five orbitals, which is accomplished through sp^3d hybridization, with a base geometry of trigonal bipyramidal. Two of the "legs" of this geometry are missing, however, because two pairs of electrons around the central atom are non-bonding. As a result, the sp^3d hybridization leads to a T-shaped geometry.

3.9 To draw this Lewis structure, I need to realize that I am working with a negative ion. The charge is 1-, indicating that I need to add one extra electron when I draw the Lewis structure:

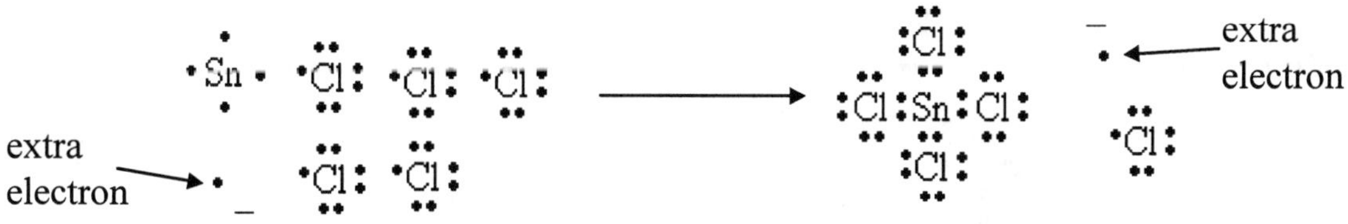

extra electron

extra electron

Even though it looks like I'm stuck, I'm really not. Since Sn has access to d-orbitals, it can have ten or 12 electrons around it. If I take the extra electron that I get from the 1- charge and put it on the tin atom, that will give me a spot to attach the last Cl atom:

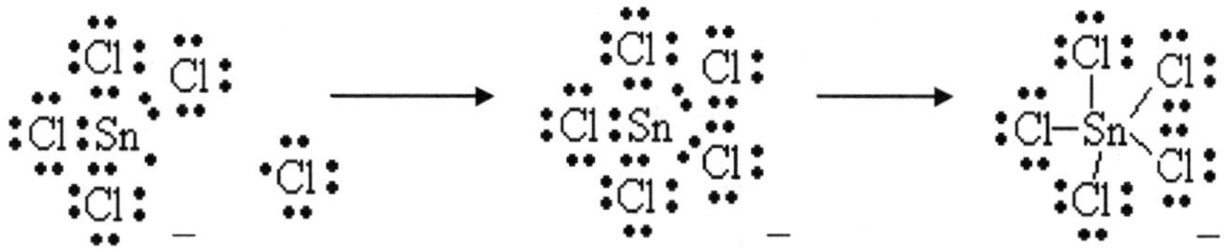

So I can now see that there are ten electrons around the central tin atom, and they are all bonding electron pairs. This means that the hybridization is sp^3d and the resulting geometry is trigonal bipyramidal.

3.10 I start by drawing a Lewis structure. Since this is a positive ion, I will need to remove an electron from the central atom to account for the 1+ charge:

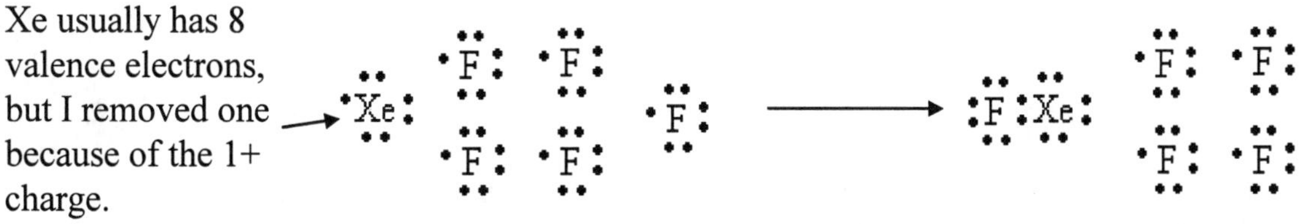

Xe usually has 8 valence electrons, but I removed one because of the 1+ charge.

I am not stuck here, because Xe has access to d-orbitals and can therefore have up to 12 electrons around it. I will split two electron pairs into 4 single electrons, providing a place for the other 4 fluorines to link up:

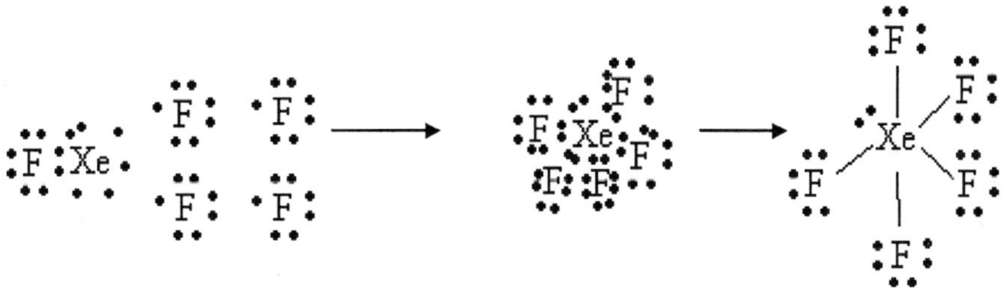

Now I can tell that there are 12 electrons around the central xenon atom, which calls for 6 hybrid orbitals. Thus, the central atom has sp^3d^2 hybridization, and the base geometry is octahedral. One of the pairs of electrons is non-bonding, however, so a "leg" from the octahedral shape is gone. As a result, the sp^3d^2 hybridization leads to a square pyramid geometry.

REVIEW QUESTIONS

1. What is a hybrid orbital?

2. What is the difference between a molecular orbital and a hybrid orbital?

3. If an atom has 5 hybrid orbitals, what type of hybridization was used?

4. If the central atom of a molecule has 3 hybrid orbitals (all of which end up forming molecular orbitals), what is the geometry of the molecule?

5. Using two dots to depict atoms that are boding together, draw the general shape of a sigma-bond and then draw another picture illustrating a pi-bond.

6. Suppose two atoms bond together by overlapping sp^2-hybrid orbitals. Did they form a pi-bond or a sigma-bond?

7. Which of the following atoms *could* have sp^3d or sp^3d^2 hybridization?

Li, Cl, C, F, S, P, N, Br

8. Two atoms in a molecule bond to one another. If they have a double bond between them, do they have a sigma-bond? Do they have a pi-bond?

9. Two atoms that bond to one another make two pi-bonds. Is this a single, double, or triple bond?

10. You go to the store and buy a red hat. What colors of light do the chemicals in the hat's dye absorb?

PRACTICE PROBLEMS

1. Urea (N_2H_4CO) is often used in fertilizers as a source of nitrogen. The Lewis structure for this molecule is shown below:

```
          ..
          O:
          ||
     ..        ..
 H — N — C — N — H
     |         |
     H         H
```

What is the orbital hybridization for the nitrogen atoms and carbon atom?

2. Hydrogen cyanide is a gas that smells like bitter almonds. Its Lewis structure is:

```
 ..
 N ≡ C — H
```

a. What is the hybridization of the nitrogen atom?

b. In what kind of orbital is the non-bonding electron pair stored?

c. How many sigma-bonds exist is the entire molecule? How many pi-bonds exist?

3. What is the hybridization of the central nitrogen atom and the resulting geometry of ClNO?

4. What is the hybridization of the central silicon atom and the resulting geometry of SiF_4?

5. What is the hybridization of the central carbon atom and the resulting geometry of F_2CO?

6. What is the hybridization of the central silicon atom and the resulting geometry of HSiP?

7. What is the hybridization of the central bromine atom and the resulting geometry of BrF_5?

8. What is the hybridization of the central iodine atom and the resulting geometry of the IF_6^+ polyatomic ion?

9. What is the hybridization of the central arsenic atom and the resulting geometry of AsF_4^-?

10. Which ion is more likely to form complex ions in solution: Li^+ or Cu^+?

Module #4: Intermolecular Forces and the Phases of Matter

Introduction

In your previous chemistry course, you were introduced to the **kinetic theory of matter**, which is also called the **kinetic-molecular theory of matter**. To remind you of what this theory says, perform the following experiment:

EXPERIMENT 4.1
The Kinetic Theory of Matter

Supplies:
From the laboratory equipment set:
- Cobalt chloride
- Two plastic test tubes
- Measuring scoop
- Two plastic eyedroppers (also called pipettes)

From around the house:
- Someone to help you
- Water
- Heat to boil water
- A small glass (like a juice glass)
- A Styrofoam coffee cup
- A white piece of paper (no lines)

Fill the juice glass with cold water from the tap. Start a pan of water boiling. You need only have enough boiling water to fill the Styrofoam cup, so you don't have to heat that much water. While you are waiting for the water to boil, carefully place a measure of cobalt chloride at the bottom of each test tube. Try not to get any of it on the sides of the test tube. Take a pipette and have your helper do the same. Now each of you take a test tube and slowly and carefully add water to it. The idea here is to add water to the cobalt chloride without stirring it up at all. This requires some patience! The best way to do it is to add the water drop-by-drop with your pipette against the side of the test tube. That way, each drop dribbles down the side of the test tube. If you have some cobalt chloride clinging to the sides of the test tube, position your pipette so that as the drops dribble down the side of the test tube, they will run into the solid clinging there and wash it down. You need to fill each test tube to the topmost mark on the tube.

Now let one test tube sit in the test tube rack. As soon as the water is boiling, fill the Styrofoam cup with it. Place one of the test tubes in the Styrofoam cup and get it to stand in the water. **Be careful! The water is hot!** Let the two tubes sit for an hour. When you come back, take each tube and place the white piece of paper behind it. Look straight through the tube, with the paper as a background. You should see that the water near the bottom of the tube is very pink, and that the water at the top of the tube is still clear. Look for the topmost part of the tube in which you see a pink color. The pink color should be higher in the test tube that was in the hot water. The color will not be as dark as it is in the other test tube, because it is dispersed over a larger volume, but it will be higher.

What did the experiment show you? Remember that the kinetic theory of matter says that the molecules or atoms which make up a substance are always moving. In gases, the molecules are far apart from each other and move quickly; in liquids they are closer together and move more slowly; and in solids they are even closer together and simply vibrate back and forth. In the experiment, you put an ionic solid at the bottom of a sample of water. As time went on, the water began to dissolve the solid, splitting it up into its respective ions. This caused the liquid to turn pink. In order for more solid to dissolve, however, the ions that were dissolved had to get out of the way to allow room for more ions. The only way that could happen was for the ions to travel up the tube. Why did they travel up the tube? Well, since the ions were in motion (after all, they were in the liquid state), they began moving about randomly. Sometimes, that random motion would take them higher in the test tube. Given enough time, this random motion would completely mix the solution, making a constant, light pink color throughout the test tube.

Why did the test tube that was in the hot water have a pink color higher up in the test tube? Well, the kinetic theory of matter tells us that as you add energy to a substance, its molecules or atoms begin to move faster. Thus, the ions that were dissolved in the test tube that was in hot water moved a lot faster than those that were in the cold water. As a result, the pink color traveled up the tube faster. Hopefully, this visual experiment has helped you remember the kinetic theory of matter.

Applying the Kinetic Theory of Matter to Phase Changes

If molecules and atoms like to move around, as the kinetic theory of matter says that they do, why isn't every substance in Creation a gas? After all, in the gas state, the molecules or atoms that make up a substance are free to move around as much as they like. Why, then, don't all molecules and atoms strive for this freedom? Because there are attractive forces that exist *between individual molecules*. These attractive forces, collectively called **van der Waals** (van dur valls) **forces**, tend to keep the molecules and atoms close to each other, limiting their ability to move. The phase that a substance finds itself in depends on the relative strength of the van der Waals forces and the energy of its constituent molecules or atoms as a result of the substance's temperature.

Consider, for example, a glass of water sitting out on a table. You know what will happen over time. The water will slowly evaporate, changing from liquid to gas. Why does this happen? Well, the first thing to realize is that even though the molecules of water in the glass are all at the same temperature, they are not moving around with the same speed. You see, the molecules of water are moving about randomly, crashing into each other, bouncing off the side of the glass, etc. Every time a collision between water molecules occurs, some energy will be exchanged. Typically, one molecule in the collision will gain more energy than the other. As this goes on, the distribution of energy becomes more uneven. Some molecules end up with more than their fair share of the energy. This makes them move faster than the molecules that have less than their fair share of energy. In the end, then, we say that at a given temperature, the molecules in a substance have a *distribution* of energy. This results in a *distribution* of speeds. Some have more, some have less. Such a distribution is illustrated by the graph in Figure 4.1.

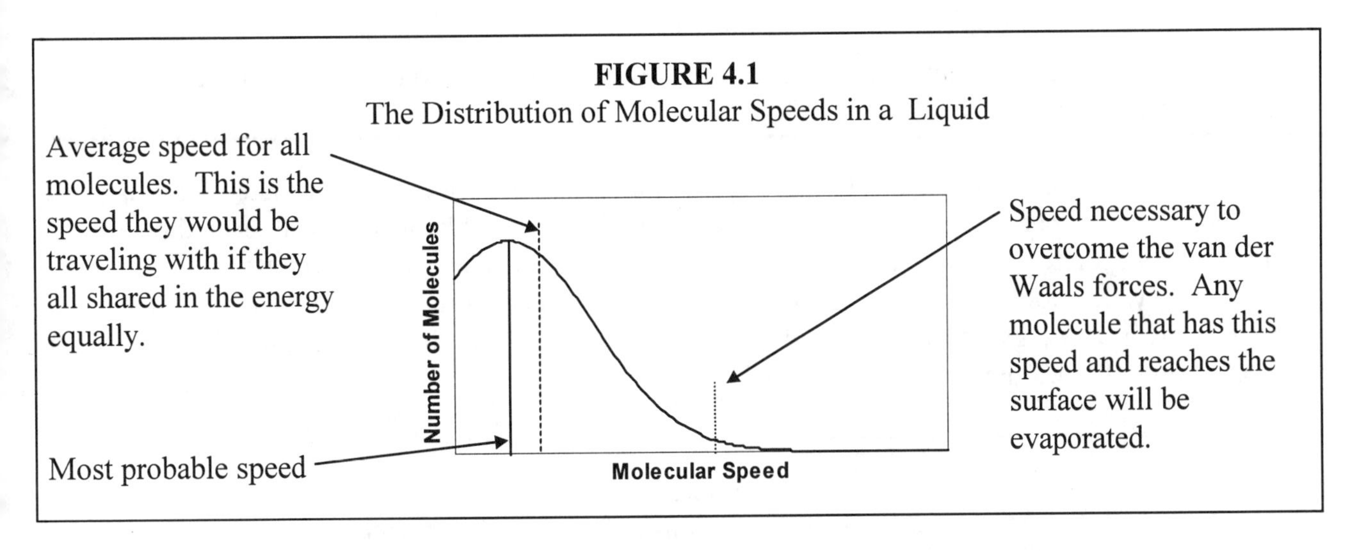

FIGURE 4.1
The Distribution of Molecular Speeds in a Liquid

Before you concentrate on the lines in the figure, make sure you understand the graph itself. The x-axis plots the speed of a molecule. The y-axis tells us how many molecules have that speed. So you see that some molecules have low speeds. As you go up in speed for a while, however, the number of molecules with that speed actually increases. At some point, you go through a maximum, which means that more molecules have that speed than any other speed. We call this the **most probable speed**, and it is shown by the solid line in the figure. As you increase in speed after that maximum, fewer and fewer molecules have that speed, until, at some point, there are essentially no molecules with very high speeds.

Now remember, this whole distribution exists because the molecules do not share the energy at their disposal equally. Some have more energy than their fair share, some have less. The dashed line near the maximum of the graph is the speed that they would all be traveling if they all have exactly their fair share of the energy. Thus, this is the **average speed** of the molecules. The other dashed line in the graph is the one relevant to the process of evaporation. Any molecule that is moving at or above that speed has enough energy to break free of the van der Waals forces that attract it to the other molecules of water. Thus, when a molecule traveling at that speed or higher reaches the surface of the liquid, it will blast out of the liquid, becoming a gas molecule.

In the end, then, a molecule will change from liquid to gas if it has enough speed to escape the pull that the other molecules of water exert on it. Now at this point you should be a little puzzled. After all, there aren't very many molecules that have the energy necessary to escape from the liquid, at least according to the figure. Why, then, will all of the water from a glass eventually evaporate away? Well, think about it. When a molecule escapes from the water, it takes its energy with it. Thus, the water actually loses energy and becomes cooler. Since the water is surrounded by room-temperature air, however, it quickly absorbs the energy it lost when the water molecule evaporates, making a new distribution that looks exactly like the original one. At that point, just as many water molecules have the speed necessary to escape as did before any evaporation took place. Thus, evaporation keeps happening. As time goes on, eventually there will be less and less water in the glass. At some point, the number of water molecules will be so small that there aren't enough molecules to exert the van der Waals forces that were being exerted, and it actually becomes *easier* for water molecules to evaporate. As this process continues, all of the water will eventually turn into a gas.

Now let me ask you a question. Suppose the water were completely insulated from the surroundings, so that it could not absorb any energy. What would happen? Well, each time a molecule evaporated, it would take some energy away with it. This would mean that the remaining water would get cooler. Since the water is insulated from the surroundings and cannot absorb any new energy, the water would get cooler and cooler. As this happens, here's what the distribution of molecular speeds would look like:

FIGURE 4.2
Molecular Speed Distributions at Different Temperatures

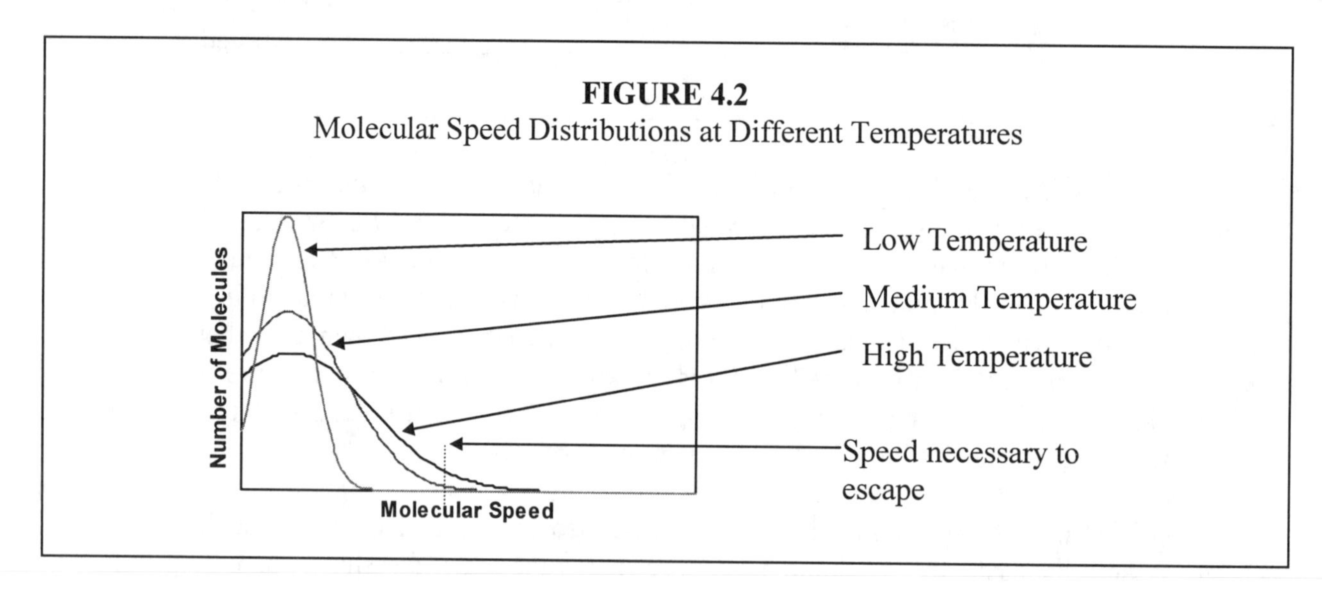

Notice what happens to the distribution as you decrease the temperature. The distribution "scrunches up" and does not extend to speeds as large as it did at higher temperatures. Now look at the distributions in relationship to the speed necessary to escape. As the water begins to cool, the number of molecules that have the speed necessary to escape decreases. Notice from the figure that only a few molecules have the speed necessary to decrease once the water cools to the medium temperature. What happens at the low temperature? At that point, no molecules have the energy necessary to escape, so the water stops evaporating. Now remember, the only reason this happens is because in our hypothetical case, the water cannot absorb energy from its surroundings. In reality, it will absorb energy from its surroundings, and thus it will continue to evaporate.

Now think for a moment what would happen to this entire situation if I increased the atmospheric pressure. How would that change our picture? Would it increase the temperature? No. But it would change something. What would it change? It would change the speed necessary to escape. Think about it. In order to escape from the water, a molecule must have speed enough to overcome the attractive van der Waals forces that hold the molecules together. What happens when I increase the atmospheric pressure? The water molecules get closer together. This *increases* the van der Waals forces. Thus, the speed necessary to escape will increase. This will slow the process of evaporation down, because fewer molecules will have the necessary speed to escape.

Now think about what happens if I change the situation entirely and talk about a solid. In a solid, the situation is different because the molecules aren't as free to move. They are held rigidly in place, as we will discuss in great detail in a later part of the module. Instead of moving freely, the molecules tend to just vibrate back and forth. Under these conditions, the distribution of molecular speeds is pretty narrow. All molecules have speeds relatively close to the average. As a result, there usually isn't a group of molecules that has the speed necessary to pull away from the other molecules. Thus, a solid will not spontaneously change into a liquid unless the temperature is above that of its freezing point.

There is an interesting effect that can occur for some solids, however. For certain substances, the temperature and pressure ranges for which the substance is a liquid become very narrow. Thus, if the temperature and pressure work out just right, it is possible for the molecules of a solid to have enough energy to escape the solid and become a gas directly. This process is called **sublimation**.

Sublimation - The process by which a solid turns into a gas without passing through a liquid phase

You can see sublimation occurring with dry ice. At normal room temperature and atmospheric pressure, carbon dioxide is a gas. If you take dry ice (which is solid carbon dioxide) out of the freezer that is keeping it cold, you will see billowing clouds of vapor leaving the ice. This is the carbon dioxide gas that has been formed by sublimation. Naphthalene is an aromatic gas at room temperature and atmospheric pressure. At pressures slightly greater than atmospheric pressure, however, naphthalene is a solid. Mothballs are made out of naphthalene and stored under slightly increased pressure. When you take a mothball out of its packaging, it slowly sublimes. That's why mothballs smell so strongly. The naphthalene vapor tends to keep moths away.

ON YOUR OWN

4.1 The following graph shows the distribution of speeds for two samples (A) and (B) of the same compound. Which sample is at the higher temperature?

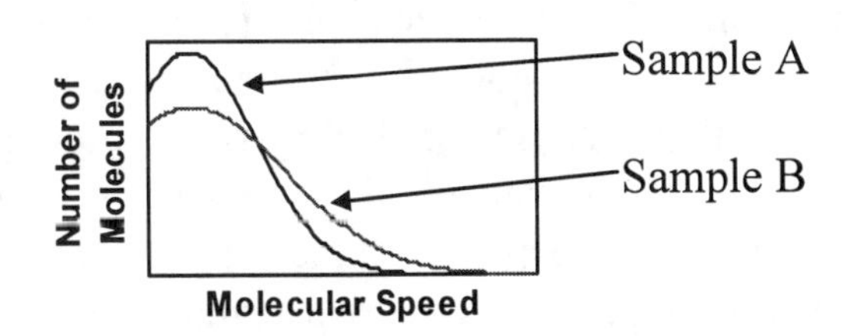

4.2 In a demonstration, a chemistry professor puts a glass of water in a big glass tank and evacuates the air from the tank. At first, you see the water boil vigorously, and then suddenly it freezes! Explain this effect in terms of the kinetic theory of matter.

The Different Types of Van Der Waals Forces

I have spent a long time talking about van der Waals forces, but you still probably do not have a good idea of what they are. Hopefully, I will now remedy this situation by describing the three basic van der Waals forces. The most prominent kind of van der Waals force is the **dipole-dipole attraction**. This attraction occurs between polar molecules. Remember from your first-year course that molecules which do not share electrons equally between the atoms end up having small charges located on the molecule. When many such molecules are together, they will orient themselves so that the small charges on one molecule are attracted to the small charges on another molecule. Consider, for example the molecule HCl. Because the chlorine atom pulls on electrons harder than does the hydrogen atom, the chlorine gets more than its fair share of electrons. This gives the chlorine a small negative charge and leaves the hydrogen with a small positive charge. When several HCl molecules get together, they arrange themselves in patterns such as this:

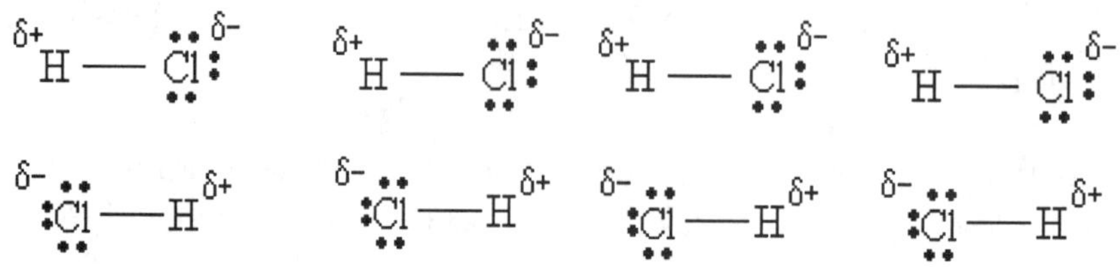

$\delta+$ H — Cl $\delta-$ $\quad$ $\delta+$ H — Cl $\delta-$ $\quad$ $\delta+$ H — Cl $\delta-$ $\quad$ $\delta+$ H — Cl $\delta-$

$\delta-$ Cl — H $\delta+$ $\quad$ $\delta-$ Cl — H $\delta+$ $\quad$ $\delta-$ Cl — H $\delta+$ $\quad$ $\delta-$ Cl — H $\delta+$

Notice that the molecules have oriented themselves such that the small positive charges are closest to the small negative charges. The resulting attraction that exists between the opposite charges attracts the molecules to one another. This is the essence of the dipole-dipole attraction.

A special case of the dipole-dipole attraction occurs between ionic molecules. In this case, full charges of the ions in one molecule are attracted to the full charges of the ions in another molecule. This resulting **ion-dipole attraction** is stronger than the dipole-dipole attraction, because the ions have more charge than do polar molecules, so the strength of the attraction between charges is greater.

The second major van der Waals force is called the **London dispersion force**, and it is more difficult to understand. This force, named after Fritz London who in 1928 was the first to explain it, helps us understand why nonpolar molecules are attracted to one another. London dispersion forces are possible because the electrons in an atom are constantly moving. Because they are constantly moving, sheer chance tells us that every now and again, more electrons will be on one side of the atom than the other. At that point, one side of the atom has excess electrons and the other side has a shortage. Thus, there is a small positive charge that exists on one side of the atom and a small negative charge on the other.

If the atom were all by itself, the electrons would quickly move and even out the charge again. However, if another atom were sitting next to this one, then the electrons in that atom would be repelled by the buildup of negative charge on the first one. This would cause the electrons in the second atom to move away, creating an imbalance of electrons in *that* atom.

Suddenly, then, each atom would have a charge imbalance. This creates an attractive force that will keep the electrons from balancing out the charge again!

Now if all of that was a bit hard to understand, try to see it based on Figure 4.3.

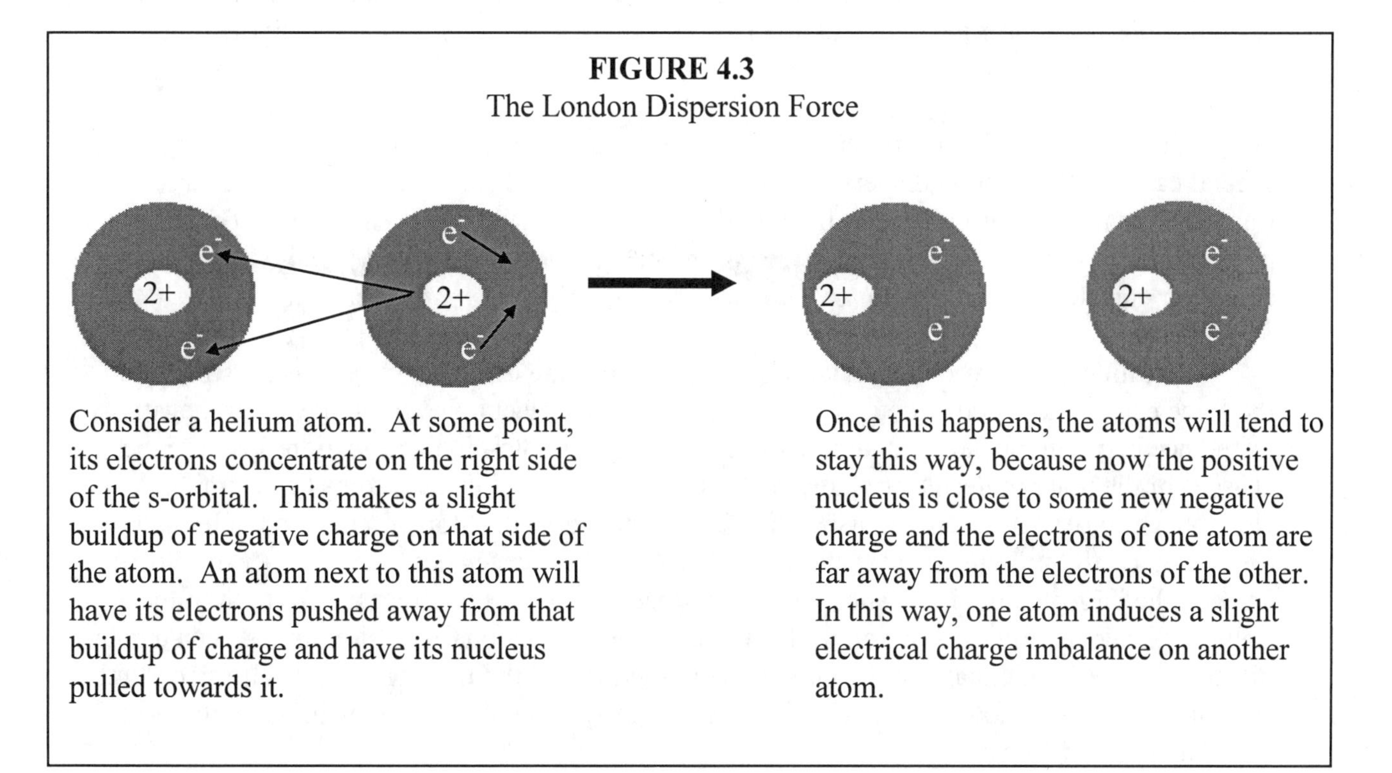

FIGURE 4.3
The London Dispersion Force

Consider a helium atom. At some point, its electrons concentrate on the right side of the s-orbital. This makes a slight buildup of negative charge on that side of the atom. An atom next to this atom will have its electrons pushed away from that buildup of charge and have its nucleus pulled towards it.

Once this happens, the atoms will tend to stay this way, because now the positive nucleus is close to some new negative charge and the electrons of one atom are far away from the electrons of the other. In this way, one atom induces a slight electrical charge imbalance on another atom.

London, who tried to be humble, called this an "induced-dipole interaction," because the process involves one atom inducing another atom to have slight electrical charge imbalance. Eventually, this force was renamed in honor of the man who first explained its origin.

Now think for a moment about the relative strengths of the ion-dipole, dipole-dipole, and London dispersion forces. Which do you think is greatest? Well, we already reasoned that the ion-dipole attraction is bigger than the dipole-dipole attraction. Where do London dispersion forces fit in? Are they stronger or weaker than the dipole-dipole attraction? Well, there is a quick way to answer this. Consider the following substances: ICl and Br_2. Both molecules are diatomic, and they actually have very similar atomic masses. They also contain atoms that all come from group 7A. As a result, you would expect these molecules to be chemically similar to one another. Well, in terms of the chemistry they both perform, they are pretty similar. There are a couple of huge differences between them, however.

First of all, you should be able to recognize that ICl is a polar molecule while Br_2 is not. If you can't remember how to tell whether or not a molecule is polar, please go back to your first year book and refresh your mind, because it is a skill you must know to be able to master the material in this module. Since ICl is a polar molecule, it experiences the dipole-dipole attraction with other ICl molecules. The Br_2 molecule is not polar, so the only van der Waals force that it can experience is the London dispersion force. Now, consider this. At 0 °C (standard temperature), ICl is a solid and Br_2 is a liquid. What does that tell you about the relative strengths of the dipole-dipole interaction and the London dispersion force? Even though these molecules are very similar, at the same temperature, one is a liquid and the other is a solid. Thus, ICl must have greater attraction to its fellow ICl molecules, because at the same temperature, the ICl molecules are closer together than the Br_2 molecules. This tells us that the dipole-dipole attraction is stronger than the London dispersion force.

Just as the ion-dipole attraction is a special case of the dipole-dipole attraction, there are special cases of the London dispersion force. For example, ions can induce a slight charge imbalance in a nonpolar molecule. In addition, a polar molecule can do the same. These interactions give rise to **ion-induced London forces** and **dipole-induced London forces**. In the end, then, this tells us that *all atoms and molecules experience London dispersion forces.*

Within a given type of van der Waals force, there are different ways that the strength of the force can be increased or decreased. For example, a molecule in which the electronegativity difference between the atoms is large will have stronger dipole-dipole attractions because the charges of the polar molecule are large compared to a molecule whose atoms have similar electronegativities. For London dispersion forces, the closer the atoms in the molecules can get to each other, the stronger the attractive force. This makes sense because the closer the atoms can get, the larger charge imbalance they can induce and hence the larger the strength of the London dispersion force. Since all molecules experience London dispersion forces, then a polar molecule, for example, has both dipole-dipole attraction and the London dispersion force. In this case, then, the polar molecule can also increase the intermolecular forces by moving closer to the other molecules in the sample.

There are two things that tend to affect how close the atoms of two different molecules can approach each other. The first is mass. The heavier the atoms, the closer they can get to one another. Thus, molecules and atoms with large masses tend to have stronger attractive forces between them than do molecules with small masses. The second is the shape of the molecule. The more linear a molecule is, the closer the atoms in one molecule can get to another molecule. If the molecules are bulky, it is hard to get more than a few of the atoms in a molecule next to the atoms of another molecule. Of these two effects, mass is by far the most important. Thus, we typically expect molecules of greater mass to have greater attractive forces between them. If the molecules are the same mass, then we tend to look at the shape of the molecule to determine the strength of the attractive forces.

There is one more van der Waals force that I need to discuss. It is rather rare, but it is stronger than all of the other van der Waals forces. It is called **hydrogen bonding**, and you are already familiar with it in some way, because hydrogen bonding gives water its unique properties. For example, you should already know that water (unlike almost any other compound in Creation) is less dense as a solid than as a liquid. This tells us that water molecules are actually closer together in their liquid phase than they are in their gas phase! This is why ice floats on liquid water.

What is hydrogen bonding? Well, remember that water is polar. The oxygen atom is much more electronegative than the hydrogen atoms to which it is bonded. This causes a deficiency of electrons on the hydrogen atoms and a surplus of electrons on the oxygen atom. Since the water molecules are polar, they will tend to align themselves so that the hydrogens (which have a slight positive charge) are near the oxygens (which have a slight negative charge). When this happens, the hydrogen atoms (which have a deficiency of electrons) find themselves very near a source for extra electrons. What source is that? The oxygen atoms! The hydrogen atoms then actually share in the extra electron density of an oxygen atom from *another molecule*. Most chemists believe that the electron density the hydrogen atom shares comes from one of the non-bonding electron pairs that are on the oxygen atom. This is illustrated in Figure 4.4.

FIGURE 4.4
Hydrogen Bonding

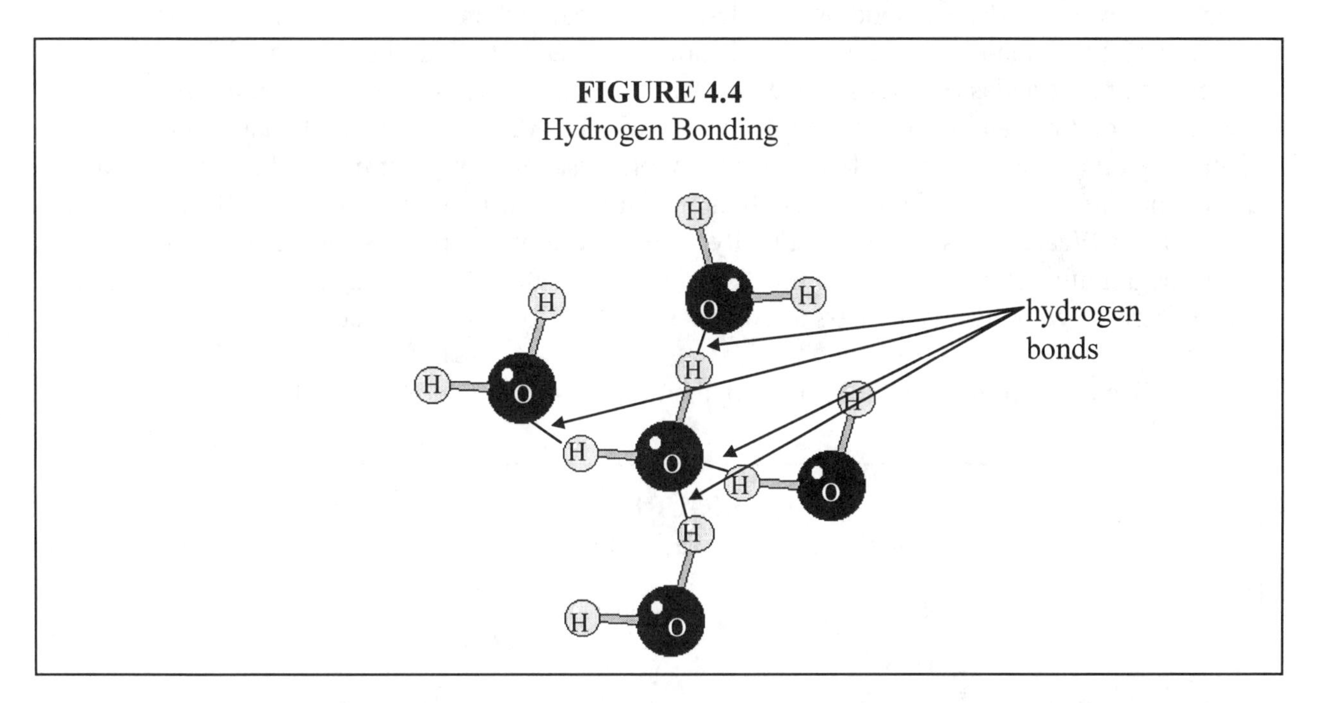

Hydrogen bonds are actually weak bonds that hold the hydrogen molecules close together. They are not as strong as a real chemical bond, because a full electron pair is not being shared. The hydrogen is just sharing part of one of the non-bonding electron pair on the oxygen atom. Even though a hydrogen bond is weak (compared to a real chemical bond), it still is stronger than any of the other van der Waals forces.

Hydrogen boding is strong enough, for example, to keep the double helix of DNA together. If you remember from biology, DNA is composed of repeating nucleotides that bond to each other in the middle of DNA's double helix. The bonds that the nucleotides use are *hydrogen bonds*. This is actually a great testament to the ingenuity of God, because in order for DNA to give instructions to the cell regarding what proteins to make, the DNA has to "unwind" its helix by breaking the bonds between the nucleotides. Once the instructions have been given, the nucleotides must bond together again, reforming the double helix. It turns out that *only hydrogen bonds* are suitable for this situation. The hydrogen bonds are strong enough to keep the double helix together, but at the same time they are weak enough to allow DNA to unwind when exposed to only a small amount of extra energy. This is important. If regular chemical bonds were used to connect the nucleotides of DNA together, the energy required to break those bonds in order to unwind the helix would be so great that it would *destroy the entire DNA molecule!* Hydrogen bonds, however, are not that strong, so they can be broken with a "safe" amount of energy.

Besides water and DNA, are there other molecules that use hydrogen bonding? Yes, there are. Hydrogen bonding is rather rare, but it is possible in any molecule where hydrogen atoms are bonded to very electronegative atoms like oxygen, nitrogen, fluorine or chlorine. The most common molecules that exhibit hydrogen bonding are water, DNA, NH_3, and HF.

Summing it all up, then, there are three basic types of van der Waals forces. In order of strength, they are: hydrogen bonding, dipole-dipole attraction, and London dispersion forces. A subclass of the dipole-dipole attraction is the ion-dipole attraction, and the London dispersion forces have two subclasses: ion-induced London forces and dipole-induced London forces. Hydrogen bonding can only exist in molecules in which hydrogen atoms are bonded to very electronegative atoms. The dipole-dipole attraction exists only in polar molecules, and London dispersion forces exist in all molecules. In nonpolar molecules, London dispersion forces are the only van der Waals forces that exist. Finally, the dipole-dipole force is stronger when the electronegativity difference between the atoms in a molecule is large, and the London dispersion forces increase with increasing mass and increasing linearity of the molecules.

If we keep all of those facts in mind, then Figure 4.5 should make a lot of sense.

FIGURE 4.5

Enthalpy of Vaporization for Certain Liquids

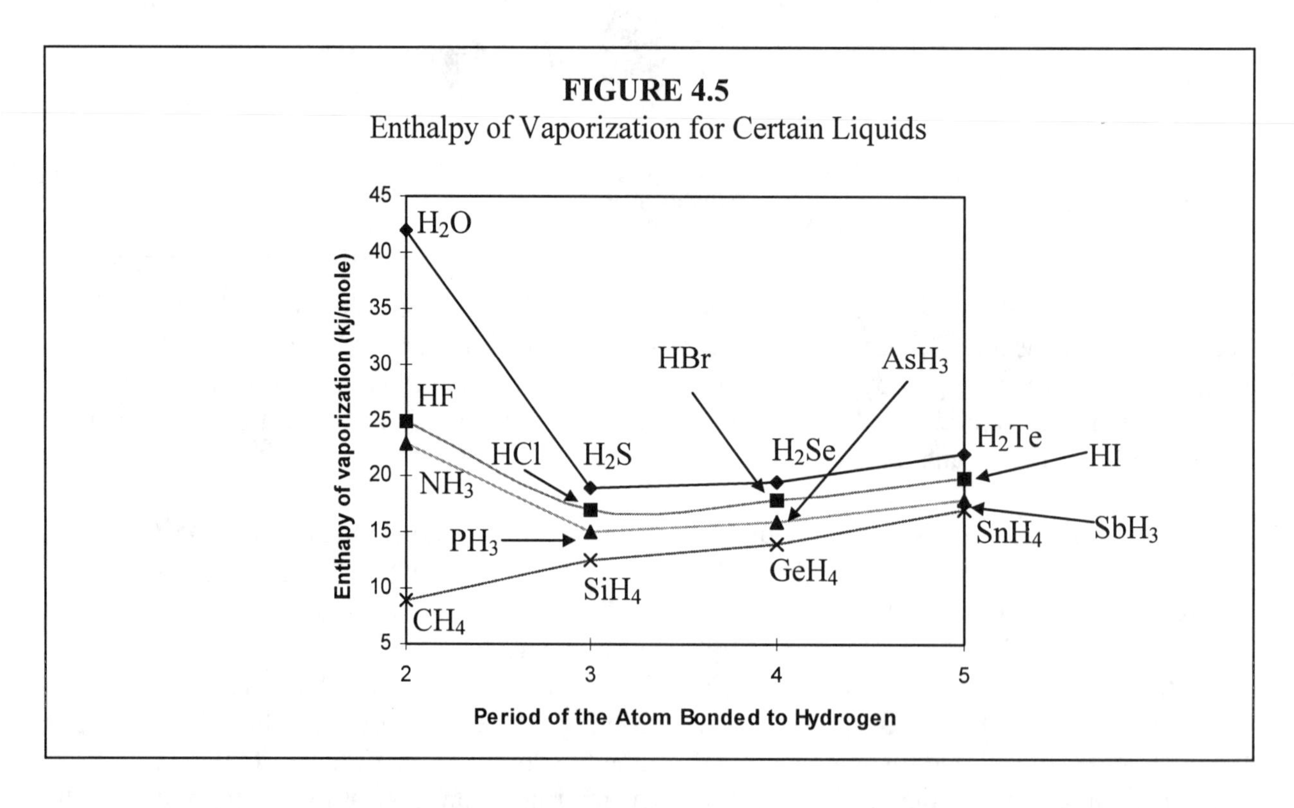

In this figure, I plot the enthalpy of vaporization for several compounds that contain hydrogen and another atom. The x-axis in the graph indicates the period (row) of the periodic chart in which you find the atom to which the hydrogen is bonded. This, in effect, tells us about the mass of the molecule. The farther to the right, the more massive the molecule. Now remember what the enthalpy of vaporization is. Enthalpy (ΔH) is the amount of energy absorbed or emitted in a process. Vaporization is the process of evaporation. Thus, the enthalpy of vaporization tells us how much energy a liquid must absorb in order to vaporize. In the end, this is a measure of the total strength of van der Waals forces in a compound. The weaker the van der Waals force, the easier it is for a molecule to become a gas; thus, the lower the enthalpy of vaporization. By the same reasoning, the stronger the van der Waals forces, the higher the enthalpy of vaporization.

Looking at the figure, we can see that for the most part, the farther to the right you go, the higher the enthalpy of vaporization. Since the x-axis is really just a mass scale, this tells us that the higher the mass, the stronger the van der Waals forces. This makes sense, since London dispersion forces are in all molecules and London dispersion forces increase with increasing mass. What about H_2O, HF, and NH_3 (ammonia)? Those three molecules buck this general trend. That's because of hydrogen bonding. In each of these molecules, hydrogens are bound to strongly electronegative atoms. This means that hydrogen bonding can occur. Since hydrogen bonding is the strongest of the van der Waals forces, this dramatically increases the enthalpy of vaporization for those compounds. Study the following example to learn this kind of reasoning.

EXAMPLE 4.1

Which has the higher melting temperature: SFCl or Cl_2?

The melting temperature is related to the strength of the intermolecular forces. The stronger the intermolecular forces, the harder it will be to pull the molecules apart, so the higher the melting temperature. The first thing to do, then, is determine what kinds of forces are at play in these compounds. As you should be able to tell, Cl_2 is nonpolar whereas SFCl is polar. This means that SFCl will have dipole-dipole attractions and Cl_2 will have London dispersion forces. Since London dispersion forces are weaker than the dipole-dipole attraction, the SFCl will have the higher melting temperature.

Which of the following molecules has the higher boiling point?

```
                                              H
                                              |
                                          H — C — H
      H   H   H   H                       H   |   H
      |   |   |   |                       |   |   |
a. H— C — C — C — C —H           b.   H — C — C — C — H
      |   |   |   |                       |   |   |
      H   H   H   H                       H   H   H
```

Both of these molecules have the same chemical formula (C_4H_{10}). Thus, they both have the same kinds of van der Waals forces. They also have the same mass. However, (a) is more linear than (b). Neither one of them are linear, of course. Each carbon is the center of a tetrahedron. Thus, both of these molecules have many tetrahedrons linked to one another. Nevertheless, (a) has all of its tetrahedrons in a line, whereas (b) has them clustered around a central carbon. Since dispersion forces increase with increasing linearity, and since the stronger the intermolecular forces, the higher the boiling point, (a) has the highest boiling point.

Which has the largest vapor pressure: $Br_2(l)$ or $I_2(l)$?

Once again, vapor pressure depends on the strength of the intermolecular forces. The weaker the intermolecular forces, the more a substance can evaporate. In your first year course, you were taught that vapor pressure is a measure of how much vapor hangs over a sample of liquid. Thus, more evaporation means a higher vapor pressure. This means that the compound with the weakest intermolecular forces will have the highest vapor pressure. Both of these compounds are nonpolar, so only London dispersion forces are at play. I_2 is much heavier than Br_2, however. This means that I_2 will have the greater London dispersion forces, which means that Br_2 (l) will have the greatest vapor pressure.

Do you see the kind of reasoning I used? First, you determine whether or not the molecule is polar. That will tell you the kind of van der Waals forces you are dealing with. If one has dipole-dipole forces and the other just London dispersion forces, then the dipole-dipole forces are stronger. If the forces are the same, then mass or linearity will tell you what forces are stronger. If hydrogen bonding is possible (an H bonded to N, F, O, or Cl), that overrides everything else. Once you determine the strength of the intermolecular forces, you can relate it to the physical observable you are asked about simply by thinking how that observable is affected by the intermolecular forces. Boiling point, for example, increases with increasing intermolecular forces while vapor pressure decreases. Try this yourself with the following "on your own" problems.

ON YOUR OWN

4.3 Which compound has the lowest vapor pressure: NH_3 (l) or CH_4 (l)?

4.4 Which compound has the highest melting point: O_3 or Cl_2?

4.5 Which is more likely to be a liquid at room temperature and which is more likely to be a gas: C_2H_6 or C_8H_{18}?

Cohesive Forces, Adhesive Forces, and Surface Tension

Before I move to the next topic, there is some terminology that needs to be taken care of. The van der Waals forces between molecules in a substance are often called **cohesive forces**, because they attract the molecules of a substance together. The stronger the van der Waals (cohesive) forces in a liquid, the harder it is to make that liquid flow. Consider the difference between the way honey flows and the way water flows. If you pour water out of a glass, it flows out right away. Honey, on the other hand, flows out slowly. This is because the van der Waals forces are stronger in honey than in water. As a result, honey does not flow as easily as water. Because the ability for a liquid to flow is important in some applications, chemists use **viscosity** to measure a liquid's resistance to flow. The higher a liquid's viscosity, the harder it is for the liquid to flow. Oil, honey, and syrup have high viscosity while water, alcohol, and gasoline have low viscosity.

High viscosity in a liquid is desirable when you want the liquid to resist high temperatures. After all, if a liquid has low viscosity, then it has weak cohesive forces. This means that when the temperature increases, the liquid will evaporate. Liquids with high viscosity have large cohesive forces, so they won't evaporate as easily. In automobiles, for example, we use high viscosity oils so that the oil will not vaporize when the engine gets really hot. Of course, the oil cannot be too viscous, or it will not be able to flow everywhere it is needed in the engine. Engines that run really hot tend to require higher viscosity oil than those that run at lower temperatures. Either way, though, when the engine first starts, the oil will not flow well because the molecules are so close together. Once the engine gets hot, the molecules get a little farther apart and the oil flows more freely. Then everything is okay. When the engine first starts, however, the oil does not flow everywhere it is needed, and as a result there is a lot of wear and tear on the engine. That's why a mechanic will tell you that the worst thing you can do to a car is to start it cold and drive it before it has a chance to warm up. Because the oil is viscous, it will not flow to where it is needed until the oil gets hot.

Cohesive forces are responsible for the phenomenon of **surface tension** in a liquid. Even though a needle is more dense than water, if you are careful, you can float a needle on water. Why? Well, there are cohesive forces attracting the water molecules together. If you think about it, a water molecule in the middle of a water sample has other water molecules pulling at it in all directions, because it is surrounded by other water molecules. What happens at the surface of the water, however? The molecules on the surface of the water have no liquid water molecules above them. Thus, they have plenty of water molecules pulling down on them with cohesive forces, but none pulling back up on them. This imbalance of force creates a tension in the surface of the water, making it hard to move those water molecules out of the way. Because of this, something which would sink if placed under the surface of the water will actually float if placed on top of the water. Such is the case with the **water strider**, an insect that walks on water. This insect is much more dense than water and should sink, but it rests its feet on the surface of the water, and the surface tension of the water is strong enough to keep the water strider from sinking.

Suppose you put a liquid into a container. The molecules are not only attracted to each other, but they are also attracted to the walls of the container. After all, if van der Waals forces can exist between two molecules in the liquid, they can also exist between a molecule in the liquid and a molecule in the container. For example, when you put water in a graduated cylinder in your first-year chemistry course, the water formed a concave meniscus at the top. This is because the water is so attracted to the glass of the graduated cylinder that the edges of the water actually "creep up" the sides of the glass. This causes a dip in the center, which we call a meniscus. The attractive forces between a liquid and a surface are called **adhesive forces**. Notice that both cohesive forces and adhesive forces are just van der Waals forces. The only difference is that cohesive forces are the van der Waals forces between the molecules in the liquid while adhesive forces are the van der Waals forces between a surface and a liquid.

Have you ever noticed that, when you wax a car and it rains shortly thereafter, the rain "beads up" on the car, forming miniature pools of water all over the car? After a while, if you do not wax the car again, that effect goes away. Why does that happen? Well, the paint on the car and water have reasonably strong adhesive forces between them. As a result, when water hits a car that has no wax on it, the water spreads out, trying to contact as much of the car as possible. Wax and water, however, have weak adhesive forces between them. Thus, when water hits a car covered in wax, it is not attracted to the surface of the car. Because of the cohesive forces between water molecules, however, water is attracted to itself. Thus, rather than spreading out, the water collects in little piles, being drawn to other water molecules rather than the car.

ON YOUR OWN

4.6 From your first-year course, you should have learned what a water meniscus looks like in a graduated cylinder. Suppose you filled a graduated cylinder with mercury. The adhesive forces between glass and mercury are weak compared to the cohesive forces. This is the opposite of how it is with water. Draw what the mercury meniscus would look like.

Phase Diagrams

Based on what you have learned so far, you should realize that phase changes are governed by three things:

1. The intermolecular forces between the atoms or molecules of the substance
2. Temperature
3. Pressure

To bring all of this together, chemists often draw **phase diagrams**. To give you an idea of what a phase diagram is, study Figure 4.6, the phase diagram for water:

FIGURE 4.6
The Phase Diagram for Water

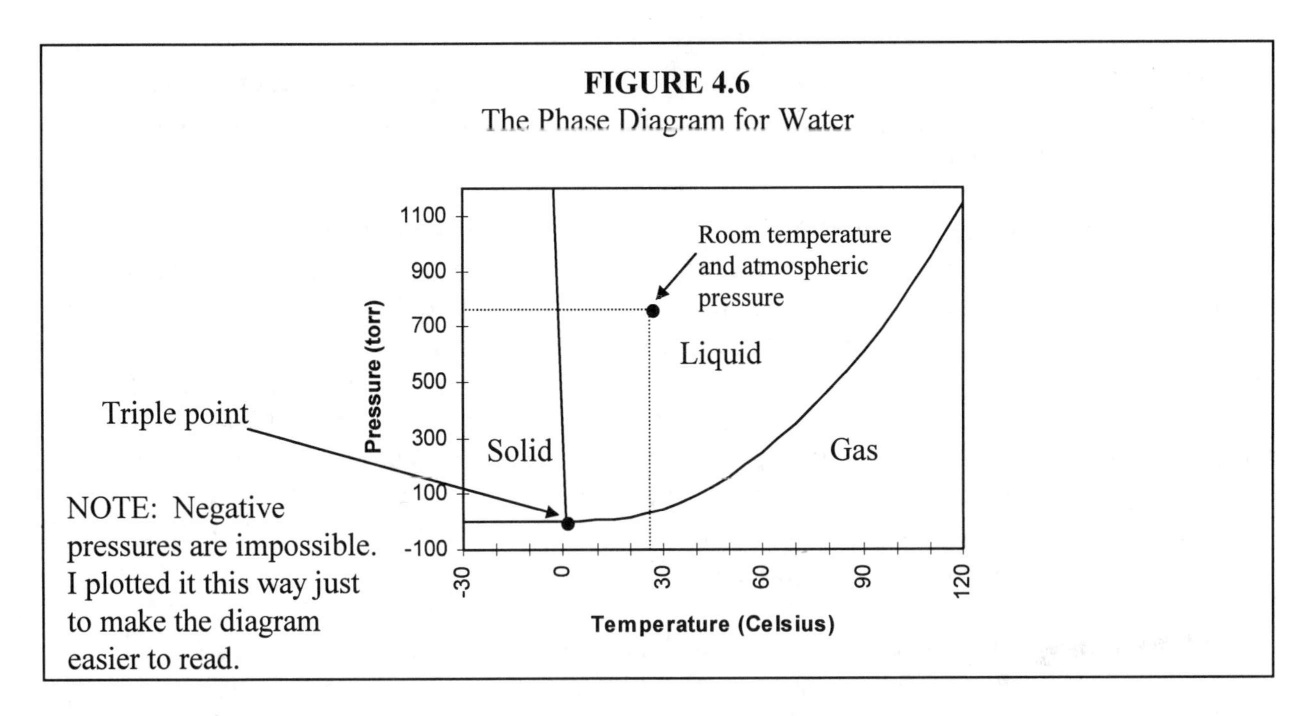

NOTE: Negative pressures are impossible. I plotted it this way just to make the diagram easier to read.

A phase diagram is used to determine the phase of a substance at any given temperature and pressure. There are three regions in a phase diagram, each corresponding to the phases of matter. You should always be able to remember which regions correspond to which phase because the x-axis is a temperature scale. Thus, the leftmost region will always be the region corresponding to the solid phase because substances always become more solid as the temperature lowers. The region corresponding to the gas phase will always be at the high temperatures, so it is the rightmost region of the phase diagram. When you are given a temperature and pressure, you can see where that point falls in the phase diagram. That will tell you what phase the substance is in under those conditions.

For example, you know that water is a liquid at room temperature (25 °C) and normal atmospheric pressure (760 torr). If you find the point on the graph above that corresponds to that temperature and pressure (as shown in the figure), you will see that it falls squarely in the liquid region of the phase diagram. This tells us that water is a liquid at that temperature and pressure. Now of course, that didn't tell us anything we didn't already know. But a phase diagram can tell us a lot of things we otherwise wouldn't know. For example, what is the phase of water at 70 °C and 100 torr? According to the phase diagram, it is a gas. That should make sense. At atmospheric pressure, water would be a liquid at 70 °C. As pressure decreases, however, the water molecules are not pushed together as strongly by the atmosphere. As a result, it doesn't take as much energy to vaporize the water.

There are a few other things about a phase diagram that need pointing out. First, what happens if a set of temperature and pressure values leads to a point that lies right on one of the lines in a phase diagram? It tells you that at those values of temperature and pressure, there is an equilibrium between the two phases separated by that line. For example, at atmospheric pressure (760 torr) and 100 °C, where do you find yourself on the phase diagram? You are on the line that separates liquid and gas, which means that an equilibrium between liquid and gas exists there. Well, 100 °C is the boiling point of water, so that makes perfect sense. As you should have learned in your first-year course, once you heat water to 100 °C, it can never heat up any more until all of the water has evaporated. At 760 torr and 100 °C, liquid water can only exist in equilibrium with water vapor.

This actually brings up a very important point. Look at the point on the phase diagram that corresponds to 500 torr and 81 °C. It might be hard for you to see exactly because the scales are so broad, but that point is once again on the line that separates liquid and gas. Once again, then, this means that an equilibrium exists between liquid and gas at that temperature. What does that tell us? Well, wherever liquid and gas exist in equilibrium, it means that the liquid is boiling. This tells us that the boiling point for water at 500 torr is only 81 °C, not 100 °C. Now this should make sense. After all, with less pressure, water molecules are not pushed together as tightly, so it takes less energy to break away from the liquid phase. However, this raises an interesting question. What is the *real* boiling point of water? Is it 100 °C or 81 °C? Well, it's both of those temperatures and a whole lot more. You see, the real definition of boiling point is:

Boiling point - The temperature at which a liquid's vapor pressure equals the current atmospheric pressure

Based on this definition, then, when the atmospheric pressure is 500 torr, the boiling point of water is 81 °C. When the atmospheric pressure is 760 torr, then the boiling point is 100 °C.

Why, then, do we say that the boiling point of water is 100 °C? Well, when we say that, we are talking about water's **normal boiling point**.

Normal boiling point - The temperature at which a liquid's vapor pressure equals normal atmospheric pressure (760 torr)

So the boiling point of a liquid changes with changing atmospheric pressure. What we generically call the boiling point of a liquid is, more precisely, its normal boiling point.

Now of course, the same can be said of melting point. Solids are not strongly affected by atmospheric pressure, but if you look at the phase diagram for water, the line that separates solid from liquid does tilt a little. This tells us that the melting point of water does change with changing pressure. Now look at that line and notice the tilt that it has. Put your pencil anywhere on that line and then draw a line straight up. What happens? Eventually, your pencil moves into the liquid region. What does that tell us? It tells us that as you increase the pressure on solid water, it will eventually turn into a liquid. Now remember, this is a unique property of water. Because of water's strong hydrogen bonding, the closer the molecules get, the more liquid they become. Thus, the more you pressurize water, the more liquid it becomes. This should immediately tell you that for virtually any other substance in Creation, the line on the phase diagram that separates solid and liquid will tilt opposite of that in Figure 4.6.

Finally, look at the point that I labeled **triple point**. There is a triple point for every substance in Creation, and it corresponds to the temperature and pressure at which all three phases of matter exist.

Triple Point - The temperature and pressure at which all three phases of matter exist together in a given substance

The triple point of water is 0.01 °C and 4.6 torr.

ON YOUR OWN

4.7 Given the following phase diagram for an unknown substance:

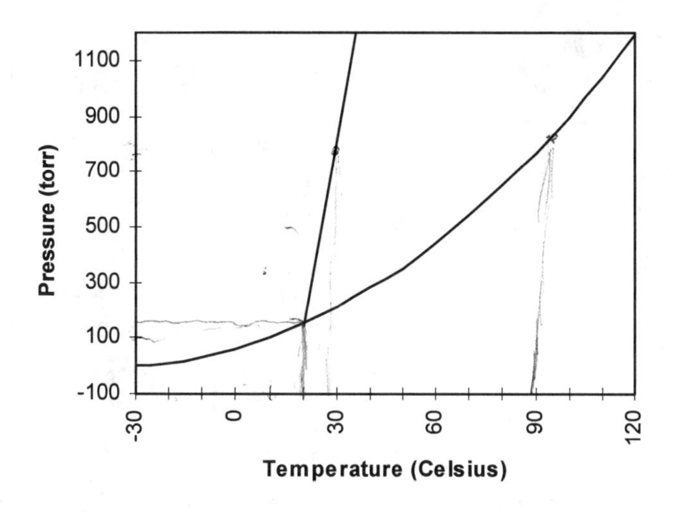

a. Determine the phase of the substance at 10 °C and 300 torr.
b. Determine the substance's triple point.
c. Determine the normal melting point and boiling point of the substance.
d. If you had a sample of the substance at 100 torr and 0 °C and suddenly decreased the pressure to near 0 torr, what would happen to the phase of the substance? What name do we give that process?

Crystals and Unit Cells

For the rest of this module, I want to concentrate on substances in their solid state. When they freeze, the vast majority of substances in Creation form **crystalline solids**. A crystalline solid is a homogeneous solid in which the atoms, ions, or molecules that make up the solid form a standard, repeating pattern. Some substances do not form crystalline solids, however. The molecules which make up these substances tend to freeze in random positions relative to each other. This results in no repeating pattern, and the solid is called an **amorphous** (uh more' fus) **solid.** Glass, candle wax, butter, and most plastics fall into this category. Since it is hard to generalize much about amorphous solids, I will stick to crystalline solids in my discussion.

As I mentioned, crystalline solids form regular, repeating patterns. For example, the following figure shows the crystalline structure of two familiar solids, sodium chloride (NaCl) and ice.

FIGURE 4.7
The Crystalline Structures of NaCl and H_2O

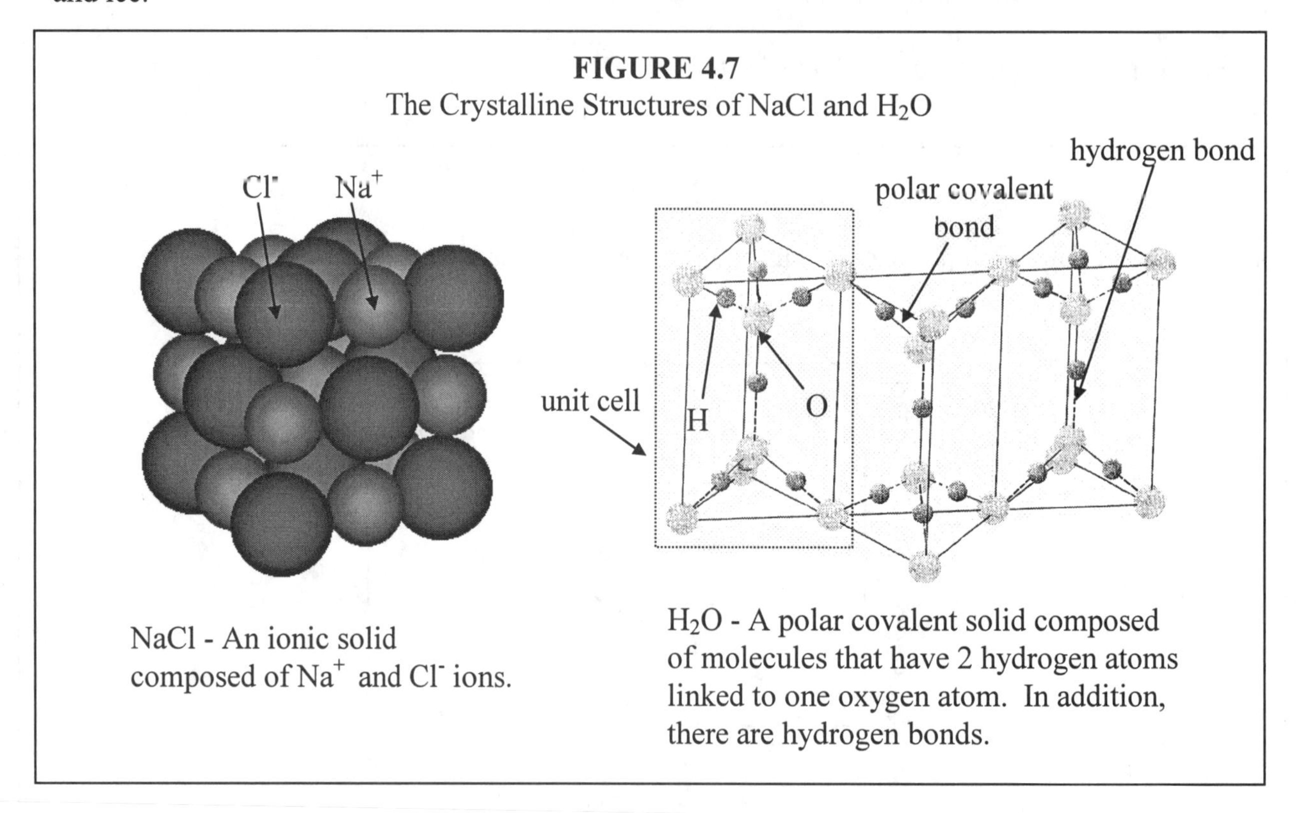

NaCl - An ionic solid composed of Na^+ and Cl^- ions.

H_2O - A polar covalent solid composed of molecules that have 2 hydrogen atoms linked to one oxygen atom. In addition, there are hydrogen bonds.

Notice that although these structure look different from each other, they both have regularly repeating patterns. That's what makes them crystalline solids.

There are three basic kinds of crystalline solids: **ionic solids**, **metallic solids**, and **covalent solids**. The difference involves how the atoms, molecules, or ions in the solid are held together. Ionic solids are formed (reasonably enough) from ionic molecules. Their constituent ions are held together by the mutual attractive force that exists between their opposite charges. Metallic solids, on the other hand, are held together by the London dispersion forces that exist between the constituent atoms in the metal. Finally, covalent solids are formed when polar covalent or purely covalent molecules freeze. In these solids, the atoms that make up each molecule are held together with covalent bonds, and the molecules themselves are held in place by the van der Waals forces which exist between them.

If you look at the water crystal in Figure 4.7, you will notice that I have outlined a portion of it called the **unit cell**. The unit cell is the smallest portion of the crystal that can generate the entire crystal if repeated over and over again in three-dimensional space. The unit cell is a very important aspect of a crystal, because it tells us a lot about the nature of the crystal itself. There are many, many different kinds of unit cells in Creation, so we will concentrate only on some of the more important ones. The unit cells are classified based on the relative lengths of the three-dimensional axes that form them and the angle that exists between those three axes.

FIGURE 4.8
Important Unit Cell Configurations in Creation

Cubic
All three axes are of equal length and are 90° from each other.

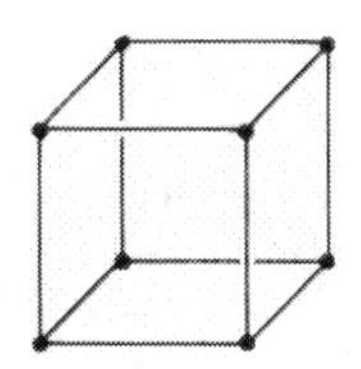

Tetragonal
One axis is different in length from the others, but all are 90° from each other.

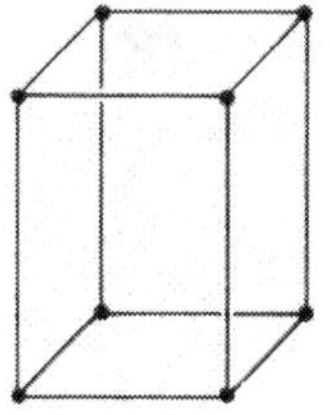

Orthorhombic
All three axes are of different length and are 90° from each other.

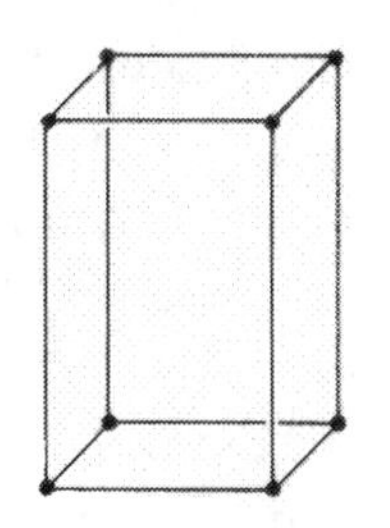

Hexagonal
One axis is different in length from the other. The two that are equal in length are 60^{0} from each other (making 6 sides rather than 4), and the other is 90^{0} from those two.

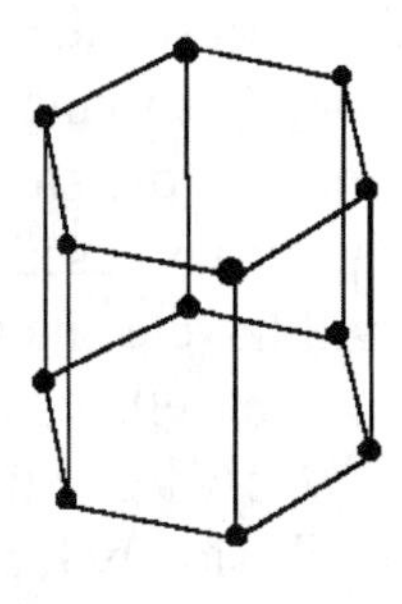

Monoclinic
All three axes are of different length. Two of them are 90° from each other but the third is at an angle other than 90^{0}

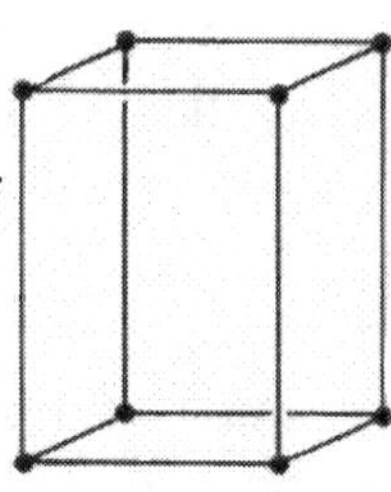

Triclinic
All three axes are of different length. All of them are at angles other than 90°

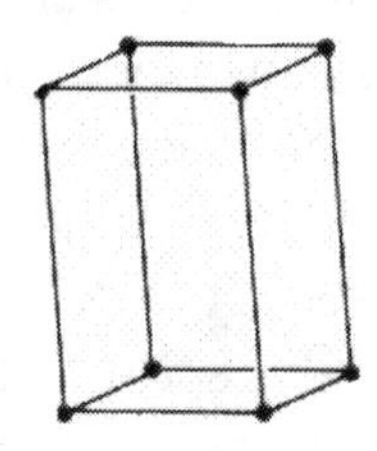

Now don't go memorizing these unit cells and their characteristics. Any good chemist can go find a table like this when he or she needs it. The point of the table is to give you the idea of what kinds of unit cells are out there. Try your hand at identifying these unit cells by performing the following experiment.

EXPERIMENT 4.2
Identifying Unit Cells

Supplies:
From the laboratory equipment set:

- Ammonium chloride
- Copper sulfate
- Sodium thiosulfate
- Sodium ferrocyanide
- Plate with all of the depressions in it

- Droppers (called pipettes)
- Measuring scoop

From around the house:
- Magnifying glass (A microscope works even better, if you have one.)

From the supermarket:
- Distilled water

The solids in your laboratory equipment set have been crushed to make them easier to use. Thus, the crystals now are rather small and hard to see. In order to make bigger crystals that are easier to see, you will form new ones. To do this, use each chemical one at a time and do the following: Add a measure of solid to one of the large depressions in your depression plate. Add ten drops of distilled water with a pipette. Take another pipette and use the end to stir the solution, dissolving as much solid as you can. Use that same pipette to draw in some of the solution and transfer 5 drops to one of the smaller depressions in your depression plate. Clean the pipette you used to stir and transfer each time you start a new chemical. When you are done, you will have 4 small depressions with 4 different solutions in them. Make sure you have a system that tells you which solution is in which depression!

Now leave the depression plate to sit for a while. As the water evaporates, crystals of solid will slowly form. Since you are not going to crush them, the crystals will be large. The slower the water evaporates, the bigger the crystals will be. Once all of the water evaporates, examine the crystals with the magnifying glass. It is best not to take the crystals out of the depressions when you observe them, because you can damage the pristine shape of the crystals when you touch them. If you do want to take them out of the depression, do so carefully. Try to identify the basic shape of the crystal's unit cell. Now, of course, you won't see a unit cell, because unit cells have only a few atoms in them. Nevertheless, since the entire crystal is built by simply repeating the same unit cell over and over again, the overall shape of the crystal should reflect the shape of its unit cell. The unit cells of each substance in this experiment are listed at the end of the answers to the "on your own" problems.

Were you correct in your identification? If not, don't worry about it. It takes some real practice to grow and identify crystals. This experiment just gave you your first shot at it!

Metallic Crystals

The easiest crystals to describe are those formed by metals. This is because, by and large, metals form crystals with a cubic unit cell. Look at the cubic cell shown in Figure 4.8 and think of a crystal that is formed by repeating that cell over and over again in three dimensional space. If I repeat the unit cell a few times, I end up getting something that looks like Figure 4.9:

FIGURE 4.9
A Simple Cubic Unit Cell Repeated in Three Dimensions

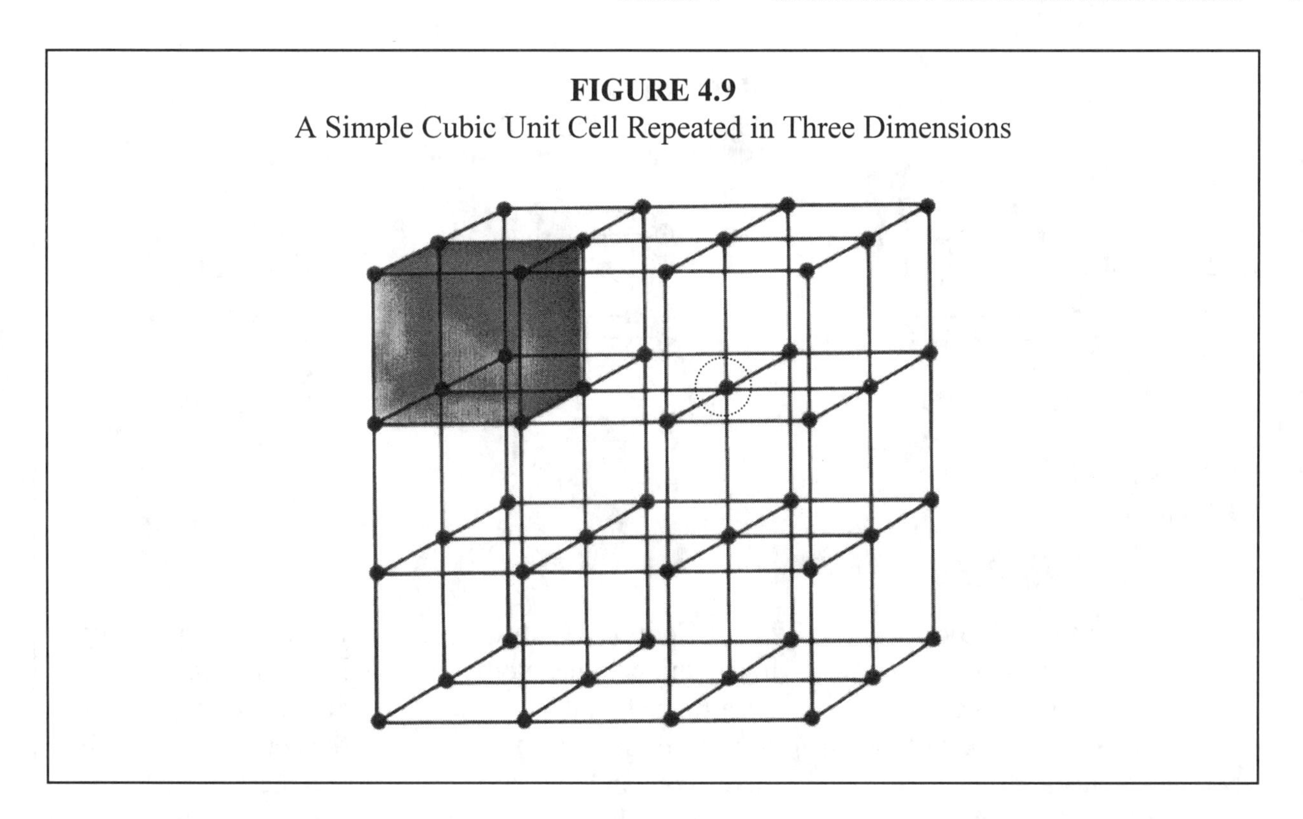

In this figure, I have shaded in the unit cell to emphasize that this system, called a **crystal lattice**, is formed by repeating that one cell over and over. Don't worry why I titled the figure with the term "simple cubic." I will get to that in a minute. First, I want you to look at the ball that I circled with the dotted circle. If that ball represented a metal atom, how many unit cells would that atom be a part of? Look closely. It sits on the intersection of a total of 8 unit cells (4 above the ball and 4 below it). It makes up the corner of each of those unit cells, so it belongs (in part) to all eight of them. This is the first thing you need to learn about unit cells. The atoms in a unit cell do not necessarily belong to that unit cell alone. They may be a part of several other unit cells, depending on where they are in the cell.

Given that fact, how many atoms are contained in one unit cell of the crystal drawn in Figure 4.9? Well, there are 8 balls in a unit cell (one in each corner). Each of those balls actually belongs to 8 unit cells, so each unit cell only owns 1/8 of the ball. Thus, a unit cell with 8 balls, each of which contributes 1/8 of itself to the unit cell, has a total of *one atom per unit cell.* Do you see how I got that? If you count the number of cells that an atom belongs in, you can divide one by that number to determine what portion of that atom belongs to a single unit cell. Then, if you add up the portions for each atom, you get the total number of atoms in the unit cell. Eight atoms, each of which contributes one eighth to the cell, gives you one atom per cell.

Why is this important? Well, most metals do not end up forming a unit cell as simple as that one. There are three kinds of cubic unit cells formed by metals. They are all illustrated in Figure 4.10.

FIGURE 4.10
The Major Cubic Unit Cells

UNIT CELL:

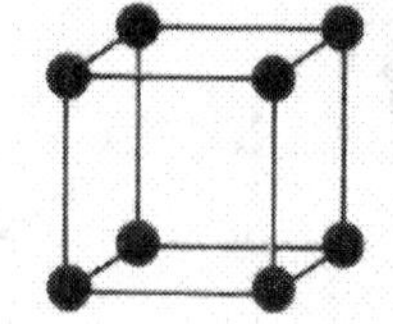

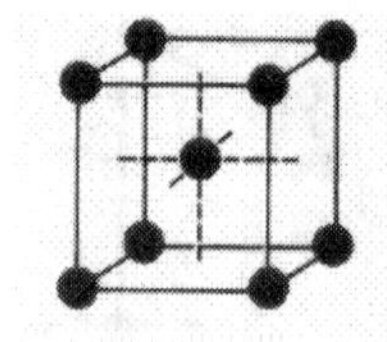

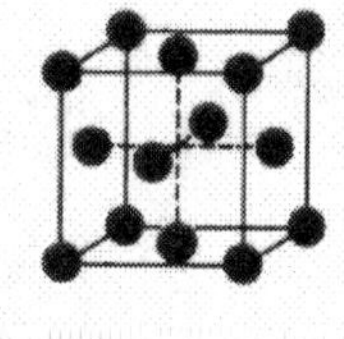

REALISTIC VERSION:

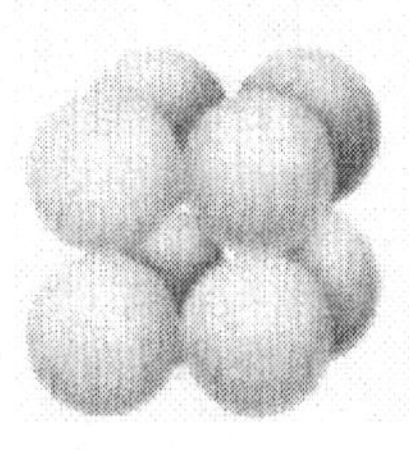

Simple Cubic
In the unit cell, the atoms are arranged on the corners of the cube.

Body-Centered Cubic
In the unit cell, the atoms are arranged on the corners of the cube, and there is one in the very center as well.

Face-Centered Cubic
In the unit cell, the atoms are arranged on the corners of the cube, and there is one at the center of each face of the cube.

Metals such as silver (Ag), aluminum (Al) and nickel (Ni) have face-centered cubic unit cells while barium (Ba), chromium (Cr), and iron (Fe) have body-centered cubic unit cells. About the only metal that crystallizes with a simple cubic unit cell is polonium (Po).

We already determined how many atoms exist in a simple cubic unit cell. How many exist in a body-centered cubic unit cell? Well, each atom on a corner contributes only 1/8 of itself to the unit cell. There are 8 atoms on the corners, so that makes one atom. Then there is that atom in the center. How many unit cells does it belong to? Well, it is right in the center of the unit cell, so there is no other cell that it can be a part of. As a result, the atom in the center of a body-centered cubic unit cell belongs completely to one unit cell. Thus, there are a total of 2 atoms in a body-centered cubic unit cell. Counting the atoms in a face-centered unit cell is left as an "on your own" problem.

Now look at the figure again and notice that in the realistic drawings which show what the unit cells really look like, the eight atoms in the simple cubic unit cell are touching each other. In the body-centered cubic cell, that is not the case. Instead, the 8 corner atoms each touch the atom in the center. Finally, in the face-centered cubic unit cell, the 8 corner atoms each touch the atoms on the face of the cube. This is a very useful thing to know, because there is an experimental technique called **x-ray diffraction** in which x-rays are bounced off of a solid and, with a little calculation, the length of the unit cell's edge can be determined. This is a useful thing to know because, with a little work, you can determine the radius of the metal atom from the length of the unit cell edge. In order to do this, however, you need to know the type of unit cell you are dealing with.

Look at the realistic drawing of the simple cubic unit cell for example:

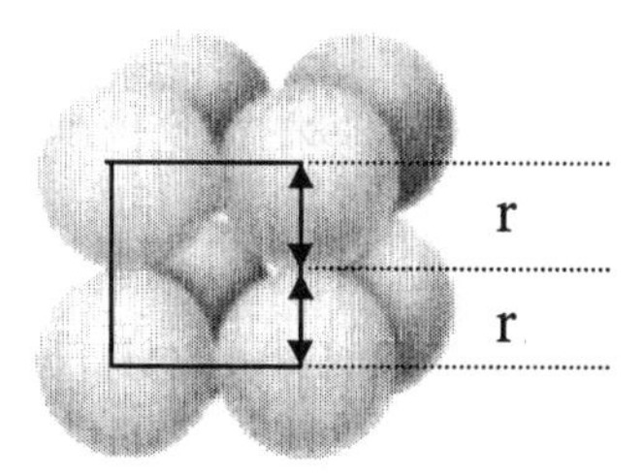

Notice that because the atoms on the corner touch each other, any edge of the simple cubic unit cell will be twice the radius of the atom. Thus, if you know the edge length, you can divide by 2 to get the radius of the atom. Now look at the face-centered cubic unit cell:

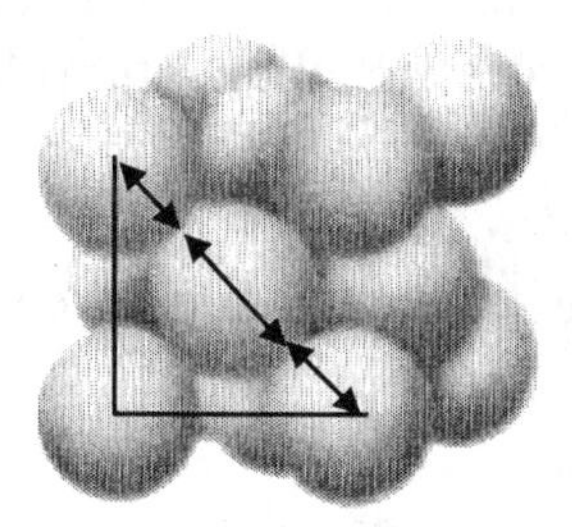

Because the corner atoms are touching the atom at the center of the face and not each other, the edge of the unit cell is larger than twice the atom's radius. We can still relate the atomic radius to the edge of the unit cell, however. Notice that a right triangle drawn on the face has its hypotenuse passing through three touching atoms. If you notice, the hypotenuse is made up of the radius from one corner atom, the diameter of the center atom, and the radius of the other corner atom. This means that the length of the hypotenuse is 4 times the length of the atomic radius. How long is the hypotenuse? The Pythagorean theorum says that the hypotenuse of that right triangle is $\sqrt{2}$ times the edge length. In the end, then,

$$\text{In a face-centered cubic unit cell: } r = \frac{\sqrt{2}\cdot(\text{edge length})}{4}$$

So even though the math is a little harder, I can still determine the radius of the atoms in a face-centered cubic unit cell if I learn the edge length of the cell from x-ray diffraction. The math is even harder for body-centered cubic unit cells, so I will just give you the result:

$$\text{In a body-centered cubic unit cell: } r = \frac{\sqrt{3}\cdot(\text{edge length})}{4}$$

Given these relationships, then, we can use the results of x-ray diffraction experiments to tell us the radius of a metal atom.

ON YOUR OWN

4.8 How many atoms are in a unit cell for a face-centered cubic crystal?

4.9 X-ray diffraction experiments indicate that the edge length of a tungsten unit cell is 3.165 Å. If tungsten has a body-centered cubic unit cell, what is the radius of tungsten?

Ionic Crystals

In Experiment 4.2, you worked exclusively with ionic crystals. Ionic crystals are typically the ones with the more exotically shaped unit cells, but there are a surprising number of ionic compounds that have cubic crystal structures. Sodium chloride, for example, exists with a simple cubic unit cell.

One of the most important things to learn about ionic crystals is the fact that they have a **lattice energy** associated with them.

Lattice energy - The energy required to separate the ions in an ionic solid

When ionic compounds crystallize, the ions are held in place by the electrical attraction between the ions that make up the ionic compound. In order for the ions in an ionic crystal to be separated, that attraction must be overcome. The energy required to do that is the lattice energy.

Chemists often write this process as a chemical equation:

$$MX\ (s) \rightarrow M^{n+}\ (g) + X^{n-}\ (g)$$

The "M" represents any metal and the "X" represents any non-metal. For sodium chloride, then, the equation would be:

$$NaCl\ (s) \rightarrow Na^{+}\ (g) + Cl^{-}\ (g)$$

The lattice energy (abbreviated as "U") for NaCl, then, would be the ΔH of this reaction.

Now it turns out that these lattice energies are quite difficult to measure. After all, how easy would it be to run the reaction above in a calorimeter so as to measure its ΔH? Since it is awfully tough to vaporize salt, it would be quite a feat to pull something like that off. In order to determine the lattice energy for an ionic solid, then, we have to go at it in some round about way. This round about way is called the **Born-Haber Cycle**.

In this experimental technique, we make use of Hess's Law. We separate the formation reaction of NaCl into several distinct steps, each of which has a ΔH that is either already known or easy to measure in the lab. Since the ΔH of formation is rather easy to calculate, the only unknown in the cycle will be the lattice energy, and thus you can calculate it. How in the world does this work?

Well, the formation reaction of NaCl (s) is:

$$Na\ (s) + \frac{1}{2}Cl_2\ (g) \rightarrow NaCl\ (s)$$

We can split this up into the following hypothetical steps:

1. $Na\ (s) \rightarrow Na\ (g)$ ΔH = the ΔH of sublimation for sodium = 109 kJ

2. $\frac{1}{2}Cl_2\ (g) \rightarrow Cl\ (g)$ $\Delta H = \frac{1}{2}$ the bond strength of Cl_2 = 122 kJ

3. $Na\ (g) \rightarrow Na^+\ (g) + e^-$ ΔH = The ionization energy of Na = 496 kJ

4. $Cl\ (g) + e^- \rightarrow Cl^-\ (g)$ ΔH = Electron affinity of Cl = -368 kJ

5. $Na^+\ (g) + Cl^-\ (g) \rightarrow NaCl\ (s)$ $\Delta H = -U$

If we sum all of these reactions up, we are back to the formation reaction for NaCl (s):

$Na\ (s) + \frac{1}{2}Cl_2\ (g) \rightarrow NaCl\ (s)$ $\Delta H = \Delta H_f$ of NaCl (s) = -411 kJ

Well, if the five equations above add up to the formation reaction of NaCl (s), then the 5 ΔH's must add up to the ΔH of the formation reaction:

$$109\ kJ + 122\ kJ + 496\ kJ + -368\ kJ + -U = -411\ kJ$$

Now we can just solve for U:

$$U = 109\ kJ + 122\ kJ + 496\ kJ + -368\ kJ + 411\ kJ = 770\ kJ$$

So we find that the lattice energy for NaCl (s) is 770 kJ.

Do you see the point here? It is almost impossible to directly measure the lattice energy of an ionic solid. Nevertheless, it is a rather important quantity to know if you are studying an ionic compound. Thus, you use Hess's Law along with a bunch of ΔH's that you already know to determine the ΔH (the lattice energy) that would have been nearly impossible to figure out otherwise. Now don't worry. I won't expect you to do Born-Haber cycle problems. I just wanted you to see an application of the version of Hess's Law that you learned back in Module #1.

ANSWERS TO THE ON YOUR OWN QUESTIONS

4.1 Sample B is at the higher temperature. There are many ways to tell this. The average energy is larger for sample B, the most probable energy is larger, and more molecules have the highest energies as compared to sample A.

4.2 As the pressure decreased because the air was being evacuated, it became easier and easier for water molecule to evaporate. Soon, the water molecules were evaporating so quickly that the water started to bubble and boil. As the water evaporated, however, it took energy away from the water that remained. This cooled the water that remained. The water ended up getting so cold this way that it froze. I have actually done this demonstration several times in class, and the water will freeze in mid-boil!

4.3 NH_3 has the lowest vapor pressure. Remember, the stronger the van der Waals forces, the harder it is for water to evaporate, so the less the vapor pressure. Since NH_3 has hydrogen bonding (it has H atoms bonded to an electronegative atom, N), it will have the strongest van der Waals forces.

4.4 Cl_2 will have the highest melting point. The stronger the van der Waals forces, the harder it will be to turn the solid into a liquid and thus the higher the melting point. Both of these molecules are nonpolar, so the only forces at play are London dispersion forces. This means the higher the mass the stronger the forces. Cl_2 has the larger mass and thus the stronger van der Waals forces.

4.5 C_8H_{18} will be a liquid and C_2H_6 will be a gas. The one with the stronger van der Waals forces will be the liquid. Since these molecules are both the same kind of molecules, they have essentially the same forces at play. Thus, the heaviest will have the strongest van der Waals forces.

4.6 The meniscus will be opposite that of water because in the case of mercury, the cohesive forces are stronger than the adhesive forces. The liquid therefore draws in on itself rather than creeping up the glass.

4.7 a. The phases aren't marked but they needn't be. The leftmost is solid, the rightmost is gas, and liquid is in between. The point at x= 10, y = 300 is squarely in the solid region, so the substance is a solid.

b. Reading from the graph might introduce some errors between you and me, but the triple point seems to be at P = 120 torr, T = 20 °C to me.

c. Once again, your numbers might be slightly different than mine. The normal melting point and boiling point are found at p = 760 torr. The point on the line separating solid and liquid at 760 torr is the normal melting temperature, and the point on the line separating liquid and gas at

760 torr is the normal boiling temperature. I read the graph as follows: The normal melting temperature is 30 °C and the normal boiling temperature is 90 °C.

d. At 100 torr and 0 °C, the substance is a solid. If the temperature remains fixed and the pressure drops to near zero, the new temperature will be 0 °C and the new pressure will be near 0 torr. At that point, the substance becomes a gas. The process of a solid going directly into the gas phase with no liquid phase in between is called sublimation.

4.8 In a face-centered cubic unit cell, there are 8 atoms on the corners, each contributing 1/8 to the cell. This makes one atom. Then, there are 6 atoms on the faces. If you think about it, an atom on the face of the unit cell can belong to only two cells: the one we are considering and the one directly adjacent. This means each of those atoms belongs half to one cell and half to the other. Since there are 6 of them and they each contribute 1/2, that makes 3 more. Thus, there are 4 atoms to a unit cell.

4.9 For this, we just need to use the formula:

$$r = \frac{\sqrt{3}\cdot(\text{edge length})}{4}$$

$$r = \frac{\sqrt{3}\cdot(3.165\ \text{A})}{4} = \underline{1.370\ \text{Å}}$$

ANSWERS TO EXPERIMENT 4.2:

Ammonium chloride - cubic
Copper sulfate - triclinic
Sodium thiosulfate - hexagonal
Sodium ferrocyanide - monoclinic

REVIEW QUESTIONS

1. If a liquid is insulated from its surroundings but is free to evaporate, what will happen to its temperature as time goes on?

2. A sample of liquid has the same temperature throughout. Does this mean that the molecules in the liquid all have essentially the same energy?

3. If a solid goes through sublimation, what phase will the substance be in?

4. What are the three main types of van der Waals forces? List them in order of increasing strength.

5. What van der Waals force exists in all substances?

6. What holds the DNA helix together?

7. What is the difference between cohesive and adhesive forces?

8. If a liquid beads up on a surface, are the cohesive forces greater than the adhesive forces, or vice-versa?

9. If a liquid's vapor pressure reaches the same value as the atmospheric pressure, what happens to the liquid?

10. What is the purpose of the Born-Haber cycle?

PRACTICE PROBLEMS

Problems 1 and 2 use the following graph of molecular speeds in a liquid:

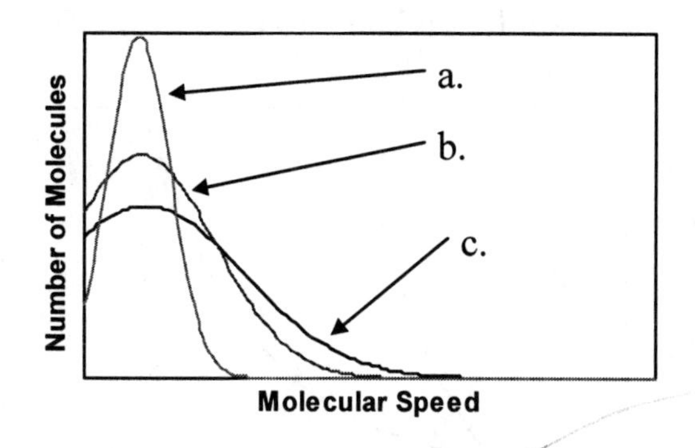

1. Which graph (a, b, or c) corresponds to the lowest temperature.

2. Which graph represents the situation in which the liquid has the highest vapor pressure?

3. Order the following in terms of increasing normal boiling point: I_2, H_2, Br_2, O_2.

4. Consider these three different Lewis structures of pentane (C_5H_{12}):

```
a.
     H   H   H   H   H
     |   |   |   |   |
 H — C — C — C — C — C — H
     |   |   |   |   |
     H   H   H   H   H

b.
             H
             |
         H — C — H
     H       |       H
     |       |       |
 H — C   —   C   —   C — H
     |       |       |
     H       |       H
         H — C — H
             |
             H

c.
                  H
                  |
              H — C — H
     H    H       |    H
     |    |       |    |
 H — C —  C  —    C  — C — H
     |    |       |    |
     H    H       H    H
```

Which has the lowest vapor pressure?

5. Arrange the following in terms of increasing melting point: N_2, CaO, NO.

6. Between the following compounds: HF and HBr, which has the strongest van der Waals forces?

7. At a certain temperature and pressure, a substance is melting. If the pressure were suddenly increased, but the temperature remained the same, what would happen to the phase of the substance?

8. Given the following phase diagram for carbon dioxide:

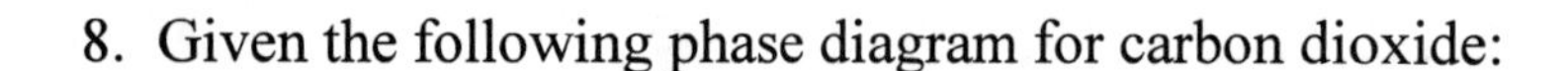

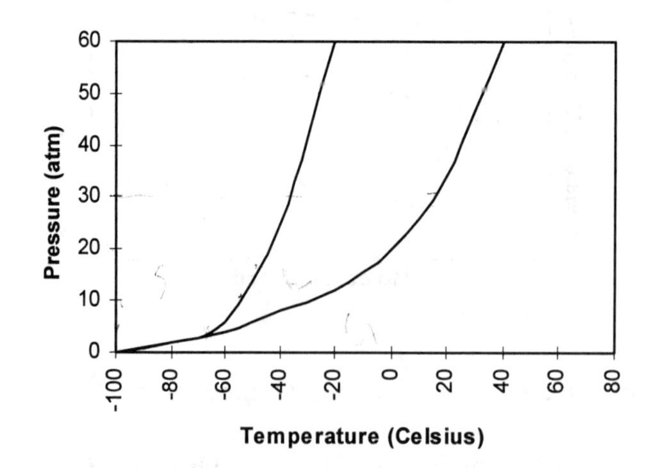

a. In what phase is CO_2 at 0 °C and 50 atm?
b. What is the triple point of CO_2?
c. What is the boiling point of CO_2 at 30 atm?
d. What is the melting point of CO_2 at 10 atm?
e. What is the highest temperature at which CO_2 can sublime?

9. Identify the type of crystalline solid (ionic, covalent, or metallic) formed by each of the following:

a. SiO_2 b. KCl c. Cu d. CO_2 e. Fe f. CaO

10. Lead crystallizes in a face-centered cubic unit cell. If lead's atomic radius is 1.75 Å, calculate the edge length of the cube.

Module #5: Solutions and Colloids

Introduction

In the last module, I concentrated on the intermolecular forces that hold solids together, keep liquid molecules close to one another, and are barely non-existent in gases. These intermolecular forces are important in the formation of solutions and colloids (kah' loyds), which I am going to discuss in this module. Since you may not know what a colloid is yet, don't worry about that term. I will get to it towards the end of this module. For right now, concentrate on the concept of solutions.

In your first-year course, you were given an introduction to the nature of solutions. If you think back on what you learned in that course in light of what you learned in the previous module, you will realize that intermolecular forces play an incredibly important role in the formation of solutions. Consider, for example, a solid dissolving in a liquid. As you learned in your previous chemistry course, the solid is called the "solute," because it is the substance that is dissolving, and the liquid is the "solvent," because it is what dissolves the solute. In order for the solvent to dissolve the solute, solvent molecules must squeeze themselves in between the ions or molecules of the solid and pull them away from each other. What will govern how effectively this happens? *Intermolecular forces!* If the van der Waals forces between the solute and solvent are large, then the solute will dissolve readily. If the van der Waals forces between the solute and solvent are small, however, the solvent will not be strongly attracted to the solute and will not get in between the solute's ions or molecules very well. Thus, the solute will not dissolve readily in the solvent.

Although intermolecular forces are important in the formation of solutions, there are other factors that play a role as well. Principal among those factors are enthalpy and entropy. Those are two words that should be familiar to you. In your previous chemistry course, you should have learned that enthalpy measures the energy of a system, while entropy measures the amount of disorder in a system. These two factors must obey certain rules of thermodynamics if a solution is going to form.

Solutions are an important part of chemistry. Most of the experiments that you did in your previous course, and most of the experiments you will do in this course deal with chemical reactions that occur in solution. This is no accident. The vast majority of the chemistry that occurs on this planet takes place in solution. Thus, the study of solutions is a very important part of your chemical education!

A Little Bit of Review

Before we go on to learning new things about solutions, there are a few concepts that we need to review. You should have learned these things before, but they might be a bit hazy in your mind. I will therefore go over them briefly so that we are all "on the same page." If you are still a bit unsure of any of these concepts after this review, please go back to your first year course. You need to be in relatively good command of the concepts presented in this section before you can proceed further.

First of all, you need to remember that there are many, many ways to measure the concentration of a solute in solution. Each measure of concentration has its own units, and each applies to a different aspect of the solution's chemistry. The units you need to review are **molarity, molality,** and **mole fraction**. Molarity has units of moles per liter, telling you how many moles of the *solute* is contained in a liter of *solution*. Molarity is typically used in stoichiometric applications, because the number of moles of a substance is critically important. Molality, on the other hand, has units of moles per kilogram, and it tells you how many moles of *solute* you have for each kilogram of *solvent*. Remember the main difference between molality and molarity. Whereas molarity measures the moles of solute per liter of *solution*, molality measures moles of solute per kilogram of *solvent*. Molality is typically used in freezing point depression and boiling point elevation problems. Finally, mole fraction has no units. It is calculated by taking the moles of *solute* and dividing by the total number of moles of *both solute and solvent*. Mole fraction is mostly used in applications involving gases.

To dust the cobwebs out of your mind on these units, please study the example below and perform the "on your own" problems that follow. If you have trouble with the "on your own" problems, you should go back to your first-year book and review these units, as we will use them all in the next few modules.

EXAMPLE 5.1

A chemist takes 5.01 grams of NaCl and dissolves it in 112.1 grams of water. The total volume of the resulting solution is 113.5 mL. Calculate the concentration of NaCl in molarity, molality, and mole fraction.

I'll start with molarity, because that's what you should remember the best. In calculating molarity, we take the number of moles of solute divided by the number of liters of solution.

$$\text{molarity} = \frac{\text{moles of solute}}{\text{liters of solution}}$$

To do this, we must convert grams of NaCl into moles of NaCl and mL of solution into L of solution.

$$\frac{5.01 \text{ g } \cancel{\text{NaCl}}}{1} \times \frac{1 \text{ mole NaCl}}{57.5 \text{ g } \cancel{\text{NaCl}}} = 0.0871 \text{ moles NaCl}$$

I will assume you can make the conversion to liters on your own. This leaves us with:

$$\text{molarity} = \frac{0.0871 \text{ moles NaCl}}{0.1135 \text{ L solution}} = \underline{0.767 \text{ M}}$$

For molality, we must take the number of moles of solute and divide by the number of kilograms of solvent. The problem states that 112.1 g of water were used, which is the same as 0.1121 kg of water. The molality, then, is:

$$\text{molality} = \frac{\text{moles of solute}}{\text{kg of solvent}}$$

$$\text{molality} = \frac{0.0871 \text{ moles NaCl}}{0.1121 \text{ kg water}} = \underline{0.777 \text{ m}}$$

For mole fraction, we must take the number of moles of solute and divide by the total number of moles of both solute and solvent. To be able to do that, we need to calculate the number of moles of water used:

$$\frac{112.1 \text{ g } H_2O}{1} \times \frac{1 \text{ mole } H_2O}{18.0 \text{ g } H_2O} = 6.23 \text{ moles } H_2O$$

Now we can calculate mole fraction (which is abbreviated as "X").

$$X_{solute} = \frac{\text{moles solute}}{\text{total moles}}$$

$$X_{NaCl} = \frac{0.0871 \text{ moles NaCl}}{6.23 \text{ moles } H_2O + 0.0871 \text{ moles NaCl}}$$

$$X_{NaCl} = \frac{0.0871 \text{ moles NaCl}}{6.32 \text{ moles total}} = \underline{0.0138}$$

ON YOUR OWN

5.1 A chemist makes a solution by dissolving 50.0 g of magnesium sulfate into 250.1 g of water. The solution's final volume is 251.1 mL. Determine the molarity, molality, and mole fraction of the solute in this solution.

There are two other things you need to remember from your previous chemistry course in order to understand the concepts presented in this module. First, remember that the amount of solute that can dissolve in a particular solvent is called the **solubility** of the solute. Typically, the solubility of a solute is measured in grams of solute per hundred grams of solvent. For example, the solubility of table salt (NaCl) in water at room temperature is 26 grams per 100 grams. This means that if you take 100 grams of water, you can dissolve up to 26 grams of NaCl in it. If you try to dissolve any more than that, the excess will not dissolve. It will simply sit at the bottom of the container.

The other thing you need to keep in mind is that solutions are not always formed with a liquid solvent and a solid solute. Although this is what we typically think of when we think of solutions, a solution can be formed between substances of any state. If you take a concentrated household cleaner and dilute it with water in order to use it with a mop, you are making a solution of two liquids. Water is the solvent in this solution while the household cleaner is the solute. Soda pop is a solution that contains many solutes. Some of the flavorings are solids, often the sweetener is a liquid syrup, and the carbonation (which makes the fizz) is a gas. Thus, soda pop is a solution with solid, liquid, and gas solutes.

Relating Units of Concentration

To extend what you learned about these units in your previous chemistry course, I want to show you how all three of these units can be related to one another with just one quantity: the density of the solution.

EXAMPLE 5.2

A solution of silver nitrate ($AgNO_3$) has a concentration of 1.2 M and a density of 1.11 g/mL. What is the molality of the solution, as well as the mole fraction of both silver nitrate and water?

Since the solution has a concentration of 1.2 M, you know that there are 1.2 moles of silver nitrate for every liter of *solution*. To determine the molality of the solution, however, we need to know the number of moles of silver nitrate per *kilogram of solvent*. How do we determine the number of kilograms of solvent? Well, we can start by using the density of the solution to convert from liters of solution to grams of solution. Since we know that there are 1.2 moles of silver nitrate for every one liter of solution, let's see how many grams there are in one liter of solution:

$$\frac{1.00 \not{L} \text{ solution}}{1} \times \frac{1 \not{mL}}{0.001 \not{L}} \times \frac{1.11 \text{ g}}{1 \not{mL}} = 1.11 \times 10^3 \text{ g}$$

Thus, a liter of solution has a mass of 1.11×10^3 g. How does that help? Well, think about what's in that one liter of solution. There are only two things: water and 1.2 moles of silver nitrate. If the total mass of one liter of solution is 1.11×10^3 g, then the mass of water plus the mass of 1.2 moles of silver nitrate must be equal to 1.11×10^3 g. Wait a minute, however. We can calculate the mass of 1.2 moles of silver nitrate:

$$\frac{1.2 \cancel{\text{moles AgNO}_3}}{1} \times \frac{169.9 \text{ g AgNO}_3}{1 \cancel{\text{mole AgNO}_3}} = 2.0 \times 10^2 \text{ g AgNO}_3$$

If the mass of water plus the mass of silver nitrate must add up to 1.11×10^3 g, then the mass of water must be:

$$\text{mass of water} = 1.11 \times 10^3 \text{ g} - 2.0 \times 10^2 \text{ g} = 9.1 \times 10^2 \text{ g}$$

Thus, we know that one liter of solution contains 1.2 moles of silver nitrate and 9.1×10^2 g of water. That's all we need to determine molality:

$$\text{molality} = \frac{\text{moles solute}}{\text{kg solvent}} = \frac{1.2 \text{ moles AgNO}_3}{0.91 \text{ kg water}} = \underline{1.3 \text{ m}}$$

Now that we know the number of grams of water in one liter of solution, it is a snap to determine the number of moles of water in one liter of solution:

$$\frac{9.1 \times 10^2 \text{ g} \cancel{\text{water}}}{1} \times \frac{1 \text{ mole water}}{18.0 \text{ g} \cancel{\text{water}}} = 51 \text{ moles water}$$

Now we know the number of moles of solute and solvent, so we can calculate the mole fraction of each:

$$X_{AgNO_3} = \frac{\text{moles AgNO}_3}{\text{total moles}} = \frac{1.2 \text{ moles AgNO}_3}{1.2 \text{ moles AgNO}_3 + 51 \text{ moles water}} = \frac{1.2}{52} = \underline{0.023}$$

$$X_{water} = \frac{\text{moles water}}{\text{total moles}} = \frac{51 \text{ moles water}}{1.2 \text{ moles AgNO}_3 + 51 \text{ moles water}} = \frac{51}{52} = \underline{0.98}$$

The two mole fractions add up to one, as they should.

Do you see what we are able to do? As long as we know one concentration unit as well as the density of the solution, we can determine any of the other concentration units. See if you truly understand this by solving the following "on your own" problems.

ON YOUR OWN

5.2 The molality of a solution of calcium chloride is 0.500 m, while the density is 1.07 g/mL. What is the molarity of the solution and the mole fraction of solute and solvent?

5.3 Which (if any) of the concentration units (molarity, molality, and mole fraction) will vary with temperature?

Solubility, van der Waals Forces, and Entropy

Now that you know about van der Waals forces, you are ready for a detailed discussion of how solutes dissolve in solvents. In your first year chemistry course, you should have seen a figure like the following.

FIGURE 5.1
Solid Sodium Chloride Dissolving in Water

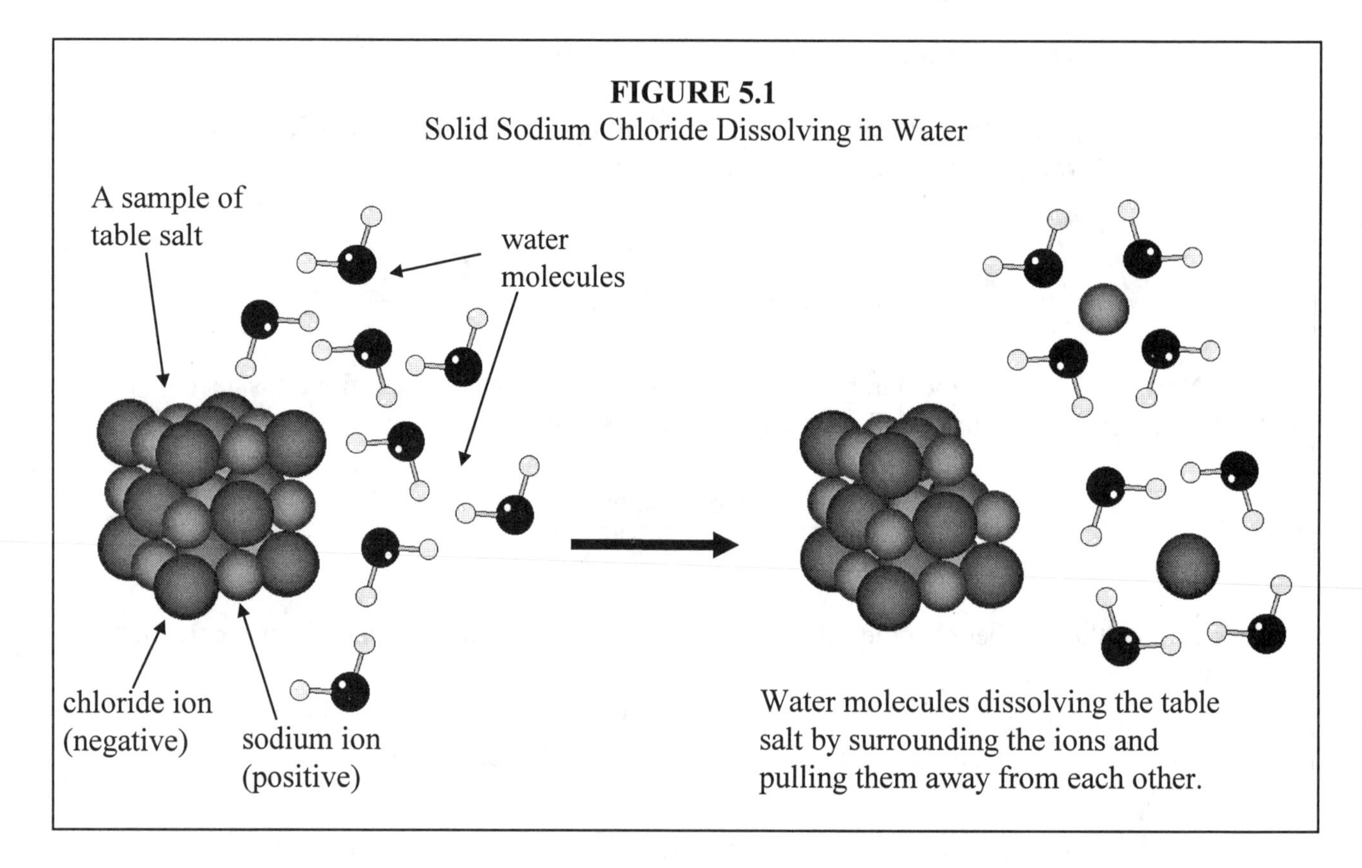

In this figure, we see that the water molecules are attracted to the solid sodium chloride. Why? Well, the water molecules are polar and sodium chloride is made up of ions, so these two compounds are attracted to one another through the ion-dipole interaction, which is a van der Waals force. This brings the water molecules close to the sodium chloride. The water molecules are so strongly attracted to the ions in the sodium chloride, however, that they do more than just get close to them. They actually shove their way in between the ions and surround them. Notice on the left-hand side of the figure that the water molecules surround the ions so that their hydrogen atoms are close to the negative chloride ions and their oxygen atoms are close to the positive sodium ions. This should make sense to you, since the polar nature of water gives the hydrogen atoms a slight positive charge and the oxygen atoms a slight negative charge.

Please remember that although Figure 5.1 is drawn specifically for the case of a solid solute dissolving in a liquid solvent, the same general arguments can be made for any type of solute dissolving in any other type of solvent. For example, consider the case of one liquid being dissolved in another. To make chocolate milk, for example, you might take chocolate syrup (a liquid) and dissolve it in milk (another liquid). In order for the milk to dissolve the chocolate syrup, the molecules in the milk must be attracted to the molecules in the syrup. They will then surround the syrup molecules and pull them away from each other, dissolving them. If you try to mix oil with water, however, the solute (oil) will not be dissolved by the solvent (water). Why? Well, oil is made up of non-polar molecules, so the only van der Waals force at their disposal is the London dispersion force. Water, on the other hand, is polar, so it is attracted to molecules that can participate in the dipole-dipole attraction. Thus, the water molecules are not attracted to the oil molecules. As a result, oil and water do not mix.

That's about where your first year chemistry course probably stopped in its explanation for how solutes dissolve in solvents. Although such an explanation is correct, it does not go nearly deep enough into all of the physical processes that are at play when a solution is made. I want to give you a more detailed understanding of how solutions form. To do that, however, you need to learn a few more things.

First, you need to learn about the energetics involved in making solutions. In your first year course, you probably learned that some solutes dissolve exothermically while others dissolve endothermically. If you dissolve solid Drano® in water, the solution gets very hot. This is because solid Drano® dissolves in water exothermically. On the other hand, you can go to a sports store and buy an instant "ice-pack" that contains ammonium nitrate (NH_4NO_3) and a small plastic bag of water. When you squeeze the bag of water hard enough to break it, the ice pack gets very cold. This is because when you break the bag of water, you cause the ammonium nitrate to dissolve in the water. Since ammonium nitrate dissolves in water endothermically, the bag gets cold.

Why do some solids dissolve endothermically and others dissolve exothermically? Well, there are many factors. First of all, you learned in the previous module that solids have lattice energies. When a solid crystal lattice forms, energy is released. When the components of the lattice are pulled apart, the lattice is destroyed. To do this, however, energy must be absorbed. Thus, when a solid's crystal lattice breaks down, it is an endothermic process.

Well then, all solids should dissolve endothermically, right? After all, the process of breaking down a crystal lattice is endothermic. Since all solids have a crystal lattice that must be broken down in order to dissolve, all solids must dissolve endothermically, right? Wrong! Don't forget the van der Waals forces! There are three sets of van der Waals forces at play in the making of a solution. The solvent molecules are attracted to each other. I will call this the "solvent-solvent attraction." In addition, the solute molecules are attracted to one another, and I will call that the "solute-solute attraction." Finally, there is a "solute-solvent attraction," caused by the van der Waals forces between the solvent molecules and the solute molecules. The relative strengths of these attractions are critical in determining whether or not a solute dissolves exothermically or endothermically.

Consider, for example, a solution in which the solute-solvent attraction is stronger than both the solute-solute attraction and the solvent-solvent attraction. This tells you that the solvent molecules would rather be close to the solute molecules than close to one another. In the same way, the solute molecules would rather be close to the solvent molecule than close to one another. As a result, as soon as the solute enters the solution, there will be a lot of activity aimed at getting the solute molecules close to the solvent molecules. This is a high energy situation. When the solute molecules do get close to the solvent molecules, they achieve the desired state. Remember that all matter desires to achieve the lowest possible energy state. Thus, the desired state of the solute molecules being close to the solvent molecules must be a low energy state. The process of dissolving the solute, then, involves going from a high energy state to a low energy state. As a result, energy is released into the surroundings. It turns out that this energy is almost always greater than the lattice energy of the solute. As a result, a net amount of energy is released. This means that **when the solute-solvent attraction is greater than the solute-solute attraction and the solvent-solvent attraction, the solute dissolves exothermically.**

There are situations in which the solute-solvent attraction is nearly identical to the solute-solute and solvent-solvent attractions. In this case, there is essentially no energy change during the process of making the solution. As a result, the solute dissolves neither exothermically nor endothermically. A solution between such a solute and solvent is called an **ideal solution**.

Ideal solution - A solution formed with no accompanying energy change

The most common type of ideal solution is formed when a gaseous solute is dissolved in a gaseous solvent. Since gas molecules are far from each other, the van der Waals forces are not very strong. As a result, there is little difference between solute van der Waals forces and solvent van der Waals forces. Thus, little or no energy change occurs when gases mix.

What happens if the solute-solvent attraction is less than the solute-solute and solvent-solvent attraction? In this case, the solute would rather stay close to other solute molecules and the solvent would rather stay close to other solvent molecules. As a result, the solute will not dissolve, right? Wrong! As you will learn in a moment, solutes can dissolve under these conditions. When they do, you can imagine that it takes energy to shove the solvent molecules close to the solute molecules. Thus, **when the solute-solvent attraction is lower than the solute-solute and solvent-solvent attraction, the solute dissolves endothermically**.

Now why in the world would a solute dissolve endothermically? If the solute molecules would just as soon stay with the other solute molecules and the solvent molecules would rather stay with other solvent molecules, why would the solute dissolve? The answer is summed up in one word: **entropy**. In your previous chemistry course, you should have learned that entropy (abbreviated as "S") is a measure of the disorder in a system. You also should have learned that the Second Law of Thermodynamics (a very misunderstood law) states that for any process which happens spontaneously (without help), the total entropy of the universe must either increase or stay the same.

The Second Law of Thermodynamics is often stated mathematically as:

$$\Delta S_{system} + \Delta S_{surroundings} \geq 0 \tag{5.1}$$

Remember, "the system" is defined as the thing we are studying, while "the surroundings" are defined as everything else. In our case, then, the system is the solution (solute and solvent), and the surroundings are the container, the air in the room, the room, etc. The Second Law of Thermodynamics, then, states that when a change occurs, the total entropy change in the thing we are studying (ΔS_{system}) plus the total entropy change of everything else ($\Delta S_{surroundings}$) must be greater than or equal to zero. This is equivalent to saying that any change which occurs in the universe must either not affect the order of the universe or it must increase the disorder of the universe. As a result, the universe is always becoming more and more disordered.

If you think about it, the process of dissolving a solute into a solvent can have a large effect on the entropy of the system. Consider, for example, the case of a solid solute dissolving in a liquid solvent. If the solute dissolves, it changes from a solid state to a liquid state. You should have learned in your previous chemistry course that the entropy of a liquid is *significantly* higher than that of a solid. In the end, then, when a solid solute dissolves, the entropy of the system increases substantially. Thus, ΔS is large and positive. How does this affect the process of forming a solution?

Remember from your previous chemistry course that there is a very simple equation which tells us whether or not a process is spontaneous:

$$\Delta G = \Delta H - T \cdot \Delta S \tag{5.2}$$

In this equation, ΔG is called the "Gibb's free energy," ΔH is the enthalpy change, and ΔS is the change in entropy of the system. If ΔG is negative, the process will happen spontaneously. If ΔG is positive, the process will not happen on its own. As you learned in your previous chemistry course, when a process is exothermic, ΔH is negative, and when it is endothermic, ΔH is positive. In the same way, when the entropy of the system increases, ΔS is positive, and when the entropy of the system decreases ΔS is negative.

In order to *really* understand how solutions form, we must think about the van der Waals forces involved and the resulting change in entropy in the light of Equation (5.2). For example, when the solute-solvent attractions are stronger than all other van der Waals forces, the solute dissolves exothermically. Thus, ΔH is negative. This will tend to make ΔG negative. In addition, if the entropy of the system increases, ΔS is positive. Looking at Equation (5.2), a positive ΔS will tend to make ΔG negative. Thus, when a solute dissolves exothermically and with a corresponding increase in disorder, ΔG will always be negative and the solute will dissolve. In the same way, if a solute dissolves endothermically and with a corresponding decrease in disorder, ΔH will be positive and ΔS will be negative. Looking at Equation (5.2), this will lead to a positive ΔG, which means that the solute will not dissolve.

Consider the case where a solute dissolves endothermically and with a corresponding increase in disorder. The ΔH will be positive, and the ΔS will also be positive. In this case, the ΔG might be positive or negative, depending on the temperature. Looking at Equation (5.2), low temperatures will result in a positive ΔG, since ΔH is positive. High temperatures, however, will make ΔG negative, because of the positive ΔS. Thus, a solute will dissolve endothermically, as long as the entropy of the system increases and the temperature is great enough to keep ΔG negative.

Now you have some idea of all the different factors necessary to determine whether or not a solute dissolves. First of all, you have to analyze the van der Waals forces to determine the ΔH of the process. Secondly, you have to look at the corresponding change in disorder. You then have to put both of those facts together with Equation (5.2) to determine how readily the solute will dissolve. Try to see how we do this by studying Example 5.3 and performing the "on your own" problems that follow.

EXAMPLE 5.3

A solute is dissolved in a solvent for which the solute-solvent attraction is small compared to the solute-solute and solvent-solvent attraction. Will this solute tend to dissolve better at high temperatures or low temperatures?

First of all, we know that the process of dissolving the solute is endothermic. We know this because the solute-solvent attraction is small compared to the solute-solute and solvent-solvent attractions. Since the solute dissolves endothermically, ΔH is positive. This tends to make ΔG positive, according to Equation (5.2). In order for the solute to dissolve, however, ΔG must be negative. The only way this can happen, according to Equation (5.2), is for ΔS to be positive. A positive ΔS can overcome the positive ΔH to make ΔG negative as long as temperature is high enough. Thus, the solute will dissolve better at high temperatures.

In the formation of an ideal solution, is ΔS positive or negative?

In an ideal solution, ΔH is zero. This means that the only factor involved in the formation of the solution is the ΔS. Since ΔG must be negative for the solution to form, and since Equation (5.2) tells us that a positive ΔS tends to make ΔG negative, then ΔS must be positive.

ON YOUR OWN

5.4 When a gas is dissolved in a liquid, is the process exothermic or endothermic?

5.5 What is the relative strength of solute-solute attractions and solvent-solvent attractions compared to solvent-solute attractions when gases are dissolved into liquids?

Temperature and Solubility

If we consider the specific case of a solid dissolving in a liquid, there is one general conclusion we can draw. If a solid dissolves, the entropy of the system should increase. In other words, the ΔS for the process of a solid dissolving in a liquid is positive. Thus, the higher the temperature, the more negative ΔG becomes. As a result, you generally expect the solubility of a solid to increase with increasing temperature. Although this is usually true, there are exceptions. Perform the following experiment to learn about one such exception.

EXPERIMENT 5.1
Solubility Curves

Supplies:

From the laboratory equipment set:
- Sodium sulfate (Na_2SO_4)
- Two test tubes with caps
- Small funnel
- Filter paper
- Clear plastic plate with depressions in it
- Plastic droppers

From around the house:
- Water
- Pan to boil water in
- Ice
- Stove
- Two Styrofoam cups
- Thermometer (It must read from 0.0 °C to 100 °C)

Procedure:

Start heating a pan of water. While you are waiting for it to boil, add ten measures of sodium sulfate to one test tube and 15 measures to the other. Fill the test tube with the least amount of solid halfway with water and fill the other one three-fourths of the way with water. Cap the test tubes and shake them vigorously. After plenty of vigorous shaking, see how much solid is left. If no solid is left, add more sodium sulfate and shake again. Continue to do so until there is visible solid left in each test tube after plenty of vigorous shaking. You now have two test tubes that each contain a saturated solution of sodium sulfate.

Fold your filter paper in a cone, as illustrated below:

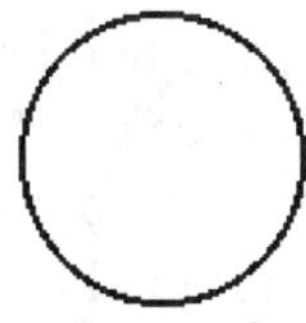

Take a large piece of circular filter paper.

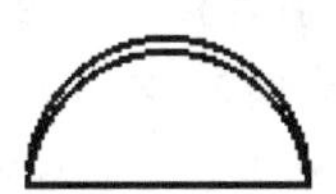

Fold it in half.

Fold it in half again so it makes a triangle.

Open the triangle to make a cone.

Place the filter paper into your funnel, and hold the funnel directly over one of the deep depressions in your depression plate. Pour a little bit of solution from one of the test tubes into the filter paper so that it filters through the paper and into the deep depression. Don't use too much solution, as you will need more later. Allow the liquid to drip into the deep depression. As the solution filters through, use your dropper to collect some of it and fill one of the small depressions in the middle of the depression plate. Because of the cohesive forces of water, the solution will form a reverse meniscus. That's okay. Measure the room's temperature with your thermometer and write it down.

Allow the funnel to drain further while you are working on the next steps. Leave the filter paper in the funnel. By now, the water should be boiling. Pour some of it into the Styrofoam cup and leave the rest boiling on the stove. Next, place the test tube that you just poured some solution from into the boiling water so that the remaining solution will heat up. Make sure the cap is NOT on the test tube. Place your thermometer into the water in the cup, NOT the solution in the test tube. Watch the temperature. It should start to decrease. Allow it to decrease to 60 degrees. If it is taking a LONG time (more than 10 minutes), you can add a little cool water to speed up the process, but the test tube needs to sit in the water for at least 3 minutes in order to equilibrate to the same temperature as the water. While you are waiting, add two measures of sodium sulfate to the filter paper. Sprinkle it around the filter paper so that when your solution is poured into the filer paper, it will hit plenty of sodium sulfate. Also, repeatedly squeeze the dropper to force out as much of the old solution as possible. Once the water is between 60 and 70 degrees, note the exact temperature and once again filter a portion (not all) of the solution through the funnel into another deep depression. Then, use the dropper to transfer solution from the deep depression to the small depression to the immediate right of the small depression you filled previously. Fill this depression to the same level that you filled the first one.

Once you have done that, pour boiling water into the other Styrofoam cup and put the same test tube you have been using into this new hot water. Put the thermometer in there as well. Let it sit for three minutes, and empty the Styrofoam cup that you used in the previous paragraph. While you are waiting, sprinkle two more measures of sodium sulfate into the filter paper again. Also, repeatedly squeeze the dropper to force out as much of the old solution as possible. Once the test tube has been sitting in the new boiling water for three minutes, fill the Styrofoam cup you just emptied with more boiling water, and transfer the test tube to that cup. Put the thermometer into the water and, after one minute, note the exact temperature. Now filter the remaining solution from the test tube into another deep depression. Once again, quickly fill a small depression from this large depression. You should use the small depression to the immediate right of the one that you filled previously.

Empty a Styrofoam cup and fill it with water from the tap. Put the other test tube (the one you haven't been using) into the water, along with the thermometer. Add a few ice cubes and stir so that they melt completely and cool the water. Get it down to about 15 degrees and let it sit for at least 3 minutes. While you are waiting, repeatedly squeeze the dropper to force out as much of the old solution as possible. Once three minutes have passed, note the exact temperature and, once again, filter the solution into another deep depression. Use the dropper to quickly fill a small depression, this time using the depression that is just to the left of all of the others that you filled.

Finally, pour a lot of water out of the cup and add more ice. Mix everything so that you have a nice mixture of ice and water. Allow the test tube and thermometer to sit in the ice water for at least 3 minutes. While you are waiting, repeatedly squeeze the dropper to force out as much of the old solution as possible. Once three minutes have passed, note the exact temperature and, once again, filter the solution into another deep depression. Use the dropper to quickly fill a small depression, this time using the depression that is just to the left of all of the others that you filled.

Now think about what you have. At each temperature, you made a saturated solution of sodium sulfate. The small depressions are filled with these saturated solutions. The one formed at the lowest temperature is the farthest to the left, and the one formed at the highest temperature is farthest to the right. The ones in between represent a steady increase in temperature. Now let all of the water evaporate, leaving behind crystals of sodium sulfate. When that is done, look at the crystals. Which depression contains the most crystals? Which contains the least? Make a qualitative graph indicating how the amount of solid changes with temperature.

In general, a solid's solubility increases with increasing temperature. However, that general trend can change under certain circumstances. To see what I mean, examine the following figure, which plots the solubility of certain compounds as a function of temperature. When these quantities are graphed together as is done in the figure, the result is called a **solubility curve**.

FIGURE 5.2
Solubility Curves for Certain Ionic Compounds

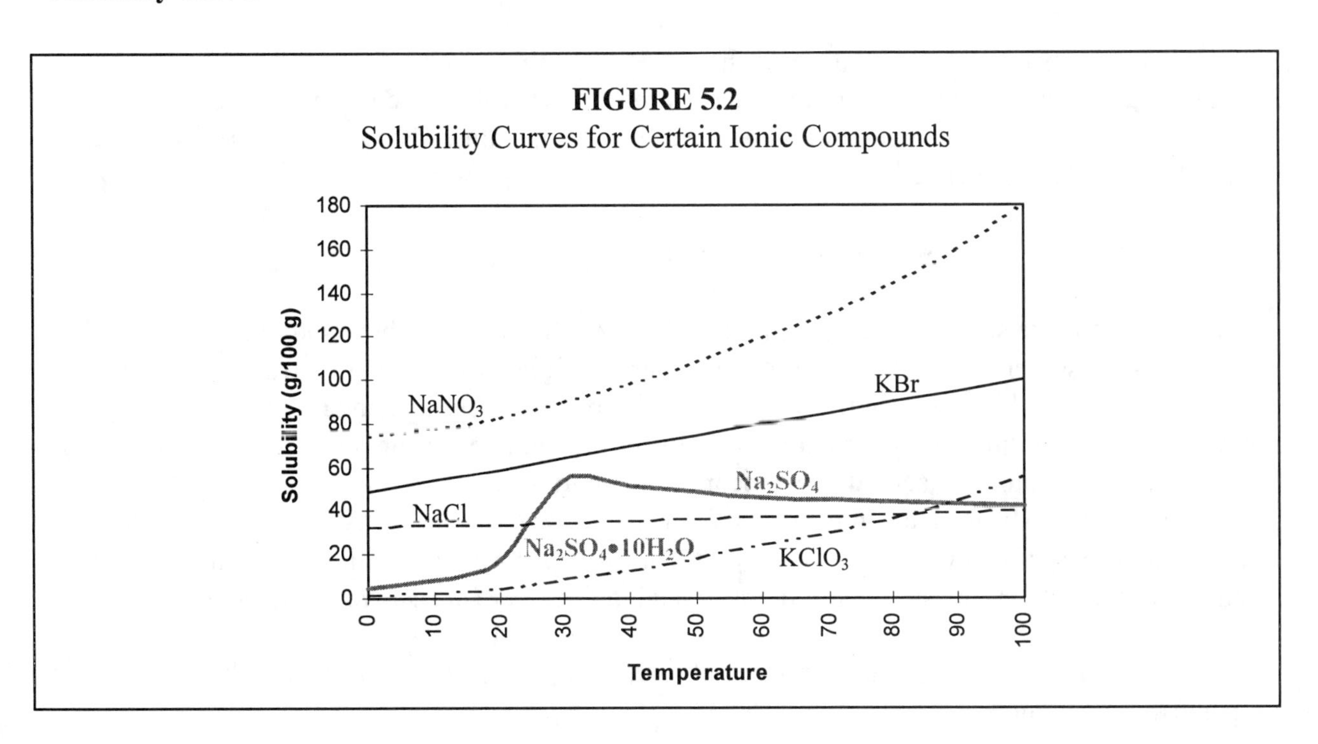

For right now, ignore all data except that illustrated with the heavy, gray line. The qualitative graph that you drew in the experiment should look something like this line. You see, while the solubility of most solids increases with increasing temperature, there are certain solids whose solubility is a more complex function of temperature.

Why? Well, first of all you have to understand that when you dissolve as much solid as you can into a solution and still have left over solid in the container, an **equilibrium** is established. For example, when you dissolve sodium chloride in water, you establish the following equilibrium:

$$NaCl\ (s) \rightleftarrows Na^+\ (aq) + Cl^-\ (aq) \quad (5.3)$$

You should have seen equilibrium equations like this in your first-year course, so it should look familiar to you. Remember what this equation tells us. It says that NaCl (s) is being broken down into its constituent ions. We call this the "forward reaction." *At the same time,* sodium ions and chloride ions are coming back together to form NaCl (s) again. We call this the "reverse reaction." When a solution becomes saturated and there is still some solid left over, the solution is in equilibrium, which means that the forward reaction and backward reactions are occurring *at the same rate*. Thus, every time an NaCl (s) is broken down into its constituent ions, an Na^+ (aq) ion reacts with a Cl^- (aq) ion to form NaCl (s).

Since the forward and reverse reactions are happening at the same rate, the overall situation does not change. Even though NaCl is constantly dissolving and re-forming, the total concentration of sodium and chloride ions in solution does not change. Likewise, the amount of left over solid in the container does not change.

How does this relate to the weird shape of the solubility curve for Na_2SO_4? Well, when you start dissolving Na_2SO_4 in solution, an equilibrium gets established. The nature of that equilibrium, however, is temperature-dependent. For temperatures above 30 °C, the following equilibrium gets established:

$$Na_2SO_4\ (s) \rightleftarrows 2Na^+\ (aq) + SO_4^{2-}\ (aq) \quad (5.4)$$

That ought to make sense to you. This is simply the equilibrium in which sodium sulfate breaks up into its constituent ions. At temperatures under 30 °C, however, the chemical nature of the sodium ions and sulfate ions (SO_4^{2-}) changes. This change causes a change in the reverse reaction. When sodium and sulfate ions react in solution at temperatures under 30 °C, they form a complex with water that actually gets incorporated into the solid crystal lattice. Na_2SO_4 (s) is no longer formed. Instead, a complex of sodium sulfate with water forms. That compound is identified as $Na_2SO_4 \bullet 10H_2O$. The "$\bullet 10H_2O$" that follows the sodium sulfate formula tells you that for every sodium and sulfate ion in the crystal lattice, there are 10 water molecules as well.

Since this new compound is formed at lower temperatures, a *different* equilibrium gets set up a low temperatures:

$$Na_2SO_4 \bullet 10H_2O\ (s) \rightleftarrows 2Na^+\ (aq) + SO_4^{2-}\ (aq) + 10H_2O\ (l) \qquad (5.5)$$

This new equilibrium has a completely different dependence on temperature as compared to the first one I discussed, so this results in a different solubility curve. Thus, the change in solubility curve that is illustrated in Figure 5.4 is due to a change in the solid that is being dissolved. At temperatures lower than 30 ^oC, the solid that is left over in the container is $Na_2SO_4 \bullet 10H_2O$. At temperatures above 30 ^oC, the solid that is left over in the container is Na_2SO_4. Since these are two fundamentally different solids, the solubility curve changes.

Now that you know why the solubility curve for sodium sulfate changes at a temperature of 30 ^oC, another question should come to mind. You should be asking yourself why the curve changes in the specific way that it does. Notice that for *every other compound in the figure*, the solubility increases with increasing temperature. When $Na_2SO_4 \bullet 10H_2O$ (s) turns into Na_2SO_4 (s) at temperatures above 30 ^oC, however, the solubility of Na_2SO_4 (s) actually decreases with increasing temperature. Why is Na_2SO_4 (s) so different than all of the other compounds in the figure?

The answer to that question comes from looking at the enthalpy of the situation. The vast majority of ionic solids dissolve endothermically. As you learned in the previous section, this means that the solute-solvent attraction is less than the solvent-solvent and solute-solute attractions. If you think about it, that ought to make sense. After all, the solute ions are held together through the mutual attraction of their opposite charges. Thus, the solute-solute attraction is an ion-ion van der Waals force. The solute-solvent attraction, however, is an attraction between a polar molecule and the ions. Thus, that is an ion-dipole van der Waals force. Which would you expect to be stronger? In most cases, the ion-ion force should be stronger, because ions have *full* electrical charges, while dipoles have only *partial* electrical charges. Thus, the ion-dipole attraction is, in general, weaker than the ion-ion attraction. As a result, most ionic solutes have a greater solute-solute attraction than solute-solvent attraction. The consequence of this is that most solutes dissolve endothermically.

Now think about what this means. If a solute dissolves endothermically, that means it *takes energy* for the solute to dissolve. Thus, we can think of energy as a *reactant* in the equilibrium. In other words, if we are considering the equilibrium that takes place in a saturated solution of sodium chloride, we could write it this way:

$$NaCl\ (s) + \text{energy} \rightleftarrows Na^+\ (aq) + Cl^-\ (aq) \qquad (5.6)$$

If sodium chloride dissolved exothermically, we would write energy as a product. However, we know that sodium chloride dissolves endothermically, so energy is a reactant.

In your first year chemistry course, you should have learned something called Le Chatelier's principle. This principle states that when an equilibrium is stressed, it tends to shift in order to relieve that stress. One way to stress an equilibrium is to add more reactants or products. You should have learned that when a substance is added to an equilibrium, the equilibrium shifts so as to decrease its concentration. When a substance is taken away, the equilibrium shifts so as to increase its concentration. You should have also learned that when using Le Chatelier's principle, we ignore solids when other, aqueous substances exist.

Using Le Chatelier's principle on the equilibrium described by Equation (5.6), then, if I added more Na^+ (aq) ions, the equilibrium would shift so as to decrease their concentration. To do this, the reverse reaction would start occurring more quickly than the forward reaction. This would result in more NaCl (s) being formed and less Cl^- (aq) in solution. Alternatively, if I took away Cl^- (aq) ions, the equilibrium would shift to make more Cl- (aq) ions. This would mean that the forward reaction would start occurring more quickly than the reverse reaction. This would result in a decrease of NaCl (s) and an increase in the concentration of Na^+ (aq) ions. If I were to add more NaCl (s), however, nothing would happen, because we ignore the effect of solids in this reaction because there are other, aqueous substances. Is Le Chatelier's principle coming back to you? If not, you might want to go back to your first year book and take a look at it again.

Now let's get back to the subject of solubility and temperature. What does Le Chatelier's principle say will happen when I increase the temperature of the equilibrium given in Equation (5.6)? Well, temperature is really just a measure of energy. Thus, increasing the temperature is the same as increasing the energy available to the reaction. According to Le Chatelier's principle, this will cause a shift in the equilibrium so as to reduce the amount of energy. Thus, the forward reaction of the equilibrium will increase in speed. This will cause more NaCl (s) to dissolve, making more Na^+ (aq) and Cl^- (aq). In other words, increasing the temperature will increase the solubility. Alternatively, if I reduce the temperature, that is the same as taking energy away from the equilibrium. In this case, Le Chatelier's principle says that the equilibrium will shift so as to make more energy. Thus, more NaCl (s) will be made and there will be less Na^+ (aq) and Cl^- (aq). In other words, lowering the temperature will decrease the solubility of NaCl!

What happens, however, if a solute dissolves exothermically? When this is the case, energy is no longer a reactant in the equilibrium. It is a product. Thus, all of the arguments reverse, making the solute less soluble at high temperatures and more soluble at low temperatures. In the end, then, the slope of the solubility curve depends on the enthalpy of the equilibrium that is established when a solute dissolves. If the solute dissolves endothermically, an increase in temperature will shift the equilibrium so as to dissolve more solute. If a solute dissolves exothermically, an increase in temperature will decrease its solubility!

If you now re-examine Figure 5.2, you can understand why the solubility curve of sodium sulfate changes the way it does. At low temperatures, the solute being dissolved is $Na_2SO_4 \bullet 10H_2O$ (s). This solute dissolves endothermically; thus, its solubility increases with increasing temperature. At higher temperatures, however, Na_2SO_4 (s) is the solute being dissolved. This solute dissolves exothermically, resulting in a decrease of solubility with increasing temperature.

ON YOUR OWN

Consider the following solubility curves:

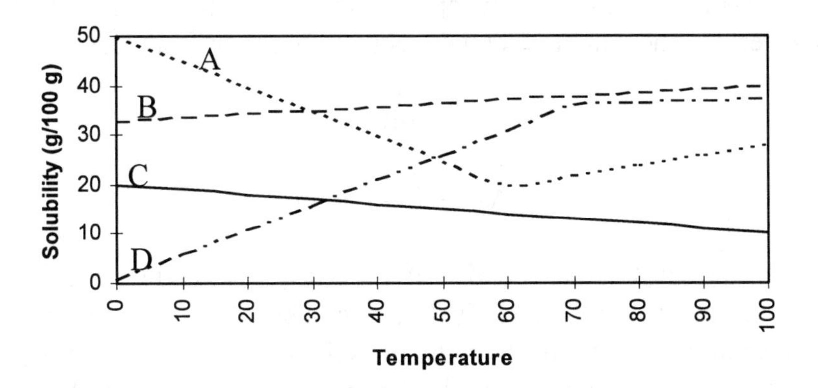

5.6 Which curve or curves (A, B, C, D) represent(s) substance(s) that change their chemical nature while dissolving over this temperature range?

5.7 Which curve or curves represent(s) solid(s) that dissolve exothermically throughout this entire temperature range?

5.8 Which curve represents a solid that dissolves endothermically over part of this temperature range and then changes to a substance that dissolves exothermically?

The Effect of a Solute on a Solvent's Phase Diagram

In Figure 4.6 of the previous module, I presented the phase diagram for water. The same phase diagram is shown here on a different scale:

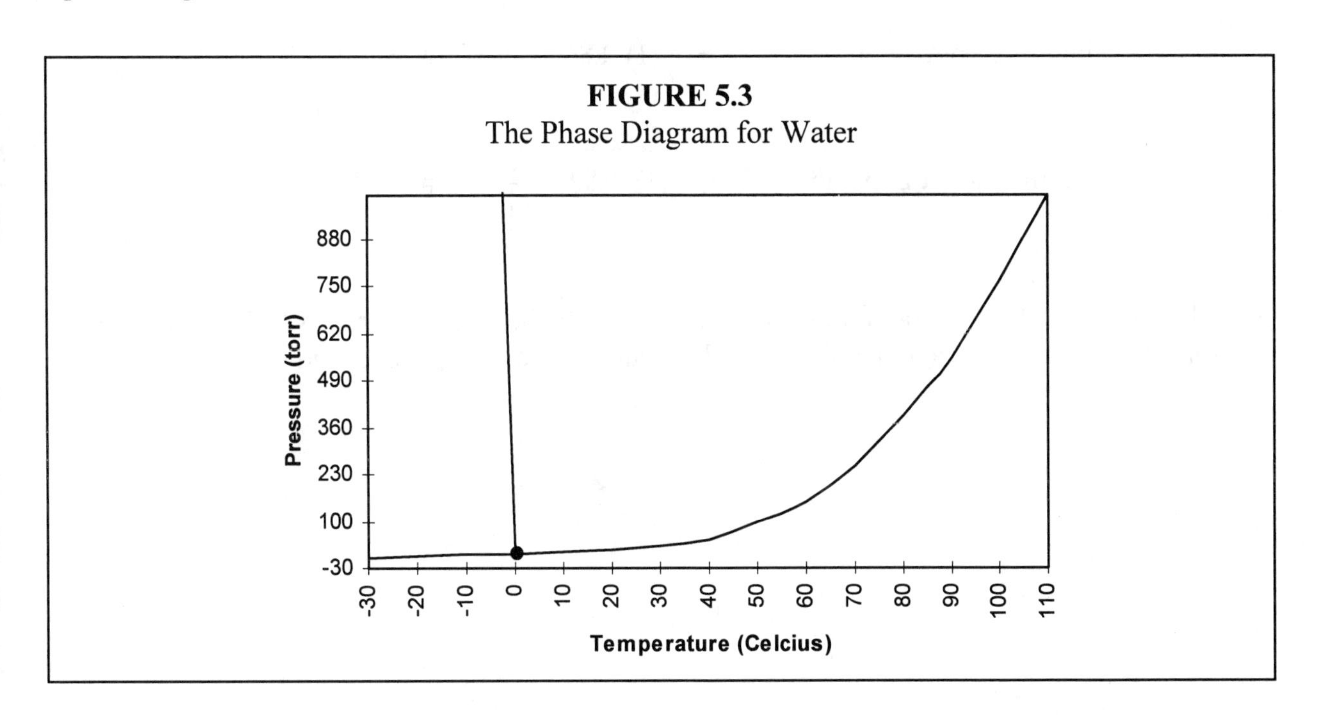

FIGURE 5.3
The Phase Diagram for Water

What happens to this phase diagram if I dissolve something in the water? Well, in your previous chemistry course, you should have learned about **freezing point depression** and **boiling point elevation**. When a solute is added to a solvent, the solvent's boiling point is raised and its freezing point is lowered. There are two equations that govern this process:

$$\Delta T = -i \cdot K_f \cdot m \quad (5.7)$$

$$\Delta T = i \cdot K_b \cdot m \quad (5.8)$$

Equation (5.7) says that when a solute is added to a solvent, its freezing point lowers by a few degrees. The change in freezing point (ΔT) is determined by taking the number of substances that the solute breaks into (i), and multiplying by the solvent's freezing point depression constant (K_f) as well as the molality of the solute (m). In the same way, Equation (5.8) says that the change in a solvent's boiling point (ΔT) can be calculated by taking the number of substances that the solute breaks into (i), and multiplying by the solvent's boiling point elevation constant (K_b) as well as the molality of the solute (m).

In general, then, the addition of a solute tends to make it more difficult to boil and freeze the solution. How will this affect the phase diagram? Well, let's look at the case of water. For water, $K_f = 1.86 \frac{^\circ C}{m}$ and $K_b = 0.521 \frac{^\circ C}{m}$. Suppose I add enough table sugar in water to make a 5.0 molal solution. We can calculate the change in the boiling point and freezing point. Remember, the quantity (i) in Equations (5.7) and (5.8) tells us the number of substances that the solute splits into when it dissolves. An ionic compound like NaCl splits up into its ions, Na^+ and Cl^-. Therefore, i=2 for NaCl. Table sugar, ($C_{12}H_{24}O_{12}$), however, is not ionic. It does not split up into ions when it dissolves. Thus, i = 1 for sugar. Therefore:

$$\text{Freezing point depression: } \Delta T = -(1) \cdot 1.86 \frac{^\circ C}{\cancel{m}} \cdot 5.0 \cancel{m} = -9.3\ ^\circ C$$

$$\text{Boiling point Elevation: } \Delta T = -(1) \cdot 0.521 \frac{^\circ C}{\cancel{m}} \cdot 5.0 \cancel{m} = 2.6\ ^\circ C$$

This tells us that the freezing point lowers by 9.3 °C and the boiling point raises by 2.6 °C. How does this affect the phase diagram? Examine the figure on the next page to find out.

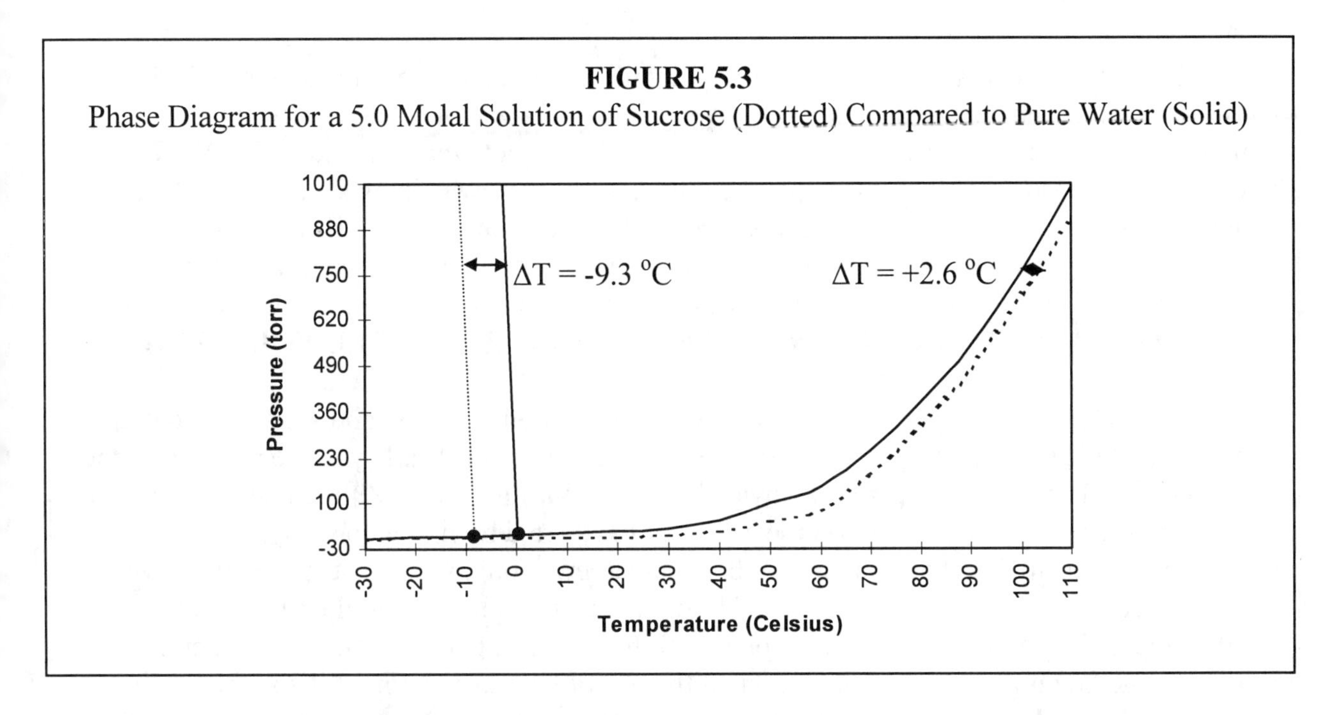

FIGURE 5.3
Phase Diagram for a 5.0 Molal Solution of Sucrose (Dotted) Compared to Pure Water (Solid)

Since the addition of a solute decreases the freezing point and increases the boiling point, a solution's phase diagram tends to have the same shape as that of the solvent, but the line separating the solid phase from the liquid phase is shifted to lower temperatures while the line separating the liquid phase from the gaseous phase is shifted to higher temperatures.

Separating Solute From Solvent in a Solution

Although a lot of interesting chemistry can be investigated with solutions, sometimes it is desirable to separate solute from solvent in a solution. Two very common ways to accomplish this are **distillation** and **chromatography** (krom uh tahg' ruh fee). We will examine each of these with an experiment.

EXPERIMENT 5.2
A Simple Distillation

Supplies:

- Water
- Salt
- Ice
- Tablespoon
- Small saucepan
- Saucepan lid or frying pan lid larger than the saucepan used
- Large bowl (It should not be plastic, as it will get hot.)
- Potholders
- Zippered plastic sandwich bag
- Stove

Fill the saucepan about three-quarters full with water. Add three tablespoons of salt to the water and stir to make as much salt dissolve as possible. Do not be concerned if you can't get it all to dissolve. Taste the saltwater you have made. Please note that you should **NEVER** get into the habit of tasting things in an experiment unless someone who knows a lot more chemistry than you do (like me) says to do so. In this case, I know that you are not at risk of poisoning yourself by tasting the saltwater you have just made. However, there may be times when you make something in an experiment which *you* think will not hurt you, but is, in fact, quite toxic. So **DO NOT TASTE THINGS IN AN EXPERIMENT UNLESS I TELL YOU TO DO SO!**

Tastes bad, doesn't it? Now set the pan of saltwater on the stove and start heating it up. Your goal is to have vigorously boiling water, so turn up the heat! While you are waiting for the saltwater to boil, take your zippered sandwich bag and fill it full of ice. Zipper it shut so that no water from the ice can leak out. Once the saltwater has started boiling vigorously, place the bowl next to the saucepan. The bowl should not be on a burner. You do not want to heat the bowl. You just want it close to the boiling water. Now use the potholder to hold the saucepan lid and put the zippered sandwich bag full of ice on top of the lid. You may have to use a finger or two from the hand holding the lid to make sure that the bag of ice stays on top of the saucepan lid. Take the lid and hold it so that one end (the one with the most ice one it) is over the saucepan and the other end is over the bowl. Tilt the lid so that it tilts toward the bowl. In the end, your setup should look like this:

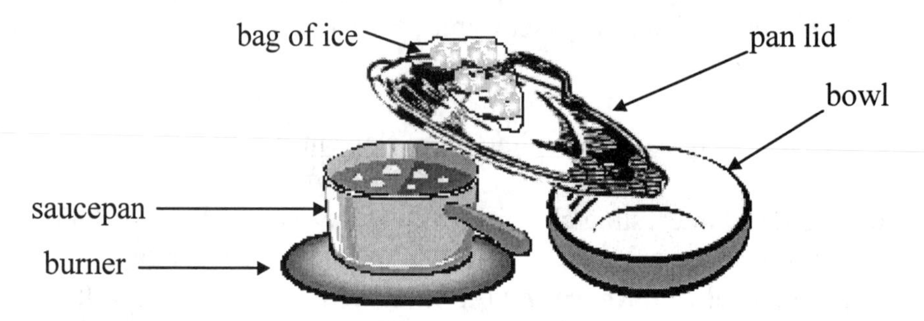

Hold the lid there for a little while, and watch what happens on the underside of the lid. **BE CAREFUL! EVERYTHING HERE IS HOT!** Notice that water droplets are forming on the underside of the lid over the saucepan, and they slowly drip down the lid towards the bowl.

If your arm gets tired, you can set the lid down so that part of it rests on the saucepan and the rest sits on the bowl. Make sure that the bowl is lower than the saucepan so that the lid still tilts towards the bowl. Eventually, you will see water dripping off of the pan lid and into the bowl. Wait until there is enough water in the bowl to be able to take a drink. Once that happens, turn off the burner and wait a moment.

Using potholders, take the lid away and put it in the sink. Pour the half-melted ice out of the bag and throw the bag away (or recycle it). **Still using potholders**, take the bowl away from the stove and set it on the counter. Empty the saucepan and put it in the sink as well. Allow the bowl to cool down completely, and then taste the water in the bowl. Once again, you can only do this because I am telling you to! Does the water in the bowl taste like saltwater?

The water in the bowl should have tasted like regular water, because there was, in fact, no salt in it. You started out with a solution of NaCl and water. When you boiled the solution, however, the water turned into vapor and escaped the pan. The temperature was not *nearly* hot enough to boil salt, however, so the salt stayed in the pan. When the water vapor hit the pan lid that was kept cool by the bag of ice, it condensed back into its liquid form. Since the pan lid was tilted towards the bowl, the water eventually trickled down into the bowl. This water had no salt in it, so it tasted like pure water. In fact, the water you tasted in the bowl was much purer than what you get from your water tap. That's because water which comes from the tap has many things dissolved in it. Most cities add chlorine to the water to destroy pathogens and fluoride to improve the dental health of the population. Even water from a well has minerals dissolved in it. The distillation you did separated the water from those dissolved substances, making it much purer.

When you go to a supermarket and purchase distilled water, you are purchasing water that has gone through a very similar process. In your experiment, you did a "simple distillation." We do not call it simple because it used only household products. Instead, we call it simple because it involved only one phase change. Water was boiled once and condensed once. To get the purest water, you need to boil and condense water several times. This is called a **fractional distillation**. In a fractional distillation, the vapor being boiled away usually travels up a tube with several prongs in it. The water tends to condense on the prongs and then vaporize again a short time later. As it travels up the tube, then, it boils and condenses several times. The vapor is condensed and collected from the top of the tube, ensuring that it has evaporated and condensed many times before it is collected. Typically, the distilled water you buy at the store has been through a fractional distillation process.

Although you might think that distillation is a method that can only be used to separate solid solutes from liquid solvents, it can also be used to separate liquid solutes from liquid solvents. After all, there is little chance of two liquids having the same boiling point. If you boil a mixture of two liquids, the one that has the lower boiling point will start boiling first. Once it starts boiling, the vast majority of the energy being put into the solution by the heat source will be used not to raise the temperature of the solution, but instead to change the phase of the liquid that is boiling. Thus, the liquid that boils first will tend to boil completely away before the other one begins to boil at all. Because of this fact, distillation can also be used to separate liquid solutes from liquid solvents.

Sometimes, however, you cannot boil your solution because the increased energy might cause chemical reactions that will alter the chemical nature of the mixture. Thus, chemists often need to separate mixtures without boiling them. One popular method for this is chromatography. The following experiment will give you some experience with this powerful chemical technique.

EXPERIMENT 5.3
Paper Chromatography

Supplies:

From the laboratory equipment set:
- Three sheets of chromatography paper (This is the rectangular, white paper that is in a bundle with a blue piece of heavier paper as the bundle's backing.)

From around the house:
- Three small glasses
- Three paper clips
- Plate
- Pencil
- Tape
- Blue food coloring
- Yellow food coloring
- Green food coloring

Take the three sheets of chromatography paper and cut one end down to a point. Make a mark with a pencil across the chromatography paper right where the point begins. Stretch out the paper clips so that they are reasonably straight. They need to be long enough to reach across the top of the glasses you are using. Now use your tape to attach the top of the chromatography paper (the end not cut down to a point) to the center of each paper clip.

Next, place a drop of each color of food coloring on the plate. Place them far enough apart from each other that they will not mix. Take a paper clip and dip it in one drop and then touch that end of the paper clip to the center of the chromatography paper, right on the pencil line. This should make a dot of food coloring on the chromatography paper In the end, the chromatography paper should look like this:

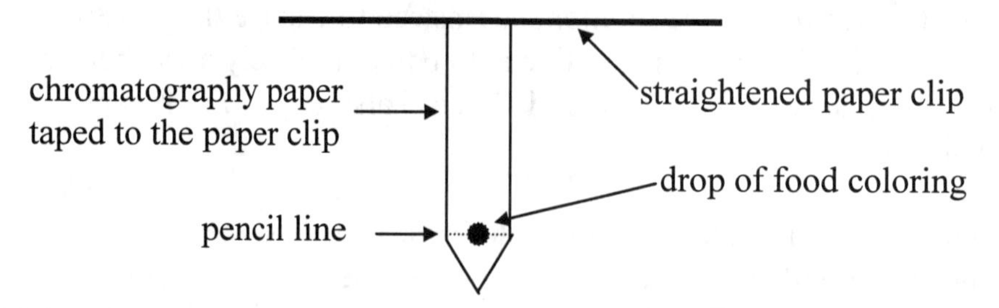

Do this so that you have a different color food coloring drop on each of the three pieces of chromatography paper.

Now fill each of the glasses with some water. Add enough water so that it looks like the pointed tip of the chromatography paper will just barely touch the water when the paper clip is suspended across the glass. **Be sure not to add too much water.** It is best to have too little water at first, as more can always be added. Too much water will mess everything up. Once you have done that, place the paper clip across the glass as shown here:

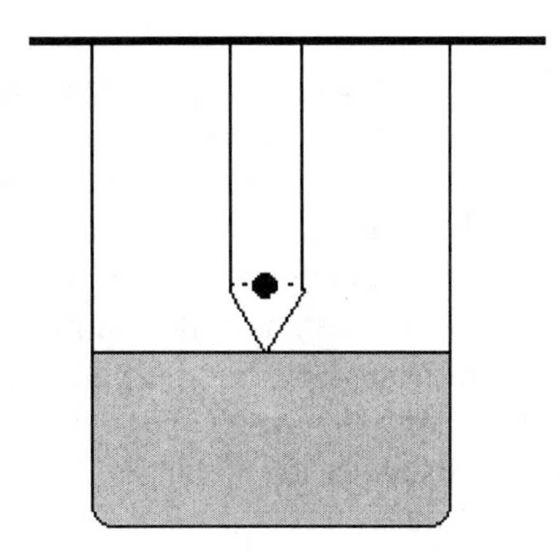

If the water touches the paper anywhere except the tip, you must start over. If the water does not touch the tip yet, carefully add more until the water level just touches the tip. The best way to do this is with one of the droppers in the laboratory equipment set:.

Set up all three glasses in this way, each with a different color of food coloring on the chromatography paper. Watch as the water begins to travel up the chromatography paper. If the paper absorbs so much water that the water level in the glass no longer touches the tip of the paper, go ahead and carefully add more water so that the paper is always in contact with the water. Wait until the water travels all the way up the paper or until it stops traveling altogether.

What did you see? Well, the water should have traveled up the chromatography paper, carrying the food coloring with it. The blue food coloring should have traveled a certain distance and the yellow food coloring another. The green food coloring should have been separated into yellow and blue, each color traveling about as far as it did on the other paper.

What explains this? There were attractive van der Waals forces between the water molecules and the molecules that make up the chromatography paper. As a result, the water was attracted to the paper. This caused the water to be pulled up the paper. As the water traveled up the paper, it encountered the food coloring. There were attractive van der Waals forces between the food coloring and the water, but there were also attractive van der Waals forces between the food coloring and the paper. These forces fought against each other. The attractive forces between the water and the food coloring tended to make the food coloring travel with the water. The forces between the paper and the food coloring tended to "drag" against this motion. Because yellow and blue food coloring are composed of different molecules, the relative strengths of these forces were different for each. As a result, the speed at which they each traveled with the water was different, resulting in different distances traveled once the water reached the top of the paper.

The green food coloring was simply a mixture of blue and yellow food coloring. As a result, these chemicals separated as they traveled up the paper. Thus, the solution formed between the two chemicals separated. That's the essence of paper chromatography. You can perform paper chromatography on other things as well. Water-soluble inks often produce very interesting results when paper chromatography is done on them. There is more chromatography paper in your kit, so feel free to investigate more!

Although paper chromatography is the easiest type of chromatography to do, there are other types as well. **Column chromatography**, for example, is done by forcing a solution to travel through a tube full of a solid that is not soluble in the solution. As the solution is forced through the tube, the components of the solution tend to travel at different speeds due to the relative strengths of the van der Waals forces involved. In the end, each component makes it to the end of the tube at different times, so they can be collected independently. **Gas chromatography** works essentially the same way, but the solution is a mixture of gases. The tube that the gases travel through must be *significantly* longer than the tube in column chromatography, however, since gas molecules are so far from each other. This relative distance decreases the effect of van der Waals forces.

ON YOUR OWN

5.9 A solution of pentane (C_5H_{12}) and decane ($C_{10}H_{22}$) is distilled. Which liquid is collected from vapor and which stays behind in the container? (HINT: Remember what you learned in Module 4!)

5.10 A mixture of two liquids is separated by paper chromatography. Liquid A rises higher on the paper than does liquid B. Which liquid has the strongest van der Waals attraction to the paper?

Colloids

A solution is a homogeneous mixture of a solvent and one or more solutes. If you try to mix an insoluble solute in a solvent (like sand in water, for example), most likely the solute will be suspended in the solvent for a brief time but will eventually settle out, forming a layer of solute and a layer of solvent. There is a type of mixture that falls in between these two. That type of mixture is called a **colloid**. A colloid is a heterogeneous mixture between a solvent and an insoluble solute. Unlike most such mixtures, however, the particles of solute are so small that they do not settle out of the solvent. Instead, they stay suspended in the solvent.

Probably the easiest example of a colloid to visualize is smoky air. When you make smoke, you are making tiny particles of ash. Unlike larger bits of ash, these ash particles can stay afloat in air; thus, they do not settle out of the air. They mix with the air but do not form a homogenous mixture, as do the nitrogen and oxygen molecules that make up 99% of the air. Compare this to a mixture of oil and water. If you shake a mixture of oil and water, the oil forms tiny droplets suspended in the water for a few moments, but eventually the oil droplets coalesce and form a layer of oil over the water. That's not a colloid, but a mixture of smoke and air is.

Colloids exhibit a phenomenon known as the **Tyndall effect**. In the Tyndall effect, light that shines through a colloid bounces off of the particles that make up the colloid. This causes the light to "scatter" to your eyes, and you see it as a beam. For example, the searchlights in the photograph below are illustrating the Tyndall effect.

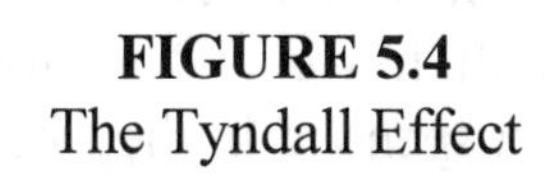

FIGURE 5.4
The Tyndall Effect

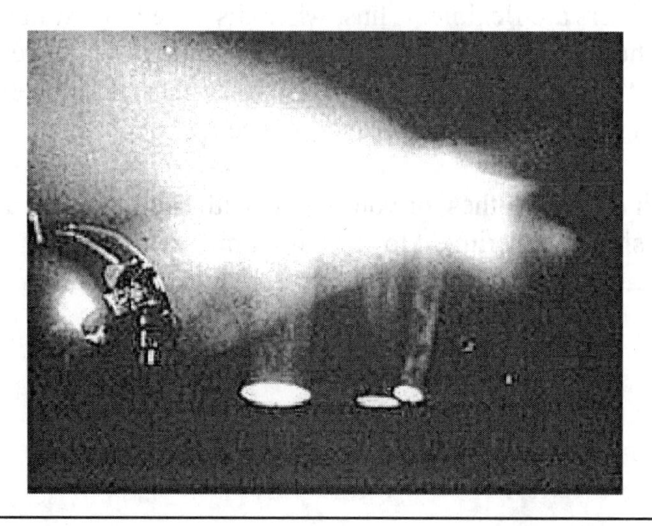

In the photograph, there is smoke in the air. The light in the searchlight beams bounces off of the smoke particles, and as a result you see the light beam. If the air were clear, without any dust, water droplets, or smoke, you would not see the beams of light. Instead, the light would travel straight from the searchlight in the direction the searchlight is pointed. You would not see the light unless you were directly in its path and facing the searchlight. Since the light is bouncing off of smoke particles in the air, however, some of it reaches your eyes even though you are not in the light's original path.

Since colloids are unique, we generally do not use the terms "solvent" and "solute" when describing them. Instead, the particles that are suspended (the smoke in smoky air) are called the **dispersed phase** while the substance in which they are suspended (the air in smoky air) is called the **dispersion medium**. Many inks and paints are actually colloids formed between a liquid and a solid. In these colloids, solid pigments are ground into a very fine powder and mixed with a liquid. The pigment is not soluble in the liquid, but the particles do stay suspended in the liquid instead of settling out. When the ink or paint dries, the liquid evaporates, but the solid stays behind.

Some colloids cannot form without the help of an **emulsifying** (ee mul' suh fie ying) **agent**. These colloids are often called **emulsions**. Milk, for example, is an emulsion of water and butterfat. Butterfat is non-polar, so it is not soluble in water. If you mix just water and butterfat, however, you will get the same effect as mixing water and oil. The butterfat will eventually coalesce and form its own layer. A specific protein (casein), however, inhibits the coalescence of butterfat. Thus, if you mix casein with water and butterfat, the casein acts as an emulsifying agent, keeping the butterfat from separating out of the water. Mayonnaise is also an emulsion between oil and vinegar, with egg yolk as the emulsifying agent.

Sometimes, the dispersed phase of a colloid will coagulate so that the whole mass, including the dispersion medium, sets into an extremely viscous body known as a **gel**. When you make gelatin, you dissolve a solid into hot water. As the water cools, the solid begins to precipitate out. The particles are so small, however, they form a colloid. The solid then sets in a complex network of fibers with liquid suspended in the fibers. The liquid is contained by the fibers of solid, but the fibers are flexible. Thus, when disturbed, the water will try to move, but the flexible fibers will hold it in place. Since the fibers are flexible, however, they move a little, causing the "jiggle" that you associate with gelatin. Other gels include jellies (made with pectin, a carbohydrate found in fruits) and the cytoplasm of a cell.

When you wash dishes, clothes, or your hands with soap, you are also forming a colloid. Perform the following simple experiment to see what I mean.

EXPERIMENT 5.4

Forming Colloidal Particles With Soap

Supplies:

- Liquid dish soap
- Vegetable oil
- Dropper
- Two bowls
- Water

Fill each bowl halfway with water and allow the water to settle. In one bowl, add a few drops of dish soap to the center of the bowl. Do not mix the solution. Just drop the soap in and leave it alone. Next, fill your dropper with vegetable oil. If you are using the a dropper from the laboratory equipment set, throw it away when you are done. If you are using a dropper that can be taken apart and washed, you will need to do that at the end of the experiment. Add a few drops of vegetable oil to the bowl with no soap in it. Not very exciting, is it? Now do the same thing to the bowl that has soap in it. What happens? Now add a few drops of soap to the bowl that originally had no soap. What happens?

The active ingredient in dish soap is sodium stearate:

```
                H H H H H H H H H H H H H H H H H O
hydrophobic     | | | | | | | | | | | | | | | | | ||              hydrophilic
    end       H-C-C-C-C-C-C-C-C-C-C-C-C-C-C-C-C-C-C-O-  Na+         end
                | | | | | | | | | | | | | | | | |
                H H H H H H H H H H H H H H H H H
```

Notice the long chain of carbon atoms. If the ions at the end of the molecule were not there, this would be a simple carbon chain, which is completely nonpolar. As a result, it would not dissolve in water at all. The ions on the other end of the molecule, however, allow the molecule to be soluble in water. The carbon chain is so long, however, that the other end of the molecule is too far away from the ions to "feel" them at all. Thus, that end of the molecule behaves as if it were nonpolar. Sodium stearate, then, has a nonpolar end which we call "hydrophobic" because it does not want to dissolve in water. The other end is called "hydrophilic" because it does want to dissolve in water.

This molecule, then, has the unique ability to be attracted to nonpolar molecules at its hydrophobic end and to polar molecules at its hydrophilic end. In your experiment, you added a nonpolar substance (vegetable oil) into water. In the water with no soap, the oil just sat on top of the water. In the bowl of water with soap, however, as soon as oil hit the water, the hydrophobic end of the sodium stearate molecules were attracted to the oil molecules. The sodium stearate molecules thus pointed their hydrophobic ends to the oil and surrounded it. The hydrophilic ends were pointed away from the oil, keeping the sodium stearate dissolved in the water. As a result, the oil piled up into drops, each drop surrounded by sodium stearate molecules!

When you added soap to the water that already had oil in it, the oil was pushed to the edge of the bowl and then formed drops. Why? As soon as you added soap, the sodium stearate spread out over the surface of the water, pushing the oil to the edge. After that, it then surrounded the oil and formed it into little drops. The drops of oil formed in your experiment are colloidal particles. They will stay as droplets, but will also travel with the water. This happens because the hydrophobic end is attracted to the oil, but the hydrophilic end is dissolved in the water. Since the oil is "stuck" to one end of the soap molecule and the other end of the soap molecule is "stuck" to the water, the oil is, in essence, "stuck" to the water, even though it is not attracted to the water. As a result, when nonpolar compounds are on dishes, clothes, or your hands, they are attracted to the hydrophobic ends of the soap molecules you are using to wash with. This pulls them off of your dishes, clothes, or hands and "sticks" them to the water. When the water rinses away, so do the particles "stuck" to it.

You should have seen one more thing in your experiment. When you initially put the soap in the water, you should have seen it sinking to the bottom. Well, you weren't actually seeing *all* of the soap sinking to the bottom. You were actually seeing everything *but the active ingredient* sinking to the bottom. You see, dish soap is a mixture of sodium stearate and a whole bunch of other chemicals that add scent, soften hands, etc. The active ingredient that does all the work, however, is just sodium stearate. When you drop the soap in water, the active ingredient stays on the surface of the water, and the rest of the stuff sinks. Thus, what you saw sinking was all of the "other stuff."

This "other stuff" is all that separates one dish soap from another. The active cleaning ingredient is the same in all of them. Thus, when one soap's advertising says it "cleans better" than another, you have to be very careful. They all use the same cleaning agent. The only thing that can really affect the soap's ability to clean is the concentration of the active ingredient. All of the other additives can dilute that concentration. Thus, the only thing to look for from dish soap to dish soap is the concentration of sodium stearate. All of the other additives will actually *dilute* the concentration of sodium stearate, making the dish soap a less powerful cleaner!

ANSWERS TO THE ON YOUR OWN PROBLEMS

5.1 To determine the molarity, we must convert grams of magnesium sulfate into moles of magnesium sulfate and mL of solution into L of solution. Remember, you should know how to take the name of an ionic molecule and determine its chemical formula. You determine the charge on each ion, then you switch the charges and treat them as subscripts. If they are equal, you ignore them. Magnesium is in group 2A; therefore, it takes on a +2 charge in an ionic compound. Sulfate is a polyatomic ion you can look up. Its formula is SO_4^{2-}; thus, its charge is negative 2. The charges are equal, so you ignore them. Magnesium sulfate, then, has a chemical formula of $MgSO_4$.

$$\frac{50.0 \text{ g } \cancel{MgSO_4}}{1} \times \frac{1 \text{ mole } MgSO_4}{120.4 \text{ g } \cancel{MgSO_4}} = 0.415 \text{ moles } MgSO_4$$

I will assume you can make the conversion to liters on your own. This leaves us with:

$$\text{molarity} = \frac{0.415 \text{ moles } MgSO_4}{0.2511 \text{ L solution}} = \underline{1.65 \text{ M}}$$

For molality, we must take the number of moles of solute and divide by the number of kilograms of solvent. The problem states that 250.1 g of water were used, which is the same as 0.2501 kg of water. The molality, then, is:

$$\text{molality} = \frac{\text{moles of solute}}{\text{kg of solvent}}$$

$$\text{molality} = \frac{0.415 \text{ moles } MgSO_4}{0.2501 \text{ kg water}} = \underline{1.66 \text{ m}}$$

For mole fraction, we must take the number of moles of solute and divide by the total number of moles of both solute and solvent. To be able to do that, we need to calculate the number of moles of water used:

$$\frac{250.1 \text{ g} \cancel{H_2O}}{1} \times \frac{1 \text{ mole } H_2O}{18.0 \text{ g} \cancel{H_2O}} = 13.9 \text{ moles } H_2O$$

Now we can calculate mole fraction.

$$X_{solute} = \frac{\text{moles solute}}{\text{total moles}}$$

$$X_{NaCl} = \frac{0.415 \text{ moles } MgSO_4}{13.9 \text{ moles } H_2O + 0.415 \text{ moles} MgSO_4}$$

$$X_{NaCl} = \frac{0.415 \text{ moles } MgSO_4}{14.3 \text{ moles total}} = \underline{0.0290}$$

5.2 Since calcium chloride is an ionic compound, you can determine its chemical formula. Calcium is in group 2A and therefore has a charge of +2 in ionic compounds. Chlorine is in group 7A and therefore has a charge of -1 in ionic compounds. Since the charges are not equal, we switch them and make them subscripts, so calcium chloride is $CaCl_2$. If the molality is 0.50 m, then we know that the solution was made by adding 0.500 moles of calcium chloride and 1.00 kg water. Mole fraction is rather easy to calculate, then. We already have the number of moles of solute, and we can calculate the number of moles of solvent:

$$\frac{1.00 \times 10^3 \text{ g} \cancel{\text{ water}}}{1} \times \frac{1 \text{ mole water}}{18.0 \text{ g} \cancel{\text{ water}}} = 55.6 \text{ moles water}$$

That's all we need to get the mole fractions:

$$X_{CaCl_2} = \frac{\text{moles } CaCl_2}{\text{total moles}} = \frac{0.500 \text{ moles } CaCl_2}{0.500 \text{ moles } CaCl_2 + 55.6 \text{ moles water}} = \frac{0.500}{56.1} = \underline{0.00891}$$

$$X_{water} = \frac{\text{moles water}}{\text{total moles}} = \frac{55.6 \text{ moles water}}{0.500 \text{ moles } CaCl_2 + 55.6 \text{ moles water}} = \frac{55.6}{56.1} = \underline{0.991}$$

Once again, the mole fractions add up to one, as they should.

How do we calculate molarity? Well, for that we need the volume of the solution. We have density, so we can get the volume if we have the mass. We can get the mass, because we know that the solution was made by adding 0.50 moles $CaCl_2$ and 1.0 kg of water. Thus, all we need to do is convert moles of $CaCl_2$ into grams:

$$\frac{0.500 \cancel{\text{ moles } CaCl_2}}{1} \times \frac{111.1 \text{ g } CaCl_2}{1 \cancel{\text{ mole } CaCl_2}} = 55.6 \text{ g } CaCl_2$$

Now we know the total mass of the solution:

$$\text{total mass} = 55.6 \text{ g} + 1.00 \times 10^3 \text{ g} = 1.06 \times 10^3 \text{ g}$$

Now that we have the mass of the solution, we can use the density of the solution to determine the volume:

$$\frac{1.06 \times 10^3 \text{ g}}{1} \times \frac{1 \text{ mL}}{1.07 \text{ g}} = 991 \text{ mL} = 0.991 \text{ L}$$

With this, we can finally calculate the molarity:

$$\text{molarity} = \frac{\text{\# moles solute}}{\text{\# Liters solution}} = \frac{0.500 \text{ moles CaCl}_2}{0.991 \text{ L}} = \underline{0.505 \text{ M}}$$

5.3 Molarity will vary with temperature. Molality relates moles and kilograms, and mole fraction deals entirely with moles. Those quantities do not change with changing temperature. Molarity, however, is moles divided by liters. The volume of a solution will change with changing temperature, so molarity varies with temperature.

5.4 You can reason this out if you think about entropy. When a gas gets dissolved, it is going from the gaseous phase to the liquid phase. This represents a *decrease* in entropy. Thus, ΔS is negative. Based on Equation (5.2), then, this tends to make ΔG positive. In order for something to be dissolved, however, the ΔG must be negative. The only way this can happen is with a negative ΔH. Thus, the process must be exothermic.

5.5 Since we already determined that gases must dissolve exothermically, we know that the solute-solvent attraction is greater than the solute-solute attraction and the solvent-solvent attraction.

5.6 Both (A) and (D) represent such situations. Remember, a dramatic change in the solubility curve occurs when the chemical nature of the solute changes. That is clearly the case for (A). This is also the case with (D). The slope of the solubility curve doesn't change sign for (D), but it clearly changes dramatically. A change in chemical nature may not cause the solubility curve to turn around and go the other way. It will, however, make a distinct difference in slope.

5.7 Curve (C) represents such a solid. Solids that dissolve exothermically decrease in solubility with increasing temperature. Curve (C) is the only one that demonstrates such behavior over the entire temperature range.

5.8 Curve (A) is such a solid. When a solid dissolves endothermically, its solubility increases with increasing temperature. When it dissolves exothermically, the solubility decreases with increasing temperature. Curve (A) exhibits both behaviors.

5.9 The liquid that boils at the lowest temperature is the one that boils away, but which is it? In module #4 you learned that when all other van der Waals forces are equivalent, the heavier molecule has the larger van der Waals forces, which will lead to the higher boiling point. These molecules are both nonpolar, but decane is much heavier. Thus, it boils at the higher temperature. This means pentane will boil off first, so pentane is boiled away.

5.10 Liquid (B) has the strongest van der Waals forces between it and the paper. The stronger the forces between the chemical and the paper, the slower the chemical travels.

REVIEW QUESTIONS

1. Which concentration unit (molarity, molality, mole fraction) varies with temperature?

2. The attractive forces between a solute and a solvent are weaker than those between the solute and other solute molecules as well as those between the solvent and other solvent molecules. Does the solute dissolve exothermically or endothermically?

3. The attractive forces between a solute and a solvent are weaker than those between the solute and other solute molecules as well as those between the solvent and other solvent molecules. At the same time, the system would decrease in entropy if the solute dissolves. Will the solute dissolve? If so, will it most likely occur at high or low temperatures, or is temperature irrelevant?

4. The attractive forces between a solute and a solvent are stronger than those between the solute and other solute molecules as well as those between the solvent and other solvent molecules. At the same time, the system would decrease in entropy if the solute dissolves. Will the solute dissolve? If so, will it most likely occur at high or low temperatures, or is temperature irrelevant?

5. When an ideal solution forms, is the entropy change positive or negative?

6. What does a dramatic change in the slope of a solubility curve indicate?

7. A solute dissolves exothermically. How does its solubility depend on temperature?

8. A chemist uses paper chromatography to separate two chemicals that have been mixed together. If chemical B travels farther up the paper than chemical A, which chemical had the weakest van der Waals attraction to the paper?

9. A colloid of water and pentane (C_5H_{10}) is distilled. Which will boil off first?

10. A speaker uses a laser pointer to point at things on the slide screen she is using. You cannot see the beam coming from the laser pointer, but you can see the dot that the laser light makes on the screen. An inconsiderate oaf near her begins to smoke. Suddenly, you see the laser beam as it travels from the pointer to the screen. Why?

PRACTICE PROBLEMS

1. What is the molarity of a solution that contains 50.0 grams of potassium hydroxide in 2.50 x 10^2 mL of solution?

2. What is the molality and mole fraction of glucose ($C_6H_{12}O_6$) in a solution made by mixing 25.0 grams of glucose with 512 grams of water?

3. A solution of potassium sulfide in water has a concentration of 5.5 M. If the density of the solution is 1.42 g/mL, what is the molality of potassium sulfide and the mole fraction of both solute and solvent?

4. The mole fraction of potassium carbonate in water is 0.0832. What is the molality of the solution?

Questions 5-8 refer to the following diagram, which is the solubility curve of the new wonder chemical, Wileium:

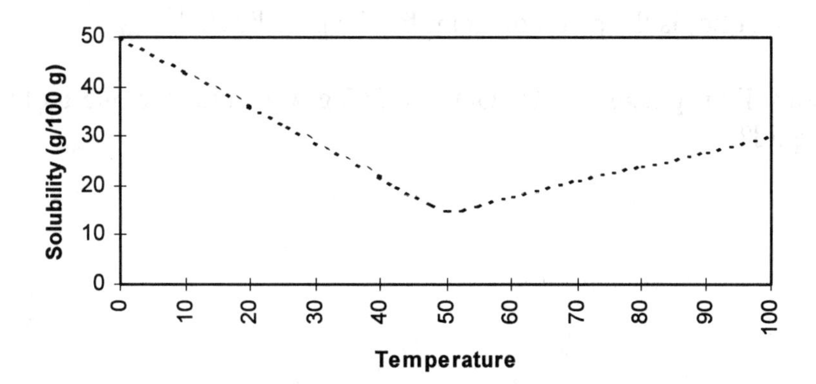

5. At what temperature does Wileium's chemical nature change in solution?

6. Over what temperature range does Wileium dissolve endothermically?

7. Over what temperature range is it possible that when Wileium dissolves there is a decrease in the disorder of the system?

8. Over what temperature range are the Wileium-water van der Waals forces stronger than the water-water and Wileium-Wileium van der Waals forces?

Questions 9 and 10 refer to the following figure, which compares two phase diagrams. One is for a liquid, the other is for that same liquid with a solid dissolved in it.

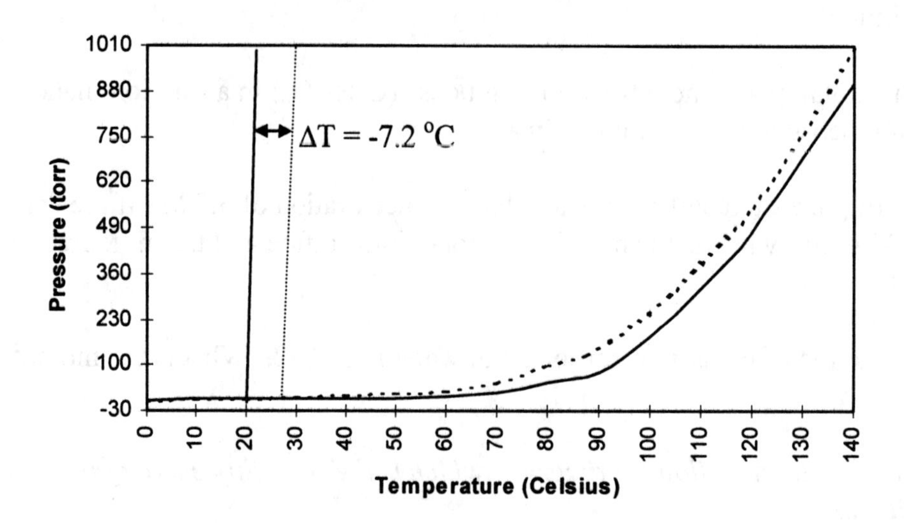

9. Which (dotted or solid) is the phase diagram for the pure liquid?

10. The dissolved solid is glucose ($C_6H_{12}O_6$), and 363 g were added to 545 g of the liquid. What is the K_f of the liquid?

Module #6: Solutions and Equilibrium

Introduction

In the previous module, I started an in-depth discussion on solutions. Along the way, I touched on viewing the process of a solute dissolving in a solvent as an equilibrium. I want to spend a great deal more time on that idea in this module. You see, using the concepts of equilibrium, we can analyze and understand solutions on a much deeper level. To do this, however, you first need to review a few things you should have learned in your first year chemistry course, and then you need to build on those concepts.

A Little Bit of Review

When you learned about chemical equilibrium in your first-year course, you should have learned about the **equilibrium constant**. The best way to remind you about the equilibrium constant is to show you the chemist's "standard reaction."

$$aA + bB \rightleftarrows cC + dD$$

This reaction is simply a general expression in which A and B represent substances that are the reactants of a chemical equilibrium, and C and D are general representations for the products. The lower-case letters (a, b, c, and d) represent the stoichiometric coefficients that multiply A, B, C, and D. For this general equilibrium, the equilibrium constant is:

$$K_{eq} = \frac{[C]^c[D]^d}{[A]^a[B]^b} \tag{6.1}$$

In the equation, "K_{eq}" is the equilibrium constant, and the square brackets are an abbreviation for "the concentration of." Thus "[A]" means "the concentration of substance A." To calculate the equilibrium constant for any equilibrium, then, we take the concentrations of the products, raise them to the power of their stoichiometric coefficients, and multiply them. We then divide that by the concentrations of the reactants, raised to their stoichiometric coefficients.

The equilibrium constant is a very important thing to know when studying an equilibrium, because for a given temperature, its value is always the same. Regardless of the *individual* concentrations of the reactants and products, as long as the reaction has achieved equilibrium, the value of the equilibrium constant is always the same. Make sure you remember how to calculate the equilibrium constant by studying the example below and performing the "on your own" problem that follows:

EXAMPLE 6.1

A chemist studies the following reaction:

$$\mathbf{N_2 (g) + 3H_2 (g) \rightleftarrows 2NH_3 (g)}$$

At equilibrium, $[N_2\ (g)] = 4.26$ M, $[H_2\ (g)] = 2.09$ M, and $[NH_3\ (g)] = 1.53$ M. What is the equilibrium constant? If the chemist then doubles the concentration of N_2 (g), what will be the value of the equilibrium constant?

If we have the equilibrium concentrations of each substance in the equilibrium, all we have to do is use Equation (6.1) to determine the equilibrium constant.

$$K_{eq} = \frac{[C]^c[D]^d}{[A]^a[B]^b}$$

$$K_{eq} = \frac{[NH_3\ (g)]^2}{[N_2\ (g)]\cdot[H_2\ (g)]^3} = \frac{(1.53\ M)^2}{(4.26\ M)\cdot(2.09\ M)^3} = \underline{0.0602\ \frac{1}{M^2}}$$

When the chemist doubles the concentration of N_2 (g), the equilibrium constant will still be $\underline{0.0602\ 1/M^2}$, because the equilibrium constant stays constant regardless of the *individual* concentrations of the substances in the equation. As Le Chatelier's principle states, however, the equilibrium will shift to the right, using up hydrogen gas and making ammonia gas in order to preserve equilibrium. If the chemist measures the new equilibrium concentrations of each reactant and then calculates the equilibrium constant, it will be exactly 0.0602 $1/M^2$ again.

ON YOUR OWN

6.1 At 2000 °C, the equilibrium constant for the following reaction:

$$N_2\ (g) + O_2\ (g) \rightleftarrows 2NO\ (g)$$

is 4.1×10^{-4}. At equilibrium, a chemist measures the concentrations of N_2 (g) and O_2 (g) to be 0.036 M and 0.0089 M, respectively. What is the concentration of NO (g)?

Before you leave this section, notice that the equilibrium constant in the example has units of "$1/M^2$" while the one in the "on your own" problem above had no units. You should recall that this is a general property of the equilibrium constant. Since the values of the exponents in Equation (6.1) depend on the stoichiometric coefficients in the specific chemical equation with which you are working, the units that result from the equation will depend on the values of those coefficients. Thus, it should not surprise you when you see different units on the equilibrium constants of different chemical reactions.

There are two more things you need to recall before I plow ahead. First, remember that the equilibrium constant is only constant for a given temperature. If the temperature changes, the equilibrium constant will as well. Second, when you use Equation (6.1) to calculate the equilibrium constant of a reaction, there are certain substances which you ignore. When there are any aqueous or gaseous substances in a reaction, you always ignore any solids or liquids in the reaction. Thus, for the following chemical equilibrium:

$$CH_3COOH\ (aq) + H_2O\ (l) \rightleftarrows H_3O^+\ (aq) + CH_3COO^-\ (aq)$$

The equilibrium constant is given by:

$$K_{eq} = \frac{[CH_3COO^-][H_3O^+]}{[CH_3COOH]}$$

Notice that even though water appears as a reactant in the equation, we ignore it in Equation (6.1) because it is a liquid in the presence of aqueous substances. Had there been a solid substance in the equation, we would have ignored that as well.

The Equilibrium Constant and Gibb's Free Energy

Until now, you have probably been told that when the ΔG for a chemical reaction is negative, the reaction will happen spontaneously. When ΔG is positive, however, the reaction will not proceed. Indeed, even in the previous module, I told you that when the ΔG for the process of dissolving a solute is negative, the solute will dissolve. If the ΔG is positive, however, the solute will not dissolve. Well, it turns out that all of that is a lie.

The value of ΔG can, indeed, be used to determine whether or not a chemical process will proceed, but using it properly is a lot more complex than what you have been told up to this point. It turns out that ΔG and the equilibrium constant are related to one another. *That's* how you can really use ΔG to determine whether or not a chemical reaction will proceed. The way in which these two quantities relate is given by Equation (6.2).

$$\Delta G = -R \cdot T \cdot \ln (K) \qquad (6.2)$$

In this equation, "ΔG" is the Gibb's free energy change for the reaction in question, "R" is the ideal gas constant, "T" is the temperature, and "K" is the equilibrium constant. Remember from algebra that "ln" refers to the natural logarithm. You will have to use the natural logarithm in this and subsequent modules, so if you are not familiar with it, go back to your algebra books to learn how to use it.

If you look at Equation (6.2) in the right way, you can see why I lied to you before about ΔG. If I rearrange Equation (6.2) to solve for K, I get:

$$K = e^{-\Delta G / RT} \qquad (6.3)$$

Now look at this equation. If ΔG is negative, the exponent for "e" will be positive. When "e" is raised to a positive power, the result is a big number. Thus, when ΔG is negative, the equilibrium constant will be large. On the other hand, when ΔG is positive, "e" will be raised to the negative power and will result in an equilibrium constant that is less than one.

What did you learn in your previous chemistry course about the relevance of value of the equilibrium constant? You should have learned that when the equilibrium constant is greater than one, the equilibrium is heavily weighted towards the products. In other words, when the equilibrium constant is greater than one, most of the reactants react to form products. On the other hand, when the equilibrium constant is less than one, the equilibrium is heavily weighted to the reactants, and very few products are formed. So, when ΔG is negative, the equilibrium constant is large and most of the reactants will react to form products. When ΔG is positive, however, the equilibrium constant is less than one and very few products are formed.

Now, consider what happens when ΔG is very large and negative. When that happens, the equilibrium constant is very large. What is the relevance of a very large equilibrium constant? When K is very large, we can assume that there are essentially no reactants left; thus, the equilibrium is not really considered an equilibrium. Instead, it is considered a complete reaction, in which all reactants react to form products. When ΔG is very large and positive, however, the equilibrium constant is an extremely small number. When K is extremely small, chemists say that there are essentially no products. In other words, when the equilibrium constant is extremely small, the reaction essentially does not happen.

You should see, then, that the general rule of thumb you learned regarding the sign of ΔG is really more of an approximation than a lie. When ΔG is large and positive, the reaction does not proceed at all, due to the very small value of K. When ΔG is large and negative, the reaction proceeds completely, with as many products formed as is allowed by the limiting reactant. This is a result of the equilibrium constant being so large. When ΔG is small, however, the result is an equilibrium. When ΔG is small and negative, the equilibrium is weighted to the products, because K is greater than one. When ΔG is small and positive, the result is an equilibrium that is weighted to the reactants, because the equilibrium constant is less than one.

Hopefully, an example will clear up any confusion that still exists in your mind.

EXAMPLE 6.2

The ΔG's for 4 reactions at 25 °C are given below. In each case, determine the equilibrium constant and state whether the reaction is an equilibrium, is a complete reaction, or does not occur.

a. ΔG = -1203 J/mole b. ΔG = 1192 J/mole c. ΔG = -167 kJ/mole d. ΔG = 321 kJ/mole

We can solve for the equilibrium constant in each case by using Equation (6.3). Now please remember that Equation (6.3) is just a rearranged version of Equation (6.2). You need not memorize both equations, as one comes from the other. The other thing you need to realize about using Equations (6.2) and (6.3) is the value of "R" to use. If you remember the ideal gas constant from your first year chemistry course, you probably remember the value 0.0821 $\frac{\text{liter}\cdot\text{atm}}{\text{mole}\cdot\text{K}}$. Although that is the value for the ideal gas constant, those units are not all that useful for our purposes here. A more useful value for the ideal gas constant is 8.314 $\frac{\text{J}}{\text{mole}\cdot\text{K}}$. Of course, both values are really the same; they just have different units. As you will see, the units in the latter value will be most useful to us. Because of these units for "R," the temperature must be expressed in Kelvin.

a. $K = e^{-\Delta G/RT} = e^{-(-1203 \frac{\cancel{J}}{\cancel{mole}})/(8.314 \frac{\cancel{J}}{\cancel{mole}\cdot\cancel{K}}\cdot 298\cancel{K})} = \underline{1.63}$

Since the equilibrium constant is just a little bit greater than one, this reaction is an <u>equilibrium that is weighted towards the products</u>.

b. $K = e^{-\Delta G/RT} = e^{-(1192 \frac{\cancel{J}}{\cancel{mole}})/(8.314 \frac{\cancel{J}}{\cancel{mole}\cdot\cancel{K}}\cdot 298\cancel{K})} = \underline{0.618}$

Since the equilibrium constant is just a little bit less than one, this reaction is an <u>equilibrium that is weighted towards the reactants</u>.

c. The ΔG is given in kJ/mole. It must first be converted to J/mole in order to be consistent with our units.

$K = e^{-\Delta G/RT} = e^{-(-1.67\times 10^{5} \frac{\cancel{J}}{\cancel{mole}})/(8.314 \frac{\cancel{J}}{\cancel{mole}\cdot\cancel{K}}\cdot 298\cancel{K})} = \underline{1.88\times 10^{29}}$

The equilibrium constant is *huge*, so this is <u>a complete reaction</u>, with essentially all reactants forming products.

d. The ΔG is given in kJ/mole. It must first be converted to J/mole in order to be consistent with our units.

$K = e^{-\Delta G/RT} = e^{-(3.21\times 10^{5} \frac{\cancel{J}}{\cancel{mole}})/(8.314 \frac{\cancel{J}}{\cancel{mole}\cdot\cancel{K}}\cdot 298\cancel{K})} = \underline{5.39\times 10^{-57}}$

This is a tiny, tiny equilibrium constant, which means that <u>the reaction doesn't really occur</u>.

Notice the results of this example. In part (a), the ΔG is negative but relatively small. The result was an equilibrium constant just a bit larger than one. This indicates that the reaction is, indeed, an equilibrium, with both products and reactants in existence. Based on the value of the equilibrium constant, however, there are more products than reactants. In part (b), the ΔG value is still relatively small, but in this case it is positive, resulting in an equilibrium constant slightly less than one. This means that the reaction is an equilibrium, with more reactants than products. In part (c), the ΔG is large and negative, resulting in a *huge* equilibrium constant. When the equilibrium constant is so large, the reaction is not even considered an equilibrium. the reaction is assumed to run to completion, using up all reactants allowed by the limiting reactant. Finally, part (d) has a large, positive ΔG. This results in an equilibrium constant so small, we assume that no products are formed.

ON YOUR OWN

6.2 What is the ΔG for a reaction with an equilibrium constant of 3.4×10^3 at 25 °C?

Now you hopefully know how to analyze chemical equations in a more detailed way. You see, one of the concepts at the heart of chemistry is that *every* chemical reaction is, in fact, an equilibrium. When the equilibrium constant is really large, however, we can *treat the reaction like it runs to completion, even though that's not really the case*. In the same way, there really aren't any reactions that do not occur at all. However, when the equilibrium constant is very small, we can *treat the reaction like it never occurs, even though that's really not the case*. To determine the size of the equilibrium constant, we usually rely on ΔG, because it is relatively easy to measure and can be related to the equilibrium constant through a reasonably simple equation.

Solubility Equilibria

So far, I have discussed solutes as only either soluble or insoluble. Vegetable oil, for example, is insoluble in water, whereas alcohol is soluble in water. In fact, many solutes are "slightly soluble" or "mostly insoluble" in a variety of solvents. What do I mean by these terms? Well, suppose I wanted to dissolve barium sulfate ($BaSO_4$) in water. If I took a sample of this ionic solid and put it in water, it would look like nothing dissolved. However, if I then filtered the water and analyzed it, I would find trace amounts of both the barium ion and the sulfate ion in the water. Thus, *some* barium sulfate dissolved, just not very much.

How much barium sulfate dissolved? Well, when I added the sample of barium sulfate to the water, the following equilibrium was established:

$$BaSO_4\,(s) \rightleftarrows Ba^{2+}\,(aq) + SO_4^{2-}\,(aq) \qquad (6.4)$$

This equilibrium is governed by the following equilibrium constant:

$$K = [Ba^{2+}(aq)] \cdot [SO_4^{2-}(aq)] \quad (6.5)$$

Remember, we ignore solids in the equilibrium constant when aqueous substances are present. That's why there is no $BaSO_4$ term in Equation (6.5).

It turns out that this equilibrium constant is given a special name. The equilibrium constant of an equilibrium involving solubility is called a **solubility product**, and it is usually given an "sp" subscript. Thus, Equation (6.5) is more properly written:

$$K_{sp} = [Ba^{2+}(aq)] \cdot [SO_4^{2-}(aq)] \quad (6.6)$$

Solubility products are relatively easy to measure, so they have been tabulated for a great many solutes. Now remember, the equilibrium constant (a solubility product is really just an equilibrium constant) is constant only for a given temperature, so the solubility product is temperature dependent. Typically, solubility products are measured for room temperature, since the majority of chemistry is done under room temperature conditions. The solubility product for barium sulfate at 25 °C, for example, is 1.08×10^{-10} M^2. Since the units on the solubility product vary with the stoichiometric coefficients involved, they are rarely listed with the solubility product. Thus, you will usually see the solubility product of barium sulfate listed simply as 1.08×10^{-10}.

What does this solubility product tell us? Well, since it is really just the equilibrium constant for the equilibrium given by Equation (6.4), and since its value is significantly smaller than one, we can say that the equilibrium is heavily weighted towards the reactants. Indeed, the equilibrium constant is so small that for all practical purposes, we could say that nothing really happens at all. Nevertheless, something *does* happen, because once a solution of $BaSO_4$ (s) and water is filtered, trace amounts of Ba^{2+} (aq) and SO_4^{2-} (aq) can be found in the water. It turns out that using the solubility product, we can actually determine what the concentration of each of these ions will be. Study the following examples to see how this is done.

EXAMPLE 6.3

What concentration of Ba^{2+} (aq) and SO_4^{2-} (aq) ions will form in a saturated solution of $BaSO_4$ (s) and water? ($K_{sp} = 1.08 \times 10^{-10}$)

To solve this problem, we have to think about the equilibrium:

$$BaSO_4(s) \rightleftarrows Ba^{2+}(aq) + SO_4^{2-}(aq)$$

The solubility product is given by:

$$K_{sp} = [Ba^{2+}(aq)] \cdot [SO_4^{2-}(aq)]$$

Since we know the value for the solubility product, we are part of the way home as far as solving for the concentration of ions present:

$$1.08 \times 10^{-10} = [Ba^{2+} (aq)]\cdot[SO_4^{2-} (aq)]$$

We can't solve this equation, however, because we don't know either concentration. Thus, we have two unknowns in the equation.

What can we do? Well, we can think about the equilibrium itself. Before the solid barium sulfate is added to the water, there is no Ba^{2+} (aq) or SO_4^{2-} (aq) in it. Thus, ANY Ba^{2+} (aq) or SO_4^{2-} (aq) that is there later will have come from the barium sulfate dissolving according to the equilibrium equation given above. Now we know that some of the barium sulfate will dissolve, we just don't know how much that will be. Well, let's say that "X" moles of barium sulfate dissolves. If "X" moles of barium sulfate dissolves, how many moles of Ba^{2+} (aq) will form? According to the stoichiometry of the reaction, for every mole of barium sulfate that dissolves, one mole of Ba^{2+} (aq) will form. Thus, if "X" moles of $BaSO_4$ (s) dissolve, "X" moles of Ba^{2+} (aq) will form. Using the same reasoning, "X" moles of SO_4^{2-} (aq) will also form.

How does this help? Well, since all of this happens in the same container, a certain number of moles will result in a certain concentration. Thus, we can use the term moles and molarity interchangeably. This means that once the barium sulfate dissolves, the concentration of Ba^{2+} (aq) will be "X," and the concentration of the SO_4^{2-} (aq) will also be "X." If we make that substitution in our equation, we get:

$$1.08 \times 10^{-10} = X\cdot X$$

$$1.08 \times 10^{-10} = X^2$$

$$1.04 \times 10^{-5} = X$$

Since "X" represents the concentration of the Ba^{2+} (aq) ions and the concentration of the SO_4^{2-} (aq) ions, then $[Ba^{2+} (aq)] = 1.04 \times 10^{-5}$ M, and $[SO_4^{2-} (aq)] = 1.04 \times 10^{-5}$ M.

Now before I move on to another example, let's make sure you see what I did here. I used the stoichiometry of the equilibrium to get an expression for the concentration of each ion in terms of "X." I then used that expression in the equation for the solubility product to solve for "X," and related "X" back to the concentration of the ions. The point is, given just the solubility product of a solute, you can determine the concentration of the ions in a saturated solution of the solute. Try studying the next example, which adds a bit of a twist to the situation.

The solubility product for calcium fluoride (CaF_2) at 25 °C is 3.9×10^{-11}. What is the concentration of calcium and fluoride ions in a saturated solution of CaF_2?

To solve this problem, we first need to think about the equilibrium which governs how CaF_2 dissolves:

$$CaF_2\ (s) \rightleftarrows Ca^{2+}\ (aq) + 2F^-\ (aq)$$

As an aside, if you don't understand why CaF_2 (s) dissolves into one Ca^{2+} (aq) and two F^- (aq) ions, you need to go back to your first year course and re-study how ionic solids dissolve. Now that we have this equilibrium, we can determine the equation for the equilibrium constant, which is the solubility product for calcium fluoride:

$$K_{sp} = [Ca^{2+}\ (aq)]\cdot[F^-\ (aq)]^2$$

Since we have the value for the solubility product, we can go ahead and stick it in there:

$$3.9 \times 10^{-11} = [Ca^{2+}\ (aq)]\cdot[F^-\ (aq)]^2$$

To solve this equation, though, we need to reduce the number of variables. Like the last example, we know that the only way calcium and fluoride ions get into the water is through the process of calcium fluoride dissolving. Thus, the concentrations of these two ions are related by the stoichiometry of the equilibrium. According to the equilibrium, if "X" moles of calcium fluoride dissolves, "X" moles of calcium ions will form, while "2X" moles of fluoride ions will form. After all, the equilibrium states that for every one mole of CaF_2 that dissolves, two moles of F^- (aq) will form. Thus, if "X" moles of CaF_2 dissolve, then "2X" moles of F^- (aq) will form.

Taking those expressions for the amount of calcium and fluoride ions, the equation becomes:

$$3.9 \times 10^{-11} = (X)\cdot(2X)^2$$

$$3.9 \times 10^{-11} = 4\cdot X^3$$

$$2.1 \times 10^{-4} = X$$

According to what I said earlier, the concentration of calcium ions will be equal to "X," so $[Ca^{2+}\ (aq)] = 2.1 \times 10^{-4}$ M. The concentration of fluoride ions, however, is "2X," so $[F^-\ (aq)] = 4.2 \times 10^{-4}$ M.

Do you see what's going on here? We use the equilibrium to develop a relationship between the concentrations of the ions so as to reduce the number of variables in our solubility product equation. This relationship comes directly from the stoichiometry of the equilibrium. Once we solve for "X," we then relate that back to the concentrations of the ions that we are interested in. I will do one more to cement this into your mind.

The solubility product of barium phosphate ($Ba_3(PO_4)_2$) at 25 °C is 1.3×10^{-29}. What is the concentration of each ion in a saturated solution of barium phosphate?

This solute dissolves according to the equation:

$$Ba_3(PO_4)_2\ (s) \rightleftarrows 3Ba^{2+}\ (aq) + 2PO_4^{3-}\ (aq)$$

The solubility product equation, then, is:

$$1.3 \times 10^{-29} = [Ba^{2+}\ (aq)]^3 \cdot [PO_4^{3-}\ (aq)]^2$$

The stoichiometry tells us that if "X" barium phosphate dissolves, "3X" Ba^{2+} (aq) and "2X" PO_4^{3-} (aq) will form. This means the equation reduces to:

$$1.3 \times 10^{-29} = (3X)^3 \cdot (2X)^2$$

$$1.3 \times 10^{-29} = 108 \cdot X^5$$

$$6.5 \times 10^{-7} = X$$

Since the concentration of Ba^{2+} (aq) was determined to be "3X," $[Ba^{2+}\ (aq)] = 2.0 \times 10^{-6}$ M.
Since the concentration of PO_4^{3-} (aq) was determined to be "2X," $[PO_4^{3-}\ (aq)] = 1.3 \times 10^{-6}$ M.

These solubility product problems are the first in a long line of equilibrium problems that use the same kind of reasoning. That's why I went through three example problems to introduce you to the concept. Please be sure that you have this kind of problem-solving technique down pat before you move on!

ON YOUR OWN

6.3 At 25 °C, the solubility product of $PbSO_4$ (s) is 1.8×10^{-8}. What is the concentration of each ion in a saturated solution of lead sulfate?

6.4 At 25 °C, the solubility product of Bi_2S_3 (s) is 7.3×10^{-91}. What is the concentration of each ion in a saturated solution of bismuth sulfide?

6.5 The concentration of lead ions in a saturated solution of PbI_2 is 1.3×10^{-3} M. What is the K_{sp} of lead iodide?

Before you leave this section, it is important to note when the solubility product plays a role in the dissolving of a solute and when it does not. Suppose, for example, you are dissolving lead sulfate into water a tiny, tiny amount at a time. The solubility product for lead sulfate is 1.8×10^{-8}. As you determined in "on your own" problem 6.3, this means that a saturated solution of lead sulfate has lead ions at a concentration of 4.2×10^{-4} M as well as sulfate ions at a concentration of 4.2×10^{-4} M. What if you added so little lead sulfate that if *all* of it dissolved, the concentration of each ion would be less than 4.2×10^{-4} M? What would happen then? Well, under those conditions, the solution would not be saturated, so *all of the lead sulfate would dissolve, and no equilibrium would be established.*

This is an important point. Dissolving a solute into a solvent cannot be thought of as an equilibrium until the solution becomes saturated. Only then will an equilibrium be set up between the solid solute and its aqueous ions. Until then, any solute that is put in solution will dissolve. The solubility product, then, is only applicable to a solute when the solution into which it is dissolving becomes saturated. This is an important thing to remember as you read the next section.

The Common Ion Effect

You might think that the usefulness of a solute's solubility product is rather limited. After all, it only applies to saturated solutions. Well, that's true, but there is more than one way to saturate a solution. To see what I mean, perform the following experiment.

EXPERIMENT 6.1

The Common Ion Effect

Supplies:

From the laboratory equipment set:

- Filter paper
- Funnel

From around the house:

- Stove
- Saucepan
- Baking soda
- Table salt
- Two glasses
- 1/3 measuring cup
- Spoons for stirring

To start the experiment, dump a large amount of baking soda into one of the glasses and a large amount of table salt into the other. Fill them both with about a cup of water. Take the glass with baking soda and water in it and stir it vigorously, making a saturated solution of baking soda. Remember, the solution is saturated when no matter how much you stir, there is still plenty of solid solute left in the glass. Add as much baking soda as needed to fill this requirement. Filter this solution like you did in Experiment 5.1, collecting the filtered liquid in the 1/3 measuring cup. Continue filtering until you have 1/3 of a cup of filtered solution. Pour the filtered solution into the saucepan and put it on the stove burner to get it boiling.

While that solution is boiling, rinse out the measuring cup. Stir the table salt solution vigorously, making another saturated solution. Filter this solution into the 1/3 measuring cup as well (you can use the same filter paper). While you are filtering, keep an eye on the boiling solution. Your goal is to boil all of the water away. When it gets close to the end, however, the solid will begin to spatter. As soon as you see solid spattering, take the pan off the heat. Almost all of the water should be gone. If a little is left, don't worry. That shouldn't affect the results of the experiment. Now you need to cool the pan down. You can do this by *gently* running cool water over the *outside* of the pan. *Make sure that no water gets into the pan!* You want to cool the saucepan, not re-dissolve the baking soda that is caked onto the pan!

Once you have 1/3 cup of filtered salt water solution, add it directly into the pan. Use a spoon to try and mix up the baking soda that is caked onto the pan and the saltwater solution. Scrape the baking soda off of the pan as you stir, trying as hard as you can to dissolve the baking soda. Can you dissolve the baking soda into the saltwater solution?

Optional Extension: There are certain water-softening salts that are made of KCl rather than NaCl. If you would like, you can repeat the experiment using one of those salts rather than table salt. The result will be different, as discussed below.

You should not have been able to dissolve anywhere near all of the baking soda in the pan. Why not? All of that baking soda dissolved in 1/3 of a cup of water. Why won't it dissolve in 1/3 of a cup of a saturated saltwater solution? The answer to that question is **the common ion effect.**

When you made a saturated solution of baking soda, you dissolved just as much baking soda as you possibly could. This was limited by baking soda's solubility product. Baking soda (whose proper name is sodium bicarbonate) dissolves according to the following equilibrium:

$$\underset{\text{(baking soda)}}{NaHCO_3\ (s)} \rightleftarrows Na^+\ (aq) + HCO_3^-\ (aq)$$

The solubility is limited by the solubility product equation:

$$K_{sp} = [Na^+\ (aq)]\cdot[HCO_3^-\ (aq)]$$

When the product of the concentrations of the sodium and bicarbonate ions reached the value of the solubility product, the solution became saturated. When you then filtered the solution, you had a solution that had just as much sodium ions and bicarbonate ions as the solubility product would allow.

When you evaporated the water, all of the sodium ions and bicarbonate ions joined up again, making baking soda. When you added the saturated sodium chloride solution and tried to dissolve the baking soda in it, a problem arose. Since table salt has a chemical formula of NaCl, it splits up into *sodium ions* and chloride ions when it dissolves. Thus, when you tried to dissolve the baking soda into the saltwater, you were trying to dissolve sodium bicarbonate into a solution that already had sodium ions in it. As soon as a little bit of sodium bicarbonate dissolved, the product of the sodium ion concentration times the bicarbonate ion concentration (the solubility product equation) was already quite large, because the sodium ion concentration was already large thanks to the salt that was dissolved in the solution. As a result, even though the concentration of bicarbonate ions was small, the *product* of the sodium concentration and the bicarbonate concentration was large enough to equal the solubility product of the sodium bicarbonate, keeping any more from dissolving.

The conclusion is rather simple. When you are dissolving a solute into a solution, the solute's solubility will be limited if that solution has a common ion. That's what we call the **common ion effect**.

Common ion effect - A change in a solute's solubility when put in the presence of a common ion

In your experiment, the common ion was sodium. Since the saltwater solution had a lot of sodium ions in it already, *any solute that contained sodium ions* would have a limited solubility in that solution. Since baking soda contains sodium ions, its solubility was limited, and you could not dissolve nearly as much baking soda into the saltwater solution as you had been able to dissolve in water.

It is important to realize that baking soda's solubility would not have been limited had it been dissolved in a saturated solution of potassium chloride, as suggested in the optional extension to the experiment. After all, the solubility of a solute depends only on the solubility product equation. In a saturated solution of potassium chloride, there are no bicarbonate ions or sodium ions. Thus, as far as baking soda is concerned, a saturated solution of potassium chloride is much the same as water. Since there is initially no sodium or bicarbonate ions in the solution, the baking soda would dissolve until the product of its ion concentrations is equal to K_{sp}, and the result would be the same solubility as if the baking soda were dissolving in pure water.

Even though the common ion effect is a relatively simple concept, chemists have a knack for taking the simple and making it as complicated as possible. This is no exception. It is not enough for you to understand that a common ion will limit the solubility of a solute. Instead, you need to be able to *calculate how much it is limited.* Study the following example to see how this is done.

EXAMPLE 6.4

In general, most ionic compounds that have the nitrate (NO_3^-) ion are very soluble in water. A student dissolves enough $Pb(NO_3)_2$ to make a lead nitrate solution whose concentration is 0.10 M. All of the lead nitrate dissolves. If the student then tries to dissolve PbF_2 (K_{sp} = 3.7 x 10^{-8}) in that solution, what will be the resulting concentration of fluoride ions?

The student starts with a solution that is 0.10 M $Pb(NO_3)_2$. This solute dissolves according to the following reaction:

$$Pb(NO_3)_2\ (s) \rightarrow Pb^{2+}\ (aq) + 2NO_3^-\ (aq)$$

Notice that I did not write this equation as an equilibrium. That is because all of the lead nitrate dissolved, so there is no equilibrium established. Since each mole of lead nitrate produces one mole of Pb^{2+} and 2 moles of NO_3^-, $[Pb^{2+}] = 0.10$ M and $[NO_3^-] = 0.20$ M. This means that when the student starts to dissolve PbF_2, the concentration of Pb^{2+} ions is *already* 0.10 M! How does that affect the solubility?

Well, when PbF_2 tries to dissolve, the following equilibrium is set up:

$$PbF_2\ (s) \rightleftarrows Pb^{2+}\ (aq) + 2F^-\ (aq)$$

We know that this is an equilibrium because the K_{sp} given is so small that even a tiny sample of the solute will not completely dissolve. Thus, the concentration of ions is governed by the solubility product equation:

$$K_{sp} = [Pb^{2+}\ (aq)]\cdot[F^-\ (aq)]^2$$

Now, the stoichiometry tells us that when "X" lead fluoride dissolves, "X" Pb^{2+} will form and "2X" F^- will form. However, there is *already* some Pb^{2+} in solution. Thus, the "X" that forms when lead fluoride dissolves will *add to* the 0.10 M that is already there. Thus, when "X" lead fluoride dissolves, the concentration of Pb^{2+} will be "0.10 + X." Since there is no fluoride ion already in solution, the only source of F^- is the lead fluoride dissolving. Thus, when "X" lead fluoride dissolves, the concentration of the fluoride ion will be "2X." This turns our equation into:

$$3.7 \times 10^{-8} = (0.10 + X)\cdot(2X)^2$$

That's a tough equation to solve! If I distribute the terms out, I will have both an X^2 term *and* an X^3 term. How do I solve that?

Well, the answer is, you don't. When faced with an equation like this, you make a simplifying assumption. Since the K_{sp} of lead fluoride is really small, "X" will probably be really small. Suppose "X" is less than 0.005. Suppose it is 0.004 M. If X = 0.004 M, what is 0.10 M + X? Well, 0.10 M + 0.004 M = 0.10 M. Remember, when adding numbers, I must report my answer to the same precision as the *least precise* number in the equation. A concentration of 0.10 M has a precision to the hundredths place, whereas 0.004 M goes out to the thousandths place. The answer, however, can only be reported to the hundredths place, because that's the precision of my least precise number. Thus, *if X is less than 0.005, it will not add to 0.10 because of the rules of precision.* Well, let's be optimistic. Let's suppose that X is less than 0.005. If that's the case, *we can simply drop it, because it does not add to 0.10!* Using this reasoning, the equation becomes:

$$3.7 \times 10^{-8} = (0.10)\cdot(2X)^2$$

That equation is pretty easy to solve!

$$0.00030 = X$$

Were we right? In order for our simplifying assumption to work, "X" had to be less than 0.005. It is, so our assumption was correct! We can therefore go on and finish the problem. Since the fluoride ion concentration is "2X," $\underline{[F^-\ (aq)] = 0.00060\ M}$.

There are two places that students typically get tripped up in these problems. The first is determining how much of the common ion exists before the slightly soluble solute is dissolved. Determining that is a simple stoichiometry exercise. Notice what I did. I just wrote out the equation for dissolving lead nitrate. Given the fact that the student was able to dissolve the solute, all I had to do was use the stoichiometry to turn the concentration of the solute into the concentration of the common ion.

The second place students tend to get tripped up is in the solution to the solubility product equation. For whatever reason, students do not like to make assumptions. Like it or not, however, assumptions are a *big* part of any science. Thus, you need to get used to making them. The important thing to learn about making assumptions, however, is that they need to be justified. In the example, once I made the assumption, I checked the answer to make sure it was consistent with that assumption. In the example, I assumed that the value for "X" had to be less than 0.005. When I solved for "X" under that assumption, it indeed was less than 0.005, which is evidence that the assumption was right. Please get used to making assumptions - but also get used to making sure that those assumptions are justified.

What would I have done if the assumption was not justified? Well, there is a technique called "successive approximations" that you will learn in an upcoming module. I could have used that technique to solve the problem had the assumption not been justified. Also, there actually is a cubic formula (like the quadratic formula) which can be used to solve cubic equations. It is a messy, messy formula, however, so chemists hate to use it. In this module, your assumption will be justified or the equation will have no larger exponent than a square, allowing you to use the quadratic formula to solve it. In the example below, I will do both kinds of problems. Study that example and then perform the "on your own" problems that follow in order to make sure you understand the common ion effect.

EXAMPLE 6.5

In general, ionic solids that have Group 1A metals in them are very soluble in water. A student makes a 0.25 M solution of NaOH. The student then tries to dissolve some cobalt hydroxide, $Co(OH)_2$ ($K_{sp} = 2.0 \times 10^{-16}$), in that solution. What will the resulting cobalt ion concentration be once the solution is saturated?

Since NaOH has the hydroxide ion as does $Co(OH)_2$, this is clearly a common ion effect problem. The first task, then, is to determine the concentration of the common ion.

$$NaOH\ (s) \rightarrow Na^{+}\ (aq) + OH^{-}\ (aq)$$

Remember, since the student can make a 0.25 M solution of NaOH, the NaOH must have dissolved fully. Thus, this is not an equilibrium. According to the equation, one mole of NaOH makes one mole of OH^-, so a 0.25 M NaOH solution has $[OH^-] = 0.25$ M. Now we can set up the solubility product equation for $Co(OH)_2$:

$$Co(OH)_2\ (s) \rightleftarrows Co^{2+}\ (aq) + 2OH^-\ (aq)$$

$$K_{sp} = [Co^{2+}\ (aq)]\cdot[OH^-\ (aq)]^2$$

When "X" amount of cobalt hydroxide dissolves, it will make "X" cobalt ion and "2X" hydroxide ion, according to the equilibrium. Since there is already 0.25 M worth of hydroxide ion, however, this means that the hydroxide ion concentration will be "0.25 + 2X."

$$2.0 \times 10^{-16} = (X)\cdot(0.25 + 2X)^2$$

As was the case in the previous example, this equation would be too hard to solve unless we simplify it. We will try to do so by assuming that "2X" is so small that it will not add to 0.25 by the rules of precision that you learned in your first year chemistry course. That will be true as long as "2X" is less than 0.005. Assuming this is the case, we can simplify the equation:

$$2.0 \times 10^{-16} = (X)\cdot(0.25)^2$$

$$3.2 \times 10^{-15} = X$$

Was our assumption valid? Well, $2X = 6.4 \times 10^{-15}$, which is quite a bit less than 0.005, so our assumption worked out okay. Since "X" amount of cobalt hydroxide makes "X" amount of cobalt ion, $\underline{[Co^{2+}] = 3.2 \times 10^{-15}\ M}$.

A student makes a 0.030 M solution of $CaCl_2$ and then tries to dissolve thallium chloride (TlCl), whose $K_{sp} = 1.9 \times 10^{-4}$, in it. When the solution is saturated, what will be the Tl^+ ion concentration?

Since the solution of $CaCl_2$ was made, all of the calcium chloride dissolved.

$$CaCl_2\ (s) \rightarrow Ca^{2+}\ (aq) + 2Cl^-\ (aq)$$

The common ion is the chloride ion, so that's what we have to worry about. The equation tells us there are twice as many Cl^-'s produced as $CaCl_2$'s that dissolve, so $[Cl^-] = 0.060$ M. Now we can set up the solubility equation:

$$TlCl\ (s) \rightleftarrows Tl^+\ (aq) + Cl^-\ (aq)$$

$$K_{sp} = [Tl^+]\cdot[Cl^-]$$

When "X" amount of TlCl (s) dissolves, it will produce "X" amount of Tl^+ (aq) and "X" amount of Cl^- (aq). The concentration of chloride ion, however, is *already* 0.060 M, so the concentration of chloride ion after the TlCl (s) dissolves will be "0.060 +X." This makes the final equation:

$$1.9 \times 10^{-4} = (X)\cdot(0.060 + X)$$

As long as "X" is less than 0.0005, it will not add anything to 0.060 because of the rules of precision. Thus, we could say:

$$1.9 \times 10^{-4} = (X)\cdot(0.060)$$

$$0.0032 = X$$

Unfortunately, this means our assumption did not work! The value for "X" is not less than 0.0005! This is bad because when 0.0032 is added to 0.060, the result according to the rules of precision is 0.063! Thus, "X" does, indeed, add to 0.060. What do we do? Well, "lucky" for us, the original equation, when distributed out, is a quadratic equation:

$$1.9 \times 10^{-4} = (X)\cdot(0.060 + X)$$

$$1.9 \times 10^{-4} = 0.060X + X^2$$

$$0 = X^2 + 0.060X - 1.9 \times 10^{-4}$$

For quadratic equations of the form $0 = aX^2 + bX + c$, the solution is:

$$X = \frac{-b \pm \sqrt{b^2 - 4\cdot a\cdot c}}{2a}$$

$$X = \frac{-0.060 \pm \sqrt{(0.060)^2 - 4\cdot(1)\cdot(-1.9\times10^{-4})}}{2\cdot(1)}$$

$$X = \frac{-0.060 \pm 0.066}{2}$$

$$X = -0.063 \text{ or } 0.003$$

Remember, all equations with squares in them have 2 solutions. In physical situations, however, these equations will have an unrealistic one that you can ignore. Since "X" gets translated into concentration, a negative number is unrealistic, so we will ignore it. Thus, X = 0.003. Since "X" is also the same as the Tl^+ concentration, $[Tl^+] = 0.003$ M.

ON YOUR OWN

6.6 What will be the resulting Cu^+ ion concentration when CuCl ($K_{sp} = 1.85 \times 10^{-7}$) is dissolved in a 0.50 M solution of NaCl?

6.7 What will be the resulting Pb^{2+} ion concentration when $PbCl_2$ ($K_{sp} = 1.7 \times 10^{-5}$) is dissolved in a 0.10 M solution of NaCl?

6.8 The ionic compound $KClO_4$ (potassium perchlorate) is made up of potassium ions and perchlorate ions (ClO_4^-). Its solubility product is 1.07×10^{-2}. What is the perchlorate ion concentration when $KClO_4$ is dissolved in a 0.10 M solution of K_2CO_3?

Precipitation From Solution

When solutions containing various ions are mixed together, it is possible to form an ion combination that makes an ionic solid which is only slightly soluble. What happens then? Perform the following experiment to find out.

EXPERIMENT 6.2
Precipitation

Supplies:

From the laboratory equipment set:
- 3 test tubes
- Test tube caps
- Test tube rack
- Potassium iodide solution
- Lime water ($Ca(OH)_2$) solution
- Copper sulfate
- Sodium sulfate
- 2 droppers
- Scoop

From around the house:
- Water

Add three measures of sodium sulfate to one test tube. Fill that test tube 3/4 of the way with water, cap it, and shake vigorously to dissolve the solid. You now have a solution of sodium sulfate. Use a dropper to fill your second test tube 1/4 of the way full of lime water. Do the same to the third test tube. Now use the other dropper to add an equal amount of potassium iodide solution to one of the test tubes that has lime water in it. Record what you see. It probably wasn't very exciting. Now add about an equal amount of the sodium sulfate solution to the other test tube that has lime water in it. You need not use a dropper here. Just pour directly from the test tube that contains the sodium sulfate solution into the test tube that contains the lime water. Once again, note anything you see.

Empty one of the test tubes that had lime water in them and rinse thoroughly. This time, add a few crystals of copper sulfate to the test tube. Fill it halfway with water, cap it with a clean cap, and shake vigorously. You may not get all of the crystals to dissolve. That's fine. As long as the solution has a nice blue color to it, you know that some of the copper sulfate dissolved. Uncap the test tube and add some potassium iodide solution with a dropper. Record what you observe.

What happened in the experiment? Well, in your first trial, you mixed two solutions. The lime water solution contained Ca^{2+} ions and OH^- ions, while the potassium iodide solution contained K^+ ions and I^- ions. Not much should have happened here. The lime water was probably a little cloudy because it was pretty much saturated with $Ca(OH)_2$. The potassium iodide was clear. When you mixed the two, you were still left with a cloudy solution.

What happened, however, when you mixed the lime water solution with the sodium sulfate solution? In that case, you should have seen a solid suspension appear. Why? Well, the lime water had Ca^{2+} ions and OH^- ions, and the sodium sulfate solution had Na^+ and SO_4^{2-} ions in it. When mixed, these ions started mingling, and a problem developed. You see, calcium sulfate ($CaSO_4$) is mostly insoluble in water ($K_{sp} = 2.4 \times 10^{-5}$). Thus, when calcium ions and sulfate ions exist together in the same solution, they cannot be very concentrated. If they are, then some of them must leave the solution. The only way that can happen is for the ions to form the solid and "fall out" of solution. When a solid "falls out" of solution, we call the process **precipitation** and we call the solid a **precipitate**. Thus, we would say that when lime water and sodium sulfate are mixed, precipitation of calcium sulfate results.

When you added potassium iodide to copper sulfate, you should have seen a yellow-brown precipitate form. This yellow-brown solid is copper iodide, which is mostly insoluble in water ($K_{sp} = 5.1 \times 10^{-12}$). Thus, when iodide ions (from the potassium iodide solution) were mixed with copper ions (from the copper sulfate solution), a copper iodide precipitate was formed. What happened to the potassium ions and sulfate ions? Nothing. They were still in solution. Potassium sulfate is very soluble in water, so those ions did not have to precipitate out. The same can be said for the sodium and hydroxide ions when lime water and sodium sulfate were mixed. Those ions just stayed in solution, because sodium hydroxide is very soluble in water.

So you see that when solutions are mixed together, the potential for forming a precipitate exists. Whether or not a precipitate forms will be determined by the combination of ions produced. If there is some combination of ions that forms a mostly insoluble solid, a precipitate will form. How do you know if a combination of ions are soluble or mostly insoluble? Well, there are some general principles of solubility. You need to be familiar with them, so they are summarized on the next page.

RULES OF SOLUBILITY IN WATER

1. **Most ionic solids made with Group 1A metals are soluble.**

2. **All ionic solids made with the ammonium ion are soluble.**

3. **Most nitrates and acetates are soluble.** (Exceptions: $AgC_2H_3O_2$, $HgC_2H_3O_2$, $Cr_2C_2H_3O_2$)

4. **Most chlorides are soluble.** (Exceptions: HgCl, AgCl, $PbCl_2$)

5. **Most sulfates are soluble.** (Exceptions: $SrSO_4$, $BaSO_4$, and $PbSO_4$)

6. **Most hydroxides (except those made with Group 1A metals and the ammonium ion) are mostly insoluble.**

7. **Most carbonates, phosphates, and sulfides (except those made with Group 1A metals and the ammonium ion) are mostly insoluble.**

It is important to know these rules. Only the information in bold-faced type need be memorized, however. These rules can be used to answer questions like the following:

EXAMPLE 6.6

A student prepares a solution of NH_4OH. If the student wants to see a precipitate form, should she mix this solution with a solution of NaCl or a solution of $Ba(NO_3)_2$?

First of all, notice that all three solutions the student starts with are made from soluble compounds. Anything made with the ammonium (NH_4^+) ion is soluble, as is anything made with Group 1A metals. Also, $Ba(NO_3)_2$ is soluble, because nitrates are soluble. Thus, the student has three solutions made with soluble ionic compounds. If she mixes the ammonium hydroxide solution with the NaCl solution, what combinations are possible? Well, the Na^+ ions could group with the Cl^- ions to make NaCl or the OH^- ions to make NaOH. Both of these are soluble, though, because they are made with a Group 1A metal. The NH_4^+ ions could group with either the OH^- ions or the Cl^- ions, making NH_4OH or NH_4Cl. Both of these are soluble, though, because ionic solids made with the ammonium ion are soluble. Thus, no precipitate forms when an ammonium hydroxide solution and a sodium chloride solution are mixed.

When the ammonium hydroxide is mixed with the barium nitrate, the ammonium ion could group with either the hydroxide ion or the nitrate ion, making NH_4OH or NH_4NO_3. Both of these are soluble, though, because ionic solids made with the ammonium ion are soluble. The barium ion could group with the nitrate or the hydroxide ions, making $Ba(NO_3)_2$ or $Ba(OH)_2$. The latter compound is mostly insoluble in water, because our rules state that hydroxides except those made with Group 1A metals or the ammonium ion are mostly insoluble. Thus, a precipitate will form when the ammonium hydroxide solution is mixed with the barium nitrate solution.

Now that's not too bad, is it? In order to answer questions like that, we just need to be familiar with how to determine the chemical formula of ionic compounds and we need to have the rules of solubility memorized. After that, it's smooth sailing. As I have stated before, however, chemists have a knack for taking the relatively simple and making it ridiculously complicated. Such is the case with this concept as well. Not only do you need to be able to determine whether or not a precipitate will form when two solutions are mixed together, you also need to determine *when* a precipitate will form.

What do I mean? Well, suppose in your experiment you took the lime water ($Ca(OH)_2$) solution and mixed it with a 0.0010 M sodium sulfate solution. Would a precipitate have formed? No, it would not have. Why not? Remember that a mostly insoluble solute can still dissolve to a certain extent in water. The concentration of calcium ions in a lime water solution is about 0.012 M. That's pretty small. A concentration of 0.0010 M for sulfate ions in a sodium sulfate solution is pretty small as well. It turns out that at those two concentrations, the calcium and sulfate ions can stay in solution without precipitating.

How do we know this? Well, we can look at the K_{sp} for calcium sulfate (2.4×10^{-5}). Remember that the K_{sp} only applies to a saturated solution. If this is the case, then, the equation for K_{sp} should tell us when a solution is saturated. For calcium sulfate, the equilibrium that will form when the solution is saturated is:

$$CaSO_4\ (s) \rightleftarrows Ca^{2+}\ (aq) + SO_4^{2-}\ (aq)$$

Which leads to a solubility product equation of:

$$K_{sp} = [Ca^{2+}]\cdot[SO_4^{2-}]$$

Now remember, this only applies to a solution that is saturated. What happens when a solution is not saturated? Well, when a solution is not saturated, it can accept more solute to dissolve right? What would happen to the product of the ion concentrations if more solute was dissolved? Well, the concentration of each ion would increase, which would cause the product of the concentrations to increase. What does that tell us? It tells us that when the solution is not saturated, the product of the ion concentrations can still get bigger. This means that *until a solution is saturated, the product of the ion concentrations is less than K_{sp}.* This is an important conclusion to remember. When a solution is not saturated, the product of the ion concentrations in the solubility product equation will be less than the K_{sp}. Once that product equals K_{sp}, no more ions can go into solution, because the solution is saturated.

What does that tell us about precipitation? Well, when we mix solutions, we can calculate the product of ion concentrations in the solubility product equation, and then compare the result to the K_{sp}. If the result is less than K_{sp}, the solution is not saturated, and no precipitate will form. If the result is equal to K_{sp}, the solution is perfectly saturated. No precipitate will form, but no more solute can be added. Finally, if the result is greater than K_{sp}, this tells us that the solution is beyond saturation, and ions must fall out as a solid precipitate!

In the hypothetical situation I was discussing, for example, a solution that had a 0.012 M concentration of calcium ions is mixed with an equal volume of a solution that has sulfate ions at a concentration of 0.0010 M. Since the solutions are mixed in equal volume, each ion is diluted to half its original concentration, because the total volume suddenly doubled. Thus, in the final solution, $[Ca^{2+}] = 0.0060$ M and $[SO_4^{2-}] = 0.00050$ M. In the solubility product equation, we multiply those concentrations to get:

$$[Ca^{2+}]\cdot[SO_4^{2-}] = (0.0060)\cdot(0.00050) = 3.0 \times 10^{-6}$$

Notice that this value is less than the K_{sp} (2.4×10^{-5}), so no precipitate will form.

Why did a precipitate form in your experiment? In your experiment, the concentration of sulfate ions in your sodium sulfate solution was probably greater than or equal 0.050 M. When the lime water was mixed with an equal volume of the sodium sulfate solution, the calcium ion concentration was once again cut in half (because the volume doubled) to 0.0060 M. The sulfate ion concentration was also cut in half to 0.025 M. Under those conditions:

$$[Ca^{2+}]\cdot[SO_4^{2-}] = (0.0060)\cdot(0.025) = 1.5 \times 10^{-4}$$

Since this result is more than the K_{sp} (2.4×10^{-5}), a precipitate did form.

So to determine whether or not a precipitate will form, all you need to do is determine what the solubility product equation is. From there, you just take the product of ion concentrations as given by that equation and compare it to K_{sp}. The size of the result relative to K_{sp} will tell you whether or not a precipitate will form. I will do one example for you to help you become confident with this kind of reasoning.

EXAMPLE 6.7

Will a precipitate form when a 0.025 M solution of $CaBr_2$ is mixed with an equal volume of a 0.010 M solution of $Pb(NO_3)_2$? ($PbBr_2$ has a K_{sp} of 6.3×10^{-6})

In this problem, we have a solution of calcium bromide and a solution of lead nitrate. When they mix, one of the possible ion combinations is $PbBr_2$, which is mostly insoluble. We know that not from the rules we learned (because none of the rules apply to lead bromide), but from the fact that we are given a K_{sp} that is less than one. To determine whether or not a precipitate occurs, then, we need to determine the concentration of each ion and use those concentrations in the solubility product equation.

What is the concentration of the bromide ion? Well, when $CaBr_2$ dissolves, it makes 2 moles of bromine ions for every one mole that dissolves. I won't write out the equation - you should see it from the chemical formula. Thus, the concentration of bromide ions in the solution is 0.050 M. However, when mixed with an equal volume of the other solution, the volume will double without any more bromide ions being added. Thus, the concentration will be cut in half. Once the two solutions are mixed, then, $[Br^-] = 0.025$ M.

What about the concentration of the lead ion? When $Pb(NO_3)_2$ dissolves, it makes 1 mole of lead ions from every 1 mole that dissolves. Thus, the concentration of lead ions in the solution is 0.010 M. When the two solutions are mixed, however, the volume doubles. This reduces the concentration by a factor of 2, resulting in $[Pb^{2+}] = 0.0050$ M.

Now we can finally get down to business. If the solution is saturated, the following equilibrium will be set up:

$$PbBr_2\ (s) \rightleftarrows Pb^{2+}\ (aq) + 2Br^-\ (aq)$$

$$K_{sp} = [Pb^{2+}]\cdot[Br^-]^2$$

In the solubility product equation, then, we multiply the concentration of lead ions by the square of the concentration of bromide ions. If this is greater than K_{sp}, a precipitate forms. If not, no precipitate forms.

$$[Pb^{2+}]\cdot[Br^-]^2 = (0.005)\cdot(0.025)^2 = 3.1 \times 10^{-6}$$

This number is less than K_{sp}, which means no precipitate forms.

So you see that K_{sp} is important when studying solutions, but so is the equation for calculating the solubility product. By calculating the product of ion concentrations and comparing that to the value of K_{sp}, we can determine whether or not a solution is saturated and whether or not a precipitate will form. Using this same reasoning, there are other questions we can answer as well. The following examples show you how to apply the techniques you have already learned to other situations.

EXAMPLE 6.8

A chemist has a 0.001 M solution of potassium chromate (K_2CrO_4). Suppose he starts adding a silver nitrate solution drop-by-drop. At what concentration of Ag^+ ions will the precipitate Ag_2CrO_4 form? (K_{sp} for Ag_2CrO_4 is 9×10^{-12}). Neglect any volume changes caused by the addition of silver nitrate.

This problem asks us to calculate what concentration of a particular ion will cause a precipitate. Well, if a precipitate forms, the following equilibrium will be established:

$$Ag_2CrO_4 \rightleftarrows 2Ag^+ + CrO_4^{2-}$$

$$K_{sp} = [Ag^+]^2\cdot[CrO_4^{2-}]$$

Now in order for a precipitate to occur, the product $[Ag^+]^2 \cdot [CrO_4^{2-}]$ must be greater than the K_{sp}, which is 9×10^{-12}. Mathematically, we could say:

$$[Ag^+]^2 \cdot [CrO_4^{2-}] > 9 \times 10^{-12}$$

We know the concentration of CrO_4^{2-}, because we are told that the original solution is 0.001 M in K_2CrO_4. When this solute dissolves, it makes one mole of CrO_4^{2-} for every mole of K_2CrO_4 that dissolves, so that tells us that $[CrO_4^{2-}] = 0.001$ M as well. The problem further tells us to ignore changes in volume due to the addition of silver nitrate, so that is the concentration for CrO_4^{2-} throughout the problem. Well, the only unknown in the inequality above is the silver ion concentration, which is what we are trying to determine.

$$[Ag^+]^2 \cdot (0.001) > 9 \times 10^{-12}$$

$$[Ag^+] > 9 \times 10^{-5}$$

Thus, <u>when the concentration of silver ions gets above 9×10^{-5} M, a precipitate will form.</u>

A chemist has a solution that is 0.10 M in *both* $CuNO_3$ and $PbNO_3$. Potassium iodide is added to this solution drop-by-drop. Which will precipitate first: CuI or PbI_2? (K_{sp} for CuI is 5.1×10^{-12} and K_{sp} for PbI_2 is 8.7×10^{-9}). Ignore any effects due to an increase in volume.

In this problem, we have two possible precipitates: copper (I) iodide and lead (II) iodide. To figure this out, let's see the concentration of iodide required for each to precipitate. For CuI, the equations are:

$$CuI\ (s) \rightleftarrows Cu^+\ (aq) + I^-\ (aq)$$

$$K_{sp} = [Cu^+] \cdot [I^-]$$

For a precipitate to form, the product of ion concentrations must be greater than K_{sp}:

$$[Cu^+] \cdot [I^-] > 5.1 \times 10^{-12}$$

$$(0.10) \cdot [I^-] > 5.1 \times 10^{-12}$$

$$[I^-] > 5.1 \times 10^{-11}$$

For PbI_2, the equations are:

$$PbI_2\ (s) \rightleftarrows Pb^{2+}\ (aq) + 2I^-\ (aq)$$

$$K_{sp} = [Pb^{2+}] \cdot [I^-]^2$$

For a precipitate to form, the product of ion concentrations must be greater than K_{sp}:

$$[Pb^{2+}]\cdot[I^-]^2 > 8.7 \times 10^{-9}$$

$$(0.10)\cdot[I^-]^2 > 8.7 \times 10^{-9}$$

$$[I^-] > 2.9 \times 10^{-4}$$

So for the CuI precipitate to form, the concentration of I^- must be greater than 5.1×10^{-11}. For the PbI_2 precipitate to form, the concentration of I^- must be greater than 2.9×10^{-4}. If the potassium iodide solution is added drop-by-drop, then, the concentration threshold for the precipitation of CuI will be hit first, because it is significantly lower than the PbI_2 precipitation threshold. Thus, CuI will precipitate first.

Notice that in both of these problems, I used the same technique you already learned. In both cases I just calculated the ion products from the solubility product equation and compared them to the K_{sp}. The only difference between the problems was the conclusions that we had to draw from them. You can see, then, why this technique is important to learn. Make sure you have it down pat by solving the following "on your own" problems.

ON YOUR OWN

6.9 Will a precipitate form when a 0.040 M solution of $MgCl_2$ is added to an equal volume of a 0.010 M solution of SrF_2? (K_{sp} for MgF_2 is 6.4×10^{-9})

6.10 A solution is 0.50 M in both $Mn(NO_3)_2$ and $Bi(NO_3)_3$. If a solution of K_2S is added drop-by-drop, which will precipitate first: MnS ($K_{sp} = 4.3 \times 10^{-22}$) or Bi_2S_3 ($K_{sp} = 7.3 \times 10^{-91}$)? Ignore any change in volume that occurs due to the addition of K_2S.

ANSWERS TO THE ON YOUR OWN QUESTIONS

6.1 This is an application of Equation (6.1). In this problem, we are given K and asked to calculate a concentration.

$$K_{eq} = \frac{[C]^c[D]^d}{[A]^a[B]^b}$$

$$K_{eq} = \frac{[NO(g)]^2}{[N_2(g)]\cdot[O_2(g)]}$$

$$4.1\times10^{-4} = = \frac{[NO(g)]^2}{(0.036\ M)\cdot(0.0089\ M)}$$

$$[NO(g)]^2 = 4.1\times10^{-4}\cdot(0.036\ M)\cdot(0.0089\ M)$$

$$[NO(g)] = \underline{3.6\times10^{-4}\ M}$$

Notice how the units worked out here. The equilibrium constant had no units, but when you worked out the math, the concentration you were solving for had units of M, as it should.

6.2 This is a direct application of Equation (6.2).

$$\Delta G = -R\cdot T\cdot \ln(K) = -(8.314\frac{J}{mole\cdot \cancel{K}})\cdot 298\ \cancel{K}\cdot \ln(3.4\times10^3) = \underline{-2.0\times10^4\ \frac{J}{mole}}$$

6.3 The first thing we have to do is come up with the equilibrium:

$$PbSO_4(s) \rightleftarrows Pb^{2+}(aq) + SO_4^{2-}(aq)$$

The solubility product equation, then, is:

$$1.8\ x\ 10^{-8} = [Pb^{2+}(aq)]\cdot[SO_4^{2-}(aq)]$$

According to the equilibrium equation, if "X" lead sulfate dissolves, "X" Pb^{2+} (aq) and "X" SO_4^{2-} (aq) will form. This means the equation reduces to:

$$1.8\ x\ 10^{-8} = X\cdot X$$

$$1.8\ x\ 10^{-8} = X^2$$

$$1.3 \times 10^{-4} = X$$

Since the concentration of each ion is given by "X," $[Pb^{2+}(aq)] = 1.3 \times 10^{-4}$ M and $[SO_4^{2-}(aq)] = 1.3 \times 10^{-4}$ M.

6.4 The first thing we have to do is come up with the equilibrium:

$$Bi_2S_3\ (s) \rightleftarrows 2Bi^{3+}\ (aq) + 3S^{2-}\ (aq)$$

The solubility product equation, then, is:

$$7.3 \times 10^{-91} = [Bi^{3+}\ (aq)]^2 \cdot [S^{2-}\ (aq)]^3$$

According to the equilibrium equation, if "X" bismuth sulfide dissolves, "2X" Bi^{3+} (aq) and "3X" S^{2-} (aq) will form. This means the equation reduces to:

$$7.3 \times 10^{-91} = (2X)^2 \cdot (3X)^3$$

$$7.3 \times 10^{-91} = 108X^5$$

$$3.7 \times 10^{-19} = X$$

Since the concentration of the bismuth ion is given by "2X," $[Bi^{3+}\ (aq)] = 7.4 \times 10^{-19}$ M. Since the concentration of the sulfide ion is given by "3X," $[S^{2-}\ (aq)] = 1.1 \times 10^{-18}$ M.

6.5 To solve any solubility problem, we must know the equilibrium that is established:

$$PbI_2\ (s) \rightleftarrows Pb^{2+}\ (aq) + 2I^-\ (aq)$$

Which means the solubility product equation is:

$$K_{sp} = [Pb^{2+}\ (aq)] \cdot [I^-\ (aq)]^2$$

Now we know that $[Pb^{2+}] = 1.3 \times 10^{-3}$ M, but what about the concentration of the iodide ion? Well, when "X" amount of PbI_2 dissolves, the equilibrium says that it will make "X" Pb^{2+} (aq) and "2X" I^- (aq). This means that the iodide ion concentration will always be double that of the lead ion concentration. This means $[I^-\ (aq)] = 2.6 \times 10^{-3}$ M. The K_{sp} can now be calculated:

$$K_{sp} = [Pb^{2+}\ (aq)] \cdot [I^-\ (aq)]^2 = (1.3 \times 10^{-3}) \cdot (2.6 \times 10^{-3})^2 = \underline{8.8 \times 10^{-9}}$$

6.6 The 0.50 M solution of NaCl has a common ion (Cl^-) with the solute. Thus, we need to determine its concentration before we can determine the CaCl's solubility. NaCl dissolves according to the equation:

$$NaCl\ (s) \rightarrow Na^{+}\ (aq) + Cl^{-}\ (aq)$$

Since there is a one-to-one correspondence between NaCl and Cl^-, $[Cl^-] = 0.50$ M. Now we can do the solubility of CuCl:

$$CuCl\ (s) \rightleftarrows Cu^{+}\ (aq) + Cl^{-}\ (aq)$$

$$K_{sp} = [Cu^{+}]\cdot[Cl^{-}]$$

When "X" amount of CuCl dissolves, it will form "X" Cu^+ and "X" Cl^-. However, there is already Cl^- in the solution. Thus, the "X" will simply add to the Cl^- that is already in there, making $[Cl^-] = 0.50 + X$. The solubility product equation thus becomes:

$$1.85 \times 10^{-7} = (X)\cdot(0.50 + X)$$

Even though this is a quadratic and we could therefore solve it exactly, we might as well make the assumption we have been making, because that makes the solution really easy. We can ignore "X" if it is less than 0.005, because it will not add to 0.50 unless it rounds up to at least 0.01. As long as that ends up to be true, we can solve the equation easily:

$$1.85 \times 10^{-7} = (X)\cdot(0.50)$$

$$3.7 \times 10^{-7} = X$$

This is much less than 0.005, so our assumption is valid, which means $[Cu^+] = 3.7 \times 10^{-7}$ M. Note that if you solved using the quadratic equation, your answer would be exactly the same.

6.7 We already determined that there is a one-to-one correspondence between the concentration of NaCl and the concentration of Cl^- in a solution of NaCl, so we can quickly say that $[Cl^-] = 0.10$ M. That's the common ion, so we can immediately start to solve the problem:

$$PbCl_2\ (s) \rightleftarrows Pb^{2+}\ (aq) + 2Cl^{-}\ (aq)$$

$$K_{sp} = [Pb^{2+}\ (aq)]\cdot[Cl^{-}\ (aq)]$$

$$1.7 \times 10^{-5} = (X)\cdot(0.10 + 2X)^2$$

We will assume that $2X < 0.005$ so that it adds nothing to 0.10. This simplifies the equation to:

$$1.7 \times 10^{-5} = (X)\cdot(0.10)^2$$

$$0.0017 = X$$

The value for “2X” is 0.0034, which is less than 0.005, so our assumption is A-okay. The equilibrium says that for every “X” of $PbCl_2$ that dissolves, “X” Pb^{2+} will be made. Thus, $[Pb^{2+}(aq)] = 0.0017$ M.

6.8 The solution has the potassium ion in common with the solute. Thus, we need to determine the concentration of that ion. When the solution was made:

$$K_2CO_3\ (s) \rightarrow 2K^+\ (aq) + CO_3^{2-}\ (aq)$$

Since the equation tells us that 2 potassium ions are made for every potassium carbonate that dissolves, $[K^+] = 0.20$ M. Now we can solve the $KClO_4$ problem:

$$KClO_4\ (s) \rightleftarrows K^+\ (aq) + ClO_4^-\ (aq)$$

$$K_{sp} = [K^+]\cdot[ClO_4^-]$$

$$1.07 \times 10^{-2} = (0.20 + X)\cdot(X)$$

To make the equation simple, we can assume that $X < 0.005$. If that is true:

$$1.07 \times 10^{-2} = (0.20)\cdot(X)$$

$$0.054 = X$$

Our assumption isn't true, so we have to solve this the hard way. Luckily, we can, because it is a quadratic equation:

$$1.07 \times 10^{-2} = (X)\cdot(0.20 + X)$$

$$0 = X^2 + 0.20X - 1.07 \times 10^{-2}$$

$$X = \frac{-b \pm \sqrt{b^2 - 4\cdot a\cdot c}}{2a}$$

$$X = \frac{-0.20 \pm \sqrt{(0.20)^2 - 4\cdot(1)\cdot(-1.07\times10^{-2})}}{2\cdot(1)}$$

$$X = \frac{-0.20 \pm .29}{2}$$

$$X = 0.05 \text{ or } -0.25$$

The negative number is not realistic, so X = 0.05. Since there is a one-to-one correspondence between the amount of potassium perchlorate that dissolves and the perchlorate ion, $[ClO_4^-] = 0.05\ M$.

6.9 To determine whether or not a precipitate occurs, we need to determine the concentration of each ion from MgF_2 and use those concentrations in the solubility product equation.

What is the concentration of the fluoride ion? Well, when SrF_2 dissolves, it makes 2 moles of fluoride ions for every one mole that dissolves. I won't write out the equation - you should see it from the chemical formula. Thus, the concentration of fluoride ions in the solution is 0.020 M. However, when mixed with an equal volume of the other solution, the volume will double. Thus, the concentration will be cut in half. Once the two solutions are mixed, then, $[F^-] = 0.010$ M.

What about the concentration of the magnesium ion? When $MgCl_2$ dissolves, it makes 1 mole of magnesium ions from every 1 mole that dissolves. Thus, the concentration of magnesium ions in the solution is 0.040 M. When the two solutions are mixed, however, the volume doubles. This reduces the concentration by a factor of 2, resulting in $[Mg^{2+}] = 0.020$ M.

Now we can finally get down to business. If the solution is saturated, the following equilibrium will be set up:

$$MgF_2\ (s) \rightleftarrows Mg^{2+}\ (aq) + 2F^-\ (aq)$$

$$K_{sp} = [Mg^{2+}]\cdot[F^-]^2$$

In the solubility product equation, then, we multiply the concentration of magnesium ions by the square of the concentration of fluoride ions. If this is greater than K_{sp}, a precipitate forms. If not, no precipitate forms.

$$[Mg^{2+}]\cdot[F^-]^2 = (0.020)\cdot(0.010)^2 = 2.0 \times 10^{-6}$$

This number is greater than K_{sp} (6.4×10^{-9}), which means a precipitate forms.

6.10 In this problem, we have two possible precipitates: manganese sulfide and bismuth sulfide. To figure this out, let's see the concentration of iodide required for each to precipitate. For MgS, the equations are:

$$MnS\ (s) \rightleftarrows Mn^{2+}\ (aq) + S^{2-}\ (aq)$$

$$K_{sp} = [Mn^{2+}]\cdot[S^{2-}]$$

For a precipitate to form, the product of ion concentrations must be greater than K_{sp}:

$$[Mn^{2+}]\cdot[S^{2-}] > 4.3 \times 10^{-22}$$

$$(0.50)\cdot[S^{2-}] > 4.3 \times 10^{-22}$$

$$[S^{2-}] > 8.6 \times 10^{-22}$$

For Bi_2S_3, the equations are:

$$Bi_2S_3\ (s) \rightleftarrows 2Bi^{3+}\ (aq) + 3S^{2-}\ (aq)$$

$$K_{sp} = [Bi^{3+}]^2\cdot[S^{2-}]^3$$

For a precipitate to form, the product of ion concentrations must be greater than K_{sp}:

$$[Bi^{3+}]^2\cdot[S^{2-}]^3 > 7.3 \times 10^{-91}$$

$$(0.50)^2\cdot[S^{2-}]^3 > 7.3 \times 10^{-91}$$

$$[S^{2-}] > 1.4 \times 10^{-30}$$

So for the MnS precipitate to form, the concentration of S^{2-} must be greater than 8.6×10^{-22}. For the Bi_2S_3 precipitate to form, the concentration of S^{2-} must be greater than 1.4×10^{-30}. If the potassium sulfide solution is added drop-by-drop, then, the concentration threshold for the precipitation of Bi_2S_3 will be hit first, because it is significantly lower than the MnS precipitation threshold. Thus, Bi_2S_3 will precipitate first.

REVIEW QUESTIONS

1. The equilibrium constant for a particular reaction is measured. What happens to the equilibrium constant when the concentration of the reactants is doubled?

2. If the ΔG for a reaction is large and negative, what is the relative size of the equilibrium constant? Is this an equilibrium or a complete reaction? If it is an equilibrium, is it weighted towards the reactants or the products?

3. If the ΔG for a reaction is small and positive, what is the relative size of the equilibrium constant? Is this an equilibrium or a complete reaction? If it is an equilibrium, is it weighted towards the reactants or the products?

4. If the solubility product for a solute is greater than one, what does that tell you about the solute's solubility?

5. Under what conditions does the solubility product apply to a solution?

6. Will $CaCl_2$ be more soluble in water or an NaCl solution?

7. Compare the solubility of $MgCl_2$ in water to the solubility of $MgCl_2$ in an NaOH solution.

8. For the following solutes, indicate whether they are soluble or mostly insoluble in water:

a. CsCl   b. BaS   c. $Fe(NO_3)_3$ d. $Ca_3(PO_4)_2$

9. A student has a solution of Na_2CO_3. If he wants to see a precipitate, should he mix it with a solution of $(NH_4)_2S$ or $Al(NO_3)_3$?

10. If you mix two solutions that each contain an ion from a mostly insoluble ionic solid, will you always see a precipitate?

PRACTICE PROBLEMS

$$(R = 8.314 \frac{J}{mole \cdot K})$$

1. What is the equilibrium constant of a reaction with $\Delta G = 1.721$ kJ/mole at 25°C? Is this a complete reaction, an equilibrium, or not really a reaction at all? If it is an equilibrium, is it weighted to the products or the reactants?

2. What is the ΔG for a reaction whose equilibrium constant is 1.12×10^{8} at 201 °C? Is this a complete reaction, an equilibrium, or not really a reaction at all? If it is an equilibrium, is it weighted to the products or the reactants?

3. At 100 °C, the solubility product of AgCl is 2.9×10^{-9}. What is the concentration of silver ions in a saturated silver chloride solution at 100 °C?

4. At 25 °C, the solubility product of $Ba_3(PO_4)_2$ is 1.3×10^{-29}. What is the concentration of each ion in a saturated solution of barium phosphate at 25 °C?

5. A chemist dissolves as much BaF_2 in pure water as possible, and the resulting concentration of fluoride ions is 1.5×10^{-2} M. What is the solubility product of BaF_2?

6. How many moles of Ag_3PO_4 will dissolve in one liter of a 0.50 M solution of Na_3PO_4? (K_{sp} for Ag_3PO_4 is 1.8×10^{-18})

7. What is the Cu^+ ion concentration when a 0.00010 M KCl solution is saturated with CuCl? ($K_{sp} = 1.85 \times 10^{-7}$ for CuCl)

8. A chemist prepares a 0.045 M solution of NaF and a 0.010 M solution of $BaCl_2$. When those solutions are mixed in equal volumes, will a precipitate form? (K_{sp} of BaF_2 is 1.7×10^{-6}).

9. A chemist prepares a 0.61 M solution of KBr. If she then adds a $Pb(NO_3)_2$ solution drop-by-drop, at what lead ion concentration will she see a precipitate? Ignore any change in volume due to the addition of lead nitrate. (K_{sp} for $PbBr_2$ is 6.3×10^{-6})

10. A chemist prepares a solution that is 0.10 M in both $Ca(NO_3)_2$ and $AgNO_3$. If a solution of Na_2SO_4 is then added drop-by-drop, which precipitates out first: $CaSO_4$ ($K_{sp} = 2.4 \times 10^{-5}$) or Ag_2SO_4 ($K_{sp} = 1.18 \times 10^{-5}$)?

Module #7: Acid/Base Equilibria

Introduction

In the previous module, I showed you that once a solution was saturated with a solute, an equilibrium was set up, and an enormous amount of information could be learned from the equilibrium constant. In this module, we will apply those same equilibrium concepts to an entire class of chemical reactions, which are called "acid/base reactions." In your previous chemistry course, you should have learned what acids and bases are, and to some extent, you should have learned how they react with one another. In this module, I will build on those concepts so that you have a much more detailed understanding of acid/base chemistry.

A Little Bit of Review

In your first-year chemistry course, you should have learned that acids are defined as substances which donate H^+ ions while bases are defined as substances which accept H^+ ions. For example, in the following reaction:

$$HCl\ (aq) + NaOH\ (aq) \rightarrow H_2O\ (l) + NaCl\ (aq) \qquad (7.1)$$

HCl acts as the acid and NaOH acts as the base. How do we know that? Well, remember that NaOH is an ionic solid. When aqueous, it splits up into its ions. Thus, I can write a more realistic version of the equation by splitting all aqueous ionic compounds into their ions:

$$HCl\ (aq) + Na^+\ (aq) + OH^-\ (aq) \rightarrow H_2O\ (l) + Na^+\ (aq) + Cl^-\ (aq) \qquad (7.2)$$

Now you can clearly see that HCl turned into Cl^-. The only way that can happen is for the HCl to lose an H^+. HCl is therefore the H^+ donor. In the same way, OH^- turned into H_2O. The only way that can happen is for OH^- to gain an H^+. Thus, the OH^- ion accepted an H^+, making it the base. Since the OH^- ion came from NaOH, however, we usually say that NaOH is the base.

Notice that in Equations (7.1) and (7.2), I used single arrows, not equilibrium signs. That's because the equilibrium constant for the reaction between hydrochloric acid (HCl) and sodium hydroxide is so large that, for all practical purposes, it is a complete reaction. You should remember from your first-year course that this is not always the case. There are a whole class of weak acids and weak bases which participate in equilibrium reactions whose equilibrium constants are smaller than one. For example, the active ingredient in vinegar is acetic acid, CH_3COOH. When acetic acid is mixed with water to make vinegar, the following equilibrium is established:

$$CH_3COOH\ (aq) + H_2O\ (l) \rightleftarrows H_3O^+\ (aq) + CH_3COO^-\ (aq) \qquad (7.3)$$

Notice that in this reaction, acetic acid is the acid, because it loses an H^+ to form CH_3COO^-, while water is the base, because it gains an H^+ to form H_3O^+.

The equilibrium constant which governs Equation (7.3) is given a special name. Since this equilibrium demonstrates how acetic acid forms ions in water, the equilibrium constant is called the **acid ionization constant**, and it is abbreviated as "K_a."

$$K_a = \frac{[H_3O^+]\cdot[OH^-]}{[CH_3COOH]} \quad (7.4)$$

Notice that because there are aqueous substances present, I must ignore the liquid water in the equilibrium. That's why the concentration of water does not appear in Equation (7.4).

Just as there are weak acids in Creation, there are also weak bases. For example, when ammonia (NH_3) is mixed with water, the following equilibrium is set up:

$$NH_3\,(aq) + H_2O\,(l) \rightleftarrows NH_4^+\,(aq) + OH^-\,(aq) \quad (7.5)$$

In this equilibrium, ammonia acts as the base because it accepts an H^+ to form NH_4^+, while water acts as the acid because it donates an H^+ to become OH^-. The fact that water can act as base in Equation (7.3) and an acid in Equation (7.5) should not surprise you, because you should have learned in your previous chemistry course that water is an **amphiprotic** substance. The equilibrium constant for this equilibrium is, not surprisingly, called the **base ionization constant**, and is given by:

$$K_b = \frac{[NH_4^+]\cdot[OH^-]}{[NH_3]} \quad (7.6)$$

Because there are weak acids and bases in Creation, and also because the strength of any acid or base depends on its concentration, chemists have developed the pH (potential hydrogen) scale. As you should already know, water has a pH of seven and is considered neutral. If a solution has a pH lower than seven, it is considered acidic. The lower the pH, the more acidic the solution. Solutions with pH greater than seven are considered **alkaline** (which means it has the characteristics of a base). The higher the pH, the more alkaline the solution.

Although it might not be review, I want to introduce some terminology before you leave this section. Look at the equilibrium given in Equation (7.5) above. As I said, the ammonia (NH_3) is acting as a base while the water is acting as an acid. When they react, they form the ammonium ion (NH_4^+) and the hydroxide ion (OH^-). Now imagine what would happen if you turned the equation around. If you did that, you would have NH_4^+ reacting with OH^- to form ammonia and water. In *that* reaction, NH_4^+ would be acting as an acid (donating an H^+ to form NH_3), and OH^- would be acting as a base (accepting an H^+ to form water). Well, *in an equilibrium, both the forward reaction and the reverse reaction occur.* That's the whole point of equilibrium. Thus, in this reaction, we really have *two* acids (H_2O in the forward reaction and NH_4^+ in the reverse reaction) and *two* bases (NH_3 in the forward reaction and OH^- in the reverse reaction). To make this more understandable, I will re-write Equation (7.5), labeling each.

$$NH_3\,(aq) + H_2O\,(l) \rightleftarrows NH_4^+\,(aq) + OH^-\,(aq) \quad (7.5)$$
(base) (acid) (acid) (base)

There is something very important to notice in this equilibrium. The ammonium ion (NH_4^+) was formed *because NH_3 acted as a base*. After all, in order for NH_4^+ to form, NH_3 must pick up an H^+. That's what a base does. What kind of substance is NH_4^+? It is an acid. In other words, **the product formed by a base after it picks up an H^+ is an acid**! Similarly, notice that when H_2O acted as an acid and gave up an H^+, the result was OH^-, which is a base. In other words, **the product formed by an acid after donating its H^+ is a base**!

In chemistry, we have a term for this. When an acid acts as an acid, we say that the product is that acid's **conjugate base**. In the same way, when a base acts as a base and accepts an H^+, the resulting product is the base's **conjugate acid**. Thus, the ammonium ion in Equation (7.5) is called ammonia's conjugate acid, because it was formed when ammonia acted as a base. In the same way, the hydroxide ion in Equation (7.5) is called water's conjugate base, because it formed when water acted like an acid. Make sure you understand this by studying the example below and solving the "on your own" problem that follows.

EXAMPLE 7.1

What is the conjugate acid for SO_4^{2-}?

The conjugate acid of a substance is formed when that substance acts as a base. In order to act as a base, SO_4^{2-} must pick up an H^+ ion. When that happens, it becomes $\underline{HSO_4^-}$. That's the conjugate acid.

What is the conjugate base of HNO_3?

The conjugate base of a substance is formed when that substance acts as an acid. In order to act as an acid, HNO_3 must lose an H^+ ion. When that happens, it becomes $\underline{NO_3^-}$. That's the conjugate base.

ON YOUR OWN

7.1 In the following equation, label the acid on the reactants side and its conjugate base on the products side. In the same way, label the base on the reactants side and its conjugate acid on the products side.

$$(CH_3)_2NH + H_2SO_4 \rightarrow (CH_3)_2NH_2^+ + HSO_4^-$$

The Real Meaning Behind the pH Scale

Although you probably learned a lot about acids and bases in your first-year chemistry course, there is a *lot* more to that subject. We will start with the pH scale. In your previous chemistry course, you probably did not learn much more about the pH scale than what I just said in the previous section. It turns out that the pH scale contains much more detailed information. In fact, from the pH, you can actually learn the precise concentration of H_3O^+ ions in a solution. That's because the pH of a solution is actually defined as follows:

$$pH = -\log([H_3O^+]) \quad (7.7)$$

We will use this equation *a lot* in this chapter, so it is important that you understand it. Remember from Algebra 2 that the function "log" refers to the base 10 logarithm, which is slightly different than the natural logarithm you used in the previous module. Make sure you understand how to use the base 10 logarithm and its inverse, because I will assume that you do! For this equation to work, concentration must always be provided in units of molarity.

EXAMPLE 7.2

What is the pH of a solution that has a hydronium ion concentration of 0.14 M?

This is a straightforward application of Equation (7.7).

$$pH = -\log([H_3O^+]) = -\log(0.14\ M) = \underline{0.85}$$

Notice that the units disappear. That's because the pH scale is defined *only* to work with molarity, so units are irrelevant. Based on the pH, this is a very strongly acidic solution.

What is the concentration of hydronium ions in a solution whose pH is 6.5?

This uses Equation (7.7), but we must first re-arrange it. I will assume you remember this kind of algebra:

$$pH = -\log([H_3O^+])$$

$$[H_3O^+] = 10^{-pH}$$

$$[H_3O^+] = 10^{-6.5} = \underline{3.2 \times 10^{-7}\ M}$$

Notice how the unit "M" just suddenly appeared. Once again, this is because the pH is defined *only* for units of molarity. Thus, concentration will *always* come out in units of molarity. The pH of the solution should have told you that this is a weakly acidic solution. This tells you the kind of hydronium ion concentration to expect from a weak acid.

Since the pH of pure water is 7.0, we can use Equation (7.7) to determine that the concentration of hydronium ions in water is 1.0×10^{-7} M. Now wait a minute. If I am talking about pure water, where did the hydronium ions come from? Well, since water can act as both an acid and a base, the following reaction occurs between water molecules, even in a sample of pure water:

$$H_2O\ (l) + H_2O\ (l) \rightleftarrows H_3O^+\ (aq) + OH^-\ (aq) \qquad (7.8)$$

This equilibrium describes what chemists call the **self-ionization of water**, and it occurs in any solution of water. The equilibrium constant, however, is rather small, indicating that there are not a *lot* of hydronium and hydroxide ions in pure water, but there are some. This equilibrium constant is important in acid/base chemistry (as you will soon see), so it is given its own name. It is called the **water ionization constant**, and it is abbreviated as "K_w."

$$K_w = [H_3O^+]\cdot[OH^-] = 1 \times 10^{-14}\ M^2 \qquad (7.9)$$

Notice that the reactants in Equation (7.9) do not appear in the equation for the water ionization constant because their phase indicates that we should ignore them when calculating the equilibrium constant. It turns out that the value of the water ionization constant is exact; thus, you can treat it as if it had an infinite number of significant figures.

The first thing that the water ionization constant tells us is that in pure water, the concentration of hydroxide ions is 1.0×10^{-7} M. After all, the pH of pure water tells us that the concentration of hydronium ions is 1.0×10^{-7} M, and the water ionization constant tells us that when the hydroxide ion concentration is multiplied by the hydronium ion concentration, the result is 1×10^{-14}. The only way this can happen is if the concentration of hydroxide ions is also 1.0×10^{-7} M.

This leads me to another important point. Since the self-ionization of water happens in all aqueous solutions, there are *always* both hydroxide ions and hydronium ions in any aqueous solution. This might sound strange at first. After all, we usually associate hydroxide ions with bases and hydronium ions with acids. However, wherever there is water, Equation (7.8) will exist, so there will always be *some* of each ion present. As a result, chemists define another scale called the "pOH scale." It is defined as follows:

$$pOH = -\log([OH^-]) \qquad (7.10)$$

This equation looks a lot like Equation (7.7), doesn't it? Well, it should. While pH tells us the concentration of hydronium ions in solution, pOH tells us the concentration of OH^- ions in solution.

If you are really familiar with logarithms, you will see that the next equation is a mathematical result of Equations (7.7), (7.9), and (7.10). If you don't see it, don't worry about it. Since Equation (7.9) tells us that there is always a relationship between the concentration of hydroxide ions and the concentration of hydronium ions in an aqueous solution, it should make sense to you that the pH of a solution and the pOH of a solution should relate to one another. They do so via the following equation:

$$pH + pOH = 14 \tag{7.11}$$

This equation holds true for any aqueous solution. Like the value for K_w, the 14 in this equation is exact. Thus, you can put as many zeros as you want to the right of the decimal.

I've thrown a lot of information at you in the last couple of pages, so I want to sum it all up with an example problem.

EXAMPLE 7.3

Aqueous solutions of ammonia are common household cleaners. A reasonably concentrated solution of household ammonia has a pH of 11.9. What is the pOH of the solution? What are the concentrations of hydronium and hydroxide ions in the solution?

There are many, many ways to solve this problem, because of the many equations that we have. I will start with calculating the pOH using Equation (7.11):

$$pH + pOH = 14$$

$$11.9 + pOH = 14$$

$$\underline{pOH = 2.1}$$

Remember, that 14 is exact. Thus, when I subtracted 11.9 from it, the limit to the digits I could report comes solely from the 11.9. Since the 11.9 goes out to tenths place, and since I am subtracting, I can report the answer to the tenths place.

I can calculate the concentration of hydronium ions from the pH:

$$pH = -\log([H_3O^+])$$

$$11.9 = -\log([H_3O^+])$$

$$\underline{[H_3O^+] = 1.26 \times 10^{-12}\ M}$$

To calculate the concentration of hydroxide ions, I could either use the pOH or the water ionization constant equation. I will choose the latter:

$$[H_3O^+]\cdot[OH^-] = 1 \times 10^{-14}\ M^2$$

$$(1.26 \times 10^{-12}\ M)\cdot[OH^-] = 1 \times 10^{-14}\ M^2$$

$$\underline{[OH^-] = 7.94 \times 10^{-3}\ M}$$

As you can see, the concentration of hydroxide ions is large compared to the concentration of hydronium ions. This is to be expected from an alkaline solution.

ON YOUR OWN

7.2 An average cup of coffee has a hydroxide ion concentration of 1.10×10^{-9} M. What is the pH, pOH, and concentration of hydronium ions in a cup of coffee? Is coffee acidic or alkaline?

Calculating the pH of a Solution of an Acid or Base

Now that you know the real definition of pH, you can calculate the pH of any acid or base solution, as long as you know the concentration of the acid or base involved. In order to do this, however, you need to know whether the acid or base is weak or strong. When an acid or base is strong, it reacts with water in a complete reaction. Weak acids, however, set up an equilibrium with water. Thus, in order to be able to analyze the acidity or alkalinity of a solution, we need to know whether the acid or base involved is strong or weak.

There's no way to know the difference between strong or weak acids except by memorization. Here is a list of the strong acids I expect you to know, along with their common names.

TABLE 7.1
Strong Acids

Formula	Name	Formula	Name
HCl	hydrochloric acid	HNO_3	nitric acid
HBr	hydrobromic acid	$HClO_4$	perchloric acid
HI	hydrogen iodide	H_2SO_4*	sulfuric acid

Notice that sulfuric acid is a diprotic acid, which means it can donate two hydrogens. We will come back to diprotic acids in a later section of this module.

What about bases? Well, the task is a bit easier here. In general, bases are only strong when they are hydroxides which are soluble in water. Remember, one quick way to recognize a base is to look for an ionic compound with the hydroxide ion. Well, if that ionic compound happens to be soluble in water, the base ends up being a strong base. Now you already should have the solubility rules down. One of those rules states that hydroxides are typically mostly insoluble, unless they are made with the ammonium ion or a Group 1A metal. Thus, NH_4OH and KOH are both strong bases because they are soluble ionic compounds that contain the hydroxide ion. However, $Ca(OH)_2$ is a weak base, because it is a mostly insoluble ionic compound that contains the hydroxide ion.

How can you calculate the pH of a solution from the concentration of the acid or base? Well, let's start with strong acids and bases, because they are a little easier to deal with.

EXAMPLE 7.4

What is the pH of a 0.10 M solution of HCl?

HCl is a strong acid, because it is on the list. This means that it reacts with water in a complete reaction.

$$HCl\ (aq) + H_2O\ (l) \rightarrow H_3O^+\ (aq) + Cl^-\ (aq)$$

How did I get this reaction? Well, I know that the only other thing in the solution is water, because that's the understood solvent. Since HCl is an acid, it reacts with water by donating an H^+ ion. When water gains an H^+ ion, it becomes the hydronium ion. After HCl lost an H^+, the Cl^- is left over. Since this is a complete reaction, we can use stoichiometry to convert from the concentration of HCl to the concentration of the hydronium ion, which will lead us to the pH.

$$\frac{0.10\ \cancel{\text{moles HCl}}}{\text{liter}} \times \frac{1\ \text{mole}\ H_3O^+}{1\ \cancel{\text{mole HCl}}} = \frac{0.10\ \text{moles}\ H_3O^1}{\text{liter}} = 0.10\ M\ H_3O^+$$

Now that I have the concentration of hydronium ions, I can calculate the pH:

$$pH = -\log([H_3O^+])$$

$$pH = -\log(0.10\ M)$$

$$pH = \underline{1.0}$$

What is the pH of a 0.500 M of NaOH?

NaOH is clearly a base, because it is an ionic compound that contains the hydroxide ion. It is also soluble, because it is made with a Group 1A metal. This means it is a soluble ionic compound. Soluble ionic compounds split up into their constituent ions when dissolved in water:

$$NaOH\ (s) \rightarrow Na^+\ (aq) + OH^-\ (aq)$$

This is not an equilibrium, since NaOH is a soluble compound. According to the stoichiometry, there is one hydroxide ion for every one sodium hydroxide molecule. Thus,

$$\frac{0.500\ \cancel{\text{moles NaOH}}}{\text{liter}} \times \frac{1\ \text{mole}\ OH^-}{1\ \cancel{\text{mole NaOH}}} = \frac{0.500\ \text{moles}\ OH^-}{\text{liter}} = 0.500\ M\ OH^-$$

Now that we have the concentration of hydroxide ions, we can use the water ionization constant equation to determine the hydronium ion concentration:

$$[H_3O^+]\cdot[OH^-] = 1 \times 10^{-14}\ M^2$$

$$[H_3O^+]\cdot (0.500) = 1 \times 10^{-14}\ M^2$$

$$\underline{[H_3O^+] = 2.00 \times 10^{-14}\ M}$$

Now we can finally calculate the pH:

$$pH = -\log([H_3O^+])$$

$$pH = -\log(2.00 \times 10^{-14}\ M)$$

$$pH = \underline{13.7}$$

So, calculating the pH of a strong acid or base solution is really just a matter of coming up with the chemical equation and then doing some pretty simple stoichiometry. What about weak acids and bases? That's a little more tricky. Study the following example to see what I mean.

EXAMPLE 7.5

What is the pH of a 1.0 M solution of acetic acid, CH_3COOH ($K_a = 1.8 \times 10^{-5}$)?

Acetic acid is a weak acid. We know that because the K_a is so small. Because it is a weak acid, then, it sets up the following equilibrium with water:

$$CH_3COOH\ (aq) + H_2O\ (l) \rightleftarrows H_3O^+\ (aq) + CH_3COO^-\ (aq)$$

We can't do stoichiometry in this case, since this is not a complete reaction. However, there is one thing we do know about the concentrations involved in this equilibrium. We know the acid ionization constant equation:

$$K_a = \frac{[H_3O^+]\cdot[CH_3COO^-]}{[CH_3COOH]}$$

We know the K_a, since it was given. How do we deal with the other unknowns, however? Well, we can do a calculation which is very similar to the ones we did with solubility. We can say that we have no idea how much acetic acid will react with water, so we will say that "X" amount of the acid reacts. The stoichiometry of the equation tells us that when "X" moles of acetic acid react with water, "X" moles of H_3O^+ and "X" moles of CH_3COO^- are produced. Thus, once equilibrium is reached, the concentrations of both products are "X." What about the acid, however? Well, the solution was made so that the concentration of the acid was 1.0 M. When "X" of the acid reacts with water, the concentration of acid will be reduced by "X." This means that the final concentration of acid is "1.0-X." This reduces the equation to:

$$1.8 \times 10^{-5} = \frac{X \cdot X}{1.0 - X}$$

This turns out to be a quadratic equation, so we could solve it. However, since the acid is weak, not much of it will react with water. Thus, the value for "X" will probably be small. Let's assume it is less than 0.05. If that's the case, we cannot subtract it from 1.0, so we can ignore it in the denominator. That makes the equation a cinch to solve:

$$1.8 \times 10^{-5} = \frac{X \cdot X}{1.0}$$

$$X = 0.0042$$

Since the concentration of hydronium ions is the same as "X," $[H_3O^+] = 0.0042$ M. We can now use that in the definition of pH:

$$pH = -\log([H_3O^+]) = -\log(0.0042\ M) = \underline{2.4}$$

Notice that the reasoning I used to determine the pH of a weak acid solution is essentially the same reasoning I used for solubility problems in the previous module. Since I do not know how much of the reactant is used, I just assume that "X" amount of it is used. From there, I relate the concentrations of all products and reactants to "X," and then plug those values into the equation for the equilibrium constant. I can then solve for "X" and relate that back to the concentration of hydronium ions, which will allow me to calculate the pH. To make sure you understand this, I will do one more example. This time I will use a weak base rather than a weak acid.

EXAMPLE 7.6

What is the pH of a 0.50 M solution of ammonia? ($K_b = 1.8 \times 10^{-5}$)

When a weak base is added to water, it forces water to act as an acid. Thus, the following equilibrium is set up:

$$NH_3\,(aq) + H_2O\,(l) \rightleftarrows NH_4^+\,(aq) + OH^-\,(aq)$$

$$K_b = \frac{[NH_4^+]\cdot[OH^-]}{[NH_3]}$$

I don't know how much NH_3 will be used up, so I will say that "X" of it will be used up. When "X" ammonia is used up, there will be only "0.50-X" ammonia left. At the same time, the stoichiometry of the reaction tells us that "X" ammonium ion and "X" hydroxide ion will form. The base ionization constant equation, then, is:

$$K_b = \frac{[NH_4^+]\cdot[OH^-]}{[NH_3]}$$

$$1.8 \times 10^{-5} = \frac{X \cdot X}{0.50 - X}$$

As long as $X < 0.005$, we can ignore it in the denominator, because it cannot add to 0.50. Thus, the equation becomes:

$$1.8 \times 10^{-5} = \frac{X \cdot X}{0.50}$$

$$X = 0.0030$$

The assumption is valid (just barely), so we can continue. As I mentioned before, the concentration of hydroxide ions will be equal to "X," so $[OH^-] = 0.0030$ M. To get pH, however, I need to know the concentration of hydronium ions. No problem:

$$[H_3O^+]\cdot[OH^-] = 1 \text{ x } 10^{-14}\ M^2$$

$$[H_3O^+]\cdot(0.0030) = 1 \text{ x } 10^{-14}\ M^2$$

$$[H_3O^+] = 3.3 \text{ x } 10^{-12}\ M$$

$$pH = -\log([H_3O^+]) = -\log(3.3 \text{ x } 10^{-12}\ M) = \underline{11}$$

ON YOUR OWN

7.3 What is the pH of a 0.235 M solution of perchloric acid ($HClO_4$)?

7.4 The pH of a solution of KOH is 13.2. What is the concentration of KOH in the solution?

7.5 What is the pH of a 0.075 M solution of HNO_2 ($K_a = 4.5 \text{ x } 10^{-4}$)?

EXPERIMENT 7.1

Calculating Concentration From pH

Supplies:

From the laboratory equipment set:

- Universal indicator solution and reference sheet for pH
- Three test tubes
- Test tube rack
- Two test tube caps
- A dropper
- Chemical scoop

From around the house:

- Baking soda
- Distilled water
- Clear soda pop (Sprite® or 7-Up®, for example)

One of the most important things to keep in mind when doing this experiment is to *avoid contamination.* Acid/base experiments are very sensitive to contamination. Thus, everything must be clean before the experiment starts and stay clean throughout the experiment. The first thing to do, then, is to wash everything out. Do not use soap, since most soaps are bases. Instead, rinse your hands *thoroughly* (this means several times) with tap water. Then rinse your test tubes, test tube caps, and chemical scoop *thoroughly* with tap water. Next, rinse your hands *thoroughly* with distilled water and do so with the test tubes, test tube caps, and chemical scoop as well. Fill one test tube 3/4 full of distilled water and use all of that distilled water to rinse out your dropper. Do this by sucking up all of the water you can in the dropper and then spraying it out of the dropper into the sink. Continue to do this until all of the distilled water is gone. Hopefully, you are now free from most contamination.

Add several measures of baking soda to another test tube and fill it halfway with distilled water. Cap and shake to dissolve the baking soda. Continue to do this until no matter how much you shake, there is still solid in the test tube. You now have a saturated solution of baking soda. Use your dropper to add 10 drops of universal indicator to the solution. Cap and shake. In the laboratory equipment set, there is a small rectangle of paper that shows the different colors that the universal indicator solution turns in the presence of different pH's. Match the color of this solution as closely as you can to a pH listed on the paper. If the color of the solution is between two colors on the sheet, then the pH is between the pH's for those two colors. Report the pH of this solution to one significant digit.

Remember from your first-year course that an acid/base indicator turns different colors in the presence of different pH's. Phenolphthalein, for example, is an indicator that is clear in acid but turns pink when the pH rises above 7. The universal indicator solution in your laboratory equipment set is actually a mixture of several different acid/base indicators, each of which have their own unique colors and pH ranges. When mixed together, they turn many different colors in many different pH's. Thus, the universal indicator is a nice way to get a quick reading on the pH of a solution.

You should have gotten a pH greater than 7. This is because baking soda is made from sodium bicarbonate ($NaHCO_3$). When it dissolves in water, it splits up into a sodium ion (Na^+) and a bicarbonate ion (HCO_3^-). Well, the bicarbonate ion is a weak base:

$$HCO_3^- (aq) + H_2O (l) \rightleftarrows OH^- (aq) + H_2CO_3 (aq)$$

$$K_b = \frac{[OH^-]\cdot[H_2CO_3]}{[HCO_3^-]}$$

We can use this information to calculate the amount of sodium bicarbonate that can exist in a saturated solution of water at room temperature. How? Think about it. When HCO_3^- reacts with water, "X" of it is consumed, making "X" OH^- and "X" H_2CO_3. We don't know what the original concentration of HCO_3^- was before the reaction, but we can call it $[HCO_3^-]_o$. The subscript "$_o$" means "initial." At equilibrium, then, the concentration of HCO_3^- will be the initial concentration minus whatever reacted with water, or "$[HCO_3^-]_o$ - X." This makes our equation:

$$K_b = \frac{X\cdot X}{[HCO_3^-]_o - X}$$

The K_b for HCO_3^- is equal to 2.3×10^{-8}. Since you measured the pH, you can actually calculate the concentration of OH^- in the solution you made, which is the same as "X." To do this, subtract the pH you measured from 14. This gives you the pOH. Use that number in Equation (7.10) and calculate the concentration of OH^-. That's your value for "X." Now that you have "X" and "K_b" in the equation above, you can solve for $[HCO_3^-]_o$.

What does this result tell you? Remember, the $[HCO_3^-]_o$ represents the initial concentration of the bicarbonate ion *before* it reacted with water. In other words, this is how much bicarbonate went into solution from dissolving the baking soda. Since each sodium bicarbonate molecule that dissolves produces one bicarbonate ion, the initial concentration of bicarbonate that you calculated is really the maximum concentration of baking soda that can dissolve in water!

I want you to do the experiment one more time, but this time use the clear soda pop you used. Soda pop gets its fizz from carbon dioxide gas that is dissolved in the soda pop. When carbon dioxide dissolves in water, however, some of it actually reacts with water to form carbonic acid.

$$CO_2 (g) + H_2O (l) \rightarrow H_2CO_3 (aq)$$

Carbonic acid can then react with water to form the following equilibrium:

$$H_2CO_3 (aq) + H_2O (l) \rightleftarrows H_3O^+ (aq) + HCO_3^- (aq)$$

$$K_a = \frac{[H_3O^+]\cdot[HCO_3^-]}{[H_2CO_3]}$$

The K_a for carbonic acid is 4.3×10^{-7}.

To do the experiment for soda pop, just pour some of the soda pop into a test tube and add 10 drops of universal indicator. Cap the test tube and shake to mix the universal indicator with the soda pop. Use the color to determine the pH to one significant digit. Now you know why the soda pop had to be clear. If it had been colored, the color would have interfered with your pH reading. Now that you have the pH, you can use Equation (7.7) to determine the concentration of hydronium ions. Use that (as well as the same reasoning that we used in the previous part of the experiment) to determine $[H_2CO_3]_o$. What does that tell you? It tells you what concentration of carbonic acid forms when carbon dioxide is dissolved in water to make soda. Your answer should be in the range of 1-3 M. The actual value is 2 M.

Amphiprotic Substances and Their Behavior

Up to this point, you know of one amphiprotic substance: water. In the presence of an acid, water acts like a base and accepts an H^+ ion. In the presence of a base, it acts as an acid and donates an H^+ ion. It turns out that there are *several* amphiprotic substances, and it is important to know *when* they act as an acid and *when* they act as a base. Take, for example, the bicarbonate ion that you used in Experiment 7.1. In that experiment, it acted like a base in water. However, in the presence of a strong base like NaOH, it acts as an acid:

$$HCO_3^- \text{ (aq)} + NaOH \text{ (aq)} \rightarrow H_2O \text{ (l)} + Na^+ \text{ (aq)} + CO_3^{2-} \text{ (aq)} \quad (7.12)$$

When dealing with the bicarbonate ion (or any other amphiprotic substance), then, how do I know whether it will act as an acid or a base? The answer is surprisingly simple:

If an amphiprotic substance is in the presence of an acid that is stronger than its own acid strength, the amphiprotic substance acts as a base. If it is in the presence of a base that is stronger than its own base strength, then it acts as an acid.

In other words, to determine whether an amphiprotic substance acts as an acid or base, compare it to the other reactant. If the other reactant is a stronger acid, then the amphiprotic substance will act as a base. If the other reactant is a stronger base, the amphiprotic substance will act as an acid.

That's why, in Equation (7.12), the bicarbonate ion acts as an acid. We know that NaOH is a strong base because it is a soluble hydroxide. Thus, since NaOH is a strong base, the bicarbonate ion is forced to act as an acid. What about the experiment, though? In the experiment, the bicarbonate ion has water as its other reactant. Water is *also* an amphiprotic substance. How do I know whether the bicarbonate ion is a weaker or stronger acid than water?

If you are unsure of whether a substance in a chemical reaction is a stronger base or acid, you can be positive by either comparing their K_a's or K_b's. Whichever substance has the larger K_a will be the stronger acid, or whichever substance has the larger K_b will be the stronger base. Now please realize that I do not have you give the K_a or K_b of a substance DIRECTLY. What do I mean by that? Well, there is a relationship between the K_a of an acid and the K_b of its conjugate base:

$$K_a \cdot K_b = K_w \quad \textbf{only for a conjugate acid/base pair} \quad (7.13)$$

As is indicated in bold-faced type, this relationship is only good for conjugate acid/base pairs. In other words, when you have the K_a of a substance, you can use Equation (7.13) to calculate the K_b for the substance's *conjugate base*. Alternatively, if you have the K_b of a substance, you can use Equation (7.13) to determine the K_a of its *conjugate acid*. This equation will, often times, allow you to calculate the K_a or K_b of a substance even when it is not given. I will show you how Equation (7.13) is used in a moment. First, however, there is another thing you need to know.

Notice that when the bicarbonate ion reacts with sodium hydroxide in Equation (7.12), the reaction is a complete reaction. When the bicarbonate ion reacts with water in the experiment, however, the reaction is an equilibrium. How do you know whether you are dealing with an equilibrium or a direct reaction when you are mixing acids and bases. Well, there is another rule:

When a strong acid or a strong base is used, the reaction will usually be a complete reaction. When both the acid and base are weak, the reaction will be an equilibrium.

This rule ought to make sense to you. After all, if an acid is strong, it will force its H^+ ion on a base, even if the base is weak. If the base is strong, it will rip the H^+ off of the acid, even if the acid is weak. If both the acid and base are weak, however, neither will be all too thrilled about doing its job, so not a lot of product will be made; thus, an equilibrium forms.

Now I will put all of this together for you in a couple of example problems, followed by some "on your own" problems.

EXAMPLE 7.7

Write an equation for the reaction that occurs between HSO_4^- and potassium hydroxide.

HSO_4^- is not on our list of strong acids, but it could be an acid because it has an "H" written in the front of the chemical formula. It could also be a base, since it is negative and therefore could accept an H^+ ion. This, then, is an amphiprotic substance. Potassium hydroxide (KOH), however, is a strong base, because it is a hydroxide made from a Group 1A metal. Thus, KOH will act as the base. Because we have a strong base, this is a complete reaction:

$$\underline{HSO_4^- + KOH \rightarrow H_2O + KSO_4^-}$$

I could have left the ions separated:

$$HSO_4^- + KOH \rightarrow H_2O + K^+ + SO_4^{2-}$$

Either answer is correct.

Write an equation for the reaction that occurs between HPO_3^{2-} and HCO_3^-. (The K_a of $H_2PO_3^-$ is 7.0×10^{-7}, and the K_a for H_2CO_3 is 4.3×10^{-7}).

We need to determine the reaction that takes place between HCO_3^- and HPO_3^{2-}. Well, since each substance is negative, they both certainly could accept an H^+ ion. Thus, they both could be bases. They also each have an "H" as the first part of the chemical formula, so they could be acids as well. Thus, these are amphiprotic substances. How strong an acid or base are they? We can determine their strengths as bases, because we are given the K_a's of their conjugate acids. Since $H_2PO_3^-$ is the conjugate acid of HPO_3^{2-}, we can determine the K_b for HPO_3^{2-} using Equation (7.13)

$$K_a \cdot K_b = K_w$$

Now remember, this equation is only valid for conjugate acid/base pairs. That's what we have here. We know the K_a of $H_2PO_3^-$, and we are looking for the K_b of its conjugate base, HPO_3^{2-}. Thus:

$$(7 \times 10^{-7}) \cdot K_b = 1 \times 10^{-14}$$

$$K_b = 1 \times 10^{-8}$$

The K_b for HCO_3^- can be determined the same way, since we are given the K_a of its conjugate acid.

$$(4.3 \times 10^{-7}) \cdot K_b = 1 \times 10^{-14}$$

$$K_b = 2.3 \times 10^{-8}$$

Which is the stronger base? The HCO_3^- is a stronger base, so it will act as the base in the equation. Both of these substances are weak, however, so the reaction is an equilibrium.

$$HPO_3^{2-} + HCO_3^- \rightleftarrows PO_3^{3-} + H_2CO_3$$

ON YOUR OWN

7.6 Write a chemical equation for the reaction between HCl and $H_2PO_4^-$.

7.7 Write a chemical equation for the reaction between NH_4OH and HCO_3^-.

7.8 Write a chemical equation for the reaction between HSO_4^- and $H_2AsO_4^-$. (The K_a for H_3AsO_4 is 4.8×10^{-3})

Diprotic and Triprotic Acids

In your first-year chemistry course, you should have learned that certain acids can donate as many as two H^+ ions. These acids are called **diprotic acids**. Acids such as carbonic acid (H_2CO_3) and sulfuric acid (H_2SO_4) are diprotic. Similarly, there are **triprotic acids** that can donate up to three H^+ ions. Arsenic acid (H_3AsO_4) and phosphoric acid (H_3PO_4) are triprotic. It is important to understand how these behave in solution.

When a diprotic or triprotic acid is put in the presence of a base, all of the acid molecules will donate their first H^+ ion before any donate the second H^+ ion. In other words, suppose I added 1 mole of NaOH to 1 mole of H_2SO_4. The reaction that would take place is:

$$H_2SO_4 + NaOH \rightarrow H_2O + Na^+ + HSO_4^- \quad (7.14)$$

This happens because each acid molecule donates one H^+, and then stops donating any more. Since there is only enough base to accept one H^+ per acid molecule, all of the acid molecules donate one H^+ and none of them donate 2 H^+ ions. If I were to then add *more* NaOH, however, the following reaction would occur:

$$HSO_4^- + NaOH \rightarrow H_2O + Na^+ + SO_4^{2-} \quad (7.15)$$

This reaction occurs because once all acid molecules have given up one H^+ ion, they then start giving up their second H^+ ion.

Now you might think that this contradicts what you learned in your first-year course. You probably learned the formula "acid plus base makes water plus salt." Thus, based on your first-year course, you would have written the reaction between sulfuric acid and sodium hydroxide as:

$$H_2SO_4 + 2NaOH \rightarrow 2H_2O + Na_2SO_4 \quad (7.16)$$

That's probably what you learned in your first-year course and it is, indeed, right. After all, if you add Equations (7.14) and (7.15) together, what do you get?

$$H_2SO_4 + 2NaOH \rightarrow 2H_2O + 2Na^+ + SO_4^{2-} \quad (7.17)$$

You should be able to tell that Equations (7.16) and (7.17) are equivalent. In Equation (7.17), the ionic compound is just split up into its ions. Since all of this happens in water, that's what would happen anyway. Thus, what you learned in your previous chemistry course was correct, it just assumed that there was plenty of NaOH in the reaction.

What happens when there isn't plenty of NaOH in the reaction? Well, study the following examples to see.

EXAMPLE 7.8

1.00 mole of sulfuric acid is dissolved in a liter of solution, and then 0.750 moles of solid NaOH are added. The solid does not affect the volume of the solution appreciably. What will the resulting pH be? (K_a of HSO_4^- is 0.012)

To solve this problem, we have to think through what happens when the NaOH is added to the solution. The first reaction that will occur is:

$$H_2SO_4 + NaOH \rightarrow H_2O + Na^+ + HSO_4^-$$

The 0.750 moles of NaOH will react with 0.750 moles of H_2SO_4 to make 0.750 moles of water, 0.750 moles of Na^+, and 0.750 moles of HSO_4^-. That, however, will leave 0.25 moles of sulfuric acid, since there was 1.00 mole of it before the NaOH was added and only 0.750 moles of it were used up in the reaction. Thus, after the solid is added, there will be 0.25 moles of H_2SO_4, 0.750 moles of water, 0.750 moles of Na^+, and 0.750 moles of HSO_4^-.

Of all of these substances floating around in solution, which will affect the pH? Well, water certainly won't, because water is perfectly neutral. The Na^+ is neither an acid nor a base, so it won't affect the pH either. The H_2SO_4 and HSO_4^- are both acids, however, so they will affect the pH. Now H_2SO_4 is a strong acid (it's on our list), but HSO_4^- is a weak acid (we are given its K_a). Since the strong acid will affect the pH most, let's start with it. The strong acid will completely react with water as per what we have already learned:

$$H_2SO_4\ (aq) + H_2O\ (l) \rightarrow H_3O^+\ (aq) + HSO_4^-\ (aq)$$

Since there are 0.25 moles of H_2SO_4 and the volume is one liter, we can do the stoichiometry:

$$\frac{0.25\ \cancel{\text{moles } H_2SO_4}}{\text{liter}} \times \frac{1\ \text{mole } H_3O^+}{1\ \cancel{\text{mole } H_2SO_4}} = \frac{0.25\ \text{moles } H_3O^1}{\text{liter}} = 0.25\ \text{M } H_3O^+$$

Thus, from the H_2SO_4 that was left over after the reaction with NaOH, the concentration of H_3O^+ will be 0.25 M.

Now we have to calculate the additional H_3O^+ concentration that will come with the HSO_4^-. Well, after the reaction with the NaOH, there were 0.750 moles of HSO_4^-. The left-over H_2SO_4, however, made an additional 0.25 moles of HSO_4^- when it reacted with water. Thus, there is 1.00 mole of HSO_4^-. This is a weak acid, so when it reacts with water, it reacts according to the equilibrium:

$$HSO_4^-\ (aq) + H_2O\ (l) \rightleftarrows H_3O^+\ (aq) + SO_4^{2-}\ (aq)$$

$$K_a = \frac{[H_3O^+]\cdot[SO_4^{2-}]}{[HSO_4^-]}$$

Now we just need to do an equilibrium problem. When "X" of the bisulfate ion reacts with water, it will make "X" H_3O^+ and "X" SO_4^{2-}. Since we started with 1.00 M HSO_4^-, after the reaction, the bisulfate ion concentration will be "1.00 - X." In the same way, before this reaction, we already had an H_3O^+ concentration of 0.25 M. Thus, after this reaction, the concentration of hydronium ions will be "0.25 + X." There were no sulfate ions before this reaction, so the SO_4^{2-} ion concentration will just be "X." This makes the acid ionization constant equation:

$$K_a = \frac{[H_3O^+]\cdot[SO_4^{2-}]}{[HSO_4^-]}$$

$$0.012 = \frac{(0.25 + X)\cdot(X)}{1.00 - X}$$

$$0.012 - 0.012\cdot X = 0.25\cdot X + X^2$$

$$X^2 + 0.26\cdot X - 0.012 = 0$$

$$X = \frac{-0.26 \pm \sqrt{0.26^2 - 4\cdot 1\cdot(-0.012)}}{2} = 0.040 \text{ or } -0.30$$

Since X = 0.040, the concentration of hydronium is 0.25 + X, or 0.29. Notice I did not use the simplifying approximation in this case. If you tried it, you would see it does not work. I already knew that (that's why I'm the teacher); thus, I just went ahead and solved the equation exactly to begin with. Now that we have the hydronium ion concentration, the pH is a snap:

$$pH = -\log([H_3O^+]) = -\log(0.29 \text{ M}) = \underline{0.54}$$

Notice that despite the fact that this was a long problem, there was really nothing new here. I first used standard stoichiometry to determine the results of a reaction between a strong acid and a strong base. Based on what was left over at the end of the reaction, I dealt with only those things that could affect the pH. One was a strong acid. Thus, I determined the $[H_3O^+]$ that resulted from it. The next was a weak acid. Given the $[H_3O^+]$ from the strong acid, I did an equilibrium calculation to determine how much more hydronium ion was added as a result of the weak acid. In the end, then, I was able to determine the pH.

When dealing with diprotic and triprotic acids, that is the procedure you must follow in order to determine the pH. Remember that once you have determined what substances will affect the pH, you must start with the strong acids and then deal with any weak acids later. I want to do another example for you so that you can really understand what's happening here.

EXAMPLE 7.9

What is the pH of a 0.50 M solution of phosphoric acid, H_3PO_4? (K_a of H_3PO_4 is 7.5×10^{-3}, K_a of $H_2PO_4^-$ is 6.3×10^{-8}.)

Notice that there are no strong acids or strong bases in this problem. Thus, the reaction is an equilibrium. Since phosphoric acid is a weak acid, the equilibrium is:

$$H_3PO_4 + H_2O \rightleftarrows H_3O^+ + H_2PO_4^-$$

We can certainly do an equilibrium calculation to see how much H_3O^+ comes from this reaction. We do not know how much phosphoric acid will react with water, so we will call it "X." When that amount of phosphoric acid reacts with water, the amount of phosphoric acid left over will be "0.50 - X." It will make "X" hydronium ion and "X" $H_2PO_4^-$. Thus,

$$K_a = \frac{[H_3O^+]\cdot[H_2PO_4^-]}{[H_3PO_4]}$$

$$7.5\times10^{-3} = \frac{X\cdot X}{0.50 - X}$$

$$0.0038 - 7.5\times10^{-3}\cdot X = X^2$$

$$X^2 + 0.0075\cdot X - 0.0038 = 0$$

$$X = \frac{-0.0075 \pm \sqrt{0.0075^2 - 4\cdot1\cdot(-0.0038)}}{2} = 0.055 \text{ or } -0.065$$

The only realistic answer is 0.055, so that's the value for "X." This means that at equilibrium, $[H_3O^+] = 0.055$ M. We are not quite ready to calculate the pH, however. We have taken care of the effect of phosphoric acid, but now another acid enters into the picture. At equilibrium, $[H_2PO_4^-] = 0.055$ M. Well, this is another acid. It is weak, so it will form the following equilibrium:

$$H_2PO_4^- + H_2O \rightleftarrows H_3O^+ + HPO_4^{2-}$$

We need to do another equilibrium calculation now, using the K_a for *this* equilibrium, which is given above. Once again, we do not know how much $H_2PO_4^-$ will react with water, so we will call it "X." Once that reacts, the concentration of $H_2PO_4^-$ will be "0.055 - X." When the reaction occurs, "X" hydronium ion will form, but the concentration of hydronium ion is *already* 0.055

M. Thus, after this reaction, the concentration of hydronium ion is "0.055 + X." There is no HPO_4^{2-} yet, so after the reaction the concentration HPO_4^{2-} will be "X." Thus, the equation becomes:

$$K_a = \frac{[H_3O^+]\cdot[HPO_4^{2-}]}{[H_2PO_4^-]}$$

$$6.3\times10^{-8} = \frac{(0.055+X)\cdot X}{0.055 - X}$$

I could go ahead and solve this equation exactly, but I won't. Remember, I told you to always try the approximation first. I haven't done that in the last couple of cases, however, since I knew that the approximation wouldn't work. It will here, however. Look how small K_a is. This means that very little product is made. Thus, I can assume $X < 0.0005$, which means I can ignore it when it is added or subtracted from 0.055. If that's the case, then

$$6.3\times10^{-8} = \frac{(0.055)\cdot X}{0.055}$$

$$X = 6.3\times10^{-8}$$

The assumption is, indeed, justified. This means that the reaction between $H_2PO_4^-$ and water is so small compared to the reaction between phosphoric acid and water, that I can just ignore the increase in hydronium ion concentration caused by the $H_2PO_4^-$. Thus, the concentration of hydronium ion is still just 0.055. Now we can calculate the pH:

$$pH = -\log([H_3O^+]) = -\log(0.055\ M) = \underline{1.3}$$

Notice that, in principle, I have another reaction I must calculate. After all, one of the products of the second equilibrium (HPO_4^{2-}) is also an acid. However, if $H_2PO_4^-$ could not add to the concentration of the hydronium ion, then this acid certainly won't. Thus, I can stop here.

The results of this example can lead you to a general rule when dealing with weak diprotic or triprotic acids:

When the conjugate of a diprotic or triprotic weak acid is 20 or more times weaker than the original acid, you need not consider it when calculating the pH.

In other words, when you are dealing with a diprotic or triprotic weak acid, you may only need to do one equilibrium calculation. If the conjugate has a K_a that is 20 or more times smaller than the K_a of the diprotic or triprotic acid, then it is simply not strong enough to add to the concentration of hydronium ion established by the original acid.

Take the example problem, for instance. The K_a of phosphoric acid is 7.5×10^{-3}. When it reacts with water, it forms the conjugate $H_2PO_4^-$, whose K_a is 6.3×10^{-8}. This K_a is more than 20 times smaller than the K_a for phosphoric acid. This tells us that even though $H_2PO_4^-$ is formed when phosphoric acid reacts with water, its acidic strength is so small compared to that of phosphoric acid that it simply cannot add to the hydronium ion concentration *already* established by phosphoric acid. Thus, when dealing with weak diprotic and triprotic acids, it is best to compare the K_a of the acid formed to that of the original acid. If it is 20 or more times smaller, then you know that you need not consider it in further calculations.

ON YOUR OWN

7.9 A chemist adds 45.0 grams of KOH to one liter of a 0.750 M solution of H_2SO_4. The solid does not add appreciably to the volume of the solution. What is the resulting pH? (K_a of HSO_4^- is 0.012)

7.10 What is the pH of a 1.2 M solution of H_2CO_3? (K_a of H_2CO_3 is 4.3×10^{-7}, K_a of HCO_3^- is 7×10^{-11})

An Alternative Definition of Acids and Bases

Before you leave this module, I want you to realize that the definition of acids and bases that you have been working with so far is not the only one which exists. The definition of acids and bases with which you are familiar was developed by the Danish chemist Johannes Brønstead and the English chemist Thomas Lowry. There is an alternative definition, developed by the American chemist G.N. Lewis, the same brilliant chemist who brought us Lewis structures.

The Lewis Definition of Acids and Bases

Acids are substances which accept a lone pair of electrons, while bases are substances which donate a lone pair of electrons.

How does the Lewis definition of acids and bases work? Well, consider the following reaction:

$$F^- + BF_3 \rightarrow BF_4^-$$

If we looked at it in terms of (you guessed it) Lewis structures, we would see:

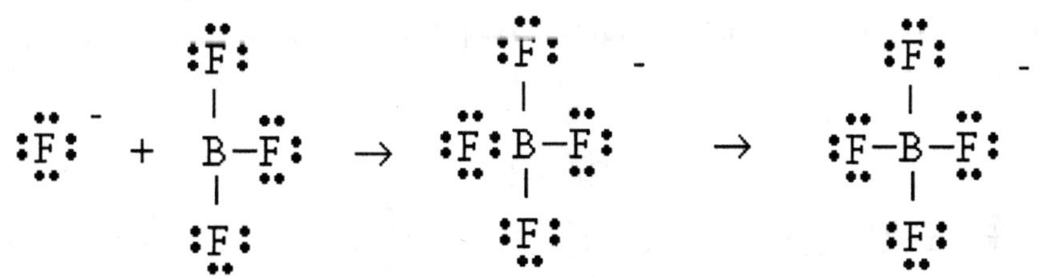

```
                ..             ..                  ..
               :F:            :F:   -             :F:   -
                |              |                   |
 ..  -          |   ..    ..   |   ..        ..    |   ..
:F:    +        B-F:  →  :F:B-F:       →    :F-B-F:
 ..             |   ..    ..   |   ..        ..    |   ..
               :F:            :F:                 :F:
                ..             ..                  ..
```

Notice what happens. The F^- has four lone pairs of electrons. The BF_3 has an empty spot for a lone pair. Thus, the F^- donates a lone pair, making it the base in the reaction. The BF_3 accepts that lone pair, making it an acid. The product, BF_4^- is usually called an **acid-base adduct** because the acid and base are stuck together.

As another example of how Lewis acids and bases work, consider the following reaction:

$$O^{2-} + SO_3 \rightarrow SO_4^{2-}$$

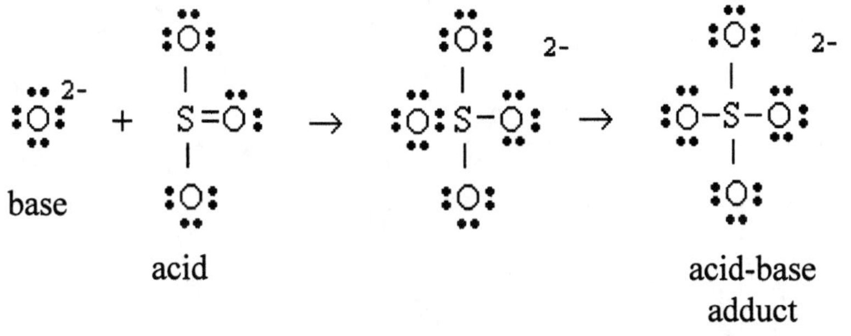

```
                ..              ..                  ..
               :O:             :O:   2-            :O:   2-
                |               |                   |
 .. 2-          |   ..     ..   |   ..         ..   |   ..
:O:     +       S=O:  →   :O:S-O:       →     :O-S-O:
 ..             |   ..     ..   |   ..         ..   |   ..
base           :O:             :O:                 :O:
                ..              ..                  ..
               acid                             acid-base
                                                 adduct
```

In this reaction, O^{2-} donates a lone pair to the sulfur atom in SO_3. This allows the sulfur atom to move one pair of electrons in the double bond back to oxygen and have 4 single bonds instead. Since the O^{2-} ion donated a lone pair of electrons, it is the base, and since SO_3 accepted that lone pair, it is an acid.

In both of these equations, you would not have characterized any of the reactants as acids or bases, because they do not conform to the definition of acids or bases that you learned up to this point. After all, no H^+ ions were transferred. What the Lewis definition of acids and bases allows us to do, then, is broaden our scope of what is an acid or base. Certain substances that you would never have identified as an acid or base using the definition you learned previously can now be identified as an acid or base using the Lewis definition. Typically, if a substance is an acid or base because it donates or accepts an H^+ ion, it is called a **Brønstead-Lowry acid or base**. On the other hand, if it is an acid or base due to the transfer of a lone pair of electrons, it is called a **Lewis acid or base.**

Now it is important to note that some substances conform to both definitions. For example, consider the reaction between HCl and NH_3:

$$HCl + NH_3 \rightarrow NH_4^+ + Cl^-$$

```
                                    H    +                       H    +
       ••          ••               ••             ••   -        |               ••   -
   H—Cl: + H—N—H   →   H—N—H   +   :Cl:   →   H—N—H   +   :Cl:
       ••           |                |             ••             |               ••
                    H                H                            H
```

Using the Brønstead-Lowry definition, HCl is an acid and NH_3 is a base. Notice, however, that you can come to the same conclusion using the Lewis definition as well. After all, the H^+ ion leaves the Cl and accepts a lone pair from the N in NH_3. Thus, NH_3 donated a pair of electrons and is therefore a Lewis base, while the H^+ ion (from HCl) accepted a lone pair, making it a Lewis acid.

ANSWERS TO THE ON YOUR OWN PROBLEMS

7.1 On the reactants side, the substance that donates an H^+ is the acid. On the products side, the substance that is left once the acid has donated its H^+ is the acid's conjugate base. In the same way, the substance that accepts an H^+ on the reactants side is the base. The substance on the products side that results from the base accepting the H^+ is the base's conjugate acid. Thus:

$(CH_3)_2NH$	+	H_2SO_4	→	$(CH_3)_2NH_2^+$	+	HSO_4^-
Base		Acid		Conjugate acid		Conjugate base

7.2 Since we have the hydroxide ion concentration, the easiest thing to calculate is the pOH:

$$pOH = -\log([OH^-])$$

$$pOH = -\log(1.10 \times 10^{-9}) = \underline{8.96}$$

Now we can calculate the pH:

$$pH + pOH = 14$$

$$pH + 8.96 = 14$$

$$pH = \underline{5.04}$$

Finally, we can calculate the concentration of hydronium ions:

$$pH = -\log([H_3O^+])$$

$$5.04 = -\log([H_3O^+])$$

$$\underline{[H_3O^+] = 9.12 \times 10^{-6} \text{ M}}$$

7.3 $HClO_4$ is a strong acid, because it is on the list. This means that it reacts with water in a complete reaction.

$$HClO_4\,(aq) + H_2O\,(l) \rightarrow H_3O^+\,(aq) + ClO_4^-\,(aq)$$

Since this is a complete reaction, we can use stoichiometry to convert from the concentration of $HClO_4$ to the concentration of the hydronium ion, which will lead us to the pH.

$$\frac{0.235 \cancel{\text{moles } HClO_4}}{\text{liter}} \times \frac{1 \text{ mole } H_3O^+}{1 \cancel{\text{mole } HClO_4}} = \frac{0.235 \text{ moles } H_3O^+}{\text{liter}} = 0.235 \text{ M } H_3O^+$$

Now that I have the concentration of hydronium ions, I can calculate the pH:

$$pH = -\log([H_3O^+])$$

$$pH = -\log(0.235 \text{ M})$$

$$pH = \underline{0.629}$$

7.4 KOH is clearly a base, because it is an ionic compound that contains the hydroxide ion. It is also soluble, because it is made with a Group 1A metal. Soluble ionic compounds split up into their constituent ions when dissolved in water:

$$KOH\ (s) \rightarrow K^+\ (aq) + OH^-\ (aq)$$

This is not an equilibrium, since KOH is a soluble compound. According to the stoichiometry, there is one hydroxide ion for every one potassium hydroxide molecule. Since we are given the pH, we can actually figure out the concentration of hydroxide ion. We can do this either by calculating the pOH and then using the definition of pOH or by calculating the concentration of hydronium ions from the pH and then using the self-ionization of water equation to determine the hydroxide ion concentration. I will do the latter:

$$pH = -\log([H_3O^+])$$

$$13.2 = -\log([H_3O^+])$$

$$[H_3O^+] = 6.31 \times 10^{-14}$$

$$[H_3O^+] \cdot [OH^-] = 1 \times 10^{-14}$$

$$(6.31 \times 10^{-14}) \cdot [OH^-] = 1 \times 10^{-14}$$

$$[OH^-] = 0.158 \text{ M}$$

Since there is a one-to-one correspondence between KOH and OH^-, <u>[KOH] = 0.158 M</u>.

7.5 Nitrous acid is a weak acid. We know that because the K_a is so small. Because it is a weak acid, then, it sets up the following equilibrium with water:

$$HNO_2\ (aq) + H_2O\ (l) \rightleftarrows H_3O^+\ (aq) + NO_2^-\ (aq)$$

We can't do stoichiometry in this case, since this is not a complete reaction. However, there is one thing we do know about the concentrations involved in this equilibrium. We know the acid ionization constant equation:

$$K_a = \frac{[H_3O^+]\cdot[NO_2^-]}{[HNO_2]}$$

We know the K_a, since it was given. We can say that we have no idea how much nitrous acid will react with water, so we will say that "X" amount of the acid reacts. The stoichiometry of the equation tells us that when "X" moles of nitrous acid react with water, "X" moles of H_3O^+ and "X" moles of NO_2^- form. Thus, once equilibrium is reached, the concentrations of both products is "X." What about the acid, however? Well, the solution was made so that the concentration of the acid was 0.075 M. When "X" of the acid reacts with water, the concentration of acid will be reduced by "X." This means that the final concentration of acid is "0.075-X." This reduces the equation to:

$$4.5\times10^{-4} = \frac{X\cdot X}{0.075 - X}$$

Let's assume that "X" is less than 0.0005. If that's the case, we cannot subtract it from 0.075, so we can ignore it in the denominator. That makes the equation a cinch to solve:

$$4.5\times10^{-4} = \frac{X\cdot X}{0.075}$$

$$X = 0.0058$$

Notice that the assumption is **not correct!** We will therefore have to solve this exactly:

$$4.5\times10^{-4} = \frac{X\cdot X}{0.075 - X}$$

$$(4.5\times10^{-4})\cdot(0.075 - X) = X^2$$

$$X^2 + (4.5\times10^{-4})\cdot X - 3.375\times10^{-5}$$

$$X = \frac{-b \pm \sqrt{b^2 - 4\cdot a\cdot c}}{2\cdot a} = \frac{-4.5\times10^{-4} \pm \sqrt{(4.5\times10^{-4})^2 - 4\cdot(1)\cdot(-3.375\times10^{-5})}}{2\cdot(1)}$$

$$X = \frac{-4.5\times10^{-4} \pm 0.012}{2} = 0.0060 \text{ or } -0.0060$$

Since the concentration of hydronium ions is the same as "X," $[H_3O^+] = 0.0060$ M. We can now use that in the definition of pH:

$$pH = -\log([H_3O^+]) = -\log(0.0060\text{ M}) = \underline{2.2}$$

7.6 First of all, HCl is a strong acid on our list, so that must be the acid in the reaction. $H_2PO_4^-$ is amphiprotic. Since it is negative, it could easily accept an H^+, but since it has H's at the beginning of its chemical formula, it can also act as an acid. In the presence of a strong acid, however, it must act as a base. Thus, the equation is:

$$\underline{HCl + H_2PO_4^- \rightarrow Cl^- + H_3PO_4}$$

Notice that the equation is not an equilibrium because a strong acid is involved.

7.7 First of all, ammonium hydroxide is a strong base because it is a soluble hydroxide. HCO_3^- is amphiprotic because it has an "H" at the beginning of its formula and it is negative. In the presence of a strong base, however, it will be forced to act as an acid:

$$\underline{NH_4OH + HCO_3^- \rightarrow H_2O + NH_4^+ + CO_3^{2-}}$$

Once again, this is a complete reaction because a strong base is involved.

7.8 Both substances are amphiprotic because they each have "H's" in front of their chemical reactions but are also negative. To determine which is the acid and which is the base, we need to compare the K_a's or K_b's. We are given the K_a for H_3AsO_4, so that means we can calculate the K_b for $H_2AsO_4^-$:

$$K_a \cdot K_b = K_w$$

$$(4.8 \times 10^{-3}) \cdot K_b = 1 \times 10^{-14}$$

$$K_b = 2.1 \times 10^{-12}$$

Now what about HSO_4^-? We have no information on that species, do we? Well, actually, we do. We know that its conjugate acid, H_2SO_4, is a strong acid. This means that the K_a is greater than one. If the K_a is greater than one, when we put it into the same equation we used above and solve for K_b, we will get a number less than 1×10^{-14}. Thus, even though we do not have an exact number for the K_b of HSO_4^-, we know that it is less than 1×10^{-14}, because we know that the K_a of its conjugate is greater than one. Thus, the K_b for $H_2AsO_4^-$ is larger, making it the base. Thus:

$$\underline{HSO_4^- + H_2AsO_4^- \rightleftarrows H_3AsO_4 + SO_4^{2-}}$$

The equation is an equilibrium because both the acid and base are weak.

7.9 The first thing that will happen is the strong acid (H_2SO_4) will react with the strong base, KOH. Since diprotic acids all give their first H^+ up before any of them give up their second H^+, we have to do our stoichiometric calculation first on the reaction in which the diprotic acid gives up its first H^+:

$$H_2SO_4 + KOH \rightarrow H_2O + K^+ + HSO_4^-$$

We have one liter of solution, and the acid concentration is 0.750 M. This means we have 0.750 moles of H_2SO_4. We need to see how many moles of KOH we have:

$$\frac{45.0 \text{ g } \cancel{\text{KOH}}}{1} \times \frac{1 \text{ mole KOH}}{56.1 \text{ g } \cancel{\text{KOH}}} = 0.802 \text{ moles KOH}$$

This tells us that we have more moles of KOH than H_2SO_4. Since one mole of acid reacts with one mole of base in the equation above, this tells us that *all* of the H_2SO_4 molecules will lose their first H^+. When they do, there will be no H_2SO_4 left, but there will be 0.750 moles of H_2O, 0.750 moles of HSO_4^-, and 0.052 moles of KOH left over. Those 0.052 moles of KOH will react with the HSO_4^- to make:

$$HSO_4^- + KOH \rightarrow H_2O + K^+ + SO_4^{2-}$$

The reaction is a complete reaction, because the base is a strong base. Once all KOH has reacted, there will only be 0.750 - 0.052 = 0.698 moles of HSO_4^-. The reaction will have added 0.052 moles to the amount of water, and there will also be 0.052 moles of SO_2^{2-}.

Now what? Well, the only thing left that can affect the pH of the solution is the HSO_4^-. It is a weak acid (as its K_a implies), so it will react with water according to the equilibrium:

$$HSO_4^- + H_2O \rightleftarrows H_3O^+ + SO_4^{2-}$$

Now we just need to do an equilibrium calculation. When "X" HSO_4^- reacts, the concentration of HSO_4^- will be "0.698 - X." This will make "X" H_3O^+ and "X" SO_4^{2-}. There is already, however, 0.052 moles of SO_4^{2-} in our liter of solution, so after the equilibrium is reached, the concentration of SO_4^{2-} will be "0.052 + X." The acid ionization constant equation, then, is:

$$K_a = \frac{[H_3O^+]\cdot[SO_4^{2-}]}{[HSO_4^-]}$$

$$0.012 = \frac{(X)\cdot(0.052 + X)}{0.698 - X}$$

$$0.00838 - 0.012\cdot X = 0.052\cdot X + X^2$$

$$X^2 + 0.064\cdot X - 0.00838 = 0$$

$$X = \frac{-0.064 \pm \sqrt{0.064^2 - 4\cdot 1\cdot(-0.00838)}}{2} = 0.065 \text{ or } -0.13$$

Since "X" is also equal to the hydronium concentration, the pH is now easy to calculate:

$$pH = -\log([H_3O^+]) = -\log(0.065\text{ M}) = \underline{1.2}$$

7.10 The acid in question is a weak acid, as is evidenced by its K_a. Thus, it reacts with water according to the following equilibrium:

$$H_2CO_3 + H_2O \rightarrow H_3O^+ + HCO_3^-$$

When "X" of the carbonic acid reacts, it will leave "1.2 - X" carbonic acid and make "X" H_3O^+ and "X" HCO_3^-.

$$K_a = \frac{[H_3O^+]\cdot[SO_4^{2-}]}{[HSO_4^-]}$$

$$4.3\times 10^{-7} = \frac{(X)\cdot(X)}{1.2 - X}$$

The K_a is so small we can probably make our assumption:

$$4.3\times 10^{-7} = \frac{(X)\cdot(X)}{1.2}$$

$$X = 7.2\times 10^{-4}$$

The assumption is fine, because X is too small to subtract from 1.2. At this point, then, we know the hydronium ion concentration is 7.2×10^{-4} M. Now the reaction also makes another weak acid, HCO_3^-. We do not have to worry about that, however, because the K_a given for that acid is more than 20 times less than the K_a for carbonic acid. Thus, this new acid does not add appreciably to the hydronium ion concentration. The pH, then, is:

$$pH = -\log([H_3O^+]) = -\log(7.2 \times 10^{-4}\text{ M}) = \underline{3.1}$$

REVIEW QUESTIONS

1. Give the conjugate bases of the following acids:

a. HCl b. CH_3COOH c. $H_2PO_4^-$ d. $HClO_4$ e. H_2SO_3

2. Give the conjugate acids of the following bases:

a. NH_3 b. OH^- c. $H_2PO_4^-$ d. SO_4^{2-} e. H_2O

3. In the following reaction, label the acid and its conjugate base, as well as the base and its conjugate acid:

$$CH_3NH_2 + H_2O \rightleftarrows CH_3NH_3^+ + OH^-$$

4. In many introductory chemistry textbooks, the pH scale is said to run from 0 to 14. Is this really true? Why or why not?

5. List the 6 strong acids that you have to know.

6. Of the following bases, which would be considered strong?

a. NH_3 b. $Ca(OH)_2$ c. KOH d. NH_4OH e. $Al(OH)_3$

7. Which of the following substances is amphiprotic?

a. H_2O b. CO_3^{2-} c. HCO_3^- d. NH_3 e. HPO_4^{2-}

8. What is the Lewis definition of acids and bases?

9. In the reactions below, determine which substance is the Lewis acid and which is the Lewis base.

a.

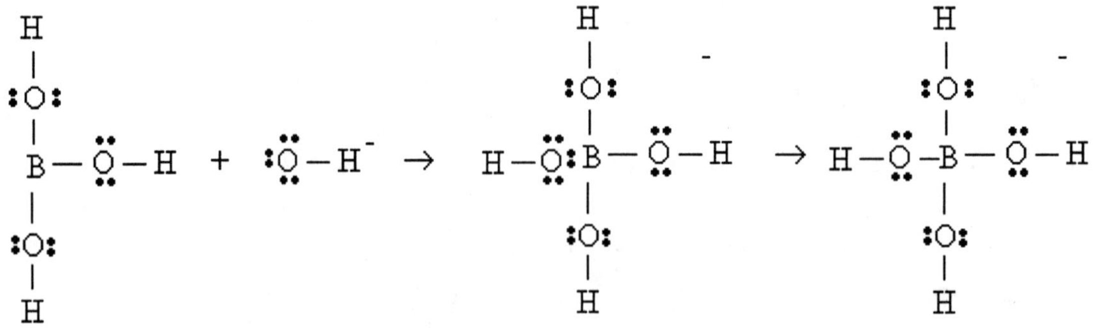

```
   H                            H         -             H         -
   |                            |                       |
  :O:                          :O:                     :O:
   |     ..          ..         |    ..              .. |    ..
   B — O — H  +  :O — H⁻  →  H—O:B — O — H   →  H—O —B — O — H
         ..          ..      ..    |    ..           ..  |    ..
   |                                |                     |
  :O:                          :O:                     :O:
   |                            |                       |
   H                            H                       H
```

b.

$$2\ H{-}\ddot{N}(H){-}H \ + \ Ag^{+} \ \rightarrow \ [H_3N{:}Ag{:}NH_3]^{+}$$

10. Are any of the acids or bases in problem #9 also Brønstead-Lowry acids or bases?

PRACTICE PROBLEMS

1. Milk of magnesia (often used as a laxative), has a hydronium ion concentration of 5.0×10^{-10} M. What is the pH, pOH, and hydroxide ion concentration for milk of magnesia? Is it an acid or base?

2. What is the pH of a 0.15 M solution of HBr?

3. What is the pH of a 0.31 M solution of LiOH?

4. What is the pH of a 1.25 M solution of hydrocyanic acid (HCN)? $K_a = 4.0 \times 10^{-10}$

5. Write a chemical equation for the reaction between HNO_3 and HSO_4^-.

6. Write a chemical equation for the reaction between KOH and $H_2AsO_4^-$.

7. Write a chemical equation for the reaction between $HC_2O_4^-$ and HPO_4^{2-}. (The K_a of $H_2PO_4^-$ is 6.3×10^{-8}, and the K_a of $H_2C_2O_4$ is 5.9×10^{-2}.)

8. A chemist takes a liter of 1.50 M NaOH and adds 0.75 moles of HCl in concentrated form. The addition of HCl does not add appreciably to the volume of the solution. What is the pH?

9. A chemist takes a liter of 0.50 M KOH and adds 0.75 moles of H_2SO_4 in concentrated form. The addition of acid does not add appreciably to the volume of the solution. What is the pH? (K_a of HSO_4^- is 0.012)

10. What is the pH of a 2.1 M solution of H_2Se? (The K_a of H_2Se is 1.7×10^{-4}, and the K_a of HSe^- is 1.0×10^{-10}.)

Module #8: More on Equilibrium

Introduction

So you thought we were done with equilibrium, didn't you? Well, you were wrong. Remember, in reality, *all* chemical reactions are equilibrium reactions. Complete reactions and reactions that do not occur are simply approximations of equilibria that have very large or very small equilibrium constants. Since all chemical reactions are, in reality, equilibrium reactions, equilibrium is clearly an important subject and cannot be glossed over lightly. Thus, we need to spend more time on this fascinating subject.

Buffer Solutions

In the last module, I talked exclusively about acid/base equilibria. Even though I will move away from acid/base concepts towards the end of this module, there is *still more* to learn about this important topic. The next thing you need to learn is best introduced with an experiment.

EXPERIMENT 8.1
The Bicarbonate Buffer

Supplies:
From the laboratory equipment set:
- Universal indicator solution and reference sheet for pH
- Two test tubes
- Test tube rack
- Two test tube caps
- Three droppers
- Chemical scoop

From around the house:
- Baking soda
- Distilled water
- Clear vinegar
- Clear soda pop (Sprite or 7-up, for example)

As was the case in Experiment 7.1, this experiment will work correctly only if you *avoid contamination*. Thus, everything must be clean before the experiment starts and stay clean throughout the experiment. Rinse your hands *thoroughly* (this means several times) with tap water. Then rinse your test tubes, test tube caps, and chemical scoop *thoroughly* with tap water. Next, rinse your hands *thoroughly* with distilled water and do so with the test tubes, test tube caps, and chemical scoop as well. Fill one test tube 3/4 full of distilled water and use all of that distilled water to rinse out your dropper. Do this by sucking up all of the water you can in the dropper and then spraying it out of the dropper into the sink. Continue to do this until all of the distilled water is gone. Repeat this for the other droppers as well. Hopefully, you are now free from most contamination.

The first thing you need to do is set up a reference test tube. Fill one of the test tubes halfway with distilled water. Add 10 drops of universal indicator solution, cap the test tube, and shake. The indicator should have turned a shade of green. Look at the reference sheet to make sure that the pH indicated by the color is, indeed, 7. If it is not, your test tube was contaminated. Clean it out and try again.

Now add 3 measures of baking soda to the second test tube and fill it 1/4 of the way with distilled water. Cap and shake. The solid may not completely dissolve, but that's okay. Next, add 10 drops of universal indicator, cap and shake. The color of the solution should be blue-green, because the bicarbonate ion will act as a base in water, making the pH slightly above 7. With the second dropper, add a few squirts of Sprite®, then cap and shake. The color should be less blue and more green now. You need to continue to add Sprite® a few squirts at a time until the color of the solution in this test tube is the same as the color in the reference test tube. In other words, you should continue to add acid (the carbonic acid in Sprite®) until the pH of the solution is 7. If you overshoot the mark, just add more baking soda. You may need to dump a little bit of solution out from time to time. That's fine. Just get to the point where the color of the solution in this test tube is the same as the color of solution in the reference test tube.

Now you have two test tubes, each with a solution that has a pH of 7. Take the reference test tube and, with the third dropper, add two drops of vinegar, cap (with the cap that you originally used on this test tube) and shake. Note the color. Record what pH corresponds to that color. Next, take the solution of Sprite® and baking soda and do the same thing. What color does it turn? In principle, once you shake it up, the solution should have had no appreciable color change. Continue to add vinegar a few drops at a time. Can you get this test tube down to the same pH as your reference test tube?

What happened in the experiment? You had two solutions with a pH of 7. The first one was distilled water and the universal indicator. With two drops of vinegar, the pH changed to nearly 3. That is expected, of course, since vinegar is an acid. With the addition of an acid, the pH of a solution should decrease. What happened in the second test tube, though? You added lots more vinegar, but the pH hardly changed. Why? Well, the second test tube contained what chemists call a **buffer solution**.

Buffer solution - A solution made up of a weak acid and its conjugate base. These solutions are resistant to changes in pH.

In the test tube, you had carbonic acid (from the Sprite) and the bicarbonate ion (HCO_3^-) from the baking soda. The carbonic acid is the weak acid, and the bicarbonate ion is its conjugate base.

Why is a buffer solution resistant to changes in pH? Think about the contents of a buffer solution. The buffer solution you used in your experiment had water, carbonic acid (from the

Sprite), sodium ions (from the baking soda), HCO_3^- ions (from the baking soda), and the universal indicator. When acetic acid (from the vinegar) was added, it could react with anything in that solution. What would it react with? Well, acetic acid wants a base with which to react. What bases were present in the solution? Water is amphiprotic, so it can act as a base. HCO_3^- is also amphiprotic, so it can act as a base as well. Those were really the only two substances in the solution that could act as a base.

What would happen if acetic acid reacted with water? It would donate an H^+ to the water, making H_3O^+.

$$CH_3COOH + H_2O \rightleftarrows H_3O^+ + CH_3COO^- \tag{8.1}$$

Notice that the hydronium ion is a product. As this reaction proceeds, the hydronium ion concentration increases, which directly lowers the pH. What would happen if the acetic acid reacted with the HCO_3^- instead? The following reaction would occur:

$$CH_3COOH + HCO_3^- \rightleftarrows H_2CO_3 + CH_3COO^- \tag{8.2}$$

Notice that this reaction does not directly change the H_3O^+ concentration, because the hydronium ion does not appear in the equation. Thus, even though acetic acid is added, the pH is not directly affected

Obviously, then, in the buffer solution that you made, the acetic acid reacted with the HCO_3^- ion, not water. This prevented the acetic acid from directly affecting the pH. Why did the acetic acid react with the bicarbonate ions rather than water? That's simple. The bicarbonate ion is a better base than water. An acid will always react with the strongest base present. Since the bicarbonate ion is a better base, acetic acid reacted with it rather than water.

Now, suppose I were to add NaOH (a base) to the buffer solution you formed in your experiment? What would the base react with? Well, it would also have the choice of reacting with water, carbonic acid, the bicarbonate ion, or the universal indicator. Just as an acid will always react with the strongest base present, a base will always react with the strongest acid present. In this case, that would be carbonic acid. Thus, when NaOH is added to your buffer solution, the following reaction would occur:

$$NaOH + H_2CO_3 \rightarrow HCO_3^- + Na^+ + H_2O \tag{8.3}$$

Notice once again that the concentration of the hydronium ion is not directly affected by this reaction, because neither the hydronium ion nor the hydroxide ion shows up in the products.

Now you know why a buffer solution is resistant to changes in pH. In order to dramatically change the pH of a solution, you need to alter the hydronium ion concentration. The easiest way to do that is to have an acid or base react with *water*. A buffer solution, however, contains a base that is stronger than water. As a result, when an acid is added, the acid will not react with water. It will, instead, react with the base, and the result is that the concentration of hydronium ions will not be directly affected. In the same way, a buffer solution contains an acid which is stronger than water. When base is added to the buffer solution, it reacts with that acid rather than with water. The result is that, once again, the hydronium ion concentration is not directly affected.

In other words, the contents of a buffer solution "protect" water. By containing an acid which is a stronger acid than water and a base which is a stronger base than water, any additional acid or base added will react with those substances rather than water. That way, the hydronium ion concentration is not directly affected and the pH does not change dramatically. In the experiment, you saw just how resistant a buffer solution was to pH changes. When you added a weak acid to water, just a couple of drops lowered the pH from 7 to almost 3. When you added the same number of drops of vinegar to the buffer, you saw no appreciable change. To see any change in pH, you had to add several times the amount of vinegar.

Now why must a buffer solution contain a weak acid and its conjugate? Think about it. A buffer solution needs an acid that is a stronger acid than water and a base which is a stronger base than water. Most weak acids are stronger acids than water, so that takes care of the first requirement. The conjugate of a weak acid is a base. Remember that an acid and its conjugate base have the following relationship between their K's:

$$K_a \cdot K_b = K_w \tag{7.13}$$

If the acid is weak, then K_a is small. When K_a is small, K_b must be larger so that their product equals K_w. Thus, the conjugate base of a weak acid is a reasonably strong base. That's what we need to "protect" water from reaction with an acid. Thus, as long as I have a weak acid and its conjugate in my buffer solution, I am assured that I will have an acid whose acidic strength is greater than water, and I will also have a base whose basic strength is greater than that of water.

It turns out that buffer solutions are found all over Creation. Like most things in science, this is another great testament to the power and wonder of the Creator. The bicarbonate buffer that you used in the experiment is actually found in the blood of mammals, including humans. You see, the enzymes and proteins that support nearly all of the chemistry in a living organism are rather sensitive to pH. Most of them have an optimal pH at which they are most effective. If the pH of their surroundings changes even slightly, they will lose their optimal performance. Since so much of life depends on millions of chemical processes all working at their peak efficiency, even a slight change from optimal pH will cause irreparable damage to a living organism. Thus, the pH of the fluids inside a living creature must be strongly regulated.

There is no better example of this than human blood. The pH of human blood plasma is 7.40. Thus "just happens" to be the pH at which the enzymes used in blood operate at their peak efficiency. If the pH of a person's blood were to lower to 7.0 or raise to 7.8, the result would be disastrous. Now think about that for a moment. A pH of 7.40 corresponds to a hydronium ion concentration of 3.98×10^{-8} M. If the pH fell to 7.0, the hydronium ion concentration would be 1.0×10^{-7} M. That's a change of only 6.02×10^{-8} M! That *small* change in hydronium ion concentration is enough to pretty much kill a human being! Thus, a person's (or any other organism's) blood must be very stable when it comes to pH!

That's where the buffer solution comes in. The same buffer solution you used in Experiment 8.1 (H_2CO_3/HCO_3^-) is used in your blood to regulate its pH. Now that's neat enough, but there's even something more incredible about this buffer. This buffer is *self-renewing* in a living organism. Why? Well, if the blood is exposed to a large amount of base, the carbonic acid will start to get used up. If it were to be completely used up, then the buffer would no longer protect against increases in the pH. That won't happen, however, because our lungs are full of carbon dioxide. When the carbonic acid concentration in our blood lowers, the body *immediately makes more* by taking that carbon dioxide and adding water:

$$CO_2\ (g) + H_2O\ (l) \rightleftarrows H_2CO_3\ (aq) \qquad (8.4)$$

Because of this fact, the buffer in our blood remains stable! It turns out that this is *the only possible buffer* that could be used to protect the blood of mammals. Aren't we "lucky" that it is there?

Of course, there are other body fluids besides blood, and they must be buffered against changes in pH as well. The fluid inside a cell, for example, usually contains a phosphate buffer which contains $H_2PO_4^-$ and the weak acid and HPO_4^{2-} as the conjugate base. Why this different buffer? Well, intracellular fluid needs a slightly different pH (7.2), so it needs a different buffer. Also, since a lot of phosphorous is transported into cells, it is relatively easy for the cell to manufacture this buffer.

Now think about that for a moment. It's hard enough just to understand how buffer systems work. In Creation, however, they are *everywhere*. Organisms regulate their pH by a host of buffer solutions all working at the pH necessary for their particular situation. Each buffer system is carefully placed to be the best possible buffer for the job. In some cases, these buffer systems are self-renewing, allowing an almost limitless protection against pH changes. This fact alone should be more than enough evidence that Creation is the result of *design*, not chance!

ON YOUR OWN

8.1 Which of the following solutions could be used as buffers?
   a. An aqueous solution of HCl and NaOH
   b. An aqueous solution of H_3PO_4 and NaH_2PO_4
   c. An aqueous solution of HCNO and NH_3
   d. An aqueous solution of CH_3COOH and KCH_3COO
   e. An aqueous solution of $NaHCO_3$ and K_2CO_3

The pH of a Buffer

If you really understand how a buffer protects a solution from pH changes, you should see an inconsistency. I have just gotten through explaining to you how a buffer protects a solution from changes in pH. Nevertheless, in your experiment, you were probably able to change the pH of the buffer solution you made as long as you added enough vinegar. Why, if a buffer protects a solution from pH changes, could you change the pH of your buffer solution? Well, notice that throughout my discussion of buffers, I used phrases like "does not directly change the hydronium ion concentration" and "does not appreciably affect pH." Why the words "directly" and "appreciably?"

The answer is rather straightforward. In Equation (8.2), for example, I showed you that when acetic acid was added to your buffer solution, it did not react with water. Instead, it reacted with the bicarbonate ion. Since the hydronium ion does not appear anywhere in Equation (8.2), I therefore told you that the hydronium ion was not "directly" affected. That's very true, but the concentration of the hydronium ion is *indirectly* affected. After all, look at the products in Equation (8.2). One of the products is *carbonic acid.* Thus, when acetic acid reacts with the bicarbonate ion, it *increases the concentration of the carbonic acid*! What happens when the concentration of an acid in solution increases? The pH lowers!

Wait a minute, though. The pH did not lower until a lot of acetic acid was added. If the concentration of carbonic acid increased each time acetic acid was added, why didn't the pH lower right away? Well, remember, carbonic acid is a *weak acid.* The effect of adding a small amount to the concentration of a weak acid is not significant, because the weak acid is in equilibrium with water. Thus, a small increase in the concentration of a weak acid will only affect the pH a tiny, tiny amount. In other words, the addition of acetic acid did affect the pH, but not directly. It affected it indirectly by slightly changing the concentration of the weak acid. Thus, when an acid is added to a buffer, the hydronium ion concentration does increase, but not directly. The increase is quite small because the effect is a secondary effect. The same is true when a base is added to a buffer. When a base is added to a buffer, the pH does change, but only marginally. Look at Equation (8.3). When a base is added to a buffer, the concentration of the conjugate base increases. This indirectly affects the hydronium ion concentration, causing a slight increase in pH.

In the end, then, buffers are not perfect. The pH of a buffer is affected by the addition of an acid or base. It is just not affected greatly. Study the figure below:

FIGURE 8.1
The pH Change in an Acetate Buffer and Water Due to the Addition of HCl

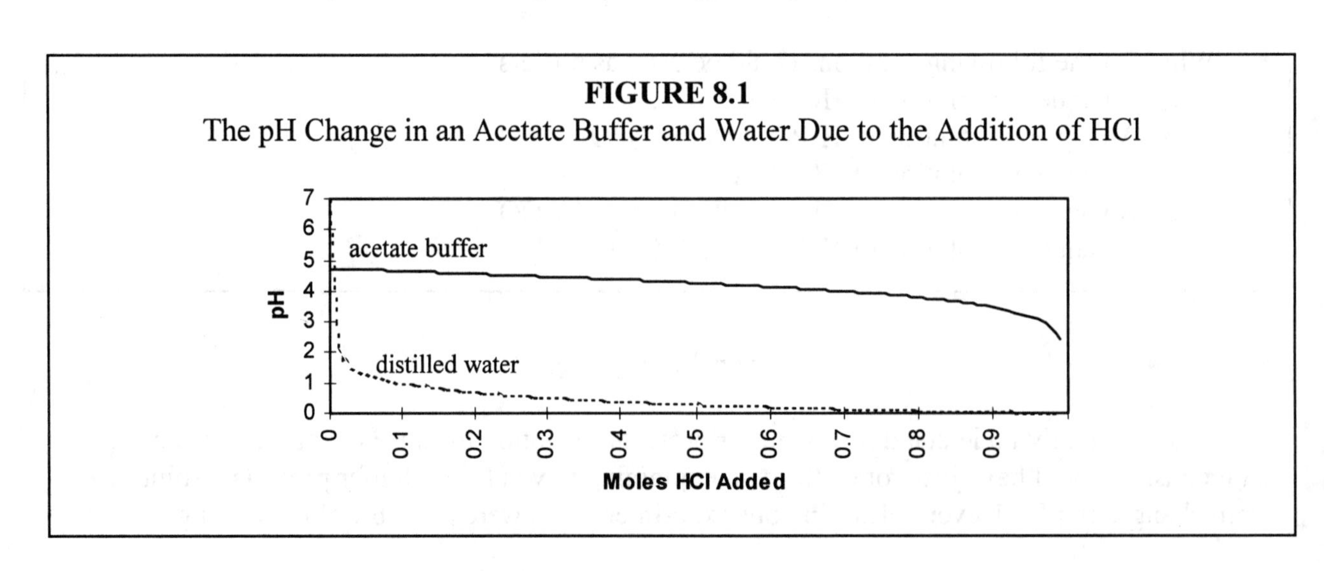

In this figure, I plot the pH change of two solutions. The first is an acetate buffer in which one mole of acetic acid (CH_3COOH) was mixed with one mole of sodium acetate ($NaCH_3COO$) in one liter of solution. The pH of the solution was initially 4.74. The other solution is distilled water, whose pH is initially 7. Notice how the pH changes as HCl is added. With the first addition of HCl, the pH of water decreases dramatically. The pH of the acetate buffer, however, does not change dramatically until about 0.8 moles of HCl are added. Nevertheless, if you look closely, you will see that the pH does decrease a bit. This is because as HCl is added, the acetate ion in the buffer reacts with it:

$$HCl + CH_3COO^- \rightarrow CH_3COOH + Cl^- \quad (8.5)$$

This reaction forms more acetic acid. Thus, the change in pH in the buffer solution is caused by the increase in concentration of acetic acid.

You should see, then, that although the pH of a buffer can change, it changes much more slowly than does a solution that is not a buffer. Now, of course, that's not enough for a chemist. You not only have to *understand* this fact, you also have to be able to *describe* it. For example, there is an equation that tells you the pH of a buffer. It is called the **Henderson-Hasselbalch equation.**

$$pH = pKa + \log\left(\frac{[A^-]}{[HA]}\right) \quad (8.6)$$

This is a very important equation in the subject of buffers, so it bears some explanation.

The first term in the equation is simply the pH of the buffer. What is "pK_a," however? Well, just as "pH" stands for "$-\log([H_3O^+])$," "pK_a" simply stands for "$-\log(K_a)$.

$$pK_a = -\log(K_a) \quad (8.7)$$

What "K_a" does this refer to? Well, a buffer is always made up of a weak acid and its conjugate base. The "pK_a" in Equation (8.6), then, refers to the negative log of the weak acid's K_a.

Now what about that log function? What does "$[A^-]$" and "[HA]" mean? Once again, remember that a buffer is made up of a weak acid and its conjugate base. One way that chemists abbreviate an acid is to use the symbol "HA." The "H" refers to the H^+ that the acid will donate, while the "A" refers to the rest of the acid. Thus, "HNO_3" could be abbreviated as "HA" where $A = NO_3$. What happens when HNO_3 donates its H^+? It forms the conjugate base NO_3^-. In our abbreviation, that's the same as "A^-." Thus, the symbol "HA" is a general symbol for an acid while "A^-" is the general symbol for that acid's conjugate base. In Equation (8.6), then, "[HA]" stands for the concentration of the weak acid in the buffer, while "$[A^-]$" stands for the concentration of the weak acid's conjugate base.

Now all of this might be a bit confusing, because there are so many new terms. The best way to clear the confusion up is with an example.

EXAMPLE 8.1

A buffer is made by taking 100.0 mL of 10.0 M cyanic acid (HCNO) and mixing it with 100.0 grams of potassium cyanate (KCNO). The resulting solution is then diluted with distilled water to a volume of 1.00 liters. What is the pH? (K_a of HCNO is 3.46×10^{-4})

This is a problem that deals with the pH of a buffer. The Henderson-Hasselbalch equation is what we use to determine a buffer's pH. Thus, we need to get all of the terms in Equation (8.6). The first term that we need is pK_a. This refers to the negative log of the K_a of the buffer's weak acid. We are given that K_a, so getting pK_a is easy:

$$pK_a = -\log(K_a)$$

$$pK_a = -\log(3.46 \times 10^{-4})$$

$$pK_a = 3.46$$

Now we need [HA] and [A$^-$]. These terms refer to the concentration of the weak acid and the concentration of its conjugate. Well, we know that 100.0 mL of 10.0 M cyanic acid was used. That tells us how many moles of cyanic acid was used:

$$\frac{10.0 \text{ moles}}{\cancel{L}} \times \frac{0.1000 \cancel{L}}{1} = 1.00 \text{ moles}$$

The solution was diluted to 1.00 liters, so the resulting concentration of cyanic acid is 1.00 M.

What about the conjugate base? Well, we know that 100.0 grams of KCNO were added. We can figure out how many moles that is:

$$\frac{100.0 \cancel{g}}{1} \times \frac{1 \text{ mole}}{81.1 \cancel{g}} = 1.23 \text{ moles}$$

That's how many moles of KCNO were added. When KCNO is dissolved, it splits up into one K^+ and one CNO^-. This tells us that for every mole of KCNO, there is 1 mole of CNO^-. Thus, once the solution is diluted to 1.00 L, the concentration of CNO^- is 1.23 M. Now we have everything we need to use the Henderson-Hasselbalch equation:

$$pH = pKa + \log\left(\frac{[A^-]}{[HA]}\right)$$

$$pH = 3.46 + \log\left(\frac{1.23 \cancel{M}}{1.00 \cancel{M}}\right) = \underline{3.55}$$

Now of course, if that was all there was to the Henderson-Hasselbalch equation, *anyone* could do buffer chemistry! It turns out that this equation is much more powerful than what I just demonstrated. You see, the Henderson-Hasselbalch equation not only allows you to calculate the initial pH of a buffer, but it allows you to calculate the pH of a buffer *as acid or base is added*! See what I mean by studying the following example.

EXAMPLE 8.2

A chemist makes one liter of an acetate buffer in which the acetic acid (CH_3COOH) concentration is 0.500 M and the acetate ion (CH_3COO^-) concentration is 0.750 M. What is the pH? If the chemist then adds 30.0 mL of 1.00 M HCl, what will be the new pH? Neglect the change in volume caused by the addition of acid. (K_a of acetic acid is 1.8×10^{-5}.)

The first part of the question is easy. It is a simple application of Equation (8.6) The concentration of the acid and conjugate base are given, and the pK_a is easily calculated:

$$pK_a = -\log(K_a) = -\log(1.8 \times 10^{-5}) = 4.7$$

Now we can plug our numbers into the Henderson-Hasselbalch equation:

$$pH = pKa + \log\left(\frac{[A^-]}{[HA]}\right)$$

$$pH = 4.7 + \log\left(\frac{0.750\ \cancel{M}}{0.500\ \cancel{M}}\right) = \underline{4.9}$$

That's the initial pH. What is the pH after some HCl is added? Well, the first thing we need to do is to understand what will happen when HCl is added. The HCl will want to react with a base. That basically means either water or the acetate ion. Since the acetate ion is the conjugate base of a weak acid, it is a stronger base than water. Thus, the HCl will react with the acetate ion as follows:

$$HCl + CH_3COO^- \rightarrow CH_3COOH + Cl^-$$

How does this affect the pH? Well, it doesn't affect it directly, because hydronium or hydroxide ions do not appear in the equation. However, it will affect it indirectly. The HCl will use up some acetate ions. That will decrease $[A^-]$ (concentration of the conjugate base) in the Henderson-Hasselbalch equation. In the same way, the reaction produces more acetic acid. This will increase [HA] (concentration of the weak acid) in the Henderson-Hasselbalch equation.

To what extent will these numbers change? We learn that from stoichiometry. The problem says the chemist added 30.0 mL of 1.00 M HCl. Let's determine how many moles that is:

$$\frac{1.00 \text{ moles}}{\cancel{L}} \times \frac{0.0300 \cancel{L}}{1} = 0.0300 \text{ moles}$$

The chemist, then, added 0.0300 moles of HCl. According to the stoichiometry of the equation, this means that the HCl will "eat up" 0.0030 moles of acetate ion and will form 0.0300 moles of acetic acid.

The problem says that we have one liter of solution. Thus, there are 0.500 moles of acetic acid in the solution and 0.750 moles of acetate ion in solution. After the HCl reaction occurs, however, there will be 0.0300 less moles of acetate ion and 0.0300 more moles of acetic acid. This means that there will be 0.720 moles of acetate ion and 0.530 moles of acetic acid. Since the problem told us to neglect the change in volume, we can still say we have one liter of solution. Thus, $[CH_3COO^-] = 0.720$ and $[CH_3COOH] = 0.530$. With those new numbers, we can use the Henderson-Hasselbalch equation to calculate the new pH:

$$pH = pKa + \log\left(\frac{[A^-]}{[HA]}\right) = 4.7 + \log\left(\frac{0.720 \cancel{M}}{0.530 \cancel{M}}\right) = \underline{\underline{4.8}}$$

So, even after 30.0 mL of 1.00 M HCl is added, the pH only changed by 0.1! Compare that to what would happen if you added 30.0 mL of HCl into one liter of distilled water. If you did that, the pH would change from 7.0 to 1.5, a change of 5.5 units!

Now don't get so wrapped up in the math that you fail to see what's going on here. In the problem, we were given a buffer. Based on the concentration of acid and conjugate base, along with the K_a of the acid, we were able to calculate the pH of the buffer. Then, the chemist added some HCl. To determine what effect that would have on pH, we determined what chemical reaction would occur. We then noticed that the resulting reaction would change the concentrations of the acid and conjugate which made up the buffer. Using stoichiometry, we could figure out how much those concentrations changed, and then we re-used the Henderson-Hasselbalch equation to determine the new pH.

I want to give you two more examples before you solve some problems on your own. The first example problem will look much like the one I just showed you. The other will look completely different, but it will use the same concepts in the solution.

EXAMPLE 8.3

A chemist makes 0.500 liters of a buffer in which the oxalic acid ($H_2C_2O_4$) concentration is 0.150 M and the oxalate ion ($HC_2O_4^-$) concentration is 0.750 M. What is the pH? If the chemist then adds 20.0 mL of 2.50 M NaOH, what will be the new pH? Neglect the change in volume caused by the addition of base. (K_a of oxalic acid is 5.9×10^{-2}.)

The initial pH is a straight application of Equation (8.6). I will assume you can determine that the pK_a in this case is 1.2.

$$pH = pKa + \log\left(\frac{[A^-]}{[HA]}\right) = 1.2 + \log\left(\frac{0.750\ \cancel{M}}{0.150\ \cancel{M}}\right) = \underline{1.9}$$

When the NaOH is added, it will want to react with the strongest acid present. Since the two choices are water and oxalic acid, it will react with oxalic acid, since it is the strongest acid:

$$H_2C_2O_4 + NaOH \rightarrow HC_2O_4^- + Na^+ + H_2O$$

This will affect the pH only because this reaction uses up $H_2C_2O_4$ and makes $HC_2O_4^-$. This changes the concentration of these two substances, changing the Henderson-Hasselbalch equation. How much are the concentrations changed? For that, we do some stoichiometry. The chemist added:

$$\frac{2.50\ \text{moles}}{\cancel{L}} \times \frac{0.0200\ \cancel{L}}{1} = 0.0500\ \text{moles}$$

of NaOH. Based on the stoichiometry of the equation, this means that 0.0500 moles of $H_2C_2O_4$ will get used up and 0.0500 moles of $HC_2O_4^-$ will get made. How much of each of these substances did we have to begin with? Well, we have 0.500 liters. Thus:

$$\frac{0.150\ \text{moles}\ H_2C_2O_4}{\cancel{L}} \times \frac{0.500\cancel{L}}{1} = 0.0750\ \text{moles}\ H_2C_2O_4$$

$$\frac{0.750\ \text{moles}\ HC_2O_4^-}{\cancel{L}} \times \frac{0.500\cancel{L}}{1} = 0.375\ \text{moles}\ HC_2O_4^-$$

Since the addition of NaOH adds 0.0500 moles to the amount of $HC_2O_4^-$ and takes 0.0500 moles of $H_2C_2O_4$ away, the new amounts are 0.425 moles $HC_2O_4^-$ and 0.0250 moles $H_2C_2O_4$. Now that we have the new amounts, we can go back to the concentrations:

$$[HC_2O_4^-] = \frac{0.425\ \text{moles}}{0.500\text{L}} = 0.850\ \text{M}$$

$$[H_2C_2O_4] = \frac{0.0250\ \text{moles}}{0.500\text{L}} = 0.0500\ \text{M}$$

Now we can finally use the Henderson-Hasselbalch equation:

$$pH = pKa + \log\left(\frac{[A^-]}{[HA]}\right) = 1.2 + \log\left(\frac{0.850\ \cancel{M}}{0.050\ \cancel{M}}\right) = \underline{2.4}$$

A chemist takes 450.0 mL of a 4.50 M solution of hydrofluoric acid (HF) and adds 25.0 grams of LiOH. What is the resulting pH? The K_a of HF is 7.2×10^{-4} and the addition of solid does not appreciably affect the volume of the solution.

Now this might not look like a buffer problem, but it really is. How do I know? Well, what will happen when LiOH is added to HF? The following reaction will occur:

$$LiOH + HF \rightarrow H_2O + Li^+ + F^-$$

We can do stoichiometry to determine how much of each substance exists after the reaction. To do this, we need to know how much of each reactant we have:

$$\frac{4.50 \text{ moles HF}}{\cancel{L}} \times \frac{0.450 \cancel{L}}{1} = 2.03 \text{ moles HF}$$

$$\frac{25.0 \text{ g} \cancel{\text{LiOH}}}{1} \times \frac{1 \text{ mole LiOH}}{23.9 \text{ g} \cancel{\text{LiOH}}} = 1.05 \text{ moles LiOH}$$

Since the reaction says each mole of HF needs 1 mole of LiOH, LiOH is the limiting reagent. This means that after the reaction is over, there will be no LiOH, 1.05 moles of Li^+, 1.05 moles of F^-, 1.05 more moles of water, and 2.03-1.05 = 0.98 moles of HF left over.

How do we calculate the pH? Well, HF is a weak acid, so we could just use the acid ionization constant equation to determine the concentration of hydronium ions and then use that to calculate pH. There is an easier way, though. Notice that now we have a solution which consists of a weak acid (HF) and its conjugate base (F^-). What is that? It is a *buffer*. Thus, we can just use the Henderson-Hasselbalch equation. Before we do that, however, we must first get back to concentration units:

$$\frac{0.98 \text{ moles HF}}{0.450 \text{ L}} = 2.2 \text{ M HF}$$

$$\frac{1.05 \text{ moles } F^-}{0.450 \text{ L}} = 2.33 \text{ M } F^-$$

$$pH = pK_a + \log\left(\frac{[A^-]}{[HA]}\right) = 3.1 + \log\left(\frac{2.33}{2.2}\right) = \underline{3.1}$$

In the end, then, these buffer pH problems involve being able to use the Henderson-Hasselbalch equation, but also being able to determine what reactions occur between acids and bases in solution as well as being able to do stoichiometry on the equations. Make sure you understand this by solving thc following "on your own" problems.

ON YOUR OWN

8.2 A chemist makes a buffer solution with $H_2PO_4^-$ and its conjugate.

a. What reaction occurs when HNO_3 is added to the buffer?
b. What reaction occurs when LiOH is added to the buffer?

8.3 A chemist makes a buffer solution by mixing 10.0 mL of 12.0 M HClO with 5.00 grams of KClO. The solution is then diluted to 0.500 liters. What is the pH? (K_a of HClO is 3.5×10^{-8}).

8.4 The buffer solution in problem 8.3 has 0.15 grams of KOH added to it. What is the new pH?

8.5 A chemist takes 500.0 mL of 1.50 M acetic acid (CH_3COOH) and adds 5.00 grams of LiOH to it. What is the pH? The addition of solid does not add appreciably to the volume of the solution. (K_a of acetic acid is 1.8×10^{-5})

The Common Ion Effect and pH

Do you remember the common ion effect you learned about in Module #6? You had better! Anyway, the phenomenon of pH plays into the common ion effect in a way you might not think about initially. Suppose, for example, you wanted to dissolve aluminum hydroxide in water. This ionic compound would split into its ions according to the equilibrium:

$$Al(OH)_3\ (s) \rightleftarrows Al^{3+}\ (aq) + 3OH^-\ (aq) \qquad (8.8)$$

The solubility product for $Al(OH)_3$ is very small (1.9×10^{-33}), indicating that this is a mostly insoluble solid.

If I asked you to calculate the concentration of aluminum ions in a saturated solution of aluminum hydroxide, what would you do? Well, you could say that when "X" aluminum hydroxide results, "X" aluminum ion and "3X" hydroxide ion would form. You could then set up the solubility product equation as follows:

$$1.9 \times 10^{-33} = [Al^{3+}]\cdot[OH^-]^3$$

$$1.9 \times 10^{-33} = (X)\cdot(3X)^3$$

$$X = 2.9 \times 10^{-9}$$

Since the aluminum ion concentration is "X," you could then say that the concentration of aluminum ions in solution is 2.9×10^{-9} M.

Now despite the fact that all of that looks perfectly reasonable, there is a problem. What's the problem? Well, think about it. There is a common ion in solution. How can I say that? Aren't we dissolving the aluminum hydroxide in water? Where's the common ion? *It's in the water*. Remember, water has a pH of 7. This means that the concentration of hydronium ions is 1×10^{-7} M. It also means that *the concentration of hydroxide ions is* 1×10^{-7} M. Thus, the self-ionization of water makes this a common ion effect problem, and it must be treated as such. Study the following example to see how this problem is done correctly:

EXAMPLE 8.4

What is the aluminum ion concentration in a saturated solution of aluminum hydroxide? (K_{sp} = 1.9×10^{-33})

To do this problem correctly, you must recognize that the self-ionization of water already results in a hydroxide ion concentration of 1.0×10^{-7}. Thus, even before the aluminum hydroxide dissolves, there are hydroxide ions in solution. Thus, dissolving aluminum hydroxide will *add* to the hydroxide concentration. Therefore, when "X" aluminum hydroxide dissolves:

$$Al(OH)_3\ (s) \rightleftarrows Al^{3+}\ (aq) + 3OH^-\ (aq)$$

the concentration of Al^{3+} will be "X," but the concentration of OH^- will be "1.0×10^{-7} +3X." The correct solubility product equation, then, is:

$$1.9 \times 10^{-33} = (X)\cdot(1.0 \times 10^{-7} + 3X)^3$$

Hopefully, "X" is small. If so, we need not worry about adding it to 1.0×10^{-7}. If so, then the equation reduces to:

$$1.9 \times 10^{-33} = (X)\cdot(1.0 \times 10^{-7})^3$$

$$X = 1.9 \times 10^{-12}$$

Our assumption was right, since $3\cdot1.9 \times 10^{-12}$ is too small to add to 1.0×10^{-7}. Thus, the aluminum ion concentration is $\underline{1.9 \times 10^{-12}\ M}$. Notice that this is more than 1,000 times smaller than the 2.9×10^{-9} M we calculated before considering the self-ionization of water.

The example shows you, then, that when you are dealing with hydroxides, the concentration of hydroxide ions can severely affect the solubility. Thus, when doing solubility (or any other equilibrium-type) problems with hydroxides, be sure to think about the self-ionization of water. Also remember that the pH affects the self-ionization of water. In general, then, you need to know the pH of a solution if you are trying to dissolve hydroxides into it. If the pH is not given, you may assume that it is 7. If it is given, be sure to use that in your calculation. Let me show you what I mean.

EXAMPLE 8.5

How many moles of $Cr(OH)_2$ can be dissolved in one liter of a solution whose pH is 8.0? ($K_{sp} = 6.7 \times 10^{-31}$)

When chromium (II) hydroxide dissolves, it does so according to the following equilibrium:

$$Cr(OH)_2\ (s) \rightleftarrows Cr^{2+}\ (aq) + 2OH^-\ (aq)$$

This equilibrium produces hydroxide, which is already present in any aqueous solution. To determine what concentration the hydroxide ion has in this solution, we look at the pH to get the hydronium ion concentration:

$$pH = -\log([H_3O^+])$$

$$8.0 = -\log([H_3O^+])$$

$$[H_3O^+] = 1.0 \times 10^{-8}$$

Now that we have the hydronium ion concentration, we can get the hydroxide ion concentration:

$$[H_3O^+]\cdot[OH^-] = 1 \times 10^{-14}$$

$$(1.0 \times 10^{-8})\cdot[OH^-] = 1 \times 10^{-14}$$

$$[OH^-] = 1 \times 10^{-6}$$

When "X" chromium (II) hydroxide dissolves, it will form "X" Cr^{2+} and "$1 \times 10^{-6} + 2X$" hydroxide. This makes the solubility product equation:

$$6.7 \times 10^{-31} = X\cdot(1 \times 10^{-6} + 2X)^2$$

Assuming that 2X is so small it cannot be added to 1×10^{-6}:

$$6.7 \times 10^{-31} = X\cdot(1 \times 10^{-6})^2$$

$$X = 6.7 \times 10^{-19}$$

This tells us the assumption is correct. Since "X" represents the number of moles of chromium (II) hydroxide that will dissolve in a liter of liquid, our answer is $\underline{6.7 \times 10^{-19}\ \text{moles}}$.

ON YOUR OWN

8.6 What is the tin ion concentration in a saturated solution of $Sn(OH)_2$ at pH = 9.0? ($K_{sp} = 5.0 \times 10^{-26}$)

The Technique of Successive Approximations

I want to use the concept of the common ion effect to introduce a very important problem-solving technique that I have put off until now. All along, the equilibrium problems I have given you can either be solved with the quadratic equation or with an assumption. It turns out that *most* equilibrium problems can be solved in this way. Not all of them can, however. Please follow the next example closely, as it will introduce the important problem-solving technique of **successive approximations**.

EXAMPLE 8.6

How many moles of $Cu(OH)_2$ can be dissolved in one liter of a solution whose pH is 7.8? ($K_{sp} = 5.6 \times 10^{-20}$)

We know that this is a common ion effect problem because when copper (II) hydroxide dissolves in water, it splits up into its ions:

$$Cu(OH)_2\,(s) \rightleftarrows Cu^{2+}\,(aq) + 2OH^-\,(aq)$$

Using the pH along with the self-ionization of water, we can learn that the hydroxide ion concentration at this pH is 6.3×10^{-7} M. We know that when "X" copper (II) hydroxide dissolves, it makes "X" Cu^{2+} and "$6.3 \times 10^{-7} + 2X$" OH^-. This makes the solubility product equation:

$$5.6 \times 10^{-20} = (X)\cdot(6.3 \times 10^{-7} + 2X)^2$$

Let's assume "X" is too small to add to 6.3×10^{-7}. If that's the case, we get:

$$5.6 \times 10^{-20} = (X)\cdot(6.3 \times 10^{-7})^2$$

$$X = 1.4 \times 10^{-7}$$

Clearly, then, the assumption will not work. We therefore cannot solve the equation this way. We cannot solve this equation exactly, either, because when you multiply it out, it contains an "X" term and an "X^3" term. We do not have the algebra skills to solve such an equation.

Are we out of luck, then? Of course not. I wouldn't have begun this problem if we were. At this point, we will employ the technique called "successive approximations." In this technique, we go ahead and accept the answer given by the assumption. Sure, it's wrong, but for now it's the best we can do. Now, we will *take that value of "X" and plug it back into the part of the equation where we ignored it before. Then we will solve for a new "X"*. Thus,

$$5.6 \times 10^{-20} = (X)\cdot(6.3 \times 10^{-7} + 2\cdot(1.4 \times 10^{-7}))^2$$

$$X = 6.8 \times 10^{-8}$$

Is this the correct answer? No, it is not. We can, however, plug *this* in the same place we plugged the previous value for "X" in and we can try again:

$$5.6 \times 10^{-20} = (X)\cdot(6.3 \times 10^{-7} + 2\cdot(6.8 \times 10^{-8}))^2$$

$$X = 1.0 \times 10^{-7}$$

Is this the correct answer? No, but it's closer. Let's plug *this* into the equation the same place we've been plugging the others in:

$$5.6 \times 10^{-20} = (X)\cdot(6.3 \times 10^{-7} + 2\cdot(1.0 \times 10^{-7}))^2$$

$$X = 8.1 \times 10^{-8}$$

I think we're almost there. Let's try again:

$$5.6 \times 10^{-20} = (X)\cdot(6.3 \times 10^{-7} + 2\cdot(8.1 \times 10^{-8}))^2$$

$$X = 8.9 \times 10^{-8}$$

Again:

$$5.6 \times 10^{-20} = (X)\cdot(6.3 \times 10^{-7} + 2\cdot(8.9 \times 10^{-8}))^2$$

$$X = 8.5 \times 10^{-8}$$

Again:

$$5.6 \times 10^{-20} = (X)\cdot(6.3 \times 10^{-7} + 2\cdot(8.5 \times 10^{-8}))^2$$

$$X = 8.7 \times 10^{-8}$$

Again:

$$5.6 \times 10^{-20} = (X)\cdot(6.3 \times 10^{-7} + 2\cdot(8.7 \times 10^{-8}))^2$$

$$X = 8.7 \times 10^{-8}$$

Notice what happened. The answer kept fluctuating, but all the while it began converging to a final number. That number did not change between 2 successive approximations. No matter how far you go now, the number will not change. Thus, we have found "X." This tells us that $\underline{8.7 \times 10^{-8}\ \text{moles}}$ of copper (II) hydroxide can dissolve in a liter of pH 7.8 water.

Although the technique of successive approximations is a lot of work, it is a powerful way to solve problems you ordinarily cannot solve. Although tedious, it really isn't that hard. All you do is make the same assumption we have always made. If the assumption works, you are done. If it doesn't work, put the answer you got for "X" back into the equation *at the same place you ignored it before*. Then, solve the problem for a new "X." As you continue to do this, you will see the values for "X" start to converge. Continue until the answer does not change. I will give you one more example to make sure you understand this.

EXAMPLE 8.7

How many moles of $Fe(OH)_3$ can be dissolved in one liter of a solution whose pH is 5.3? (K_{sp} = 1.1×10^{-36})

At this pH, the hydroxide ion concentration is 2.0×10^{-9} M. You can calculate this fact from the pH and the self-ionization of water. When iron (III) hydroxide dissolves, it makes hydroxide ions, so this is clearly a common ion effect problem. The equilibrium is:

$$Fe(OH)_3 \text{ (s)} \rightleftarrows Fe^{3+} \text{ (aq)} + 3OH^- \text{ (aq)}$$

When "X" iron (III) hydroxide is dissolved, it will make "X" Fe^{3+} and "$2.0 \times 10^{-9} + 3X$" OH^-. The solubility product equation, then, is:

$$1.1 \times 10^{-36} = (X)\cdot(2.0 \times 10^{-9} + 3X)^3$$

We will start with the assumption, which you can bet won't work:

$$1.1 \times 10^{-36} = (X)\cdot(2.0 \times 10^{-9})^3$$

$$X = 1.4 \times 10^{-10}$$

The assumption is clearly not valid. We cannot solve this problem exactly, however, because the algebra is beyond our means. Thus, we will employ successive approximations. Let's take that value and plug it in for "X" where we ignored "X" before:

$$1.1 \times 10^{-36} = (X)\cdot(2.0 \times 10^{-9} + 3\cdot(1.4 \times 10^{-10}))^3$$

$$X = 7.8 \times 10^{-11}$$

All right, let's do it again:

$$1.1 \times 10^{-36} = (X)\cdot(2.0 \times 10^{-9} + 3\cdot(7.8 \times 10^{-11}))^3$$

$$X = 9.9 \times 10^{-11}$$

Again:

$$1.1 \times 10^{-36} = (X)\cdot(2.0 \times 10^{-9} + 3\cdot(9.9 \times 10^{-11}))^3$$

$$X = 9.1 \times 10^{-11}$$

Again:

$$1.1 \times 10^{-36} = (X)\cdot(2.0 \times 10^{-9} + 3\cdot(9.1 \times 10^{-11}))^3$$

$$X = 9.4 \times 10^{-11}$$

Again:

$$1.1 \times 10^{-36} = (X)\cdot(2.0 \times 10^{-9} + 3\cdot(9.4 \times 10^{-11}))^3$$

$$X = 9.3 \times 10^{-11}$$

Again:

$$1.1 \times 10^{-36} = (X)\cdot(2.0 \times 10^{-9} + 3\cdot(9.3 \times 10^{-11}))^3$$

$$X = 9.3 \times 10^{-11}$$

Finally! The solution did not change from one step to the next, so we are done. This tells us that 9.3×10^{-11} moles of iron (III) hydroxide will dissolve in a liter of water at pH = 5.3.

ON YOUR OWN

8.7 What is the manganese ion concentration in a saturated solution of $Mn(OH)_2$ if the initial pH is 9.9? ($K_{sp} = 4.5 \times 10^{-14}$)

Other Equilibrium Situations

Although I have spent the majority of my discussions on equilibrium focusing on solubility equilibria and acid/base equilibria, it is important to be aware that many other chemical reactions form equilibria which can be analyzed using the same techniques you have applied to solubility equilibria and acid/base equilibria. Study the following example to see what I mean.

EXAMPLE 8.8

Some nail polish removers use the solvent ethyl acetate ($CH_3CO_2C_2H_5$) to dissolve dried nail polish. This substance can be made via the following equilibrium:

$$CH_3COOH\ (l) + C_2H_5OH\ (l) \rightleftarrows CH_3CO_2C_2H_5\ (l) + H_2O\ (l)$$

whose equilibrium constant is 4.00. If a chemist starts with 200 mL of 12.3 M CH_3COOH and 100 mL of 11.2 M C_2H_5OH and then dilutes the mixture to one liter, what will be the equilibrium concentrations of all substances in the reaction?

We can determine the initial concentration of the reactants by using the concentration and volumes given in the problem to determine the number of moles of each reactant used:

$$\frac{12.3 \text{ moles } CH_3COOH}{\cancel{L}} \times \frac{0.200 \cancel{L}}{1} = 2.46 \text{ moles } CH_3COOH$$

$$\frac{11.2 \text{ moles } C_2H_5OH}{\cancel{L}} \times \frac{0.100 \cancel{L}}{1} = 1.12 \text{ moles } C_2H_5OH$$

Since we have one liter of solution, this means $[CH_3COOH] = 2.46$ M and $[C_2H_5OH] = 1.12$ M.

Now what happens in the reaction? If the reaction were a complete reaction, C_2H_5OH would be the limiting reactant. Now, of course, it isn't a complete reaction. Nevertheless, we learned in stoichiometry to always do things in terms of the limiting reactant, so that's what we'll do here. Although we do not know how much C_2H_5OH reacts, we can say that "X" amount of it reacts. When that happens, it will decrease the CH_3COOH and C_2H_5OH concentrations by "X", and it will increase the concentration of each product by "X". This means that, at equilibrium, the concentration of CH_3COOH will be "2.46 - X," the concentration of C_2H_5OH will be "1.12-X," the concentration of $CH_3CO_2C_2H_5$ will be "X," and the concentration of water will be "X."

Now that we have the concentrations of each substance at equilibrium, we can use the equation for the equilibrium constant of this reaction:

$$K = \frac{[CH_3CO_2C_2H_5]\cdot[H_2O]}{[CH_3COOH]\cdot[C_2H_5OH]}$$

$$4.00 = \frac{X \cdot X}{(2.46 - X)\cdot(1.12 - X)}$$

Note that we could not ignore water, because all substances were liquids. We only ignore water when other substances are aqueous. How do we solve this equation? We could assume that "X" is small, but that won't work. The equilibrium constant is greater than one. Thus, this equilibrium forms mostly products. Thus, "X" will be large! This is a quadratic equation, however, so we can solve it exactly.

$$4.00\cdot(2.46-X)\cdot(1.12-X) = X^2$$

$$4.00\cdot(2.76-3.58\cdot X+X^2) = X^2$$

$$3\cdot X^2 - 14.3\cdot X + 11.0 = 0$$

$$X = \frac{14.3 \pm \sqrt{14.3^2 - 4\cdot 3\cdot 11.0}}{2\cdot 3} = 3.80 \text{ or } 0.967$$

Wait a minute here. Both answers are positive. Which do we choose? Well, despite the fact that both answers are positive, there is still only one physically meaningful choice. Look at how much limiting reactant we started with. There are only 1.12 moles of C_2H_5OH. The variable "X" represents *how many* of those moles react. If we start with 1.12, can we ever have 3.80 react? Of course not! Thus, 0.967 is the only physically meaningful result. That's "X." Using our definitions of concentration in terms of "X," we find that $[CH_3COOH] = 1.49$ M, $[C_2H_5OH] = 0.15$ M, $[CH_3CO_3C_2H_5] = 0.967$ M, and $[H_2O] = 0.967$ M.

Notice that although this is neither a solubility or acid/base problem, we have used no new techniques here. We had to determine the limiting reactant to figure out what reactant to relate to "X," but after that, the problem was solved just like all of the other problems we have solved in reference to equilibrium. Thus, although this is a new situation, no new techniques were used.

Another thing that you must learn when it comes to equilibrium is that equilibria which are composed entirely of gases usually have their own, unique kind of equilibrium constant. You should recall from your previous chemistry course that the pressure of a gas is directly proportional to its concentration. The more concentrated a gas is, the higher its pressure will be. As a result, an equilibrium that has only gases in it can have an equilibrium constant expressed in terms of pressure instead of concentration. Typically, we denote such equilibrium constants with a "p" subscript, to make it clear that the equilibrium constant is based on pressure, not concentration.

Thus, if our "standard" chemical reaction contained only gases:

$$aA\,(g) + bB\,(g) \rightleftarrows cC\,(g) + dD\,(g)$$

the equilibrium constant could be expressed as follows:

$$K_p = \frac{p_C{}^c\cdot p_D{}^d}{p_A{}^a\cdot p_B{}^b}$$

In this equation, "p_A" represents the pressure of gas "A," and so on for each gas in the equation. This, of course, looks just like the normal expression we see for the equilibrium constant except for the fact that the pressure of the gases are used instead of their concentrations. Other than that, this equilibrium constant has all of the properties of the equilibrium constant you are used to using. Here is an example of how you can use this kind of equilibrium constant.

EXAMPLE 8.9

At low temperatures, the reaction between nitrogen gas and hydrogen gas to make ammonia is rather small. For example, at 100^0 C:

$$N_2\,(g) + 3H_2\,(g) \rightleftarrows 2NH_3\,(g) \qquad K_p = 0.00412$$

If a chemist fills a vessel with 1.10 atms of nitrogen gas and 4.10 atms of hydrogen gas, what will be the pressures of each gas at equilibrium?

Although this problem uses K_p instead of the equilibrium constant based on concentration, we can still solve it using the same techniques that we have learned up to now. Since the equation tells us we need three times as much hydrogen as nitrogen, and since the problem says we have nearly 4 times as much hydrogen as nitrogen, the nitrogen will clearly run out first. Thus, the nitrogen is the limiting reagent. When "X" atms of nitrogen react, they will use up "3X" atms of hydrogen and produce "2X" atms of ammonia. Thus, at equilibrium, there will be "1.10-X" atms of nitrogen, "4.10-3X" atms of hydrogen, and "2X" atms of ammonia. The equilibrium constant expression, then, is:

$$K_p = \frac{p_{NH_3}^{\ 2}}{p_{N_2} \cdot p_{H_2}^{\ 3}}$$

$$0.00412 = \frac{(2X)^2}{(1.10 - X)\cdot(4.10 - 3X)^3}$$

We can solve this by hoping that "X" is too small to add to 1.10 and that "3X" is too small to add to 4.10. If that were the case, then:

$$0.00412 = \frac{(2X)^2}{(1.10)\cdot(4.10)^3}$$

$$X = 0.279$$

The assumption is not correct. We cannot solve the equation exactly, because there would be an X^4 term in it. Thus, we will have to do successive approximations. Remember, in successive approximations, you put the previous answer into the equation *everywhere* you ignored "X." We ignored "X" in two different places, so we need to put 0.279 in for "X" in two places:

$$0.00412 = \frac{(2X)^2}{(1.10-0.279)\cdot(4.10-3\cdot 0.279)^3}$$

$$X = 0.171$$

Now we start the tedious business of continually replacing "X" with the previous answer until the answers do not change:

$$0.00412 = \frac{(2X)^2}{(1.10-0.171)\cdot(4.10-3\cdot 0.171)^3}$$

$$X = 0.210$$

$$0.00412 = \frac{(2X)^2}{(1.10-0.210)\cdot(4.10-3\cdot 0.210)^3}$$

$$X = 0.196$$

$$0.00412 = \frac{(2X)^2}{(1.10-0.196)\cdot(4.10-3\cdot 0.196)^3}$$

$$X = 0.201$$

$$0.00412 = \frac{(2X)^2}{(1.10-0.201)\cdot(4.10-3\cdot 0.201)^3}$$

$$X = 0.199$$

$$0.00412 = \frac{(2X)^2}{(1.10-0.199)\cdot(4.10-3\cdot 0.199)^3}$$

$$X = 0.200$$

$$0.00412 = \frac{(2X)^2}{(1.10-0.200)\cdot(4.10-3\cdot 0.200)^3}$$

$$X = 0.199$$

The value just repeated what it was two times ago. Thus, I am done. The value for "X" is 0.199. At equilibrium, then, the pressure of nitrogen is 0.90 atms, the pressure of hydrogen is 3.50 atms, and the pressure of ammonia is 0.398 atms.

This example should have shown you two things. First, it should have shown you how pressure can be used in place of concentration for gases in equilibrium calculations. Also, it should have shown you yet another type of problem that can use successive approximations as a problem-solving technique. There is one more thing you need to see before you are done with this module. You may or may not have noticed that in all of the equilibrium calculations we have done so far, the equilibrium constants (be they solubility products, acid ionization constants, or regular equilibrium constants) have all been rather small. In fact, the largest equilibrium constant used in any example or problem I have given you was 4.00. Of course, equilibrium constants can be much, much larger than that. When dealing with a large equilibrium constant, however, you must apply a very important trick to be able to solve the problem. Study the following example to see what that trick is.

EXAMPLE 8.10

For the reaction

$$2NO\ (g) + Cl_2\ (g) \rightleftarrows 2NOCl\ (g)$$

$K_p = 2.5 \times 10^3$. What pressure of each gas exists at equilibrium if the reaction started with 4.00 atms of NO and 1.90 atms of Cl_2?

This looks like a standard equilibrium-type problem. There is, however, a subtle difference. The equilibrium constant is rather large. This is a problem. You see, if we set this up the same way we set up all other problems, here is the equation we would get:

$$2.5 \times 10^3 = \frac{(2X)^2}{(4.00 - 2X)^2 \cdot (1.90 - X)}$$

If we assume that "X" is too small to add to 1.90 and "2X" is too small to add to 4.00, we would solve the equation to find that $X = 1.4 \times 10^2$. Obviously, then, we cannot assume that "X" is small. We can't solve the equation exactly, either, because it has an X^3 term in it. Thus, we try successive approximations, right? Wrong! Successive approximations only work if the value for "X" is smaller than the numbers from which it is being subtracted. Thus, we can't make our approximation; we can't solve the equation exactly; and we can't use successive approximations. What do we do?

Well, all of our problem-solving techniques besides solving the equation exactly depend on the variable for which we are solving being small. Thus, we need to find a way to get "X" to be small. How do we do that? Well, the equilibrium constant is so large, that it is almost like this reaction is a complete reaction. Thus, let's first assume it is. If the reaction is complete, the stoichiometry tells us that the 1.90 atms of Cl_2 will run out first. When it does, it will have used up 3.80 atms of NO and it will have made 3.80 atms of NOCl. Thus, after the complete reaction, there will be no Cl_2, 0.20 atms of NO, and 3.80 atms of NOCl, right?

Now of course, we know this is not a complete reaction. It is an equilibrium. Let's suppose, however, that rather than starting with 4.00 atms of NO and 1.90 atms of Cl_2, we actually started with the mixture we just determined. Let's assume that the equilibrium *starts* with no Cl_2, 0.20 atms of NO, and 3.80 atms of NOCl. How does this help? Well, watch and see. If the equilibrium started like this, then "X" NOCl will break apart to form "X" NO and "0.5X Cl_2." If you don't like "0.5X," look back at the equation. The equation says that for every 2 moles of NOCl, there is one mole of Cl_2. Thus, there will be half as much Cl_2 formed as NOCl that breaks up. Thus, if "X" NOCl breaks up, it will form "0.5X" Cl_2.

When that happens, the NO concentration will increase from 0.2 atms to "0.20 + X" atms. The Cl_2 concentration will increase from zero atms to "0.5X" atms, and the NOCl concentration will decrease from 3.8 atms to "3.80 - X" atms. Now the equilibrium constant equation becomes:

$$2.5\times10^{3} = \frac{(3.80 - X)^{2}}{(0.5\cdot X)\cdot(0.20 + X)^{2}}$$

Now this might not look much better, but it really is. Look what happens when we *now* assume that "X" is too small to subtract from 3.8 or add to 0.2:

$$2.5\times10^{3} = \frac{(3.80)^{2}}{(0.5\cdot X)\cdot(0.20)^{2}}$$

$$X = 0.29$$

Now X is small. It is not too small to subtract from 3.8, however, so we will still have to do successive approximations, but that's not too bad:

$$2.5\times10^{3} = \frac{(3.80-0.29)^{2}}{(0.5\cdot X)\cdot(0.20+0.29)^{2}}$$

$$X = 0.041$$

$$2.5\times10^3 = \frac{(3.80-0.041)^2}{(0.5\cdot X)\cdot(0.20+0.041)^2}$$

$X=0.19$

$$2.5\times10^3 = \frac{(3.80-0.19)^2}{(0.5\cdot X)\cdot(0.20+0.19)^2}$$

$X=0.069$

$$2.5\times10^3 = \frac{(3.80-0.069)^2}{(0.5\cdot X)\cdot(0.20+0.069)^2}$$

$X=0.15$

$$2.5\times10^3 = \frac{(3.80-0.15)^2}{(0.5\cdot X)\cdot(0.20+0.15)^2}$$

$X=0.087$

.
.
.

If you carry this on, you will find that the answers start to fluctuate between X = 0.12 and X = 0.10. Thus, the answer is X = 0.11. This tells me that at equilibrium, the pressure of Cl_2 (g) is 0.055 atms, the pressure of NO is 0.33 atms, and the pressure of NOCl is 3.69 atms.

Now this might have seemed like a long, complicated problem, but it really wasn't. Think about it. All we did was recognize that since the equilibrium constant was really large, the reaction was almost a complete reaction. Thus, we treated it as a complete reaction and determined the pressures of the gases assuming that all of the limiting reagent was used. *At that point,* we then decided to treat it as an equilibrium. We used *those new values of pressure* and treated the reaction as if it ran in reverse. This helped us because the techniques that we use to solve the equation when it cannot be solved exactly rely on "X" being small. Since the reaction is nearly a complete reaction, the "X" that you solve for in that way will be small.

This problem-solving technique is the one you will want to use whenever you see a large (greater than 100) equilibrium constant. To solve such problems, just run the reaction to completion and then do an equilibrium calculation with those pressures (or concentrations). This little "trick" completes your repertoire of equilibrium problem-solving techniques.

ON YOUR OWN

8.8 A chemist studies the following reaction

$$PCl_5\ (g) \rightleftarrows PCl_3\ (g) + Cl_2\ (g)$$

at a temperature for which $K_p = 0.0211$. What is the pressure of each gas at equilibrium if the chemist started with 5.00 atms of PCl_5?

8.9 For the following equilibrium at a certain temperature:

$$2Cl_2\ (g) + 2H_2O\ (g) \rightleftarrows 4HCl\ (g) + O_2\ (g)$$

$K_p = 0.051$. What are the pressures of all gases at equilibrium if the reaction started with 1.00 atms of both H_2O and Cl_2?

8.10 At low temperatures, the following equilibrium:

$$2NO\ (g) + Br_2\ (g) \rightleftarrows 2NOBr$$

has a K_p of 3.5×10^4. What is the pressure of each gas at equilibrium if 1.00 atm of Br_2 and 2.50 atm of NO are placed in a low-temperature container?

You've learned a lot of problem-solving techniques in this chapter, but I don't want them to obscure the chemistry you've learned. Remember, *all chemical reactions are equilibrium reactions*. Whether we are talking about acid/base reactions, solubility processes, or other types of chemical reactions, they are *all* equilibria. Now if the K of an equilibrium is incredibly large, we can approximate and say that it is a complete reaction. This is not true, but the equilibrium is so heavily weighted towards products that it might as well be a complete reaction. Likewise, even though we might say that a reaction does not happen, it does to some extent. The equilibrium constant is just so small that, for all practical purposes, it is like no products are formed. That's why equilibrium is so important. It applies to *all* chemical reactions. That's why I spent 3 modules of the course talking about it.

Even the information I gave you in these 3 modules does not encompass all of what you need to know about equilibrium. In your first year course, you should have learned how to use LeChatelier's principle to predict which way an equilibrium will shift in response to stress. You should also be able to determine whether or not a reaction is at equilibrium by looking at the concentrations of all substances. If the reaction is not at equilibrium, you should be able to predict which way it will shift in order to achieve equilibrium. Obviously, then, there are a *lot* of things you need to know about equilibrium! Between this course and your first-year course, however, you should have covered them all.

ANSWERS TO THE ON YOUR OWN PROBLEMS

8.1 a. This is not a buffer. HCl is a strong acid and NaOH is not the conjugate.

b. This is a buffer. H_3PO_4 is not on our list of strong acids, so it must be weak. In solution, NaH_2PO_4 will split up into the Na^+ ion and the $H_2PO_4^-$ ion. $H_2PO_4^-$ is the conjugate base of H_3PO_4. Thus, we have a weak acid (H_3PO_4) and its conjugate ($H_2PO_4^-$).

c. This is not a buffer. HCNO is a weak acid and NH_3 is a base, but the base must be the conjugate of the weak acid.

d. This is a buffer. CH_3COOH (acetic acid) is a weak acid and KCH_3COO splits up in solution into K^+ and CH_3COO^-. We therefore have a weak acid and its conjugate.

e. This is a buffer. In solution, $NaHCO_3$ will split up into Na^+ and HCO_3^-. In the same way, K_2CO_3 will split into $2K^+$ and CO_3^{2-}. HCO_3^- is a weak acid and CO_3^{2-} is its conjugate base.

8.2 The conjugate of $H_2PO_4^-$ is HPO_4^{2-}. Those are the two compounds in the buffer.

a. When HNO_3 (a strong acid) is added, it will want to react with the strongest base. Water, $H_2PO_4^-$, and HPO_4^{2-} are all possible bases, but HPO_4^{2-} is the strongest, because it is the conjugate of the weak acid. Thus, HNO_3 will react with it:

$$\underline{HNO_3 + HPO_4^{2-} \rightarrow H_2PO_4^- + NO_3^-}$$

b. When LiOH (a strong base) is added, it will want to react with the strongest acid. Water, $H_2PO_4^-$, and HPO_4^{2-} are all possible acids, but $H_2PO_4^-$ is the strongest, because it is the weak acid in the buffer. Thus, LiOH will react with it:

$$\underline{LiOH + H_2PO_4^- \rightarrow HPO_4^{2-} + Li^+ + H_2O}$$

8.3 The solution in question is a buffer, because it is a weak acid (HClO) and its conjugate base (ClO^-, which comes from KClO). The Henderson-Hasselbalch equation is what we use to determine a buffer's pH. Thus, we need to get all of the terms in that equation. The first term that we need is pK_a. This refers to the negative log of the K_a of the buffer's weak acid. We are given that K_a, so getting pK_a is easy:

$$pK_a = -\log(K_a) = -\log(3.5 \times 10^{-8}) = 7.5$$

Now we need [HA] and [A^-]. These terms refer to the concentration of the weak acid and the concentration of its conjugate. Well, we know that 10.0 mL of 12.0 M HClO was used. That tells us how many moles of acid are in the solution:

$$\frac{12.0 \text{ moles}}{\cancel{L}} \times \frac{0.0100\ \cancel{L}}{1} = 0.120 \text{ moles}$$

The solution was diluted to 0.500 liters, so the resulting concentration of acid is:

$$\frac{0.120 \text{ moles}}{0.500 \text{ L}} = 0.240 \text{ M}$$

What about the conjugate base? Well, we know that 5.00 grams of KClO were added. We can figure out how many moles that is and then get the concentration:

$$\frac{5.00 \text{ g}}{1} \times \frac{1 \text{ mole}}{90.6 \text{ g}} = 0.0552 \text{ moles}$$

$$\frac{0.0552 \text{ moles}}{0.500 \text{ L}} = 0.110 \text{ M}$$

Since one KClO will split up into one K^+ ion and one ClO^- ion in solution, this means that the concentration of ClO^- (the conjugate base) is 0.110 M as well. That's all we need to calculate pH.

$$pH = pKa + \log\left(\frac{[A^-]}{[HA]}\right) = 7.5 + \log\left(\frac{0.110 \cancel{M}}{0.240 \cancel{M}}\right) = \underline{7.2}$$

8.4 The buffer solution in problem 3 has an acid concentration of 0.240 M and a conjugate base concentration of 0.110 M. There is also 0.500 L of it. This means that there are 0.120 moles of acid and 0.0552 moles of conjugate base in the solution.

When KOH is added, it will react with the strongest acid available, which is HClO:

$$HClO + KOH \rightarrow ClO^- + K^+ + OH^-$$

This means that the addition of KOH will lower the acid concentration and raise the base concentration. How much? That's what stoichiometry will tell us:

$$\frac{0.15 \text{ g } \cancel{KOH}}{1} \times \frac{1 \text{ mole KOH}}{56.1 \text{ g } \cancel{KOH}} = 0.0027 \text{ moles KOH}$$

This tells us that the amount of acid will lower by 0.0027 moles, bringing it to 0.120 - 0.0027 = 0.117. It also tells us that the amount of conjugate base will increase to 0.0552 + 0.0027 = 0.0579. This changes the acid and base concentrations to:

$$\text{acid: } \frac{0.117}{0.500} = 0.234 \text{ M}$$

$$\text{base: } \frac{0.0579}{0.500} = 0.116 \text{ M}$$

That changes the pH to:

$$pH = pKa + \log\left(\frac{[A^-]}{[HA]}\right) = 7.5 + \log\left(\frac{0.116 \cancel{M}}{0.234 \cancel{M}}\right) = \underline{7.2}$$

Notice, then, despite the fact that a significant amount of base was added, the pH did not change!

8.5 When LiOH is added to CH_3COOH, the reaction will be:

$$LiOH + CH_3COOH \rightarrow CH_3COO^- + Li^+ + H_2O$$

We can determine the number of moles of each reactant to do the stoichiometry:

$$\frac{1.50 \text{ moles } CH_3COOH}{\cancel{L}} \times \frac{0.500 \cancel{L}}{1} = 0.750 \text{ moles } CH_3COOH$$

$$\frac{5.00 \text{ g } \cancel{LiOH}}{1} \times \frac{1 \text{ mole LiOH}}{23.9 \text{ g } \cancel{LiOH}} = 0.209 \text{ moles LiOH}$$

Since the reaction says each mole of CH_3COOH needs 1 mole of LiOH, LiOH is the limiting reagent. This means that after the reaction is over, there will be no LiOH, 0.209 moles of Li^+, 0.209 moles of CH_3COO^-, 0.209 more moles of water, and 0.750-0.209 = 0.541 moles of CH_3COOH left over.

Notice that we now have a weak acid (CH_3COOH) and its conjugate base (CH_3COO^-). Thus, the addition of LiOH to acetic acid resulted in a buffer solution. We can now calculate pH with the Henderson-Hasselbalch equation:

$$\frac{0.541 \text{ moles } CH_3COOH}{0.500 \text{ L}} = 1.08 \text{ M } CH_3COOH$$

$$\frac{0.209 \text{ moles } CH_3COO^-}{0.500 \text{ L}} = 0.418 \text{ M } CH_3COOH^-$$

$$pH = pK_a + \log\left(\frac{[A^-]}{[HA]}\right) = 4.7 + \log\left(\frac{0.418}{1.08}\right) = \underline{4.3}$$

8.6 When tin dissolves, it does so according to the following equilibrium:

$$Sn(OH)_2\ (s) \rightleftarrows Sn^{2+}\ (aq) + 2OH^-\ (aq)$$

Since this is water, there are already hydroxide ions present. The exact number can be calculated from the pH and the self-ionization of water:

$$pH = -\log([H_3O^+])$$

$$9.0 = -\log([H_3O^+])$$

$$[H_3O^+] = 1.0 \times 10^{-9}$$

$$[H_3O^+]\cdot[OH^-] = 1 \times 10^{-14}$$

$$(1.0 \times 10^{-9})\cdot[OH^-] = 1 \times 10^{-14}$$

$$[OH^-] = 1.0 \times 10^{-5}$$

When "X" amount of tin hydroxide dissolves, then, the concentration of tin ions will be "X," and the concentration of hydroxide ions will be "$1.0 \times 10^{-5} + 2X$." This makes the solubility equation:

$$5.0 \times 10^{-26} = (X)\cdot(1.0 \times 10^{-5} + 2X)^2$$

Assuming that 2X is too small to add to 1.0×10^{-5}:

$$5.0 \times 10^{-26} = (X)\cdot(1.0 \times 10^{-5})^2$$

$$X = 5.0 \times 10^{-16}$$

The assumption is correct. This tells us that the tin ion concentration is $\underline{5.0 \times 10^{-16}\ M}$.

8.7 The initial pH leads to a hydroxide ion concentration of 7.9×10^{-5} M. When $Mn(OH)_2$ dissolves, it will add to that concentration through the following equilibrium:

$$Mn(OH)_2\ (s) \rightleftarrows Mn^{2+}\ (aq) + 2OH^-\ (aq)$$

When "X" manganese (II) hydroxide dissolves, it will lead to a manganese ion concentration of "X" and a hydroxide ion concentration of "$7.9 \times 10^{-5} + 2X$."

$$4.5 \times 10^{-14} = X\cdot(7.9 \times 10^{-5} + 2X)^2$$

Assuming that 2X is simply too small to add to 7.9×10^{-5}:

$$4.5 \times 10^{-14} = X\cdot(7.9 \times 10^{-5})^2$$

$$X = 7.2 \times 10^{-6}$$

This number is not too small to add to 7.9×10^{-5}. We cannot solve the equation exactly, so we will have to resort to successive approximations. To do this, we take the value of "X" that comes from the assumption and put it back in the equation where we ignored it before:

$$4.5 \times 10^{-14} = X \cdot (7.9 \times 10^{-5} + 2 \cdot (7.2 \times 10^{-6}))^2$$

$$X = 5.2 \times 10^{-6}$$

Now we continue the process until the value for "X" does not change:

$$4.5 \times 10^{-14} = X \cdot (7.9 \times 10^{-5} + 2 \cdot (5.2 \times 10^{-6}))^2$$

$$X = 5.6 \times 10^{-6}$$

$$4.5 \times 10^{-14} = X \cdot (7.9 \times 10^{-5} + 2 \cdot (5.6 \times 10^{-6}))^2$$

$$X = 5.5 \times 10^{-6}$$

$$4.5 \times 10^{-14} = X \cdot (7.9 \times 10^{-5} + 2 \cdot (5.5 \times 10^{-6}))^2$$

$$X = 5.6 \times 10^{-6}$$

This is the end. Notice that this is the same answer I got 2 steps ago. Thus, I will keep it. The manganese ion concentration, then, is $\underline{5.6 \times 10^{-6}\text{ M}}$.

8.8 This is a standard equilibrium calculation. When "X" atms of PCl_5 react, it will form "X" of each product. At equilibrium, then, the pressure of PCl_5 will "5.00-X," and the pressure of the other two gases will each be "X." This makes the equilibrium constant equation:

$$0.0211 = \frac{X \cdot X}{5.00 - X}$$

If you try to assume that "X" is too small to subtract from 5.00, you would find that you were wrong. Thus, I can either solve this equation exactly or with successive approximations. I will choose the exact method, because it is less tedious.

$$0.106 - 0.0211 \cdot X = X^2$$

$$X^2 + 0.0211 \cdot X - 0.106 = 0$$

$$X = \frac{-0.0211 \pm \sqrt{0.0211^2 - 4 \cdot 1 \cdot (-0.106)}}{2} = 0.315 \text{ or } -0.336$$

At equilibrium, then, the pressure of PCl_5 will be 4.69 atms and the pressure of each product will be 0.315 atms.

8.9 In this equilibrium, there is no limiting reagent since both reactants exist at equal pressures and the reaction requires the same number of moles of each reactant. When "X" of either reactant reacts, it will form "2X" HCl and "0.5X" O_2. Notice how I got that. The reaction says 2 moles of each reactant make 4 moles of HCl. That means twice as many moles of HCl form as the number of moles of each reactant. Thus, if "X" reacts, the reaction will form "2X" (twice as much) HCl. In the same way, when 2 moles of each reactant react, only 1 mole of O_2 is formed. That means the reaction forms only half as much O_2 as what reacts. Thus, when "X" reactants react, "0.5X" (half as much) O_2 forms. At equilibrium, then, the pressure of Cl_2 is "1.00-X," the pressure of H_2O is "1.00-X," the pressure of HCl will be "2X," and the pressure of O_2 will be "0.5X." This makes the equation:

$$0.051 = \frac{(2 \cdot X)^4 \cdot (0.5 \cdot X)}{(1.00 - X)^2 \cdot (1.00 - X)^2}$$

We can't solve this exactly. Let's make our assumption:

$$0.051 = \frac{(2 \cdot X)^4 \cdot (0.5 \cdot X)}{(1.00)^2 \cdot (1.00)^2}$$

$$X = 0.36$$

Our assumption does not work, and we cannot solve the equation with algebra, so we have to do successive approximations:

$$0.051 = \frac{(2 \cdot X)^4 \cdot (0.5 \cdot X)}{(1.00 - 0.36)^2 \cdot (1.00 - 0.36)^2}$$

$$X = 0.25$$

$$0.051 = \frac{(2 \cdot X)^4 \cdot (0.5 \cdot X)}{(1.00 - 0.25)^2 \cdot (1.00 - 0.25)^2}$$

$$X = 0.29$$

$$0.051 = \frac{(2 \cdot X)^4 \cdot (0.5 \cdot X)}{(1.00 - 0.29)^2 \cdot (1.00 - 0.29)^2}$$

$$X = 0.28$$

$$0.051 = \frac{(2 \cdot X)^4 \cdot (0.5 \cdot X)}{(1.00 - 0.28)^2 \cdot (1.00 - 0.28)^2}$$

$$X = 0.28$$

Since X = 0.28, the pressures of Cl_2 and H_2O are 0.72 atm; the pressure of HCl is 0.56 atm; and the pressure of O_2 is 0.14 atm.

8.10 Look at the equilibrium constant. It is huge. Thus, this is an equilibrium that is best treated first as a complete reaction. After we have done that, we can then allow the reaction to approach equilibrium. Thus, assuming this is a complete reaction, we see that we need twice as much NO as Br_2. We have more than that. Thus, once the Br_2 runs out, there will still be NO left. This means that Br_2 is the limiting reagent. Based on the stoichiometry, 1.00 atm of Br_2 will react with 2.00 atm of NO to make 2.00 atm of NOBr. This will leave 0.50 atm of NO left over.

After complete reaction, then, there are 2.00 atms of NOBr and 0.50 atms of NO. Now we can do an equilibrium calculation. When the reaction shifts backward to achieve equilibrium, "X" of the NOBr will react. That will form "X" NO and "0.5X" Br_2. At equilibrium, then, the concentration of NOBr will be "2.00-X," while the concentrations of NO and Br_2 will be "0.50 + X" and "0.5X," respectively. The equilibrium constant equation, then, is:

$$3.5 \times 10^4 = \frac{(2.00 - X)^2}{(0.50 + X)^2 \cdot (0.5 \cdot X)}$$

Now we will make the assumption that "X" is too small to add to 0.50 and subtract from 2.00:

$$3.5 \times 10^4 = \frac{(2.00)^2}{(0.50)^2 \cdot (0.5 \cdot X)}$$

$$X = 0.00091$$

Our assumption is correct. Since X = 0.00091, the pressure of NOBr is 2.00 atm; the pressure of NO is 0.50 atm; and the pressure of Br_2 is 0.00046 atm.

REVIEW QUESTIONS

1. How do you make a buffer solution?

2. What is the unique property of a buffer solution?

3. What buffer is in human blood?

4. If you have 1.0 moles of a weak acid and add less than 1.0 mole of KOH, why does that form a buffer solution?

5. Which of the following solutions can be used as a buffer?

   a. An aqueous solution of HF and NH_3
   b. An aqueous solution of HNO_3 and $NaNO_3$
   c. An aqueous solution of $H_2PO_4^-$ and K_2HPO_4
   d. An aqueous solution of HClO and KClO

6. A buffer is made with CH_3COOH and its conjugate. What chemical reaction occurs when HNO_3 is added to the buffer?

7. A buffer is made with $H_2AsO_4^-$ and $HAsO_4^{2-}$. What chemical reaction occurs when LiOH is added to the buffer?

8. Is $Ca(OH)_2$ more soluble in a solution with pH=2 or a solution with pH=12?

9. When should you use the method of successive approximations?

10. How do we solve an equilibrium problem when K is very large (>100)?

PRACTICE PROBLEMS

1. A chemist takes 200 mL of 11.0 M CH_3COOH and adds 150.0 g of KCH_3COO. The chemist then dilutes the solution to 300 mL. What is the pH? ($K_a = 1.8 \times 10^{-5}$)

2. A chemist wants to make a pH = 7.0 buffer using carbonic acid (H_2CO_3) and its conjugate. If the carbonic acid concentration is 0.50 M, what concentration must the base have in order to get pH = 7.0? ($K_a = 4.3 \times 10^{-7}$)

3. A H_2S/HS^- buffer is made with the concentration of H_2S being 0.15 M and the concentration of HS^- being 0.10 M. What is the pH? If 1.0 gram of KOH is added to 500.0 mL of the buffer, what is the new pH? ($K_a = 1.0 \times 10^{-7}$)

4. A chemist makes a buffer by adding 100.0 mL of 1.00 M $H_2PO_4^-$ ($K_a = 6.3 \times 10^{-8}$) to 100.0 mL of 0.500 M NaOH and then diluting the result to 300.0 mL. What is the pH of the buffer?

5. What is the lead ion concentration in a saturated solution of $Pb(OH)_2$ if the initial pH of the water is 13.0? ($K_{sp} = 2.8 \times 10^{-16}$)

6. What would the answer to #5 be if the pH were 9.2?

7. In solution, I_3^- partially decomposes into iodine and the iodide ion:

$$I_3^- (aq) \rightleftarrows I_2 (aq) + I^- (aq)$$

The equilibrium constant is 0.00129. If a solution is made with the initial I_3^- concentration equal to 0.10 M, what is the concentration of each substance at equilibrium?

8. The weak acid HF is not 100% stable in its gaseous state. It can decompose to some extent into hydrogen and fluorine gas:

$$2HF (g) \rightleftarrows H_2 (g) + F_2 (g)$$

with $K_p = 0.0087$. A chemist vaporizes HF thinking that it will make 3.0 atms of HF. How much HF pressure will there really be?

9. At a certain temperature, the K_p for forming ammonia from nitrogen gas and hydrogen gas is 0.0034. If a chemist starts with 4.0 atm of hydrogen and 1.0 atm of nitrogen, how many atms of ammonia will exist at equilibrium?

10. The equilibrium constant for the reaction

$$2NO (g) + F_2 (g) \rightleftarrows 2NOF (g)$$

is 1.34×10^4 at 200 °C. If 3.0 atm of NO and 1.0 atm of F_2 are placed in a reaction vessel at 200 °C, what is the pressure of each gas at equilibrium?

Module #9: Electrochemistry - Part 1

Introduction

In your first-year chemistry course, you should have learned about **reduction/oxidation reactions**. These reactions, also called **redox reactions**, form the basis of the field known as **electrochemistry**. This fascinating field of chemistry gives us batteries, sanitary drinking water, aluminum cans, rust, and much, much more. Electrochemical reactions run the muscles in our bodies and are the basis of our nervous system. In short, electrochemistry in an important and useful field of chemistry. We will study electrochemistry in great detail. First, however, you need to recall some things from your first-year course.

A Little Bit of Review

The key to understanding redox reactions is being able to identify the **oxidation numbers** of atoms within a molecule. You should have learned about oxidation numbers in your previous chemistry course, but I will go ahead and review them because they are critically important. The best way to remind you about oxidation numbers is to talk about a polar molecule such as HCl.

The HCl molecule is made up of a hydrogen atom and a chlorine atom sharing one pair of electrons, as shown in the following Lewis structure:

$$H:\ddot{\underset{\cdot\cdot}{Cl}}:$$

Even though these atoms share a pair of electrons, the Cl is much more electronegative than the H. As a result, the Cl gets more than its fair share of that electron pair. When we calculate oxidation numbers, we are asking what charges would develop if the Cl just took that electron pair and stopped sharing it. If that happened, the Cl would suddenly have 8 electrons, and H would have none. Since Cl usually has 7 electrons, the fact that it would now have 8 indicates that it would have a -1 charge. In the same way, a hydrogen atom usually has 1 electron. If the hydrogen atom were left with no electrons, it would have a +1 charge. We say, then, that the oxidation number of hydrogen in this molecule is +1 while the oxidation number of Cl is -1.

By calculating oxidation numbers, you are actually calculating *possible* charges on the atoms in a molecule *assuming* that the electrons were not shared, but were transferred to the most electronegative of the atoms. That idea leads to the following definition:

Oxidation number - The charge that an atom in a molecule would develop if the most electronegative atoms in the molecule took the shared electrons from the less electronegative atoms

Thus, oxidation numbers treat all molecules as ionic, attempting to assign charges to all atoms within a molecule. Now remember, oxidation numbers **aren't real**. In our example above, H

does not have a charge of +1 and Cl **does not** have a charge of -1. Oxidation numbers are a way of saying "what if?" They don't have any significance in the real world.

If oxidation numbers don't have any real physical or chemical meaning, why do we bother to learn them? Well, it turns out that oxidation numbers are an excellent bookkeeping technique that will allow us to determine where electrons travel during a chemical reaction. As we will see later, knowing where electrons go during a chemical reaction is a major key to understanding the reactions we will study in the next two modules. Thus, even though oxidation numbers have no real physical or chemical significance to them, they are a powerful *tool* that we can use in analyzing chemical equations. Thus, we need to learn how to calculate and use them.

The first rule to learn in calculating oxidation numbers is that, in any compound, the sum of all oxidation numbers (times the number of atoms that have that oxidation number) must equal the electrical charge on the compound. For example, we determined the oxidation numbers for H and Cl in the HCl molecule. The H has an oxidation number of +1 and the Cl has an oxidation number of -1. They add to zero, which is the overall charge of the molecule.

The sum of all oxidation numbers in a molecule must equal the charge of that molecule.

Since most molecules have no net electrical charge, most of the time the oxidation numbers in a molecule add to zero. When you are dealing with an ion of some sort, however, you must make sure that the oxidation numbers in that ion add up to the charge of the ion.

Well, it's nice to know what the sum of all oxidation numbers has to equal, but we need to be able to calculate the oxidation number of each individual atom in a molecule. How do we do that? There are some rules we can develop for this task.

1. When a substance has only one type of atom in it (F_2, O_3, or Mg^{2+} for example) the oxidation number for that atom is equal to the charge of the substance divided by the number of atoms present.

2. Group 1A metals (Na, K, Rb, Cs, and Fr) always have oxidation numbers of +1 in molecules that contain more than one type of atom.

3. Group 2A metals (Be, Mg, Ca, Sr, Ba and Ra) always have oxidation numbers of +2 in molecules that contain more than one type of atom.

4. Fluorine always has a -1 oxidation number in molecules that contain more than one type of atom.

What's nice about these three rules is that they are **always true**. The problem is, after these four rules, things get a little fuzzy, because there are a few exceptions to the next rules. In general, though, the next rules usually apply:

5. When it groups with just one other atom that happens to be a metal, H has an oxidation number of -1. In *all other cases* in which it is grouped with other atoms, H has an oxidation number of +1.

6. Oxygen has an oxidation number of -2 in molecules that contain more than one type of atom. H_2O_2 is an exception. Oxygen has an oxidation number of -1 in H_2O_2.

In fact, there are exceptions to both rule number 5 and rule number 6. However, you will be happy to know that for this course, we will ignore those exceptions, except for the one listed in rule #6. Thus, as far as you are concerned, rules 1-6 always are true. After that, though, things get even more fuzzy. If you are really stuck and none of these rules apply, you can follow this general (and often not true) guideline:

7. If all else fails, assume that the atom's oxidation number is the same as what it would take on in an ionic compound. The atoms that are most likely to follow this rule are in groups 3A, 6A, and 7A.

This means that you might be able to assume that chlorine has a -1 oxidation number because it is in group 7A and those atoms take on a -1 charge in ionic compounds. Similarly, it might be a good guess that S has a -2 oxidation number because all group 6A atoms take a -2 charge in ionic compounds. This general principle, however, has so many exceptions that it should only be used as a last resort!

So, in the end, we have 6 rules that you can assume will always work, and one general principle that has many exceptions but is a good guess when all else fails. How do we use these rules to determine oxidation numbers? In general, we end up trying to determine the oxidation numbers of all atoms using rules 1-6. If that doesn't work, before we resort to rule 7, we see if rules 1-6 apply to all but one of the atoms in the molecule. If they do, then we can determine the oxidation number of the last atom by using the condition that all oxidation numbers have to add up to the charge on the molecule. Thus, you use rule #7 *only* when rules 1-6 do not apply to all or all but one of the atoms in the molecule. This may sound really complex right now, but some practice will make you a virtual master at this!

EXAMPLE 9.1

What is the oxidation number of each atom in the following molecules?

a. H_2

According to rule #1, H's oxidation number is 0 divided by 2 which is <u>0</u>.

b. Fe^{2+}

According to rule #1, Fe's oxidation number is +2 divided by 1 which is <u>+2</u>.

c. $Ca(OH)_2$

In this molecule, we can apply our rules to every atom. Ca is a group 2A metal, so its oxidation number is +2 (rule #3). Rule #6 tells us that oxygen has an oxidation number of -2. Finally, since H is not grouped with just one other atom, its oxidation number is +1. As a check, we need to make sure that these numbers add up to the total charge (0) of the molecule. There are one Ca, two O's, and two H's in the molecule. Adding the oxidation numbers:

$$(+2) + 2\cdot(-2) + 2\cdot(+1) = 0.$$

Thus, the oxidation numbers are: Ca: +2, O: -2, H: +1.

d. F_2CO

Rule # 4 tells us that F is always -1. Rule #6 says that oxygen is always -2. We have no rule for C, but that's okay. Since we know all but one of the atoms, we can figure out C by making sure that the oxidation numbers sum to the overall charge of the molecule (0). There are two F's and one O, so the total oxidation number so far is $2\cdot(-1) + (-2) = -4$. To make all of the numbers add up to 0, the C must have an oxidation number of +4. Thus, the oxidation numbers are: F: -1, C: +4, O: -2.

e. $NaSO_4^-$

Rule #2 tells us that Na's oxidation number is +1, since it is a group 1A metal. Rule #6 says that O must have a -2 oxidation number. Once again, we have no rule for S, but we can figure out its oxidation number because the sum of all oxidation numbers must equal the charge of the molecule (-1). Since Na is +1 and O is -2, we have $(+1) + 4\cdot(-2) = -7$. In order for the sum of oxidation numbers to be -1, S must have an oxidation number of +6. Thus, the oxidation numbers are Na: +1, S: +6, O: -2.

f. C_2H_4

According to rule #5, when H is grouped with just one other atom that happens to be a metal, its oxidation number is -1, otherwise it is +1. In this molecule, H is grouped with just one other atom, but that atom is not a metal. Thus, H's oxidation number is +1. We have no rules for C, but we know that the sum of oxidation numbers must equal the overall charge (0). Since H is +1, we have a +4 sum so far. Thus, in order for them to add to zero, the two C's must account for a sum of -4. Thus, each individual C must have a -2 oxidation number. In the end, then, the oxidation numbers are C: -2, H: +1.

g. $MgCoBr_4$

In this molecule, we know that Mg is +2 because it is a group 2A metal (rule #3). For the other two, rules 1-6 do not apply. Thus, we are stuck with rule #7. We will apply this to Br for two reasons. First, the rule states that group 7A is a group that likely follows the rule, and Br is in group 7A. Also, Co is not in an "A" group, and our rules for determining charge apply only to groups 1A - 8A. Thus, Br is the only atom we can apply this rule to. Since Br is in group 7A, it develops a -1 charge when in ionic compounds. That leaves us to determine Co's oxidation number by making use of the fact that all oxidation numbers add up to the overall charge (0). Thus, the oxidation numbers are Mg: +2, Co: +2, Br: -1.

Try this on your own to make sure you can calculate oxidation numbers.

ON YOUR OWN

9.1 What are the oxidation numbers of all atoms in the following molecules?

a. $LiNH_2$ b. N_2H_2 c. $Ca(NO_2)_2$ d. CO_2 e. BF_4^- f. PO_4^{3-}
g. ClNO h. S_8

Analyzing Redox Reactions

Now that you remember oxidation numbers, I want you to perform the following quick experiment so that you remember how to use them in analyzing redox reactions.

EXPERIMENT 9.1
A Redox Reaction Between Copper and Zinc

Supplies:
From the laboratory equipment set:
- Depression plate
- Copper (II) sulfate
- Measuring scoop
- Copper wire
- Zinc wire (The shiny wire that looks like aluminum.)
- Dropper

From the supermarket:
- Distilled water

Use your dropper to fill two of the deep wells in your depression plate halfway with distilled water. Add a few crystals of copper (II) sulfate to each and stir with your dropper. Next, put one end of the zinc wire into one of the wells and put one end of the copper wire into the other. Wait for five minutes. Take each wire out and examine it. What do you see?

Keep both wires in your kit, because you will use them again in another experiment. Clean up your depression plate (not in the kitchen sink, remember) and put everything away.

In the experiment, you should have seen nothing interesting on the copper wire. However, the zinc wire should have been another story. The part of the wire that was submerged should be tarnished. Despite the fact that nothing seemed to happen, something nevertheless did. Here's what happened:

$$Zn\,(s) + CuSO_4\,(aq) \rightarrow ZnSO_4\,(aq) + Cu\,(s) \qquad (9.1)$$

When you placed the zinc wire in the solution, the aqueous copper sulfate reacted with it, forming zinc sulfate and solid copper. The tarnish that you see on the zinc wire is the solid copper that formed.

Now remember, when you see aqueous ionic compounds, you are allowed to split them up into their ions. In electrochemistry, you should *always* do that. As a result, the equation becomes:

$$Zn\,(s) + Cu^{2+}\,(aq) + SO_4^{2-}\,(aq) \rightarrow Zn^{2+}\,(aq) + SO_4^{2-}\,(aq) + Cu\,(s) \qquad (9.2)$$

Notice that the sulfate ion appears on each side of the equation. Thus, we can cancel it:

$$Zn\,(s) + Cu^{2+}\,(aq) + \cancel{SO_4^{2-}}\,(aq) \rightarrow Zn^{2+}\,(aq) + \cancel{SO_4^{2-}}\,(aq) + Cu\,(s) \qquad (9.3)$$

$$Zn\,(s) + Cu^{2+}\,(aq) \rightarrow Zn^{2+}\,(aq) + Cu\,(s) \qquad (9.4)$$

This is called the **net ionic equation** for the chemical reaction in (9.1), because it deals with only those substances that changed. The ions in the reaction which did not change at all (the sulfate ion in this case) have been canceled out.

What does this equation tell you? It tells you that in the reaction, zinc metal turned into zinc 2+. How can it do that? By losing two electrons. In addition, Cu^{2+} turned into Cu. How can that happen? The only way to get rid of a 2+ charge is to gain two electrons. In this reaction, then, zinc gave up two electrons to become Zn^{2+}, and those two electrons went to Cu^{2+} so that it could become Cu. Thus, the reaction occurred because *electrons were transferred from one reactant to another.*

When electrons are transferred from one reactant to another, the reaction is classified as a reduction/oxidation reaction. In these reactions, we say that the reactant that loses electrons is **oxidized** and the reactant that gains electrons is **reduced**. You can keep these definitions straight with the mnemonic "LEO says GER." "LEO" abbreviates "lose electrons oxidation" and "GER" abbreviates "gain electrons reduction." To make the terminology even a little more confusing, the reactant that is oxidized is called the **reducing agent**, because it causes the other reactant to be reduced. In the same way, the reactant that is reduced is called the **oxidizing agent**, because it causes the other reactant to be oxidized.

In reactions like the one represented in Equation (9.4), it is rather easy to see where the electrons went. After all, the zinc went from a charge of 0 to a charge of 2+. Thus, zinc lost two electrons. This means it was oxidized (lose electrons oxidation) and is called the reducing agent.

Copper, on the other hand, went from a charge of +2 to 0, which means it gained two electrons. That tells us that copper was reduced (gain electrons reduction) and is therefore called the oxidizing agent.

In other reactions, it's a bit harder to tell that electrons were exchanged between reactants. Consider, for example, the following reaction:

$$H_2O_2\ (aq) + Cl_2\ (g) \rightarrow 2HCl\ (aq) + O_2\ (g) \qquad (9.5)$$

Although you might not be able to see it, this is a redox reaction as well. Study the following example to see how you can tell.

EXAMPLE 9.2

Show that the reaction represented by Equation (9.5) is, in fact, a redox reaction. Determine the oxidizing and reducing agents.

Let's look at the oxidation numbers of each atom. In H_2O_2, we can use rule #5 to determine that H has an oxidation number of +1 This is the exception listed in rule #6, so O has an oxidation number of -1. In Cl_2, rule #1 tells us that Cl has an oxidation number of 0.

Looking at the products now, rule #5 tells us that H has a +1 oxidation number in HCl, leaving Cl with a -1 oxidation number. By rule #1, O has a zero oxidation number in O_2. Now, look what happened in the reaction. Cl started with an oxidation number of zero and ended with an oxidation number of -1. Thus, we can say that Cl gained an electron. This means is was reduced, which means Cl_2 (the source of Cl) is the oxidizing agent. H had an oxidation number of +1 before and after the reaction. This tells us that H neither gained nor lost electrons. Finally, O went from an oxidation number of -1 to one of 0. This means it lost electrons. Thus, it was oxidized. This tells us that H_2O_2 (the source of O) is the reducing agent.

This is why you need to learn oxidation numbers. It allows you to really track where electrons go in a chemical reaction.

ON YOUR OWN

9.2 Determine the oxidizing and reducing agents in the following reactions:

a. $Ca\ (s) + Cl_2\ (g) \rightarrow CaCl_2\ (s)$

b. $3IF_5\ (aq) + 2Fe\ (s) \rightarrow 2FeF_3\ (aq) + 3IF_3\ (aq)$

Galvanic Cells

In your first-year chemistry course, you should have seen an application of redox reactions: **Galvanic cells**. Since a redox reaction involves the exchange of electrons, if we arrange things properly, we can make use of them. If we put each reactant in a different container and then connect the containers with a wire, the electrons will travel from the reducing agent to the oxidizing agent through that wire. The result is a source of electricity, which chemists call a Galvanic cell.

As you should have learned last year, things are a bit more complicated than that. Figure 9.1 illustrates a typical Galvanic cell, to remind you of what exactly is involved.

FIGURE 9.1
A Galvanic Cell

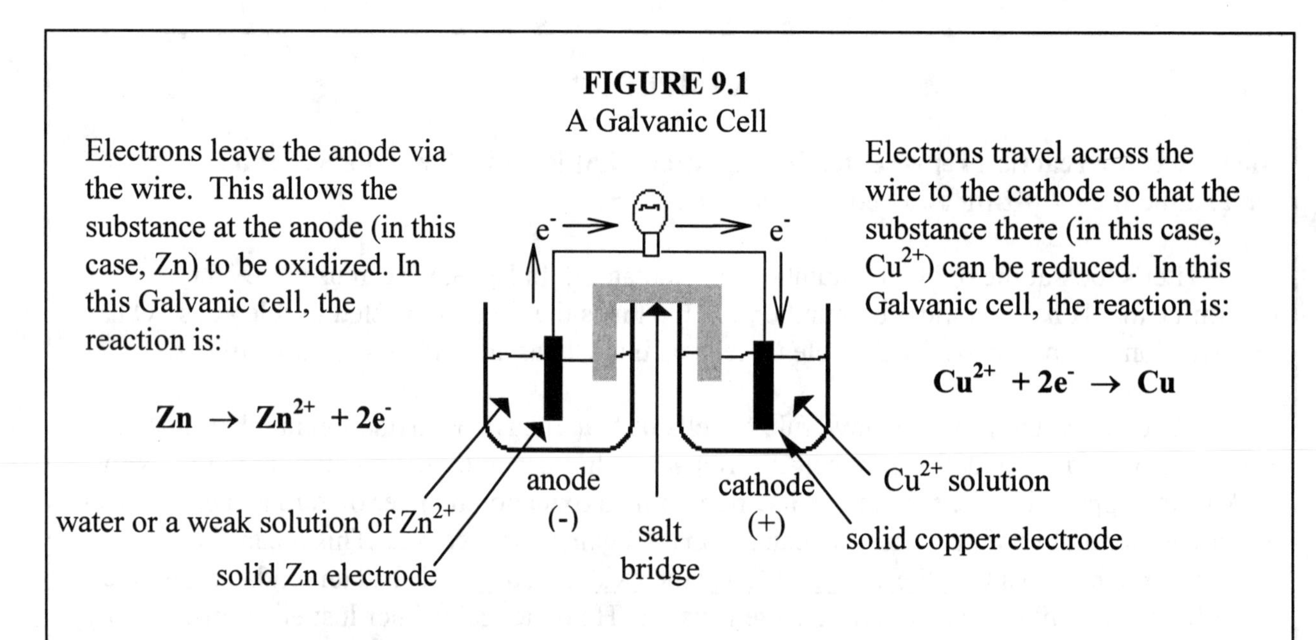

As time goes on and this Galvanic cell is allowed to operate, the copper electrode will grow and grow, because new copper is forming on it. At the same time, the Cu^{2+} concentration at the cathode will decrease, because Cu^{2+} is being used up. At the anode, the zinc electrode will get smaller and smaller, because zinc is being used up, and the Zn^{2+} concentration will increase.

Notice that the two containers are connected not only by the wire, but also by a salt bridge. The salt bridge contains ions which travel opposite the electron flow. This balances out the charges, keeping each container electrically neutral. Notice also that each container has a special name. The container in which reduction occurs is called the **cathode** and the container in which oxidation occurs is called the **anode**. In a Galvanic cell, electrons always flow from the anode to the cathode.

This is, in essence, what a battery is. The reactants of a redox reaction are separated so that the only way they can exchange electrons is for those electrons to travel down a wire. As the electrons travel across the wire, their energy can be used to do work. For example, in the Galvanic cell shown above, the light bulb will glow due to the electrons running through it.

A battery, of course, is rated by its voltage. If you took physics, you know that voltage represents a potential energy difference through which electrons travel. If you didn't take physics, you can think about voltage as a "push" that propels the electrons through a wire. The higher the voltage, the harder the push. Given that concept of voltage, what do you think determines the voltage of a Galvanic cell? Since electrons travel down a wire in a Galvanic cell, it must have a voltage. What determines how large or small that voltage is?

Since the electrons travel through the wire as a result of a need to complete the reaction, the voltage must therefore depend on the energetics of the reaction itself. If, for example, the reactant that is being oxidized *really* "wants" to lose electrons, it will push those electrons down the wire with a strong force. If the reactant that is being reduced *really* "wants" electrons, it will pull them down the wire with a strong force. Thus, the voltage of a Galvanic cell depends on the chemical nature of the reactants. The more the oxidized reactant "wants" to lose electrons, the higher the voltage. Likewise, the more the reduced reactant "wants" to gain electrons, the higher the voltage.

It turns out that chemists can actually measure how strongly a substance will pull electrons towards itself or push them away from itself. These strengths have been tabulated as **standard reduction potentials**. Several of them are listed in the table below:

TABLE 9.1
Standard Reduction Potentials

Half-Reaction	E° (Volts)	Half-Reaction	E° (Volts)
$K^+ + e^- \rightarrow K$	-2.925	$Sn^{4+} + 4e^- \rightarrow Sn$	0.15
$Ba^{2+} + 2e^- \rightarrow Ba$	-2.90	$AgCl + e^- \rightarrow Ag + Cl^-$	0.222
$Ca^{2+} + 2e^- \rightarrow Ca$	-2.87	$Hg_2Cl_2 + 2e^- \rightarrow 2Hg + 2Cl^-$	0.27
$Na^+ + e^- \rightarrow Na$	-2.714	$Cu^{2+} + 2e^- \rightarrow Cu$	0.337
$Mg^{2+} + 2e^- \rightarrow Mg$	-2.37	$NiO_2 + 2H_2O + 2e^- \rightarrow Ni(OH)_2 + 2OH^-$	0.49
$Al^{3+} + 3e^- \rightarrow Al$	-1.66	$I_2 + 2e^- \rightarrow 2I^-$	0.5355
$Zn(OH)_2 + 2e^- \rightarrow Zn + 2OH^-$	-1.245	$MnO_4^- + 2H_2O + 3e^- \rightarrow MnO_2 + 4OH^-$	0.588
$Mn^{2+} + 2e^- \rightarrow Mn$	-1.18	$Fe^{3+} + 3e^- \rightarrow Fe$	0.771
$Fe(OH)_2 + 2e^- \rightarrow Fe + 2OH^-$	-0.877	$Hg_2^{2+} + 2e^- \rightarrow 2Hg$	0.789
$Zn^{2+} + 2e^- \rightarrow Zn$	-0.763	$Ag^+ + e^- \rightarrow Ag$	0.7991
$Cr^{3+} + 3e^- \rightarrow Cr$	-0.74	$Br_2 (l) + 2e^- \rightarrow 2Br^- (aq)$	1.0652
$Fe^{2+} + 2e^- \rightarrow Fe$	-0.440	$Pt^{2+} + 2e^- \rightarrow Pt$	1.20
$Cd^{2+} + 2e^- \rightarrow Cd$	-0.403	$O_2 + 4H_3O^+ + 4e^- \rightarrow 6H_2O$	1.23
$PbSO_4 + 2e^- \rightarrow Pb + SO_4^{2-}$	-0.356	$Cl_2 + 2e^- \rightarrow 2Cl^-$	1.3595
$Co^{2+} + 2e^- \rightarrow Co$	-0.277	$Au^{3+} + 3e^- \rightarrow Au$	1.50
$Ni^{2+} + 2e^- \rightarrow Ni$	-0.250	$MnO_4^- + 8H_3O^+ + 5e^- \rightarrow Mn^{2+} + 12H_2O$	1.51
$Sn^{2+} + 2e^- \rightarrow Sn$	-0.136	$PbO_2 + SO_4^{2-} + 4H_3O^+ + 2e^- \rightarrow PbSO_4 + 6H_2O$	1.685
$Pb^{2+} + 2e^- \rightarrow Pb$	-0.126	$F_2 + 2e^- \rightarrow 2F^-$	2.87
$2H_3O^+ + 2e^- \rightarrow H_2 + 2H_2O$	**0.000**		

What do the contents of this table mean? Well, as I stated before, the voltage of a Galvanic cell depends on the strength with which one reactant pulls on the electrons *and* the strength with which the other reactant pushes on the electrons. In a sense, then, there are two "halves" to the Galvanic cell: the reactant that is trying to get the electrons and the reactant that is trying to get rid of them. These standard reduction potentials allow you to calculate how strong each half of the Galvanic cell is. That's why they are called **half-reactions**. They represent what goes on in one half of a Galvanic cell. Notice that all of the half-reactions listed involve reactants gaining electrons. What do we call it when a reactant gains electrons? We call it reduction. That's why Table 9.1 is called a **standard reduction potential** table, because all potentials listed are for reduction half-reactions.

Let's start with the half-reactions listed on the right side of the table. In the top half-reaction, Sn^{4+} is gaining 4 electrons to become Sn. The number next to that chemical equation is the reduction potential for that particular half-reaction. In other words, Sn^{4+} pulls on electrons with the equivalent strength of 0.15 Volts. If you go eight lines down, you will see that Fe^{3+} pulls on electrons with a strength of 0.771 Volts. What does this tell you? It tells you that Fe^{3+} "wants" electrons more than Sn^{4+}, because it pulls on them harder.

Now look at the left side of the table. In the top half-reaction, for example, K^+ is gaining an electron to become K. The potential for this reaction is -2.925. What does that mean? It means that K^+ does not want to gain electrons. In fact, it means that as written, the half-reaction is not very likely to proceed. However, if the half-reaction were reversed to $K \rightarrow K^+ + e^-$, it would have a potential of +2.925 Volts. That's a very strong potential, indicating that K tends to push electrons away from itself. In other words, K would like to be oxidized. Thus, the potential for K to be oxidized is high. A negative reduction potential, then, means that the half- reaction is much more favorable if it is written in reverse, making it an oxidation half-reaction. All half-reactions but one on the left side of the table, then, will be more likely to run if they were turned around.

Notice I said "all reactions but one." Which reaction on the left-hand side of the table will not run better if it is turned around? The last one. Notice that the potential for that reaction is zero. This means that the reaction really has no preference for the direction it runs. This particular half-reaction is called the **standard hydrogen electrode**. Since its reduction potential is zero, it is the standard against which other half-reactions are judged.

So what does all of this mean, anyway? Well, let's look at the Galvanic cell shown in Figure 9.1. At the cathode of that Galvanic cell, Cu^{2+} is gaining electrons to become Cu. According to Table 9.1, that half-reaction has a potential of 0.337 Volts. This means that the cathode is pulling on the electrons with a strength of 0.337 Volts. At the anode, Zn is losing two electrons to become Zn^{2+}. That reaction does not appear in Table 9.1, but the *reverse* of it does. On the left-hand side of the table, the reaction $Zn^{2+} + 2e^- \rightarrow Zn$ appears, and it has a potential of -0.763 Volts. The reverse reaction (the one we are interested in) must therefore have a potential of +0.763 Volts. This tells us that the anode is pushing on the electrons with a strength of 0.763 Volts.

So, the cathode is pulling on the electrons with a strength of 0.337 Volts and the anode is pushing on them with a strength of 0.763 Volts. In the end, then, the total potential that the electrons feel is 0.337 Volts + 0.763 Volts, or 1.100 Volts. Thus, the voltage of our Galvanic cell is 1.100 Volts. In other words, if a battery were made from these two half-reactions, the voltage of the battery would be 1.100 Volts.

That's what standard reduction potentials tell us. Provided we know the half-reactions that go on in a Galvanic cell, they allow us to calculate the voltage of the Galvanic cell. You are going to see a couple of examples of this in a moment, but I want to point out one more thing to you before I get to that. Suppose you use the standard reduction potentials to calculate the voltage of a Galvanic cell, and suppose the result is negative. What does that tell you? It tells you that there is no force to push the electrons across the wire. As a result, the Galvanic cell will not work.

Galvanic cells with negative voltage do not work.

The voltage of a Galvanic cell, then, tells us how strongly the electrons are pushed through the wire to complete the reaction. If the voltage ends up negative, you know that no electrons will pass through the wire, and the Galvanic cell does not work.

EXAMPLE 9.3

Determine the voltage of a Galvanic cell that uses the following half-reactions:

Cathode: $Ag^+ + e^- \rightarrow Ag$
Anode: $Ni \rightarrow Ni^{2+} + 2e^-$

In this Galvanic cell, one container has a solution and some solid nickel. The nickel gives up two electrons to become Ni^{2+}. Those electrons travel down the wire to the other container, which holds Ag^+ ions. These ions accept the electrons that come down the wire, forming solid silver. To determine how much voltage the Galvanic cell has, we need to look at the standard reduction potentials. For the cathode half-reaction, Table 9.1 says that the standard potential is 0.7991. Table 9.1 has no oxidation half-reactions, so how do we find the reaction that is listed for the anode? Well, we look for its reverse. On the left-hand side of the table, the reaction $Ni^{2+} + 2e^- \rightarrow Ni$ is listed with a reduction potential of -0.250 Volts. Since the anode has the reverse of this reaction, it means the anode's potential is 0.250 Volts. In the end, then, we have a cathode that pulls electrons with a strength of 0.7991 Volts and an anode that pushes them with a strength of 0.250 Volts. We can add the two to get the total voltage, which is <u>1.049 Volts</u>.

To determine the voltage of a Galvanic cell, then, we determine the potential for each half-reaction and add them.

Now please remember that a Galvanic cell is just a special way of running a chemical reaction. In a Galvanic cell, we separate the reactants and connect them with a wire.

Nevertheless, the end result is simply a chemical reaction. How do we determine what that chemical reaction is? Well, a Galvanic cell is described with two *half-reactions*. To get the overall chemical reaction that occurs in a Galvanic cell, we just add up the two half-reactions. There is a trick, though. When we add up half-reactions, we need to make sure that the electrons cancel out. Study the following example to see what I mean.

EXAMPLE 9.4

Determine the overall reaction for a Galvanic cell that runs with the following half-reactions:

$$\textbf{Cathode: } Au^{3+} + 3e^{-} \rightarrow Au$$
$$\textbf{Anode: } Sn \rightarrow Sn^{4+} + 4e^{-}$$

What is the voltage of this Galvanic cell?

To get the overall reaction, all we need to do is add up the half-reactions. There is a problem, however. In our final equation, there cannot be any electrons. Thus, we need to add these equations so that the electrons cancel. Here's how we do that:

$$4\cdot[Au^{3+} + 3e^{-} \rightarrow Au]$$
$$\underline{+\,3\cdot[\,Sn \rightarrow Sn^{4+} + 4e^{-}]}$$

Remember, this is legal. I can do anything to chemical equations that I can do to algebraic equations. If I now distribute the 4 and the 3 out, I get:

$$4Au^{3+} + 12e^{-} \rightarrow 4Au$$
$$\underline{+\,3Sn \rightarrow 3Sn^{4+} + 12e^{-}}$$
$$4Au^{3+} + \cancel{12e^{-}} + 3Sn \rightarrow 4Au + 3Sn^{4+} + \cancel{12e^{-}}$$

The answer, then, is:

$$\underline{4Au^{3+} + 3Sn \rightarrow 4Au + 3Sn^{4+}}$$

What is the voltage? For that, we look at Table 9.1. The cathode half-reaction is listed in the table with a potential of 1.50 Volts. The anode half-reaction is not in the table, but its reverse is, listed with a potential of 0.15 Volts. This means that to get our anode half-reaction potential, we have to take the negative, giving us a potential of -0.15 Volts. The voltage, then, is 1.50 Volts + -0.15 Volts = 1.35 Volts.

Notice what happened in this case. The anode half-reaction as written has a negative potential. This means that the anode half-reaction would normally not proceed. However, the cathode reaction has such a large potential, that the total voltage of the Galvanic cell is still positive. This tells us that although the anode half-reaction is not all that favorable, the cathode reaction pulls on the electrons so strongly that the Galvanic cell still works. Thus, the individual anode and cathode half-reactions need not be favorable in order for the Galvanic cell to still work. The only thing that is important is the *combination* of the two.

Before I leave this section, there is one more thing that I need to point out. In order to condense the notation for Galvanic cells, chemists have developed a shorthand. The Galvanic cell in Example 9.4, for example, is written in shorthand as follows:

$$Sn \mid Sn^{4+} (1M) \parallel Au^{3+} (1M) \mid Au$$

There are several specific things that are important here. First of all, the "||" represents the salt bridge in the Galvanic cell. In this notation, it separates the anode (to the left) from the cathode (to the right). For each half-reaction, the substances are listed. For the anode, the electrode is usually the first thing listed. After that, the rest of the substances at the anode (products and reactants) are listed. For every change of phase, a "|" is given. If a substance is aqueous, its concentration is given. Water is usually ignored in this notation.

In the above notation, then, the anode has an electrode made of Sn. There are no more solids, so the "|" indicates that the next substance is in a different phase. We know that it must be aqueous, because a concentration is given. That's the end of the anode. If we look at Table 9.1, there is only one half-reaction that has Sn^{4+} in it. Thus, that is the half-reaction for the anode. Remember, however, that *oxidation* occurs at the anode. Thus, the correct anode half-reaction is the *reverse* of what's listed in Table 9.1. On the other side of the "||" we find the cathode. The last thing listed on the cathode side is usually the electrode. In this case, then, we have a gold electrode (pretty fancy). The electrode is separated from the other substance because the electrode is solid and Au^{3+} is aqueous. The concentration of it is given. Since there is only one reaction with Au^{3+} listed in Table 9.1, that's the cathode half-reaction. Since reduction is what occurs at the cathode, we know that the half-reaction is correct as written in Table 9.1.

Please realize that the electrodes are not always a part of the half-reactions. They are sometimes there just to give the electrons something to move in. You will see how this all works in the next example, which sums up everything you need to know from this section.

EXAMPLE 9.5

Determine the overall reaction and voltage for the following Galvanic cell:

$$\mathbf{Fe \mid Fe^{2+} (1M) \parallel MnO_4^- (1M); Mn^{2+} (1M); H_3O^+ (1M) \mid Pt}$$

Draw a representation of this Galvanic cell as well, and describe what will happen at each electrode.

The first thing we need to do is determine the half-reactions. On the left of the "||," we see Fe followed by a "|." This means that the electrode is Fe. This may or may not be involved in the actual half-reaction. The next item (Fe^{2+}), however, is definitely a part of the reaction.

Since the only reaction in Table 9.1 that has Fe^{2+} in it is $Fe^{2+} + 2e^- \rightarrow Fe$, that's the reaction. Since this is the anode, however, oxidation occurs here. The reactions in Table 9.1 are reduction half-reactions, so we need to *reverse* the reaction for this cell. Since we reverse the reaction, we also take the negative of the reduction potential. Thus:

$$\text{anode: } Fe \rightarrow Fe^{2+} + 2e^- \quad E^o = 0.440 \text{ Volts}$$

Now what about the cathode. The "Pt" written all the way at the end is the electrode. That may or may not be in the reaction. Thus, we can't use it in our search of Table 9.1. The other substances are in the reaction, however. There is only one equation that has all of those substances in it: $MnO_4^- + 8H_3O^+ + 5e^- \rightarrow Mn^{2+} + 12H_2O$. Since this is the cathode, a reduction half-reaction is just what we need, so we do not reverse the reaction.

$$\text{cathode: } MnO_4^- + 8H_3O^+ + 5e^- \rightarrow Mn^{2+} + 12H_2O \quad E^o = 1.51$$

Remember, the notation ignores water, so don't be worried that water is in this half-reaction but not in the notation.

To get the voltage, we just add up the potentials. Thus, <u>the voltage is 1.95 Volts</u>. To get the overall reaction, we also add up the equations, but we do so in order to make the electrons cancel. Since one equation has 2 electrons and the other has 5, the only way I can get the electrons to cancel is to multiply the anode reaction by 5 and the cathode reaction by 2. When I do that, I get:

$$5Fe \rightarrow 5Fe^{2+} + 10e^-$$
$$+2MnO_4^- + 16H_3O^+ + 10e^- \rightarrow 2Mn^{2+} + 24H_2O$$

$$2MnO_4^- + 16H_3O^+ + 5Fe + \cancel{10e^-} \rightarrow 2Mn^{2+} + 24H_2O + 5Fe^{2+} + \cancel{10e^-}$$

When the electrons are canceled, we are left with:

$$\underline{2MnO_4^- + 16H_3O^+ + 5Fe \rightarrow 2Mn^{2+} + 24H_2O + 5Fe^{2+}}$$

Now for the picture. The Galvanic cell has an iron electrode at its anode and a platinum electrode at its cathode. Electrons always travel from the anode to the cathode in a Galvanic cell, and I already know the half-reactions. Thus:

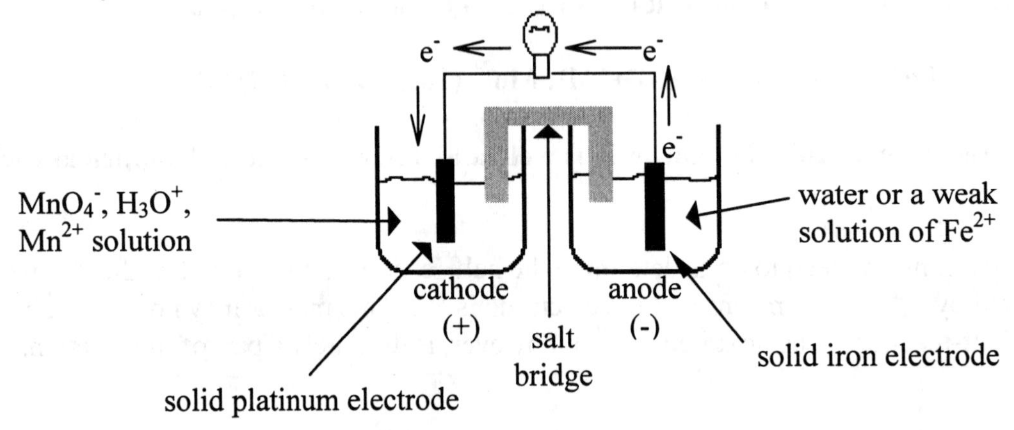

Now let's think about the chemistry involved here. At the cathode, MnO_4^- and H_3O^+ are reacting to form Mn^{2+} and water. As time goes on, then, the concentration of Mn^{2+} will increase in the cathode solution, and the concentration of MnO_4^- and H_3O^+ will decrease. On the anode side, iron is being used up and Fe^{2+} is being formed. Thus, the iron electrode will slowly disappear and the concentration of Fe^{2+} will increase.

Now that's a lot to know, but at least some of it should be review from your first-year course. The first thing you do is use Table 9.1 to determine the half-reactions. Determine the potential for each half-reaction by looking at Table 9.1 and be sure to take the negative of what's listed for the anode reaction. Add those two numbers together to get the total voltage. Then, add the half-reactions together (being sure to multiply if necessary to cancel electrons) to get the overall reaction. From all of that, you should be able to draw a picture of the cell. Finally, think through the chemistry of what's going on to determine what happens as the cell runs.

One point of confusion for students comes when they add the half-reactions together to get the total reaction. Some students think that if they must multiply the half-reaction by a number to ensure that electrons cancel, they must also multiply the half-reaction potential by that number as well. This is a natural thing to think, since you had to do that with ΔH in Hess' Law problems. However, this is not Hess' Law and half-reaction potentials are not ΔH's. Determining the voltage of the cell and determining its overall reaction are two separate steps. They are not related. Never multiply these half-reaction potentials by a number! There is never any reason to. Try this on your own to make sure you have it.

ON YOUR OWN

9.3 Draw a picture of the following Galvanic cell and discuss what happens as time goes on. Also, give the cell voltage and the overall reaction.

$$Cd \mid Cd^{2+}\ (1\ M)\ \|\ Pb^{2+}\ (1M) \mid Pb$$

9.4 Draw a picture of the following Galvanic cell and discuss what happens as time goes on. Also, give the cell voltage and the overall reaction.

$$Pb;\ PbSO_4 \mid SO_4^{2-}\ (1M)\ \|\ H_3O^+\ (1M) \mid H_2\ (1\ atm) \mid Pt$$

9.5 A Galvanic cell runs on the following overall reaction:

$$Mn\ (s) + Cd^{2+}\ (aq) \rightarrow Mn^{2+}\ (aq) + Cd$$

Assuming that both electrodes are made of substances present in each half-reaction, write the Galvanic cell's condensed notation.

EXPERIMENT 9.2
Making Your Own Galvanic Cell

Supplies:
From the laboratory equipment set:
- Copper sulfate
- Copper wire
- Zinc wire
- Two test tubes
- One test tube cap
- Chemical scoop
- Test tube rack

From around the house:
- Paper towel
- Scissors
- Table salt
- Water
- Aluminum foil
- Small glass
- Voltmeter (optional)

Use the chemical scoop to put 5 measures of copper sulfate in a test tube. Fill the test tube 3/4 full of water, cap and shake, trying to dissolve as much copper sulfate as possible. There will be some left over. That's okay. Uncap the test tube. Fill the other test tube 3/4 of the way with water. Stand them next to each other in the test tube rack. Next, make a saturated table salt solution in the small glass. You don't need a lot. Don't worry if there is extra salt at the bottom, that's fine. Once you have the saturated salt solution, use the scissors to cut the paper towel along its width. In the end, you want a 1-inch wide strip that is as long as the paper towel was wide. Soak the strip of paper towel in the saturated salt solution and then roll it into a long, thin roll. Stick one end of the roll in one test tube and the other end in the other test tube. Play around with the positioning until each end of the paper towel strip stays in each test tube. The paper towel can be as far immersed in each test tube as you need it to be in order to stay stable.

You now have the makings of a Galvanic cell. The test tubes are your anode and cathode, and the paper towel is your salt bridge. Next, stick the copper wire in the test tube that has the copper sulfate solution, and stick the zinc wire in the test tube that has only water. If you have a Voltmeter, use it now by touching one lead to the zinc wire and the other lead to the copper wire. You should read a voltage. That shows you the Galvanic cell is acting like a battery. Unfortunately, the current produced by this cell is so low that it can't be used to light a light bulb or anything, but you can see that there is a reaction going on by performing the procedure in the next paragraph.

Cut a thin strip of aluminum foil and use it to make an electrical connection between the copper and zinc wires. Be sure that the aluminum foil touches each wire. You can do this by wrapping the aluminum foil around each end of the wire. Now just wait. In five to ten minutes, you should see flakes appearing in the copper sulfate solution and on the copper wire. In this battery, the cathode half-reaction is $Cu^{2+} + 2e^- \rightarrow Cu$. The flakes you see are the copper being formed in this reaction. Thus, electrons flowed from the zinc wire, through the aluminum foil, into the copper wire, and then reacted with the Cu^{2+} (from the copper sulfate) and formed Cu. After a while, take the copper wire out and examine it. You should see plenty of new copper on the wire. There should also be some in the solution. Also, look at the solution. It should be a lot less blue. That's because the Cu^{2+} that gave the solution its blue color has reacted and is thus no longer in solution. Take the zinc wire out and examine it. It will be harder to see any difference, but the zinc wire should be different, because the anode reaction of this Galvanic cell is $Zn \rightarrow Zn^{2+} + 2e^-$. Thus, the zinc wire lost zinc, and the solution now has Zn^{2+} in it.

The Nernst Equation

In the previous section, every time I listed the concentration of an aqueous substance or the pressure of a gas, notice that I used 1 M or 1 atm, respectively. There's a reason for that. Notice the title of Table 9.1: "Standard Reduction Potentials." As you might expect, the potential of a half-reaction depends on the concentration of the substances involved in the half-reaction. The "standard" concentration of all substances in a Galvanic cell is 1 M in the case of aqueous substances and 1 atm in the case of gases. If the concentration of substances in a Galvanic cell are at these "standard" levels, then the voltage will be exactly what you calculate from Table 9.1.

For example, in the Galvanic cell you made in Experiment 9.2, you can calculate that the voltage should be 0.763 + 0.337 = 1.100 Volts. If you had a Voltmeter for Experiment 9.2, however, you certainly did not measure 1.10 Volts. That's because in order to get that voltage, the concentration of Cu^{2+} at the cathode and the concentration of Zn^{2+} at the anode would both have to have been 1 Molar. In your experiment, the concentration of Zn^{2+} was quite small, because you initially used tap water at the anode. As the reaction progressed, Zn^{2+} was made, but it never reached a concentration of 1 Molar. Although there certainly was Cu^{2+} at the cathode to begin with, the concentration was probably not 1 Molar. Even if it were briefly, it would not stay that way for long, because Cu^{2+} was used up in the reaction. Thus, your Galvanic cell was certainly not at standard conditions.

How do we calculate the voltage of a Galvanic cell if it is not at standard conditions? We use the **Nernst equation**.

$$E_{cell} = E^{o}_{cell} - \frac{0.05916}{n} \cdot \log(Q) \qquad (9.6)$$

In this equation (developed by German chemist W. H. Nernst), the "E^{o}_{cell}" term represents the voltage of the cell that you calculate from Table 9.1. In other words, it represents the voltage of the cell under standard conditions. The "n" represents how many electrons are transferred per reaction, and "Q" is called the **reaction quotient**. Do you remember how the equilibrium constant is defined? Well, "Q" is defined in the same way. Since we aren't dealing with equilibrium, however, we cannot call it the equilibrium constant, so we call it the reaction quotient. The following example ought to clear up any confusion you have on the Nernst equation.

EXAMPLE 9.6

Determine the cell potential (voltage) of the following Galvanic cell:

$$Cd \mid Cd^{2+}\ (2.00\ M) \parallel Pb^{2+}\ (0.0010\ M) \mid Pb$$

First we have to get through the notation. The cell has a Cd electrode at the anode. The only half-reaction with Cd^{2+} is $Cd^{2+} + 2e^- \rightarrow Cd$ with a potential of -0.403 Volts. Since this is for the anode, however, the equation must be reversed, because oxidation occurs at the anode. This means we must take the negative of the potential, making the anode half-reaction potential 0.403 Volts. The cathode has a lead electrode and Pb^{2+}. Referring to Table 9.1, then, the cathode half-reaction must be $Pb^{2+} + 2e^- \rightarrow Pb$, with a potential of -0.126 Volts. The standard cell potential, therefore, is 0.403 + -0.126 = 0.277 Volts. Since we are not under standard conditions, however, we need to use the Nernst equation.

Well, we just calculated E^o_{cell}. It is equal to 0.277 Volts. What about the other components in the Nernst Equation? The "n" refers to the number of electrons transferred per reaction. How do we get that? We determine how many electrons cancel out of the overall equation when we add the two equations together. That's "n." Thus:

$$Cd \rightarrow Cd^{2+} + 2e^-$$
$$+\ Pb^{2+} + 2e^- \rightarrow Pb$$

$$Cd + Pb^{2+} + \cancel{2e^-} \rightarrow Cd^{2+} + \cancel{2e^-} + Pb$$

$$Cd + Pb^{2+} \rightarrow Cd^{2+} + Pb$$

Since 2 electrons canceled out on both sides of the equation, n=2.

It's good that we had to add the equations together, because we need the overall equation to get "Q." Remember, the reaction quotient has the same definition as the equilibrium constant, it just applies to non-equilibrium situations. Thus, we take the products and raise them to their stoichiometric coefficients and divide by the reactants raised to their stoichiometric coefficients. However, we need to ignore liquids and solids. What are the solids in this case? Well, look at the notation. Both Cd and Pb are separated from the ions with a "|." Remember, that means a change in phase. Thus, Cd and Pb are both solids. The reaction quotient, then, is given by:

$$Q = \frac{[Cd^{2+}]}{[Pb^{2+}]}$$

That makes the Nernst equation:

$$E_{cell} = E^{o}_{cell} - \frac{0.05916}{n} \cdot \log(Q)$$

$$E_{cell} = 0.277 \text{ Volts} - \frac{0.05916}{2} \cdot \log\left(\frac{2.00}{0.0010}\right) = 0.277 \text{ Volts} - 0.0976 \text{ Volts} = \underline{0.179 \text{ Volts}}$$

Noticc that I did not bother to put the units in when I calculated "Q." That's because the last term of the Nernst equation is *defined* to come out in the units Volts as long as you *always* use units of either molarity or atmospheres when calculating it. Thus, I didn't bother to put the molarity units in. Also, since it is defined that way, I simply stuck the unit "Volts" on after I determined that the last term equaled 0.0976.

That's how you use the Nernst equation. You determine the overall cell reaction so that you can determine the values for "n" and "Q." As long as your units in "Q" are only moles or atmospheres, your answer will come out in the proper units: Volts.

ON YOUR OWN

9.6 What is the cell potential of the following Galvanic cell?

$$Zn \mid Zn^{2+} (0.0100 \text{ M}) \parallel Cu^{2+} (1.00 \text{ M}) \mid Cu$$

9.7 What is the cell potential of the following Galvanic cell?

$$Pt \mid Br_2 (l); Br^- (0.450 \text{ M}) \parallel Cl^- (0.0500 \text{ M}) \mid Cl_2 (g) (0.900 \text{ atm}) \mid Pt$$

Electrolytic Cells

Throughout this entire module, I have concentrated on Galvanic cells. In such cells, the cell potential is positive, and the result is that electrons flow freely between the anode and cathode. I already told you that if the cell potential turns out to be negative, then the Galvanic cell will not work. That's because there is no force pushing the electrons from anode to cathode, and so the electrons simply stay where they are.

There is another type of cell in chemistry called the **electrolytic cell**. These cells use electricity (often produced in Galvanic cells) to force redox reactions that would normally not occur. To get an idea of what I mean, perform the following experiment.

EXPERIMENT 9.3
The Electrolysis of Copper Sulfate

Supplies:
From the laboratory equipment set:
- Copper sulfate
- Depression plate
- 9-Volt battery cap (It has black and red wires and clamps on top of a 9-Volt battery.)
- Chemical scoop
- Dropper

From around the house
- 9-Volt battery
- Distilled water

Add a measure of copper sulfate to one of the deep wells in your depression plate. Now use your dropper to add distilled water to fill the well. Stir with your dropper so as to dissolve as much copper sulfate as possible. You will not be able to dissolve it all. That's okay. Next, place the cap on the 9-Volt battery and hold the wires so that the ends of the wires do not touch. Immerse the ends of both wires into the copper sulfate solution, making sure that they don't touch each other. Observe what happens. Watch the reaction for a few minutes. Note what you see. In particular, note what you see on each color of wire.

What happened in the reaction? Before you immersed the wires in the solution, the copper sulfate simply dissolved in the water and stayed there. Thus, there were Cu^{2+} ions, SO_4^{2-} ions, and water, all happily coexisting with one another. When you immersed the ends of the battery, you supplied the solution with some electrical energy. That caused the constituents of the solution to do something they would ordinarily not do: react. On the black wire, you should have noticed a blackish orange solid growing right on the wire. That was copper. You see, the black wire attaches to the negative side of the battery. Thus, the black wire was a source of electrons. Those electrons participated in the following half-reaction:

$$Cu^{2+} + 2e^{-} \rightarrow Cu$$

The copper ions in solution took electrons from the black wire, forming solid copper. Since the copper ions had to travel to the wire to get the electrons, that's where the solid copper formed.

On the red wire, you saw no solid forming, but you should have seen a lot of bubbles. That's because the red wire was attached to the positive end of the battery, which attracts electrons to itself. In the end, the water molecules were not holding onto their electrons strongly enough, and the energy with which the positive side of the battery pulled on them was enough to tear some electrons away from the water via this half-reaction:

$$6H_2O \rightarrow O_2 + 4H_3O^{+} + 4e^{-}$$

The bubbles which you saw on the red wire were the oxygen forming in the process.

The overall reaction that you witnessed, then, was the sum of the two half-reactions:

$$2\cdot[Cu^{2+} + 2e^- \rightarrow Cu]$$
$$+ 6H_2O \rightarrow O_2 + 4H_3O^+ + 4e^-$$

$$6H_2O + 2Cu^{2+} + \cancel{4e^-} \rightarrow 2Cu + O_2 + 4H_3O^+ + \cancel{4e^-}$$

Once the electrons are canceled, this equation becomes:

$$6H_2O + 2Cu^{2+} \rightarrow 2Cu + O_2 + 4H_3O^+$$

Now look at this equation for a moment. The reactants (water and copper ions) were always in the solution. They never reacted, however, because the ΔG for this reaction is positive. Under normal conditions, then, this reaction just wouldn't go. However, the battery forced the issue, forcing electrons on Cu^{2+} and ripping them away from water.

That's the difference between Galvanic cells and electrolytic cells. Electrolytic cells use the energy from a source of electricity to force electrons to move where they otherwise would not. Galvanic cells, on the other hand, simply facilitate the movement of electrons in ways that the electrons would normally move. In other words, Galvanic cells use spontaneous reactions resulting in the flow of electrons, whereas electrolytic cells use the flow of electrons to force non-spontaneous reactions to actually happen.

The electrolytic cell that you made in Experiment 9.3 is diagrammed as follows:

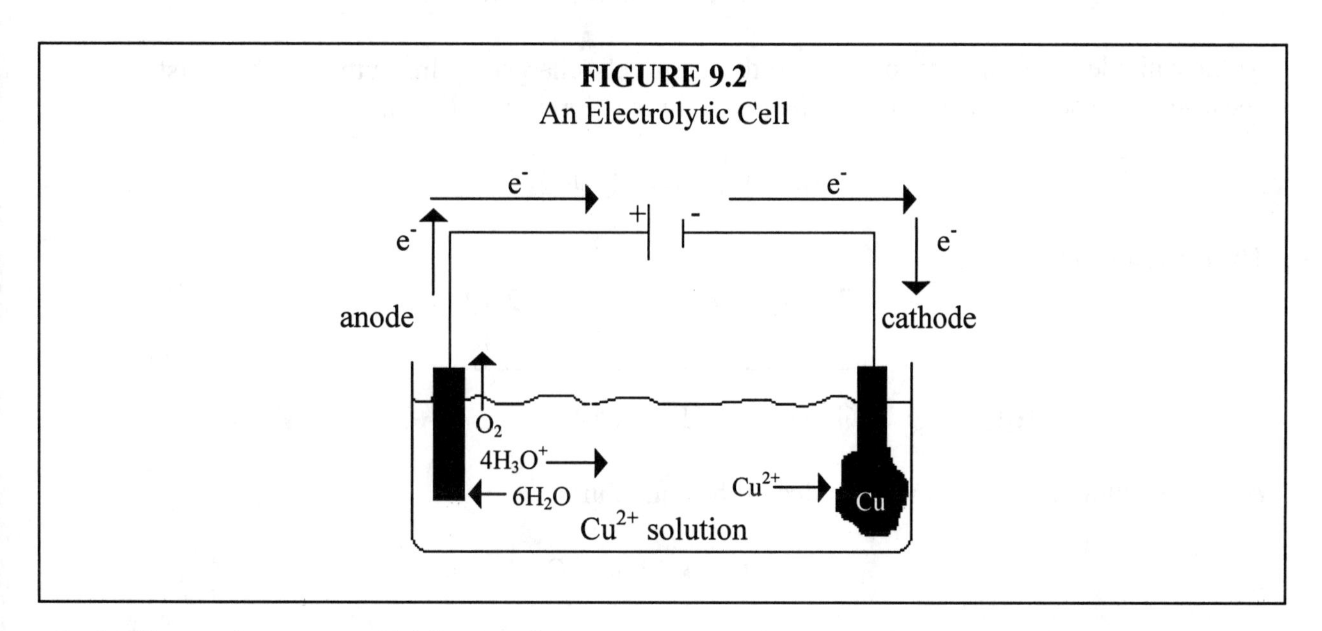

FIGURE 9.2
An Electrolytic Cell

In the figure, the two parallel lines at the top represent a battery. The longer line is the positive side and the shorter line is the negative side. At the anode, electrons are being ripped away from water molecules. Those electrons then travel to the positive side of the battery. Notice that 6 water molecules are being shown going towards the anode. This indicates that water is the molecule whose electrons are being taken. The products of the anode half-reaction (oxygen and the hydronium ion) are shown leaving the anode. The oxygen is shown rising out of solution, and the hydronium ion is shown going away from the anode into solution. Electrons are flowing from the negative side of the battery to the cathode. These electrons attract the Cu^{2+} ions and, when the ions hit the cathode, they are turned into solid copper, making the size of the cathode grow.

Notice one other difference between an electrolytic cell and a Galvanic cell. In a Galvanic cell, we take great pains to separate the reactants. After all, if we mixed the reactants together, they would immediately begin exchanging electrons, completing the reaction. Under those circumstances, we would not be able to use the electrons, because they would go from one reactant to another without bothering to pass through any wires. Thus, we separate the reactants and connect them with a salt bridge and wires. The wires give the electrons a path to travel, and the salt bridge keeps the charges in balance. In an electrolytic cell, none of that is necessary. Since the reactants *won't* react without the help of a battery, they can be mixed right in the same container. That way, we need no salt bridge. Thus, an electrolytic cell is much easier to set up than a Galvanic cell!

There are a couple of specific electrolytic cells that are worth mentioning. One of the classic electrolytic cells is made from dilute sulfuric acid. When such a solution is exposed to the electrical energy of the battery, you see gas bubbles forming on both electrodes. The anode, which rips electrons away from the reactants of an electrolytic cell, takes electrons away from the water molecules in the solution, just like it did in the electrolytic cell that you made.

$$6H_2O \rightarrow O_2 + 4H_3O^+ + 4e^-$$

At the cathode, electrons are forced onto the positively-charged hydronium ions that exist because of the acid in the solution. The resulting reaction is as follows:

$$2H_3O^+ + 2e^- \rightarrow H_2 + 2H_2O$$

The overall reaction, then, is:

$$2\cdot[2H_3O^+ + 2e^- \rightarrow H_2 + 2H_2O]$$
$$+ 6H_2O \rightarrow O_2 + 4H_3O^+ + 4e^-$$

$$4H_3O^+ + 4e^- + 6H_2O \rightarrow 2H_2 + 4H_2O + O_2 + 4H_3O^+ + 4e^-$$

After canceling like terms on both sides of the equation:

$$2H_2O \rightarrow 2H_2 + O_2$$

In the end, then, this particular electrolytic cell causes water to be broken down into hydrogen and oxygen.

This electrolytic cell is often used as a demonstration in introductory chemistry courses. When done in a special apparatus called a **Hoffmann apparatus**, the hydrogen and oxygen gas are collected in separate containers. Since the mole ratio of gases in the reaction is two hydrogens to one oxygen, and since the volume of gases is proportional to the number of moles of gas, this electrolytic cell always produces twice as much hydrogen as oxygen. This can be a visual demonstration of the fact that a water molecule is composed of twice as much hydrogen as oxygen.

FIGURE 9.3
The Electrolysis of Water

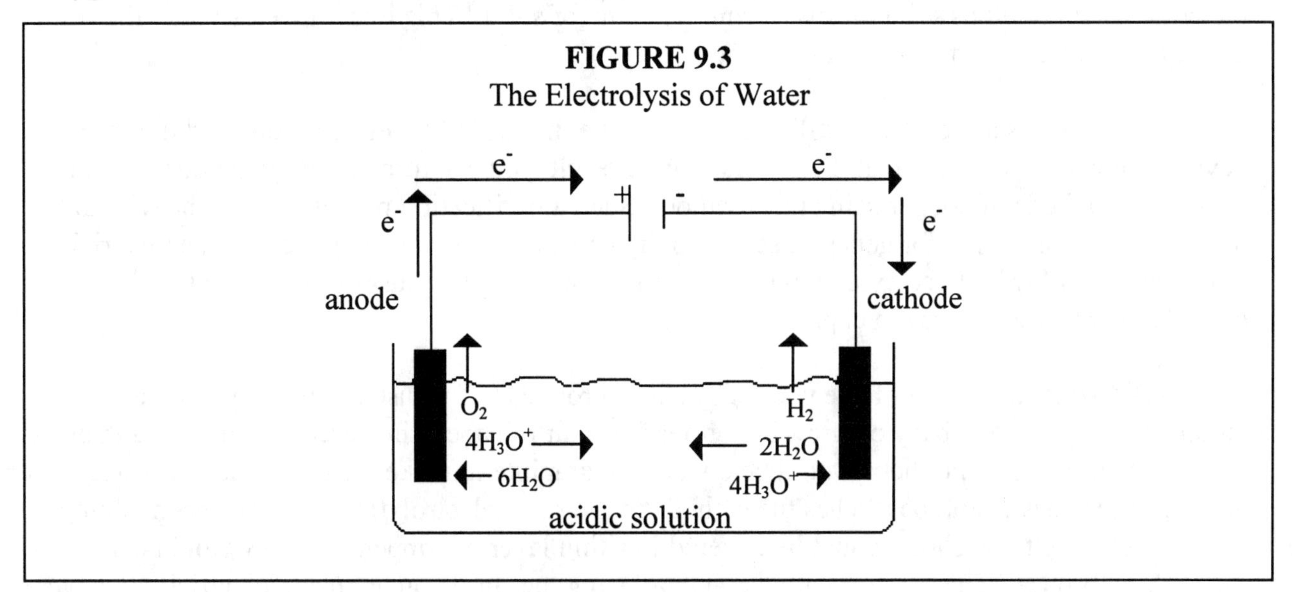

The last specific electrolytic cell I want to mention is used in the purification of copper. Interestingly enough, I can make an electrolytic cell that involves only copper and copper ions. If I connect an impure rod of copper to the positive side of a battery and I connect a pure rod of copper to the negative side of a battery and then immerse them into a solution of salt water, the rod of impure copper will slowly disintegrate, and the rod of pure copper will have more pure copper formed on it. The electrolytic cell is shown below.

FIGURE 9.4
Purifying Copper With Electrolysis

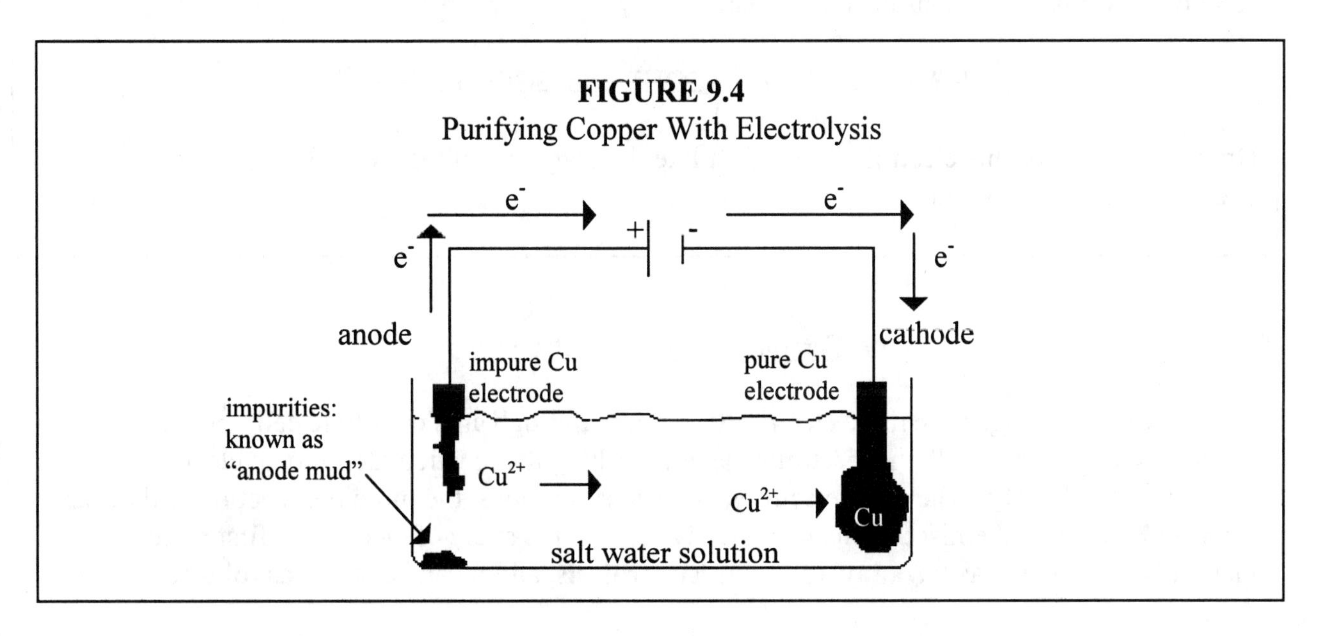

In this electrolytic cell, the anode pulls electrons away from the copper electrode, making copper ions. The cathode then adds electrons to those copper ions again, making them solid copper. Now why do such a crazy thing? Why waste electricity disintegrating a copper rod just to have the copper form again on another copper rod? Well, natural sources of copper are always contaminated with other metals. These contaminating metals, however, do not give up their electrons as easily as copper does. Thus, only the copper in the impure copper rod disintegrates. The other metals stay solid. As the copper rod disintegrates, then, these impurities simply sink to the bottom of the cell as solid metals, forming a lump of solid. This lump appears under the anode and is often referred to as "anode mud."

Thus, this is a means of purifying copper. The impurities never form ions and thus are never formed back into metal at the cathode. As a result, you can turn an impure anode of copper into a pure cathode of copper using this method. Such a purification process is used heavily in the production of copper for general use. Also, it can be used to cover any conductive material with copper. If I want to cover a metal with copper, for example, I just use that metal as the cathode. It will get covered in copper.

Did your parents ever have your baby shoes bronzed? This used to be very popular among parents. After a baby outgrew his or her first pair of shoes, the parents would send them away to be bronzed. The shoes would be covered in graphite to make them conduct electricity, and then the shoes would be used as the cathode in the same electrolytic cell I have been talking about. As a result, the shoes would be covered in a thin layer of copper, and they would come out looking bronze. This is often called **electroplating**, because you are using electricity to cover something with a "plate" of metal.

ON YOUR OWN

9.8 Two platinum electrodes are connected to a battery and immersed in a sodium chloride solution. Two half-reactions occur:

$$2H_2O + 2e^- \rightarrow H_2 + 2OH^- \quad \text{and} \quad 2Cl^- \rightarrow Cl_2 + 2e^-$$

Draw a diagram for this electrolytic cell, just like the diagrams in Figures 9.3 - 9.4. What is the overall reaction of this cell?

Faraday's Law of Electrolysis

Not surprisingly, the more electricity you pass through an electrolytic cell, the more products you get. After all, the electrons that are pushing the reaction along in an electrolytic cell come from the battery. The more electrons the battery supplies, the more the reaction will occur. Michael Faraday, a chemist that made great advances in the study of electricity, first noticed this fact. It is now known as **Faraday's Law of Electrolysis**, and it can be stated as follows:

Faraday's Law of Electrolysis - The number of moles of products in an electrolytic cell is directly proportional to the current supplied and the time over which it is supplied.

It turns out that we can actually use Faraday's law to predict how much product is produced in an electrolytic cell, providing we know the current of the battery and the time that the cell runs.

How can we do such a thing? It starts with the most important relationship in electrochemistry: the **Faraday**.

1 mole of electrons = 96,485 Coulombs of charge

The Faraday is sometimes expressed as a constant: 96,485 Coulombs per mole of electrons. Whether you express it as a constant or as a relationship, its importance cannot be ignore. The Faraday tells us how much charge it takes to make a mole of electrons.

Why is this important? Well think about it. Batteries are rated by the amount of charge they can deliver in a second. This is called the **current** of the battery, and the unit of current is called the **Amp**. An Amp is defined as a Coulomb of charge per second. If a battery can deliver 5 Amps, then, it can produce 5 Coulombs of charge every second. What does that do for us? Well, if I know the Amps that come from a battery and I know how long the battery runs, I can calculate the total number of Coulombs the battery delivers. Using the Faraday, I can then convert that to *moles* of electrons. At that point, I can use the number of moles of electrons in stoichiometry! Study the following example to see what I mean.

EXAMPLE 9.7

An electrolytic cell is used to purify copper. At the cathode, Cu^{2+} is reduced to copper. If the battery can deliver 12 Amps of current, how many grams of copper will be plated out in an hour?

When Cu^{2+} is reduced to copper, it must gain two electrons. Thus, the cathode half-reaction must be:

$$Cu^{2+} + 2e^{-} \rightarrow Cu$$

In these problems, we always assume that the electrons are the limiting reagent in the reaction. Thus, if I can figure out how many moles of electrons were supplied by the battery, I can figure out how many moles of copper were formed.

Well, it turns out that I *can* determine how many moles of electrons the battery supplied. The current and time can be used to determine how much charge the battery produced. Remember, Amps are defined as Coulombs per second. Thus, if I take Amps times seconds, I will get Coulombs:

$$12\ \frac{\text{C}}{\cancel{\text{sec}}} \times 3600\ \cancel{\text{sec}} = 43{,}000\ \text{C}$$

Now that I have the number of Coulombs, I can use the Faraday to convert to *moles* of electrons:

$$43{,}000\ \cancel{\text{C}} \times \frac{1\ \text{mole of electrons}}{96{,}485\ \cancel{\text{C}}} = 0.45\ \text{moles of electrons}$$

Now that I have moles of electrons, this becomes a stoichiometry problem. The half-reaction tells me it takes 2 moles of electrons to make one mole of Cu.

$$0.45\ \cancel{\text{moles}\ e^-} \times \frac{1\ \text{mole Cu}}{2\ \cancel{\text{moles}\ e^-}} = 0.23\ \text{moles Cu}$$

All I have to do now is convert from moles of Cu to grams of Cu.

$$0.23\ \cancel{\text{moles Cu}} \times \frac{63.5\ \text{g Cu}}{1\ \cancel{\text{mole Cu}}} = \underline{15\ \text{g Cu}}$$

Notice what I did here. Using current and time, I calculated the charge delivered by the battery. Then, using the Faraday, I converted charge into the number of moles of electrons. From there, I was able to do a simple stoichiometry problem to figure out how much product was produced. I want to do one more example to make sure you understand this. To add a twist, I will do the problem the other way around.

EXAMPLE 9.8

While in college, the author of this text made extra money by electroplating silver from silver nitrate solutions that were being disposed of by a local hospital. The cathode of his electrolytic cell reduced Ag^+ from the silver nitrate solution to solid Ag. He then sold this silver to jewelry dealers. If the author had an 11 Amp battery and needed to produce 100.0 grams of silver, how long would he have to run his electrolytic cell?

We know how much product will be produced. The question asks how much time it will take. Well, that depends on the number of moles of electrons that are needed. We can determine that using stoichiometry. The cathode reaction is:

$$Ag^+ + e^- \rightarrow Ag$$

Using this reaction in a stoichiometry problem:

$$100.0\,\text{g}\,\cancel{\text{Ag}} \times \frac{1\,\text{mole Ag}}{107.9\,\text{g}\,\cancel{\text{Ag}}} = 0.9268\,\text{moles Ag}$$

$$0.9268\,\cancel{\text{moles}}\,\cancel{\text{Ag}} \times \frac{1\text{ mole electrons}}{1\,\cancel{\text{mole}}\,\cancel{\text{Ag}}} = 0.9628\text{ moles electrons}$$

So we need 0.9628 moles of electrons. We can convert that to Coulombs using the Faraday, and then use that to determine how long it would take for 11 Amps to deliver that charge.

$$0.9628\ \cancel{\text{moles}\,e^-} \times \frac{96{,}485\,\text{C}}{1\,\cancel{\text{mole}\,e^-}} = 9.290 \times 10^4\ \text{C}$$

$$9.290 \times 10^4\ \cancel{\text{C}} \times \frac{\text{sec}}{11\ \cancel{\text{C}}} = \underline{8445\,\text{sec}}$$

Thus, it takes a little over 2 hours to make 100 grams.

ON YOUR OWN

9.9 The cathode of an electrolytic cell is used to reduce Au^{3+} into solid gold. If the battery can deliver 5.00 Amps, how many grams of gold will be made in 2.00 hours?

9.10 A Hoffmann apparatus uses a battery that delivers 5.00 Amps. How long will it take for the Hoffmann apparatus to make 15.0 grams of hydrogen gas?

ANSWERS TO THE ON YOUR OWN PROBLEMS

9.1 a. Rule #2 says Li has an oxidation number of +1. Rule #5 tells us that in this molecule, H has an oxidation number of +1. We have no rule for N, but we can figure it out because all oxidation numbers need to add up to the total charge (0). The oxidation numbers from Li and H sum up to +3. Thus, N must be -3. The oxidation numbers, then, are: Li: +1, N: -3, H: +1.

b. Rule #5 tells us that, since N is not a metal, H's oxidation number is +1. Thus, to make all oxidation numbers add up to the total charge (0), N must be -1. The oxidation numbers, then, are: N: -1, H: +1.

c. Rule #3 says that Ca is +2. Rule #6 tells us that O is -2. There is one Ca and 4 O's, so the total oxidation number so far is $(+2) + 4 \cdot (-2) = -6$. There are two N's, and they must combine to make +6 in order for the oxidation numbers to sum up to the total charge (0). Thus, each N must have a +3. The answers, then, are: Ca: +2, N: +3, O: -2.

d. Rule #6 tells use that O has a -2 oxidation number. Since the total charge is zero, C must have a +4 oxidation number. C: +4, O: -2

e. Rule #4 tells us that F is always -1. In order for the oxidation numbers to add up to the total charge (-1), that means B must be +3. B: +3, F: -1

f. Rule #6 tells us that O is -2. Since the oxidation numbers must add up to the total charge (-3), P must be +5. P: +5, O: -2

g. Rule #6 tells us that O is -2. We have no ironclad rules for N or Cl. So now we must resort to rule #7. This works best on groups 3A, 6A, and 7A. Since Cl is in 7A, we will use it on Cl. Thus, Cl has an oxidation number of -1. In order to get all oxidation numbers to add up to the overall charge (0), N must be +3. Cl: -1, N: +3, O: -2

h. Rule #1 says that when a molecule is made up of only one type of atom, the oxidation number is the charge (0) divided by the number of atoms (8). S: 0

9.2 a. On the reactants side, the oxidation numbers are as follows: Ca: 0, Cl: 0. On the products side, the oxidation numbers are: Ca: +2, Cl: -1. Since Ca went from 0 to +2, electrons were lost. This means that Ca was oxidized; therefore, Ca is the reducing agent. On the other hand, Cl went from 0 to -1, indicating that electrons were gained. This means that Cl was reduced, which tells us that Cl is the oxidizing agent.

b. On the reactants side, the oxidation numbers are: F: -1, I: +5, Fe: 0. On the products side, the oxidation numbers are Fe: +3, F: -1 (in both molecules), I: +3. Since Fe went from 0 to +3, it lost 3 electrons. This means that Fe was oxidized and is therefore the reducing agent. On the other hand, I went from +5 to +3, indicating that it gained 2 electrons. Thus, I was reduced, which means I is the oxidizing agent.

9.3 The first thing we need to do is determine the half-reactions. On the left of the "||," we see Cd followed by a "|." This means that the electrode is Cd. This may or may not be involved in the actual half-reaction. The next item (Cd^{2+}), however, is definitely a part of the reaction. Since the only reaction in Table 9.1 that has Cd^{2+} in it is $Cd^{2+} + 2e^- \rightarrow Cd$, that's the reaction. Since this is the anode, however, oxidation occurs here. The reactions in Table 9.1 are reduction half-reactions, so we need to *reverse* the reaction for this cell. Since we reverse the reaction, we also take the negative of the reduction potential. Thus:

$$\text{anode: } Cd \rightarrow Cd^{2+} + 2e^- \quad E^o = 0.403 \text{ Volts}$$

Now what about the cathode? The "Pb" written all the way at the end is the electrode. That may or may not be in the reaction. Thus, we can't use it in our search of Table 9.1. The other substance (Pb^{2+}) is in the reaction, however. There is only one equation that has Pb^{2+} in it: $Pb^{2+} + 2e^- \rightarrow Pb$. Since this is the cathode, a reduction half-reaction is just what we need, so we do not reverse the reaction.

$$\text{cathode: } Pb^{2+} + 2e^- \rightarrow Pb \qquad E^o = -0.126$$

To get the voltage, we just add up the potentials. Thus, the voltage is 0.277 Volts. To get the overall reaction, we also add up the equations, but we do so in order to make the electrons cancel. Since each equation has 2 electrons in it, we can just add them.

$$
\begin{array}{r}
Cd \rightarrow Cd^{2+} + 2e^- \\
+ Pb^{2+} + 2e^- \rightarrow Pb \\
\hline
\end{array}
$$

$$Cd + Pb^{2+} + \cancel{2e^-} \rightarrow Cd^{2+} + \cancel{2e^-} + Pb$$

When the electrons are canceled, we are left with:

$$\underline{Cd + Pb^{2+} \rightarrow Cd^{2+} + Pb}$$

Now for the picture. The Galvanic cell has a Cd electrode at its anode and a Pb electrode at its cathode. Electrons always travel from the anode to the cathode in a Galvanic cell, and I already know the half-reactions. Thus:

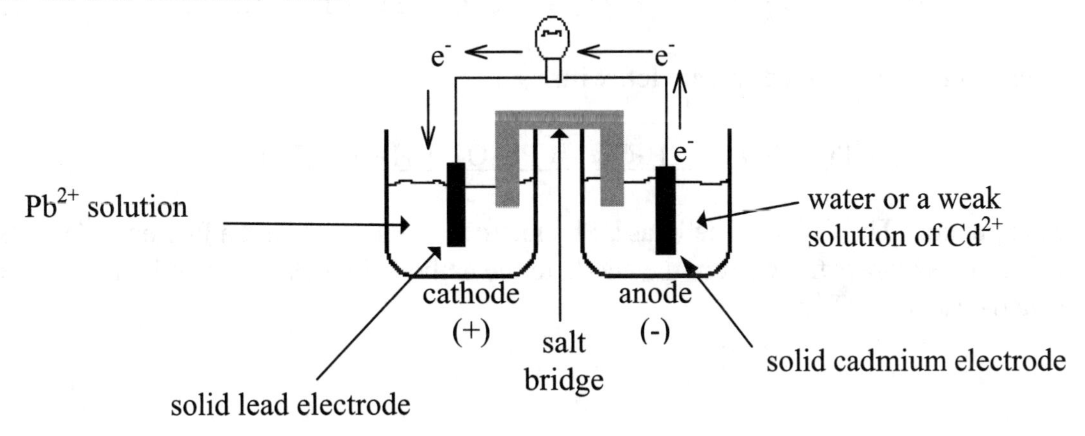

Now let's think about the chemistry involved here. At the cathode, Pb^{2+} ions react to form solid lead. As time goes on, then, the lead electrode will slowly grow. On the anode side, cadmium is being used up and Cd^{2+} is being formed. Thus, the cadmium electrode will slowly disappear and the concentration of Cd^{2+} will increase.

9.4 On the left of the "||," we see Pb followed by a semicolon and then $PbSO_4$. This means there are two solids at the anode: lead and lead sulfate. The anode's electrode is always the first thing written, though, so the electrode is lead. After the "|," the sulfate ion is listed. Thus, we are looking for a reaction that has lead sulfate and the sulfate ion. The only possible reaction, then, is $PbSO_4 + 2e^- \rightarrow Pb + SO_4^{2-}$. There is another reaction that has lead sulfate and the sulfate ion, but it also has PbO_2 in it. That's not listed in the notation, so that can't be the anode reaction. We reverse the anode reaction and take the negative of the potential, so

$$\text{anode: } Pb + SO_4^{2-} \rightarrow PbSO_4 + 2e^- \quad E^o = 0.356 \text{ Volts}$$

Now what about the cathode? The "Pt" written all the way at the end is the electrode. That may or may not be in the reaction. Thus, we can't use it in our search of Table 9.1. The other substances (H_3O^+ and H_2) are in the reaction, however. There is only one equation that has those substances in it: $2H_3O^+ + 2e^- \rightarrow H_2 + H_2O$. Remember, we ignore water in the notation, that's why water can be in the reaction. Since this is the cathode, a reduction half-reaction is just what we need, so we do not reverse the reaction.

$$\text{cathode: } 2H_3O^+ + 2e^- \rightarrow H_2 + 2H_2O \qquad E^o = 0.000$$

To get the voltage, we just add up the potentials. Thus, the voltage is 0.356 Volts. To get the overall reaction, we also add up the equations, but we do so in order to make the electrons cancel. Since each equation has 2 electrons in it, we can just add them.

$$\begin{array}{r} Pb + SO_4^{2-} \rightarrow PbSO_4 + 2e^- \\ + 2H_3O^+ + 2e^- \rightarrow H_2 + 2H_2O \\ \hline \end{array}$$

$$Pb + SO_4^{2-} + 2H_3O^+ + \cancel{2e^-} \rightarrow PbSO_4 + \cancel{2e^-} + H_2 + 2H_2O$$

When the electrons are canceled, we are left with:

$$\underline{Pb + SO_4^{2-} + 2H_3O^+ \rightarrow PbSO_4 + H_2 + 2H_2O}$$

Now for the picture. The Galvanic cell has a Pb electrode at its anode and a Pt electrode at its cathode. Electrons always travel from the anode to the cathode in a Galvanic cell, and I already know the half-reactions. Thus:

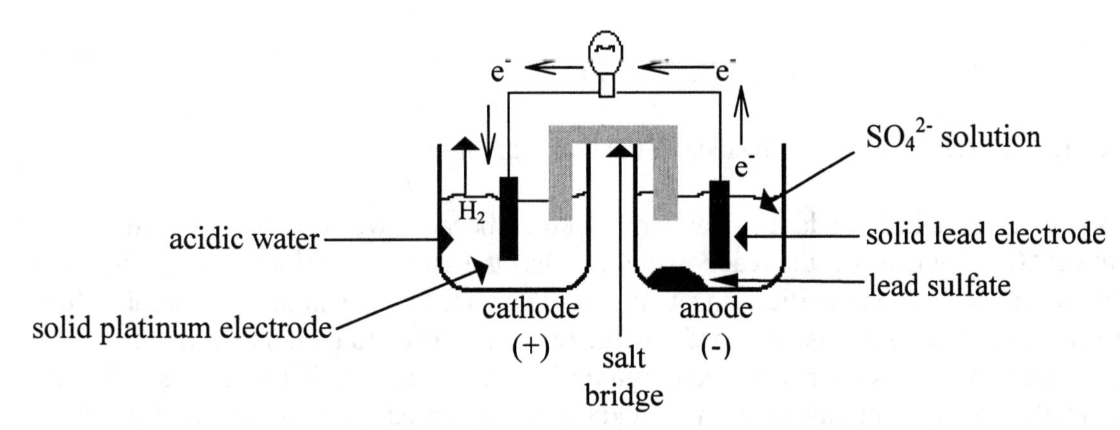

Now let's think about the chemistry involved here. At the cathode, hydronium ions react to form hydrogen gas. As time goes on, then, bubbles will appear at the cathode, and the water will get less acidic. On the anode side, lead and sulfate ions are being used up and lead sulfate is being formed. Thus, the lead electrode will slowly disappear and the concentration of SO_4^{2-} will decrease. At the same time, a pile of lead sulfate will increase in size.

9.5 We need to look at the chemical reaction and determine what's going on. Mn is losing electrons to become Mn^{2+}. The substance at the anode loses electrons, so Mn is at the anode. We are told to assume that the electrode is a part of the reaction, so the electrode at the anode is made of Mn. The other thing going on is that Cd^{2+} is gaining electrons to make Cd. Things at the cathode gain electrons, so this is the cathode half-reaction. Once again, the problem tells us to assume that the electrode is a part of the reaction, so the electrode is Cd. We will assume that the concentrations are 1 molar. Thus, the notation is:

$$\underline{Mn \mid Mn^{2+}\ (1\ M)\ \|\ Cd^{2+}\ (1\ M) \mid Cd}$$

9.6 First we have to get through the notation. The cell has a Zn electrode at the anode. The only half-reaction with Zn^{2+} is $Zn^{2+} + 2e^- \rightarrow Zn$ with a potential of -0.763. Since this is for the anode, however, the equation must be reversed , because oxidation happens at the anode. This means we must take the negative of the potential, making the anode half-reaction potential 0.763. The cathode has a copper electrode and Cu^{2+}. Referring to Table 9.1, then, the cathode half-reaction must be $Cu^{2+} + 2e^- \rightarrow Cu$, with a potential of 0.337. The standard cell potential, therefore, is 0.763 + 0.337 = 1.100 Volts. Since we are not under standard conditions, however, we need to use the Nernst equation.

Well, we just calculated E^o_{cell}. It is equal to 1.100 Volts. What about the other things in the Nernst equation? The "n" refers to the number of electrons transferred per reaction. How do we get that? We determine how many electrons cancel out of the overall equation when we add the two equations together. That's "n." Thus:

$$Zn \rightarrow Zn^{2+} + 2e^-$$
$$+\ Cu^{2+} + 2e^- \rightarrow Cu$$

$$Zn + Cu^{2+} + \cancel{2e^-} \rightarrow Zn^{2+} + \cancel{2e^-} + Cu$$

$$Zn + Cu^{2+} \rightarrow Zn^{2+} + Cu$$

Since 2 electrons canceled out on both sides of the equation, n=2.

It's good that we had to add the equations together, because we needed the overall equation to get "Q." Remember, the reaction quotient has the same definition as the equilibrium constant, it just applies to non-equilibrium situations. Thus, we take the products and raise them to their stoichiometric coefficients and divide by the reactants raised to their stoichiometric coefficients. Remember, however, we need to ignore liquids and solids. What are the solids in this case? Well, look at the notation. Both Zn and Cu are separated from the ions with a "|." Remember, that means a change in phase. Thus, Zn and Cu are both solids. Thus:

$$Q = \frac{[Zn^{2+}]}{[Cu^{2+}]}$$

The Nernst equation, then, is:

$$E_{cell} = E^{o}_{cell} - \frac{0.05916}{n} \cdot \log(Q)$$

$$E_{cell} = 1.100 \text{ Volts} - \frac{0.05916}{2} \cdot \log\left(\frac{0.0100}{1.00}\right) = 1.100 \text{ Volts} + 0.05916 \text{ Volts} = \underline{1.159 \text{ Volts}}$$

9.7 First we have to get through the notation. The cell has a Pt electrode at the anode. The only half-reaction with Br_2 (l) and Br^- is $Br_2 (l) + 2e^- \rightarrow 2Br^-$ with a potential of 1.0652 Volts. Since this is for the anode, however, the equation must be reversed , because oxidation happens at the anode. This means we must take the negative of the potential, making the anode half-reaction potential -1.0652 Volts. The cathode has a platinum electrode Cl^-, and Cl_2. Referring to Table 9.1, then, the cathode half-reaction must be $Cl_2 + 2e^- \rightarrow 2Cl^-$, with a potential of 1.3595. The standard cell potential, therefore, is -1.0652 + 1.3595 = 0.2943 Volts. Since we are not under standard conditions, however, we need to use the Nernst equation.

Well, we just calculated E^{o}_{cell}. It is equal to 0.2943 Volts. What about the other things in the Nernst Equation? The "n" refers to the number of electrons transferred per reaction. How do we get that? We determine how many electrons cancel out of the overall equation when we add the two equations together. That's "n." Thus:

$$\begin{array}{r} 2Br^- \rightarrow Br_2 (l) + 2e^- \\ + \; Cl_2 + 2e^- \rightarrow 2Cl^- \\ \hline \end{array}$$

$$2Br^- + Cl_2 + \cancel{2e^-} \rightarrow Br_2 (l) + \cancel{2e^-} + 2Cl^-$$

$$2Br^- + Cl_2 \rightarrow Br_2\ (l) + 2Cl^-$$

Since 2 electrons canceled out on both sides of the equation, n=2.

It's good that we had to add the equations together, because we need the overall equation to get "Q." Remember, the reaction quotient has the same definition as the equilibrium constant, it just applies to non-equilibrium situations. Thus, we take the products and raise them to their stoichiometric coefficients and divide by the reactants raised to their stoichiometric coefficients. Remember, however, we need to ignore liquids and solids. What are the solids in this case? Well, look at the notation. Both electrodes are platinum and are not in the equation, so they are irrelevant. The only thing we ignore this time is Br_2 (l), because we ignore liquids in "Q."

$$Q = \frac{[Cl^-]^2}{[Cl_2]\cdot[Br^-]^2}$$

The Nernst equation, then, is:

$$E_{cell} = E^\circ_{cell} - \frac{0.05916}{n}\cdot \log(Q)$$

$$E_{cell} = 0.2943 \text{ Volts} - \frac{0.05916}{2}\cdot \log\left(\frac{0.0500^2}{(0.900)\cdot(0.450)^2}\right) = 0.2943 \text{ Volts} + 0.0551 \text{ Volts} = \underline{0.3494 \text{ Volts}}$$

9.8 In an electrolytic cell, the anode rips away the electrons. Thus, the substances which lose electrons are at the anode. This means that Cl^- gravitates to the anode and Cl_2 bubbles off around it. At the cathode, electrons are added to the reactants. That means the water will go to the cathode, because it has electrons added to it. The result is that OH^- will be formed at the cathode, and hydrogen will bubble off there.

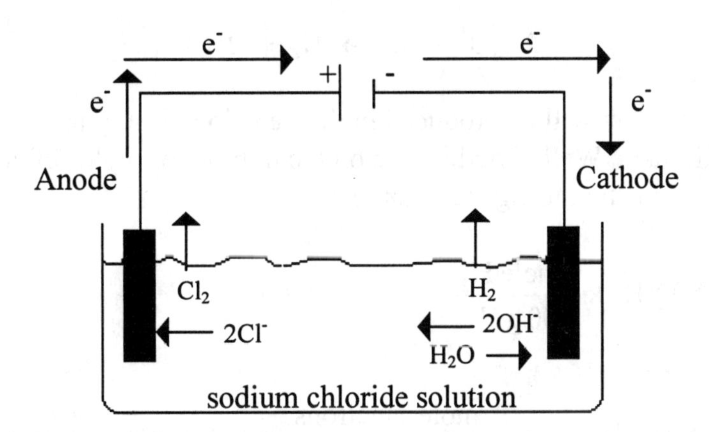

The overall reaction is simply the sum of the individual reactions. Since they each have the same number of electrons, they electrons cancel right away:

$$\underline{2H_2O + 2Cl^- \rightarrow H_2 + 2OH^- + Cl_2}$$

9.9 When Au^{3+} is reduced to gold, it must gain three electrons. Thus, the cathode half-reaction must be:

$$Au^{3+} + 3e^- \rightarrow Au$$

The current and time can be used to determine how much charge the battery produced. Remember, Amps are defined as Coulombs per second. Thus, if I take Amps times seconds, I will get Coulombs;

$$5.00\ \frac{C}{sec} \times 7200\ sec = 3.60 \times 10^4\ C$$

Now that I have the number of Coulombs, I can use the Faraday to convert to *moles* of electrons:

$$3.60 \times 10^4\ C \times \frac{1\ \text{mole of electrons}}{96{,}485\ C} = 0.373\ \text{moles of electrons}$$

Now that I have moles of electrons, this becomes a stoichiometry problem. The half-reaction tells me it takes 3 moles of electrons to make one mole of Au.

$$0.373\ \text{moles}\ e^- \times \frac{1\ \text{mole Au}}{3\ \text{moles}\ e^-} = 0.124\ \text{moles Au}$$

All I have to do now is convert from moles of Au to grams of Au.

$$0.124\ \text{moles Au} \times \frac{197.0\ \text{g Au}}{1\ \text{mole Au}} = \underline{24.4\ \text{g Au}}$$

9.10 In a Hoffmann apparatus, hydrogen is made at the cathode, where the following reaction occurs (see text):

$$2H_3O^+ + 2e^- \rightarrow H_2 + 2H_2O$$

We know how much product will be produced in this reaction (15.0 grams). The question asks how much time it will take. Well, that depends on the number of moles of electrons that are needed. We can determine that using stoichiometry.

$$15.0\ \text{g}\ H_2 \times \frac{1\ \text{mole}\ H_2}{2.00\ \text{g}\ H_2} = 7.50\ \text{moles}\ H_2$$

$$7.50\ \text{moles}\ H_2 \times \frac{2\ \text{mole electrons}}{1\ \text{mole}\ H_2} = 15.0\ \text{moles electrons}$$

So we need 15.0 moles of electrons. We can convert that to Coulombs using the Faraday, and then use that to determine how long it would take for 5.00 Amps to deliver that charge.

$$15.0\ \cancel{\text{moles}\,e^-} \times \frac{96{,}485\,\text{C}}{1\ \cancel{\text{mole}\,e^-}} = 1.45 \times 10^6\ \text{C}$$

$$1.45 \times 10^6\ \cancel{\text{C}} \times \frac{\text{sec}}{5.00\ \cancel{\text{C}}} = \underline{2.90 \times 10^5\ \text{sec}}$$

Thus, it takes a little over 80 hours to make 100 grams.

REVIEW QUESTIONS

1. How do we use oxidation numbers in chemistry?

2. In the following equation, is Cl_2 oxidized or reduced?

$$Cl_2 + 2e^- \rightarrow 2Cl^-$$

3. In the following equation, is K oxidized or reduced?

$$K \rightarrow K^+ + e^-$$

4. Why must we keep the reactants in separate containers in order for a Galvanic cell to produce electricity?

5. In a Galvanic cell, what kind of reactions (oxidation or reduction) happen at the anode? What kind of reactions happen at the cathode?

6. Do electrons flow from the cathode to the anode or from the anode to the cathode in a Galvanic cell?

7. What does the half-reaction potential tell you?

8. Looking at the values in Table 9.1, which is more readily reduced: Cu^{2+} or Mg^{2+}?

9. If the cell potential of a Galvanic cell is negative, will the cell produce a current of electrons?

10. What are the differences between a Galvanic cell and an electrolytic cell?

PRACTICE PROBLEMS

(You may use Table 9.1 to solve these problems)

1. Give the oxidation numbers of all atoms in the following substances:

a. MnO_2 b. H_2SO_4 c. CO_3^{2-} d. $MgCl_2$ e. KNO_3 f. SF_6
g. $IrCl_6^{3-}$ h. VS

2. Identify the oxidizing agent and the reducing agent in each of the following reactions:

a. $NiO (s) + Cd (s) \rightarrow CdO (s) + Ni (s)$
b. $2VO_3^- (aq) + Zn (s) + 8H^+ (aq) \rightarrow 2VO^{2+} (aq) + Zn^{2+} (aq) + 4H_2O (l)$
c. $2Na (s) + Cl_2 (g) \rightarrow 2NaCl (s)$

3. Draw a picture of the following Galvanic cell and discuss what happens as time goes on. Also, give the cell voltage and the overall reaction.

$$Ni \mid Ni^{2+} (1 M) \parallel Br^- (1M) \mid Br_2 (l) \mid Pt$$

4. A Galvanic cell runs on the following overall reaction:

$$Mn + 2AgCl \rightarrow Mn^{2+} + 2Cl^- + 2Ag$$

What is the voltage of the Galvanic cell?

5. A Galvanic cell runs on the following overall reaction:

$$Mn (s) + 2Ag^+ (aq) \rightarrow Mn^{2+} (aq) + 2Ag$$

Assuming that both electrodes are made of substances present in each half-reaction, write the Galvanic cell's condensed notation.

6. A Galvanic cell runs on the following overall reaction:

$$Zn + I_2 \rightarrow 2I^- + Zn^{2+}$$

Write the individual half-reactions, indicating which occurs at the anode and which occurs at the cathode.

7. What is the cell potential and overall reaction of the following Galvanic cell?

$$Al \mid Al^{3+} (0.200 M) \parallel Co^{2+} (0.00100 M) \mid Co$$

8. A Galvanic cell runs on the following overall reaction:

$$Mn\ (s) + Cu^{2+}\ (aq) \rightarrow Cu\ (s) + Mn^{2+}\ (aq)$$

If the concentration of copper ions is 0.500 M and the concentration of manganese ions is 0.0100 M, what is the cell potential?

9. An electrolytic cell runs the following reaction:

$$Fe + Cu^{2+} \rightarrow Fe^{2+} + Cu$$

Write the half-reactions that occur and indicate whether they run at the anode or cathode.

10. For the above electrolytic cell, how many grams of copper can be produced in an hour if the battery can supply 25.0 Amps of current?

Module #10: Electrochemistry - Part 2

Introduction

In the previous module, you learned a lot about Galvanic cells, electrolytic cells, oxidation numbers, and cell potentials. There is a lot more to the subject of electrochemistry, however. Although the last module concentrated on how electrochemistry applies to Galvanic and electrolytic cells, electrochemistry can take place with or without a cell. Whenever an atom changes its oxidation number, a redox reaction has occurred. *That's* electrochemistry. Because you are probably used to thinking of redox reactions as occurring either in Galvanic or electrolytic cells, you might forget that a redox reaction can occur like any other chemical reaction, as long as the reactants are allowed to interact with each other.

After all, when chemists make a Galvanic cell, they purposefully separate the reactants so as to force the electrons that need to be transferred to travel along a wire. That way, we can use the energy of those electrons as they move from one reactant to another. If you were to simply mix the two reactants together, the reaction will proceed at a quick pace. Since you are not forcing the electrons to travel down a wire, however, you cannot use their energy as they travel from one reactant to another. Thus, Galvanic cells are very artificial. They are a means of manipulating what naturally occurs in Creation so as to make it suit our needs.

In this module, I will move away from viewing electrochemistry in terms of Galvanic or electrolytic cells and discuss redox reactions in general. I will show you how to balance redox reactions (*not* an easy task), how to determine whether or not they are spontaneous, and how to relate the electrical potential of a redox reaction to things with which you are more familiar, such as the equilibrium constant and ΔG.

Balancing Redox Reactions - The Half-Reaction Method

Do you think that balancing chemical equations is easy? You probably think of yourself as an expert at that by now. However, you have been shielded from the "real world" of chemistry just a bit. You have probably never had to balance redox reactions before. That's because, unlike balancing other types of chemical reactions, balancing redox reactions can be *quite tough*! Thus, most courses only teach it to the more advanced students. Since you are advanced enough to be taking a second year of chemistry, it is time for you to be exposed to it.

There are two ways to balance redox reactions: the **half-reaction method** and the **change in oxidation number method**. Both yield the same result. However, I consider the half-reaction method to be much more physically illustrative than the change in oxidation number method. Thus, I will concentrate on the half-reaction method. In the next section, I will show you the other method, and you are free to use it if you prefer. However, my solutions will all use the half-reaction method.

Why is balancing redox reactions so difficult? Well, consider the following redox reaction:

$$Fe^{2+} + Cl_2 \rightarrow Fe^{3+} + Cl^- \qquad (10.1)$$

Now you might think that just putting a "2" in front of the "Cl^-" would balance the equation:

$$Fe^{2+} + Cl_2 \rightarrow Fe^{3+} + 2Cl^- \qquad (10.2)$$

That's not right! Although the atoms are, indeed, balanced (1 iron on each side and 2 chlorines on each side), look instead at the charges. On the reactants side, the total charge is 2+. On the products side, the total charge is 1+ (a plus 3 and 2 negative ones makes a plus 1). That's not legal! A chemical reaction must be balanced with respect to charge as well as with respect to atoms. This is important:

A chemical reaction must be balanced with respect to charge as well as with respect to atoms.

Why do I say that? Well, remember why a chemical equation must be balanced with respect to atoms. If there are more atoms on one side of the equation than on the other, it would imply that atoms were either made or destroyed in the reaction. That would violate the law of conservation of matter.

Well, what causes charge? An excess of protons or electrons. Thus, if there is more charge on one side of the equation than on the other, that would mean that protons or electrons are either made or destroyed in the reaction. That would also violate the law of conservation of matter. Thus, just as a chemical equation must be balanced with respect to atoms, it must also be balanced with respect to charge. How do we do that in redox reactions? Well, we have to remember that in redox reactions, electrons are given up by the reducing agent and accepted by the oxidizing agent. Thus, we must keep track of those electrons. Here's how to do it using the half-reaction method:

EXAMPLE 10.1

Balance the following redox reaction:

$$\mathbf{Fe^{2+} + Cl_2 \rightarrow Fe^{3+} + Cl^-}$$

You first have to notice that this is a redox reaction. Iron +2 goes to iron +3, which means it lost an electron. On the other hand, chlorine went from an oxidation number of zero to an oxidation number of -1. This means it gained an electron. Now suppose we were to make this into a Galvanic cell. What would the half-reactions be?

Well, let's look at iron first. Iron 2+ changed into iron 3+ and therefore lost an electron. In chemical equation notation, that would be:

$$Fe^{2+} \rightarrow Fe^{3+} + e^-$$

Now let's look at the chlorine. It went from oxidation number of zero to negative one. This means it gained an electron.

$$Cl_2 + e^- \rightarrow Cl^-$$

What's wrong here? Well, first, the Cl's are not balanced. There are 2 Cl's on one side and one on the other. Thus, we must multiply the Cl^- by 2:

$$Cl_2 + e^- \rightarrow 2Cl^-$$

That's still not enough. Now the charges are not balanced. There are 2 negatives on the products side and only one on the reactants side. To balance charges, then, I need to multiply e^- by 2:

$$Cl_2 + 2e^- \rightarrow 2Cl^-$$

Now both the charges and the atoms are balanced. Notice also that unlike the other two versions of the equation, this one actually makes sense. After all, a Cl_2 molecule has 2 Cl atoms in it. In order for both of them to get a negative one charge, I need 2 electrons. Thus, this makes physical sense.

In the end, then, I have two half-reactions:

$$Fe^{2+} \rightarrow Fe^{3+} + e^-$$
$$Cl_2 + 2e^- \rightarrow 2Cl^-$$

To get the total equation, all I have to do is add them up. I must do this, however, in such a way as to make the electrons cancel.

$$2\cdot[Fe^{2+} \rightarrow Fe^{3+} + e^-]$$
$$+ \; Cl_2 + 2e^- \rightarrow 2Cl^-$$

$$2Fe^{2+} + Cl_2 + \cancel{2e^-} \rightarrow 2Fe^{3+} + 2Cl^- + \cancel{2e^-}$$

$$\underline{2Fe^{2+} + Cl_2 \rightarrow 2Fe^{3+} + 2Cl^-}$$

Notice that this equation is balanced with respect to both atoms and charge. Thus, it is the correct chemical equation.

You can balance redox equations, then, by first splitting them into their half-reactions. You do that by observing which atoms lost and gained electrons. You then balance the individual half-reactions with respect to both atoms and charge. Once you are through with all of that, you add the half-reactions back together so that the electrons cancel and the result is a balanced redox reaction.

Now you might be thinking to yourself that this is an awfully large amount of work just to balance an equation. After all, if you "fiddled" with the stoichiometric coefficients in Equation (10.1), you could probably come up with an answer through a little bit of trial and error. Why resort to all of the work I just did? Well, most redox reactions are a little more involved than this simple reaction. Thus, trying to find an answer with trial and error is a lot less likely to succeed. Also, there is a wrinkle in all of this that I haven't mentioned yet. I will get to that soon. For right now, solve the following "on your own" problems exactly like I solved the example. You will see why you need to do it that way in a little bit:

ON YOUR OWN

10.1 Balance the following redox equation:

$$Fe^{2+} + Na \rightarrow Na^{+} + Fe$$

10.2 Balance the following redox equation:

$$Al^{3+} + Zn \rightarrow Zn^{2+} + Al$$

Okay, now it is time to get to more difficult redox reactions as well as add the wrinkle that I mentioned earlier. What is the wrinkle? Well, it stems from the laziness of chemists. In a chemical reaction, there are often times reactants or products that are so common that chemists tend not to list them. Typically, water is one of those chemicals. Since the vast majority of chemistry takes place in solution, water is everywhere in chemistry. Lazy chemists often do not bother to write water out, because it is simply understood. Also, since the water in a chemical reaction is often acidic or basic, the hydronium or hydroxide ion is usually present as well. Once again, however, lazy chemists often do not list them. In a redox reaction, then, there might be water, hydronium ions, or hydroxide ions that are simply not listed as a part of an unbalanced redox equation. You have to know whether or not to add them in as you go.

Now this might sound very confusing, but it is only mildly confusing. Study the following example to see how all of this works.

EXAMPLE 10.2

Balance the following redox reaction, which happens in acidic solution:

$$Cr_2O_7^{2-} + Fe^{2+} \rightarrow Cr^{3+} + Fe^{3+}$$

There are clearly a lot of things wrong with this equation. There are oxygens on the reactants side and not on the products side. The chromiums are not balanced, and neither are the charges. What do we do? Well, first let's see what happened from an oxidation number perspective. Iron went from +2 to +3, so it lost an electron. That's a pretty easy reaction to write:

$$Fe^{2+} \rightarrow Fe^{3+} + e^-$$

Now, what about the reduction that happened? Chromium went from +6 to +3, so it gained electrons. We can start, therefore, by writing that much:

$$Cr_2O_7^{2-} + e^- \rightarrow Cr^{3+}$$

There are still a lot of problems with this equation, but let's attack them one at a time. First, let's balance the atoms that were reduced. There are 2 chromiums on one side and one on the other. This means that I need to multiply the Cr^{3+} by 2.

$$Cr_2O_7^{2-} + e^- \rightarrow 2Cr^{3+}$$

Now, what about electrons? The chromiums were both +6 before the reaction and both turned into +3. Thus, there are 2 atoms, each gaining 3 electrons. As a result, we need a total of 6 electrons:

$$Cr_2O_7^{2-} + 6e^- \rightarrow 2Cr^{3+}$$

Now what? Well, here's where you have to *really think*. We need to balance oxygen atoms. How can we do that? Remember, chemists often do not mention water, the hydronium ion, or the hydroxide ion. Since we are in solution, however, we can add water anywhere we want. That will give us some oxygens to put somewhere. Also, since the problem says we are in acidic solution, we can also add hydronium ions anywhere that we want. So, if we add some water and hydronium ions, we can balance oxygen atoms and hydrogen atoms.

Where do we add which, and how much of each do we add? That's the thinking part. Look at the charges. On the reactants side, there is a total charge of -8 (-2 from the ion and -6 from the electrons). On the products side, there is a total charge of +6. In the end, that must balance. If we have water and H_3O^+ to "play with," which should go where? If we add H_3O^+ to the products side, it will make the charge *even more positive*. Thus, the charges will *never balance*. If we added H_3O^+ to the reactants side, however, it will cancel the negative charges. If we add enough H_3O^+, we can actually make the reactants side as positive as the products side, balancing the charges in the equation.

How many H_3O^+'s must we add? The total charge on the reactants side is -8. It needs to be +6 to make the charges balance. Thus, we need to add 14 hydronium ions:

$$Cr_2O_7^{2-} + 6e^- + 14H_3O^+ \rightarrow 2Cr^{3+}$$

You might think that things are even worse now, because the reactants side now has both H's and O's, while the product's side has none! Well, you are forgetting that we can throw water in anywhere we want. If we add water to the products side, it is easy to balance the H's and O's:

$$Cr_2O_7^{2-} + 6e^- + 14H_3O^+ \rightarrow 2Cr^{3+} + H_2O$$

There are 21 oxygens and 42 hydrogens on the reactants side, thus, we need to multiply water by 21:

$$Cr_2O_7^{2-} + 6e^- + 14H_3O^+ \rightarrow 2Cr^{3+} + 21H_2O$$

Notice what we have now. We have a completely balanced half-reaction. The atoms and charges both balance, and the proper number of electrons exist to take account of the change in oxidation number. Now we can add this half-reaction to the other, being sure to cancel electrons:

$$Cr_2O_7^{2-} + 6e^- + 14H_3O^+ \rightarrow 2Cr^{3+} + 21H_2O$$
$$+\ 6\cdot[\ Fe^{2+} \rightarrow Fe^{3+} + e^-]$$

$$Cr_2O_7^{2-} + \cancel{6e^-} + 14H_3O^+ + 6Fe^{2+} \rightarrow 2Cr^{3+} + 21H_2O + 6Fe^{3+} + \cancel{6e^-}$$

The final answer, then, is

$$\underline{Cr_2O_7^{2-} + 14H_3O^+ + 6Fe^{2+} \rightarrow 2Cr^{3+} + 21H_2O + 6Fe^{3+}}$$

Do you think you could have gotten that one by trial and error?

That was pretty complex for "just" balancing an equation. Let's review how it is done. First, you use the oxidation numbers to determine what was oxidized and what was reduced. When you do, start the half-reactions. Look at the atom that was oxidized or reduced and balance just that atom. Then, figure out how many electrons are necessary to make the oxidation number change. Once you have that, if the half-reaction is still not balanced, use water and hydronium ions (if you are in acidic solution) or hydroxide ions (if you are in basic solution) to balance the charge as well as the oxygen and hydrogen atoms. Since the hydronium ions or hydroxide ions are charged there is only one place you can put them in order to get the charge to balance.

Now like I said, that's a pretty complex process. I therefore want to do two more examples for you, just so you can get a better feel for all of this. After that, of course, you will need to do this on your own.

EXAMPLE 10.3

Balance the following redox reaction, which occurs in acidic solution:

$$\mathbf{H_2O_2 + Ag \rightarrow H_2O + Ag^+}$$

Notice that water *is* in this reaction. That's because water contains the atom that is reduced. Thus, water, hydronium, and hydroxide are *usually* not written, but sometimes they are. Anyway, let's see what happened in terms of oxidation number. Silver went from an oxidation number of 0 to +1. Thus, one half-reaction is:

$$Ag \rightarrow Ag^+ + e^-$$

This equation is already balanced in terms of atoms and charge, so we are done with this half-reaction.

In H_2O_2, the oxidation number of oxygen is -1. In water, the oxidation number of oxygen is -2. Thus, oxygen gained electrons. Since water and H_2O_2 contain this atom, they need to be in this half-reaction:

$$H_2O_2 + e^- \rightarrow H_2O$$

We always start balancing the half-reaction by working with the atom that was reduced:

$$H_2O_2 + e^- \rightarrow 2H_2O$$

The electron is not right now. After all, both O's went from -1 to -2. That takes 2 electrons. Thus:

$$H_2O_2 + 2e^- \rightarrow 2H_2O$$

At this point, charge is not balanced, and neither are the hydrogens. We next balance the charge. Since the reaction is in acidic solution, we can balance the charge with H_3O^+. Since the reactants side has a charge of -2 and the products side has a charge of 0, the hydronium ions must be added to the reactants side.

$$H_2O_2 + 2H_3O^+ + 2e^- \rightarrow 2H_2O$$

Now the charges are balanced, but the oxygens and hydrogens are not. We can always balance the hydrogens and oxygens by adding water. Don't worry that there is already water there. Remember the protocol. After using hydronium (or hydroxide if the reaction is in basic solution) to balance the charge, you add water to balance the hydrogens and oxygens. Thus:

$$H_2O_2 + 2H_3O^+ + 2e^- \rightarrow 2H_2O + 2H_2O$$

We can combine those waters:

$$H_2O_2 + 2H_3O^+ + 2e^- \rightarrow 4H_2O$$

That's the balanced half-reaction. Now we can add the two half-reactions, making sure that the electrons cancel:

$$H_2O_2 + 2H_3O^+ + 2e^- \rightarrow 4H_2O$$
$$+ 2\cdot[Ag \rightarrow Ag^+ + e^-]$$

$$H_2O_2 + 2H_3O^+ + \cancel{2e^-} + 2Ag \rightarrow 4H_2O + 2Ag^+ + \cancel{2e^-}$$

$$\underline{H_2O_2 + 2H_3O^+ + 2Ag \rightarrow 4H_2O + 2Ag^+}$$

Balance the following redox reaction, which occurs in basic solution:

$$\mathbf{Fe + MnO_4^- \rightarrow Fe(OH)_2 + MnO_2}$$

In this reaction, iron is going from an oxidation number of 0 to an oxidation number of +2. This means it lost two electrons. We take both molecules that contain iron and put those in the oxidation half-reaction:

$$Fe \rightarrow Fe(OH)_2 + 2e^-$$

The iron atoms are balanced, and since those are the atoms that were oxidized, that's all we worry about for now. The charge is not balanced, however. For that we use hydroxide ions, since the reaction occurs in basic solution. Since the charge of the products side is -2 and the charge of the reactants side is 0, we need to add hydroxide ions to make the reactants side more negative:

$$Fe + 2OH^- \rightarrow Fe(OH)_2 + 2e^-$$

Now all the atoms are balanced as well. Thus, the oxidation half-reaction is ready. What about reduction? Well, Mn goes from an oxidation number of +7 to an oxidation number of +4. This means it gained 3 electrons:

$$MnO_4^- + 3e^- \rightarrow MnO_2$$

Mn was the atom reduced, and it is balanced, so we use hydroxide ions to balance the charge. The reactants side has a charge of -4, so we need to add 4 hydroxide ions to the products side:

$$MnO_4^- + 3e^- \rightarrow MnO_2 + 4OH^-$$

Now we can balance the rest of the atoms with water:

$$MnO_4^- + 3e^- + 2H_2O \rightarrow MnO_2 + 4OH^-$$

Since the two half-reactions are now balanced, we can add them together, making sure the electrons cancel.

$$2\cdot[MnO_4^- + 3e^- + 2H_2O \rightarrow MnO_2 + 4OH^-]$$
$$+\ 3\cdot[Fe + 2OH^- \rightarrow Fe(OH)_2 + 2e^-]$$

$$2MnO_4^- + \cancel{6e^-} + 4H_2O + 3Fe + 6OH^- \rightarrow 2MnO_2 + 8OH^- + 3Fe(OH)_2 + \cancel{6e^-}$$

$$\underline{2MnO_4^- + 4H_2O + 3Fe + 6OH^- \rightarrow 2MnO_2 + 8OH^- + 3Fe(OH)_2}$$

ON YOUR OWN

10.3 Balance the following redox reaction, which occurs in acidic solution:

$$H_2S + Hg_2^{2+} \rightarrow Hg + S$$

10.4 Balance the following redox reaction, which occurs in basic solution:

$$MnO_4^- + NO_2^- \rightarrow MnO_2 + NO_3^-$$

10.5 Balance the following redox reaction, which occurs in basic solution:

$$Al + [Sn(OH)_4]^{2-} \rightarrow [Al(OH)_4]^- + Sn$$

10.6 Balance the following redox reaction, which occurs in acidic solution:

$$NO_3^- + I_2 \rightarrow IO_3^- + NO_2$$

Balancing Redox Reactions - The Change in Oxidation Number Method

Remember that I told you there were two techniques for balancing redox reactions. The half-reaction method is, in my opinion, the best of the two, but I do want to show you the other method so that you know that both exist. You can use either method, but I guarantee you that you are more likely to make mistakes with this method than you are with the half-reaction method. In the change in oxidation number method, you never split up the reaction into half-reactions. Instead, you use the change in oxidation number of the oxidizing agent and reducing agent to determine how to balance the electrons. Study the following example to see what I mean:

EXAMPLE 10.4

Use the change in oxidation number method to balance the reaction between antimony and chlorine:

$$\mathbf{Sb + Cl_2 \rightarrow SbCl_3}$$

You should notice right away that this is an "easy" redox equation to balance, because there are no hydrogens or oxygens in the equation. Thus, there is no worry about adding water and hydronium or hydroxide ions. In the change in oxidation number method, you look at the oxidation number of the atoms involved. In this equation, both Sb and Cl start with zero oxidation number. We cannot use any of our hard and fast rules to see what the oxidation numbers of the products are, but since Cl is in group 7A, we can assume its oxidation number is -1 unless there is another atom that follows the rules better. There is no other such atom, so Cl has an oxidation number of -1 and Sb has an oxidation number of +3.

This means that in the reaction, each Cl gained an electron and each Sb lost 3 electrons. When we use the change in oxidation number method, we see how many of each reactant went through the process of oxidation and reduction, and then try to balance the electrons transferred. In this reaction, 2 Cl's each gained one electron. Thus, there are 2 electrons required for this process. Only 1 Sb lost 3 electrons, so 3 electrons are lost when Sb is oxidized. There is an imbalance here. The electrons that the Cl's gain come from the Sb. There are 3 electrons given up by Sb and only 2 gained by the Cl. This will not work, because that would result in an extra electron "floating around" somewhere. Thus, we need to balance electrons. The common multiple of 2 and 3 is 6. Thus, if we require 2 antimonies to start the reaction, they will each lose 3 electrons for a total of 6. If we require 3 Cl_2's to start the reaction, each Cl atom will gain an electron, for a total of 6 electrons gained. Thus, the only way to ensure the fact that every electron lost by Sb is picked up by a Cl atom is to make sure there are 2 Sb's and 3 Cl_2's in the reactants:

$$2Sb + 3Cl_2 \rightarrow SbCl_3$$

Now we just balance out the products:

$$\underline{2Sb + 3Cl_2 \rightarrow 2SbCl_3}$$

This might have seemed easy, but remember, this is not the challenging kind of redox reaction. Try the next one to see the level of difficulty involved.

Balance the following redox reaction, which occurs in acidic solution:

$$\mathbf{MnO_4^- + Cl^- \rightarrow Mn^{2+} + Cl_2}$$

This is a more complicated equation. The oxygens present tell us we will have to mess with water and, since it is in acidic solution, hydronium ions as well. We start by determining the oxidation numbers that changed. The Mn went from +7 to +2, so it gained 5 electrons. The Cl^- lost a single electron to go from -1 to 0. To get the electrons to balance, then, I will have to multiply Cl^- by 5.

$$MnO_4^- + 5Cl^- \rightarrow Mn^{2+} + Cl_2$$

Once I do that, I now know that every electron gained by Mn comes from Cl^-. Now I balance the atoms involved in oxidation or reduction. The Mn's are balanced. We don't worry about the O's yet because they were neither oxidized nor reduced. The Cl's were oxidized, however, so I must balance them. To do that, I have to multiply Cl_2 by 5/2:

$$MnO_4^- + 5Cl^- \rightarrow Mn^{2+} + \frac{5}{2}Cl_2$$

Remember from your first-year course that using fractions to balance an equation is all right as long as you get rid of them in the end. Now we have an equation in which electrons are balanced as are the atoms involved in oxidation and reduction. Now we need to balance charge. We are in acidic solution, so we will use H_3O^+ ions. The reactants side has a charge of -6, while the products side has a charge of +2. The only way to balance that with hydronium ions is to add them to the reactants side. If we add 8 hydronium ions, the reactants side will have a total charge of +2, just like the products side.

$$MnO_4^- + 5Cl^- + 8H_3O^+ \rightarrow Mn^{2+} + \frac{5}{2}Cl_2$$

Now we just have to balance the H's and O's. There are 24 H's on the reactants side and 12 O's. We can balance both of them by adding 12 waters to the products side:

$$MnO_4^- + 5Cl^- + 8H_3O^+ \rightarrow Mn^{2+} + \frac{5}{2}Cl_2 + 12H_2O$$

This equation is balanced, but we now have to multiply everything in it by 2 in order to get rid of the fraction. Thus, the final balanced equation is:

$$\underline{2MnO_4^- + 10Cl^- + 16H_3O^+ \rightarrow 2Mn^{2+} + 5Cl_2 + 24H_2O}$$

Of course, had I balanced these two equations with the half-reaction method, the final answer would have been the same. Thus, you can use either method you like. They essentially involve the same steps, but the half-reaction method is less prone to mistakes. On the other hand, the change in oxidation number method involves less writing. You can therefore choose whichever method you wish. Since I consider the half-reaction method superior, however, all of the solutions I give will be using that method.

The Strengths of Oxidizing and Reducing Agents

Now that you know how to balance redox reactions, you need to know how to analyze them in the same way that you analyze other chemical reactions. For example, you need to know whether or not a redox reaction occurs. Actually, you already know how to determine that. In the previous module, I told you that Galvanic cells work with spontaneous reactions. Thus, any redox reaction that can run a Galvanic cell will happen spontaneously if the reactants are simply mixed together. What else do you know about Galvanic cells? Their electrical potential (their voltage) is always positive. Thus, you can determine the E^o of any redox reaction, and, if it is positive, the reaction will be spontaneous. If the E^o is negative, however, the reaction will not be spontaneous.

How do you determine E^o? You did that in the previous module. You determine each half-reaction, look up each of their potentials in Table 9.1, and add them together. The only twist is to remember that if the half-reaction is reversed, its potential is the negative of that listed in Table 9.1. To remind you of how this is done, I will show you a quick example. Since this example (and many of the rest of the problems in this module) use Table 9.1, it is best to have a copy of it handy. The table that accompanied your Module #9 test would work really well for this purpose.

EXAMPLE 10.5

Is the following redox reaction spontaneous?

$$2MnO_4^- + 10Cl^- + 16H_3O^+ \rightarrow 2Mn^{2+} + 5Cl_2 + 24H_2O$$

The trick to this is first determining what atoms were oxidized and reduced, and then looking for the proper half-reaction for each of them. In this reaction, Mn went from +7 to +2. Thus, it was reduced (it gained electrons). If you look in Table 9.1, there is only one half-reaction in which electrons are added to MnO_4^- to make Mn^{2+}:

$$MnO_4^- + 8H_3O^+ + 5e^- \rightarrow Mn^{2+} + 12H_2O \qquad E^o = 1.51 \text{ Volts}$$

Now you might think that this is wrong because the stoichiometric coefficients are off. In the reaction of interest, there are 2 MnO_4^-'s, 16 hydronium ions, and 24 waters. That's not the case in the half-reaction. Remember, however, that in order to cancel electrons when the half-reactions are added, the half-reactions might have to be multiplied by a factor. Thus, you can't look at the stoichiometric coefficients in order to pick the appropriate half-reaction. You simply look for a half-reaction in which the proper reactant either gains or loses electrons to become the proper product. That's all.

The other half-reaction must involve Cl atoms giving up electrons, because Cl went from -1 to 0. Looking at Table 9.1, we can find the reverse of such a reaction. If we reverse what's in Table 9.1, then, we will have our other half-reaction:

$$2Cl^- \rightarrow Cl_2 + 2e^- \qquad E^o = -1.3595 \text{ Volts}$$

Once again, the stoichiometric coefficients are off, but that's okay. That's just an issue of canceling electrons when the half-reactions are added together. This reaction has the right reactant losing electrons to become the right product. That's all we are interested in.

If we add these two half-reaction potentials together, we get 0.15 Volts. That's a positive number, so the reaction is spontaneous.

Now think about the example you just read. Suppose the oxidation half-reaction had a potential that was 0.16 Volts more negative. Would the overall reaction have been spontaneous? No. At that point, the sum of the half-reaction potentials would have been negative. When analyzing a redox reaction, then, the potentials of the half-reactions are very important. If both half-reactions have a positive potential, then the reaction is definitely spontaneous, because 2 positive numbers will always add to make a positive number. If both potentials are negative, however, the reaction will definitely not be spontaneous, because the sum of two negative numbers is always negative.

What happens when one half-reaction has a positive potential and the other has a negative one? Well, then it depends on the relative size of the two potentials. If the positive number is larger than the absolute value of the negative number, the sum of the two potentials will be positive, and the reaction will be spontaneous. If the absolute value of the negative number is larger than the positive number, then their sum will be negative, and the reaction will not be spontaneous. In other words, it is possible for one half-reaction to "force" another half-reaction to proceed, even though it would "rather" not.

Consider the reaction shown in Example 10.5. In that reaction, the oxidation half-reaction ($2Cl^- \rightarrow Cl_2 + 2e^-$) is not spontaneous, because its potential is negative. That should make sense from what you learned in your previous chemistry course. Cl atoms strive to gain an extra electron so that they have the same electron configuration as a noble gas. Thus, once they have that extra electron, they will not give it up easily. Even though that is the case, the overall reaction in Example 10.5 is spontaneous, because the reduction half-reaction ($MnO_4^- + 8H_3O^+ + 5e^- \rightarrow Mn^{2+} + 12H_2O$) is so strong that it rips the electrons away from Cl^-, despite the fact that the Cl^- doesn't "want" to give the electrons up.

In chemistry, we have terminology that describes such situations. We say that MnO_4^- is such a powerful oxidizing agent, it will force Cl^- to be oxidized, even though Cl atoms are more stable as Cl^- than as Cl_2. Thus, a powerful oxidizing agent can force an atom to give up its electrons, even though the atom doesn't "want" to. In the same way, a powerful reducing agent can force its electrons on an atom, even if the atom doesn't "want" to take them. Zinc metal, for example, is more stable as Zn^{2+} than it is as Zn. Nevertheless, when mixed with magnesium, Zn^{2+} will accept two electrons from magnesium to become Zn. The reason is that magnesium is a powerful reducing agent. Thus, even though Zn^{2+} doesn't "want" any electrons, magnesium forces them on the Zn^{2+}, and the result is that Zn^{2+} becomes the less stable Zn metal.

How do we determine whether a substance is a powerful reducing or oxidizing agent? We look at Table 9.1. For example, the last reaction in the table is $F_2 + 2e^- \rightarrow 2F^-$. This is a reduction half-reaction, which has a potential of 2.87 Volts. Think about what that means. The potential is so large and positive that it can be added to potentials as high as -2.86 Volts and the result would still be positive. Thus, if F_2 were added to a substance whose oxidation potential is as low as -2.86 Volts, F_2 would still force the substance to be oxidized. Thus, F_2 must be a powerful oxidizing agent. After all, it oxidizes substances whose oxidation half-reaction has a negative potential. This means it can force a substance to be oxidized, even if the substance doesn't "want" to be oxidized.

In the same way, look at the first reaction in Table 9.1. If we were to reverse it, we would get $K \rightarrow K^+ + e^-$. This is an oxidation half-reaction with a potential of 2.925 Volts. This means that it can participate in a redox reaction with a substance whose reduction potential is as low as -2.924 Volts. After all, the sum would still be positive, indicating a spontaneous reaction. Thus, K has such a strong oxidation potential that it can force another substance to be reduced, even if that substance doesn't "want" to be reduced. As a result, we would say that K is a strong reducing agent.

In the end, then, the potential of a substance's oxidation or reduction half-reaction tells us how strong a reducing or oxidizing agent it is.

When a substance has an oxidation potential that is large and positive, it is a strong reducing agent.

When a substance has a reduction potential that is large and positive, it is a strong oxidizing agent.

Now don't get so wrapped up in the terminology that you get confused. If a substance has a large and positive oxidation potential, it means that the substance "wants" to be oxidized. In order to be oxidized, however, it must find another substance that can be reduced. After all, in order to be oxidized, a substance must lose electrons. Those electrons must go somewhere. They must be gained by another substance. Thus, every time oxidation takes place, reduction must take place as well. So, if a substance "wants" to be oxidized, it must find something to be reduced. If the substance "really wants" to be oxidized, it can force another substance to be reduced, even if the other substance doesn't "want" to be reduced. Therefore, a substance that has a large oxidation potential will force other substances to be reduced. As a result, a substance with a large oxidation potential will be a strong reducing agent.

It works the same way for substances with large reduction potentials. In order to be reduced, a substance needs to take the electrons from another substance. Thus, a substance with a large reduction potential can force other substances to be oxidized. We therefore say that substances with large reduction potentials are strong oxidizing agents. See how all of this plays out in real life by studying the following example.

EXAMPLE 10.6

Which is the strongest reducing agent: Zn^{2+}, Au, Ca, Mn, or Pb?

Remember what a reducing agent is: a substance that gets oxidized. The stronger the oxidation potential of the substance, the stronger a reducing agent it is. To determine the oxidation potential, we need to find each substance in Table 9.1 and see if it can be oxidized. If so, we need to determine its oxidation potential. The substance with the largest oxidation potential will be the strongest reducing agent.

The first substance, Zn^{2+} is found in Table 9.1 in $Zn^{2+} + 2e^- \rightarrow Zn$. Of course, Table 9.1 lists *reduction* half-reactions, not oxidation half-reactions. Thus, to get an oxidation half-reaction, we must turn the reaction around. This gives us $Zn \rightarrow Zn^{2+} + 2e^-$. Now we see that Zn^{2+} *cannot be oxidized*! When we turn the reaction around to make it an oxidation half-reaction, Zn^{2+} is a *product*. It must be a reactant in an oxidation half-reaction in order to be oxidized. Thus, Zn^{2+} cannot even be a reducing agent, because it cannot be oxidized.

The next substance, Au can be found in Table 9.1. After we reverse its reaction in order to make it an oxidation reaction, we find that Au is, indeed a reactant: $Au \rightarrow Au^{3+} + 3e^-$. Since we had to reverse the reaction, we take the negative of the potential, so Au's oxidation potential is -1.50 Volts. Since it is negative, Au is obviously not much of a reducing agent. The rest of the substances can also be found in Table 9.1. When we reverse their reactions to make them oxidation half-reactions and we take the negative of their potentials, we get:

$$Ca \rightarrow Ca^{2+} + 2e^- \qquad E^\circ = 2.87 \text{ Volts}$$
$$Mn \rightarrow Mn^{2+} + 2e^- \qquad E^\circ = 1.18 \text{ Volts}$$
$$Pb \rightarrow Pb^{2+} + 2e^- \qquad E^\circ = 0.126 \text{ Volts}$$

The substance with the highest oxidation potential is Ca, so Ca is the strongest reducing agent.

In chemistry, it is important to be able to determine what substances are strong reducing or oxidizing agents, because when a substance is strong at reduction or oxidation, it can be hazardous. For example, chlorine gas is a reasonably strong oxidizing agent. We can tell this because the reduction half-reaction for chlorine gas ($Cl_2 + 2e^- \rightarrow 2Cl^-$) has one of the largest reduction potentials listed on Table 9.1. If you breathe in chlorine gas at a significant concentration, then, the chlorine gas will try to oxidize anything it comes into contact with, so that the chlorine itself will be reduced. Thus, all of the cells in your lungs and respiratory system will come under attack, as the chlorine gas forces substances within the cells to oxidize, even though they ordinarily would never oxidize. The result can be respiratory disease or failure. We would expect, then, that chlorine gas is toxic to work with.

What about Cl^- ions, however? You might think that since chlorine gas is hazardous, chloride ions would be as well, since they can come from chlorine gas. But look at Table 9.1. Cl^- cannot be an oxidizing agent, because Cl^- has no reduction half-reaction. However, if we turn the equation around, we see that Cl^- can be oxidized ($2Cl^- \rightarrow Cl_2 + 2e^-$). The reaction, however, has a potential of -1.3595 Volts. This means that although Cl^- can be a reducing agent, it is not a good one. As a result, from an oxidation/reduction point of view, Cl^- is not toxic, but Cl_2 gas is. That's because Cl_2 is a strong oxidizing agent, but Cl^- is a poor reducing agent. On the other hand, potassium metal (K) is extremely toxic because its oxidation half-reaction ($K \rightarrow K^+ + e^-$) has a large potential (2.925 Volts). The potassium ion (K^+), however, is non-toxic, because it's reduction half-reaction ($K^+ + e^- \rightarrow K$) has a potential of -2.925 Volts. Thus, potassium is a strong reducing agent, but K^+ is a poor oxidizing agent.

There is another reason that the strengths of reducing agents and oxidizing agents are important to know. It allows us to determine what reactants to use in a specific chemical reaction. Study the following examples to see what I mean.

EXAMPLE 10.7

A chemist wants to turn platinum (Pt) into its 2+ ion (Pt^{2+}). If she has the following substances "in stock," which should she use: MnO_4^-, I_2, or Pb^{2+}?

The chemist wants to oxidize Pt. This means she wants the following half-reaction to occur:

$$Pt \rightarrow Pt^{2+} + 2e^- \qquad E^o = -1.20 \text{ Volts}$$

The potential of this half-reaction indicates that it is not spontaneous. The only way she can make it spontaneous is to react it with something whose reduction potential is greater than 1.20 Volts. That way, when the two half-reactions are added together, the overall potential will still be positive. Each of the substances listed can be found in Table 9.1, and the reduction half-reactions they participate in are as follows:

$$MnO_4^- + 8H_3O^+ + 5e^- \rightarrow Mn^{2+} + 12H_2O \qquad E^o = 1.51 \text{ Volts}$$
$$I_2 + 2e^- \rightarrow 2I^- \qquad E^o = 0.5355 \text{ Volts}$$
$$Pb^{2+} + 2e^- \rightarrow Pb \qquad E^o = -0.126 \text{ Volts}$$

Of these three, only MnO_4^- has a reduction potential great enough so that when the two half-reactions are added, the overall potential is still positive. Thus, the chemist must use MnO_4^-. If you are worried that water and hydronium ions are also in the reaction, remember that chemists add water, hydronium ions, and hydroxide ions freely in redox reactions. Thus, they can also be ignored if they are not central to the problem. What's central to the problem here is what substances are oxidized and what ones are reduced.

A chemist wants to reduce Cr^{3+} to Cr. Of the following chemicals, which should he use in the reaction: Pt^{2+}, Mn, or Ni?

In this case, a chemist wants to reduce a substance. Thus, the half-reaction that he wants is:

$$Cr^{3+} + 3e^- \rightarrow Cr \qquad E^o = -0.74 \text{ Volts}$$

To accomplish this, the chemist has three alternatives. The first one is particularly bad. If we look for Pt^{2+} in Table 9.1, we find it being reduced ($Pt^{2+} + 2e^- \rightarrow Pt$). We can't use a reduction reaction however, however. Since we are trying to reduce Cr^{3+}, we must find an oxidation reaction to go along with it. When we turn the Pt^{2+}-containing reaction around, we get $Pt \rightarrow Pt^{2+} + 2e^-$. At this point, Pt^{2+} is no longer a reactant. This tells us that Pt^{2+} *cannot even be oxidized*. Thus, Pt^{2+} cannot be used to reduce something else.

The other two chemicals can be used as reducing agents. We can find them in Table 9.1 and, when we reverse their reactions, they end up as reactants. Thus, they can both be oxidized. Their oxidation half-reactions are as follows:

$$Mn \rightarrow Mn^{2+} + 2e^- \qquad E^o = 1.18 \text{ Volts}$$
$$Ni \rightarrow Ni^{2+} + 2e^- \qquad E^o = 0.250 \text{ Volts}$$

Looking at the oxidation potentials, only Mn has a large enough oxidation potential so that when the two half-reactions are added together, the overall potential is still positive. Thus, the chemist must use Mn.

In the end, then, the relative strength of an oxidizing or reducing agent can tell us a lot about the reactions in which it can be involved. Make sure you understand the reasoning that goes on here by performing the following "on your own" problems.

ON YOUR OWN

10.7 List the following in order of increasing strength as an oxidizing agent: Sn^{4+}, Cu, Na^{+}, Sn^{2+}, and Fe^{2+}. If it cannot be an oxidizing agent, leave it out of the list entirely.

10.8 A chemist wants to turn Cl^{-} into Cl_2. Which of the following reagents can she use: I_2, Fe, AgCl, or F_2?

Now that you are familiar with analyzing the strength of reducing and oxidizing agents, perform the following experiment to see how this works in the real world.

EXPERIMENT 10.1

Supplies:
From the laboratory equipment set:
- Zinc wire
- Ferric ammonium sulfate
- Copper sulfate
- Lime water
- Depression plate
- Dropper
- Chemical scoop

From the supermarket
- Distilled water

From the hardware store
- Steel wool, sandpaper, or some other form of abrasive

In this experiment, you will try to react solid zinc (from the zinc wire) with aqueous Fe^{3+} (from the ferric ammonium sulfate), Cu^{2+} (from the copper sulfate), or Ca^{2+} (from the lime water). In other words, you will try to perform the following reactions:

$$Zn + Fe^{3+} \rightarrow ????$$
$$Zn + Cu^{2+} \rightarrow ????$$
$$Zn + Ca^{2+} \rightarrow ????$$

Using the skills you have just learned (as well as Table 9.1), you should be able to determine which of these reactions are spontaneous. Make those predictions *before* you do the experiment. Write them down.

Once you have made the predictions, try to run the reactions. First, use the chemical scoop to fill a deep well on the depression plate 3/4 of the way full with copper sulfate. Use the dropper to fill that well the rest of the way with distilled water. This will make a saturated solution of copper sulfate. Next, use the steel wool or other abrasive to shine up the end of the zinc wire. Then soak that shined up end in the saturated solution of copper sulfate. Allow it to soak for 5 minutes. Then pull the zinc out and see if it has been discolored. If it has, that means copper has covered the zinc metal. This indicates that copper ions were reduced and zinc ions were oxidized. This means that the reaction occurred.

Now repeat that procedure with ferric ammonium sulfate. Once again, if the zinc is discolored, it means that iron ions were reduced and zinc ions were oxidized, which means the reaction did occur. If the zinc is not discolored, a reaction did not occur. Finally, repeat the experiment with lime water. The lime water is already in solution, so there is no need to add water. Just fill a deep well with the lime water solution and soak the shined-up end of the zinc wire in it. Once again, if the zinc wire is discolored, calcium ions were reduced and zinc was oxidized. This means a reaction occurred. If the zinc wire is not discolored, a reaction did not occur.

Check your predictions with the outcome to see if you were right. You can look at the end of the answers to the "on your own" problems to see what the correct predictions would have been, as well as to see what the correct experimental results should have been.

Relating Redox Potential to ΔG and the Equilibrium Constant

If you think about it, the potential of a redox reaction must, in some way, relate to ΔG. After all, ΔG changes sign depending on whether or not a reaction is spontaneous. If a reaction is spontaneous, ΔG is negative. If a reaction is not spontaneous, ΔG is positive. Well, if a redox reaction is spontaneous, E^o is positive. If a redox reaction is not spontaneous, E^o is negative. Thus, there must be a relationship between these two quantities. There is:

$$\Delta G^o = -n \cdot F \cdot E^o \qquad (10.3)$$

Now remember what the "^{o}" superscript means. It means "under standard conditions." In this equation, then, the "ΔG^o" is the ΔG of the reaction under standard conditions. The "n" refers to the number of electrons transferred in the reaction; the "F" refers to the Faraday, a constant I introduced in the previous module; and "E^o" refers to the potential of the reaction under standard conditions.

Although I introduced the Faraday in the previous module, I want to introduce it a little differently in this module. This way that will make the math a little easier. In the previous module, I told you that the Faraday tells us how much charge exists in a mole of electrons. That's true, but it also tells us other things as well. We can express the Faraday as a physical constant:

$$F = 96{,}485 \ \frac{J}{V \cdot mole}$$

This really means the same thing as how much charge exists in a mole of electrons, but as you will see in a moment, it will make the math in this module simpler.

Now before I go on, I want to say a few words about what ΔG really means. You already know that ΔG is negative for a spontaneous reaction and positive for a non-spontaneous reaction. There is something else that ΔG tells us, however. I have put off telling you this because the discussion is easiest when applied to redox reactions. However, it is applicable to *all* chemical reactions, so please remember that as you read along.

Remember what ΔG is called. It's called the "Gibbs free energy." This is, in fact, what it means. It tells us how much energy that is "free" to be used. If the Gibbs free energy is negative, this means that there is so much free energy that the reaction emits it, which means that the reaction can be used to do work. If the Gibbs free energy is positive, however, that means the reaction must *use* energy in order to occur. Thus, something must *work on* the reaction in order for it to run. That's why we say a positive ΔG tells you a reaction is not spontaneous. If something must work on a reaction in order for it to occur, the reaction certainly cannot be spontaneous.

Now how can we take a redox reaction that is not spontaneous and make it happen? We can work on it by making an electrolytic cell. The electricity source (the battery, for example) works on the reaction, supplying the energy that ΔG says the reaction *must* receive in order to progress. So what happens with spontaneous redox reactions? We can use them *as a source for work.* If we take a redox reaction and make it run a Galvanic cell, the free energy that ΔG tells us about can then be used to do work, like light a light bulb or run a radio.

The point, then, is that ΔG tells us the maximum amount of work you can get out of a chemical reaction. In a redox reaction, this tells you how much work the Galvanic cell can do. Thus, if you calculate the ΔG for a redox reaction that runs a Galvanic cell, then, you are calculating the maximum amount of work that the Galvanic cell can do. If the redox reaction runs your muscles, the ΔG tells you the maximum amount of work that your muscles can do.

Please remember, that this discussion applies to all chemical reaction, not just redox reactions. The problem is, it is harder to talk about the work a non-redox chemical reaction can do. Nevertheless, if you can access the energy that the chemical reaction is emitting, you can get that chemical reaction to do work. If you can do this, then the ΔG of the reaction tells you the maximum amount of work that you can get out of the system.

You also need to remember something else about ΔG. It relates to the equilibrium constant. In Module #6, I presented the following equation:

$$\Delta G = -R \cdot T \cdot \ln(K) \qquad (6.2)$$

Using the ideal gas constant (R) and the temperature (T), you can calculate the equilibrium constant (K) from ΔG. Since Equation (10.3) allows us to determine ΔG from E^o, you can, in the end, determine the equilibrium constant for a redox reaction from its potential. Study the following example to see how all of this comes together.

EXAMPLE 10.8

Calculate the potential, the ΔG, and the equilibrium constant for the following reaction:

$$2Ag^+ (aq) + Zn (s) \rightarrow 2Ag (s) + Zn^{2+} (aq)$$

To determine the potential, we need to determine the half-reactions. That's easy. Silver ions are turning into solid silver. That reaction is in Table 9.1:

$$Ag^+ + e^- \rightarrow Ag \qquad E^o = 0.7991 \text{ Volts}$$

In reaction to this reduction, Zn is oxidized to Zn^{2+}. We can get this reaction by reversing one that we see in Table 9.1:

$$Zn \rightarrow Zn^{2+} + 2e^- \qquad E^o = 0.763 \text{ Volts}$$

The total potential of the redox reaction is the sum of these half-reaction potentials. Thus, $\underline{E^o = 1.562 \text{ Volts}}$. We can now relate that to ΔG:

$$\Delta G^o = -n \cdot F \cdot E^o$$

We already know what "F" is, but what is "n?" Well, it's the number of electrons exchanged. We can get that by adding the half-reactions together in such as way as to cancel electrons. By looking at the half-reactions, you can tell that the silver reduction reaction needs to be multiplied by 2 in order to get electrons to cancel. Thus, each half-reaction deals with 2 electrons, so n=2.

$$\Delta G = -2 \cdot 96{,}485 \frac{J}{\cancel{V} \cdot \text{mole}} \cdot 1.562 \cancel{V} = -3.014 \times 10^5 \frac{J}{\text{mole}}$$

So $\underline{\Delta G = -3.014 \times 10^5 \text{ J/mole}}$. The value of ΔG here tells us that a Galvanic cell which runs on this reaction could do slightly more than 300,000 J of work for every mole of reactants.

Finally, since ΔG can be related to K, we can calculate the equilibrium constant for this reaction:

$$\Delta G = -R \cdot T \cdot \ln(K)$$

Remember, when we use this equation, it is best to use the value of 8.314 $\frac{J}{mole \cdot K}$ for the ideal gas constant. Also, what do we use for "T?" Well, we are doing things under standard conditions. That's what the "o" superscript means. For redox reactions, standard conditions are at concentrations of 1 M, pressure of 1 atm, and temperatures of 25 °C, or 298 K. That's what we use in the equation:

$$\Delta G = -R \cdot T \cdot \ln(K)$$

$$-3.014 \times 10^5 \frac{J}{mole} = -8.314 \frac{J}{mole \cdot K} \cdot 298\,K \cdot \ln(K)$$

$$\ln(K) = \frac{-3.014 \times 10^5 \frac{\cancel{J}}{\cancel{mole}}}{-8.314 \frac{\cancel{J}}{\cancel{mole} \cdot \cancel{K}} \cdot 298\,\cancel{K}} = 121.6$$

$$\underline{K = 6.460 \times 10^{52}}$$

Make sure you can do this on your own.

ON YOUR OWN

10.9 What is the maximum amount of work that a Galvanic cell can do if it runs on the following reaction?

$$3I_2 + 2Al \rightarrow 2Al^{3+} + 6I^-$$

10.10 What is the equilibrium constant for the following redox reaction?

$$2H_3O^+ + Pb \rightarrow H_2 + 2H_2O + Pb^{2+}$$

Corrosion

Before I close this module out, I want to tell you about one of the biggest problems caused by redox reactions: corrosion. This phenomenon is best illustrated with iron. In the presence of oxygen and water, iron is oxidized to Fe^{2+}:

$$Fe \rightarrow Fe^{2+} + 2e^-$$

Iron gives up its electrons because oxygen and water rip them away according to the following half-reaction:

$$O_2 + 2H_2O + 4e^- \rightarrow 4OH^-$$

These two reactions add together to give us:

$$2Fe + O_2 + 2H_2O \rightarrow 2Fe^{2+} + 4OH^-$$

The iron ions and hydroxide ions formed in this reaction immediately join to form $Fe(OH)_2$. Thus, the same reaction is often written as:

$$2Fe + O_2 + 2H_2O \rightarrow 2Fe(OH)_2$$

Often, $Fe(OH)_2$ is called "rust." Although it is reasonable to say that, $Fe(OH)_2$ is not really rust. In the presence of water and oxygen, $Fe(OH)_2$ is immediately oxidized again to $Fe_2O_3 \cdot H_2O$. That is the "real" chemical formula for rust. The reason I put "real" in quotes is because the chemical composition of rust actually changes throughout a sample of rust. As a result, $Fe_2O_3 \cdot H_2O$ is just an approximation of rust's chemical formula.

The problem with rust is that it takes iron from its metal form and puts it into an ionic compound. This, of course, changes the physical characteristics of the iron. As a result, the rust flakes away from the iron, slowly destroying it. The destructive nature of corrosion cannot be overemphasized. Literally billions of dollars of damage is caused by the corrosion of iron used in construction materials every year!

Since corrosion is such a problem, there are many methods used to attempt to prevent it. If you cover iron with a waterproof seal of paint or lacquer, for example, water can never contact the iron, and thus it cannot react with the iron to form rust. Sometimes, an even more durable ceramic coating is used to make sure that no water gets to the iron. For example, your refrigerator is made out of iron that has such a coating. Also, if you form a solid solution (an alloy) of iron, chromium, and nickel, the redox reaction is inhibited, and no rust will form. This solid solution is often called "stainless steel."

Please remember that corrosion happens to many other metals as well. Basically, any metal whose oxidation potential is greater than -1.23 Volts will undergo the same chemical process, which is generally called **corrosion**. After all, the potential of the reduction half-reaction involving water and oxygen is 1.23 Volts. Thus, any metal whose oxidation potential is greater than -1.23 Volts will spontaneously react with oxygen and water, because the sum of the half-reaction potentials will be positive.

There are a few exceptions to this general rule. Aluminum, for example, has an oxidation potential of 1.66 Volts. Nevertheless, it does not corrode. Why? Well, if aluminum metal is exposed to water and oxygen, it will, indeed, corrode. The aluminum metal that you buy from a store, however, is protected by a waterproof seal of aluminum oxide. This prevents water and oxygen from getting to the aluminum metal under the coating, thus no corrosion occurs.

Interestingly enough, the aluminum oxide coating that protects aluminum is not put on by some manufacturing process. As soon as aluminum is exposed to air (which always has water in it), the aluminum begins to corrode according to the process I have just discussed. It turns out, however, that the corrosion process itself forms the waterproof seal of aluminum oxide! In this case, then, the corrosion process itself protects the rest of the aluminum metal in an aluminum product!

ANSWERS TO THE ON YOUR OWN PROBLEMS

10.1 In this equation:

$$Fe^{2+} + Na \rightarrow Na^{+} + Fe$$

Iron ions gain 2 electrons to become iron atoms:

$$Fe^{2+} + 2e^{-} \rightarrow Fe$$

Sodium atoms lose an electron to become sodium ions:

$$Na \rightarrow Na^{+} + e^{-}$$

Both reactions are balanced with respect to atoms and charge, so we just add them together in such a way as to make the electrons cancel:

$$\begin{array}{c} Fe^{2+} + 2e^{-} \rightarrow Fe \\ \underline{+\ 2\cdot[\,Na \rightarrow Na^{+} + e^{-}]} \end{array}$$

$$Fe^{2+} + \cancel{2e^{-}} + 2Na \rightarrow 2Na^{+} + \cancel{2e^{-}} + Fe$$

$$\underline{Fe^{2+}\quad 2Na \rightarrow 2Na^{+} + Fe}$$

10.2 By looking at this reaction:

$$Al^{3+} + Zn \rightarrow Zn^{2+} + Al$$

We can see that the following half-reactions occurred:

$$\begin{array}{c} Al^{3+} + 3e^{-} \rightarrow Al \\ Zn \rightarrow Zn^{2+} + 2e^{-} \end{array}$$

Both are balanced with respect to charge and atoms, so we just need to add them together so that the electrons cancel:

$$\begin{array}{c} 2\cdot[Al^{3+} + 3e^{-} \rightarrow Al] \\ \underline{+3\cdot[Zn \rightarrow Zn^{2+} + 2e^{-}]} \end{array}$$

$$2Al^{3+} + \cancel{6e^{-}} + 3Zn \rightarrow 2Al + 3Zn^{2+} + \cancel{6e^{-}}$$

$$\underline{2Al^{3+} + 3Zn \rightarrow 2Al + 3Zn^{2+}}$$

10.3 Now the problems start getting a little tougher. In this reaction:

$$H_2S + Hg_2^{2+} \rightarrow Hg + S$$

We can see that S is going from -2 to 0. This means it lost electrons. Thus, the oxidation half-reaction starts with:

$$H_2S \rightarrow S + e^-$$

First, we see that the S's are balanced. Second, we need to realize that to go from -2 to 0, sulfur had to lose 2 electrons. Thus:

$$H_2S \rightarrow S + 2e^-$$

Now we need to balance charge. Since we are in acidic solution, we can do this with H_3O^+. The reactants side has a total charge of 0 and the products side has a total charge of -2. The only way to balance charge with hydronium ions, then, is to add them to the products side:

$$H_2S \rightarrow S + 2e^- + 2H_3O^+$$

To balance the H's and O's, we can use water:

$$H_2S + 2H_2O \rightarrow S + 2e^- + 2H_3O^+$$

In the reduction half-reaction, Hg goes from +1 to 0. This means it gained electrons:

$$Hg_2^{2+} + e^- \rightarrow Hg$$

In order to balance Hg atoms, we need to multiply the product by 2:

$$Hg_2^{2+} + e^- \rightarrow 2Hg$$

Now we need to realize that since 2 Hg's each went from +1 to 0, there need to be two electrons:

$$Hg_2^{2+} + 2e^- \rightarrow 2Hg$$

The half-reactions are now both balanced. To get the overall reaction, then, we just add them so as to make their electrons cancel:

$$\begin{array}{l} H_2S + 2H_2O \rightarrow S + 2e^- + 2H_3O^+ \\ \underline{+\ Hg_2^{2+} + 2e^- \rightarrow 2Hg} \end{array}$$

$$H_2S + 2H_2O + Hg_2^{2+} + \cancel{2e^-} \rightarrow S + \cancel{2e^-} + 2H_3O^+ + 2Hg$$

$$\underline{H_2S + 2H_2O + Hg_2^{2+} \rightarrow S + 2H_3O^+ + 2Hg}$$

10.4 In this reaction:

$$MnO_4^- + NO_2^- \rightarrow MnO_2 + NO_3^-$$

Mn goes from +7 to +4, while N goes from +3 to +5. Thus, Mn gained electrons and N lost them. Starting with Mn:

$$MnO_4^- + e^- \rightarrow MnO_2$$

The Mn's are balanced, so we move on to considering electrons. Since Mn went from +7 to +4, it gained 3 electrons:

$$MnO_4^- + 3e^- \rightarrow MnO_2$$

Now we balance charge with OH^-, since the solution is basic. Since there are 4 negatives on the reactants side, there need to be 4 negatives on the products side:

$$MnO_4^- + 3e^- \rightarrow MnO_2 + 4OH^-$$

Now we balance H's and O's with water:

$$MnO_4^- + 3e^- + 2H_2O \rightarrow MnO_2 + 4OH^-$$

The reduction half-reaction is now balanced. On to the oxidation half-reaction:

$$NO_2^- \rightarrow NO_3^- + e^-$$

Since N goes from +3 to +5 and there is only one on each side of the equation, the nitrogen must have lost 2 electrons:

$$NO_2^- \rightarrow NO_3^- + 2e^-$$

Now we need to balance charge with OH^-:

$$NO_2^- + 2OH^- \rightarrow NO_3^- + 2e^-$$

We finish by balancing H's and O's with water:

$$NO_2^- + 2OH^- \rightarrow NO_3^- + 2e^- + H_2O$$

Now we can add the equations so as to make the electrons cancel:

$$2\cdot[MnO_4^- + 3e^- + 2H_2O \rightarrow MnO_2 + 4OH^-]$$
$$+\,3\cdot[NO_2^- + 2OH^- \rightarrow NO_3^- + 2e^- + H_2O]$$

$$2MnO_4^- + 6e^- + 4H_2O + 3NO_2^- + 6OH^- \rightarrow 2MnO_2 + 8OH^- + 3NO_3^- + 6e^- + 3H_2O$$

$$\underline{2MnO_4^- + H_2O + 3NO_2^- \rightarrow 2MnO_2 + 2OH^- + 3NO_3^-}$$

10.5 In this reaction:

$$Al + [Sn(OH)_4]^{2-} \rightarrow [Al(OH)_4]^- + Sn$$

Al goes from 0 to +3, while Sn goes from +2 to 0. Thus, Sn gained electrons and Al lost them. Starting with Sn:

$$[Sn(OH)_4]^{2-} + e^- \rightarrow Sn$$

The Sn's are balanced, so we move on to considering electrons. Since Sn went from +2 to 0, it gained 2 electrons:

$$[Sn(OH)_4]^{2-} + 2e^- \rightarrow Sn$$

Now we balance charge with OH^-, since the solution is basic. Since there are 4 negatives on the reactants side, there need to be 4 negatives on the products side:

$$[Sn(OH)_4]^{2-} + 2e^- \rightarrow Sn + 4OH^-$$

There is no need to balance with water, as the reduction half-reaction is now balanced. On to the oxidation half-reaction:

$$Al \rightarrow [Al(OH)_4]^- + e^-$$

Since Al goes from 0 to +3 and there is only one on each side of the equation, the Al must have lost 3 electrons:

$$Al \rightarrow [Al(OH)_4]^- + 3e^-$$

Now we need to balance charge with OH^-:

$$Al + 4OH^- \rightarrow [Al(OH)_4]^- + 3e^-$$

Once again, the equation is balanced without the need of adding water. Now we can add the equations so as to make the electrons cancel:

$$3\cdot[[Sn(OH)_4]^{2-} + 2e^- \rightarrow Sn + 4OH^-]$$
$$+ 2\cdot[Al + 4OH^- \rightarrow [Al(OH)_4]^- + 3e^-]$$

$$3[Sn(OH)_4]^{2-} + \cancel{6e^-} + 2Al + \cancel{8OH^-} \rightarrow 3Sn + \overset{4}{\cancel{12}}OH^- + 2[Al(OH)_4]^- + \cancel{6e^-}$$

$$\underline{3[Sn(OH)_4]^{2-} + 2Al \rightarrow 3Sn + 4OH^- + 2[Al(OH)_4]^-}$$

10.6 In this reaction

$$NO_3^- + I_2 \rightarrow IO_3^- + NO_2$$

N goes from +5 to +4 (it gained electrons) while I goes from 0 to +5 (it lost electrons).

Starting with NO_3^-:

$$NO_3^- + e^- \rightarrow NO_2$$

The N's are balanced, so we move on to considering electrons. Since N went from +5 to +4, it gained 1 electron. Thus, the reaction is okay as written. Now we balance charge with H_3O^+, since the solution is acidic. Since there are 2 negatives on the reactants side and none on the products side, we must add hydroniums to the reactants side.

$$NO_3^- + e^- + 2H_3O^+ \rightarrow NO_2$$

We now balance with water:

$$NO_3^- + e^- + 2H_3O^+ \rightarrow NO_2 + 3H_2O$$

On to the oxidation half-reaction:

$$I_2 \rightarrow IO_3^- + e^-$$

The I's are not balanced. We need to do that first, since they are the atoms participating in oxidation.

$$I_2 \rightarrow 2IO_3^- + e^-$$

Since I goes from 0 to +5 and there are two on each side of the equation, there must be a total of 10 electrons involved:

$$I_2 \rightarrow 2IO_3^- + 10e^-$$

Now we need to balance charge with H_3O^+:

$$I_2 \rightarrow 2IO_3^- + 10e^- + 12H_3O^+$$

Now we balance by adding water.

$$I_2 + 18H_2O \rightarrow 2IO_3^- + 10e^- + 12H_3O^+$$

Now we can add the equations so as to make the electrons cancel:

$$10\cdot[NO_3^- + e^- + 2H_3O^+ \rightarrow NO_2 + 3H_2O]$$
$$+ I_2 + 18H_2O \rightarrow 2IO_3^- + 10e^- + 12H_3O^+$$

$$10NO_3^- + 10e^- + 20H_3O^+ + I_2 + 18H_2O \rightarrow 10NO_2 + 30H_2O + 2IO_3^- + 10e^- + 12H_3O^+$$

$$\underline{10NO_3^- + 8H_3O^+ + I_2 \rightarrow 10NO_2 + 12H_2O + 2IO_3^-}$$

10.7 In asking about a substance's strength as an oxidizing agent, we need to know its reduction potential. Remember, the substance that is *reduced* is called the *oxidizing agent*. Reduction potentials are in Table 9.1.

Notice that of the substances discussed, Cu does not have a reduction reaction in Table 9.1. It can be found in a reduction reaction, but in that reaction, it is a *product*, not a reactant. Thus Cu can be oxidized, but not reduced. Since it cannot be reduced, it cannot be an oxidizing agent. Thus, it is not in the list. The others all have reduction half-reactions in which they are reactants, so we just list them in the order of their reduction potentials:

$\underline{Na^{+}, Fe^{2+}, Sn^{2+}, Sn^{4+}}$

10.8 To turn Cl^- into Cl_2, you have to get the following half-reaction in Table 9.1 to proceed:

$$2Cl^- \rightarrow Cl_2 + 2e^- \qquad E^o = -1.3595 \text{ Volts}$$

Thus, we must add to it an oxidizing agent whose potential is greater than 1.3595 Volts, so that the total potential is positive. Of the following agents: I_2, Fe, AgCl, or F_2, Fe cannot be reduced. Thus, it is not even a choice. The others all have reduction half-reactions in which they participate as reactants, so they are all potential oxidizing agents. F_2 has the only reduction potential large enough (2.87 Volts), so <u>the chemist must use F_2.</u>

10.9 The maximum work a chemical reaction can do is given by ΔG. Thus, this question really asks us to calculate ΔG. To do that, we need E^o. This reaction involves Al going to Al^{3+}. This oxidation, according to Table 9.1, has a potential of 1.66 Volts. It also involves I_2 turning into I^-. According to Table 9.1, this reduction half-reaction has a potential of 0.5355 Volts. The overall potential is the sum of the two, or 2.20 Volts. To use Equation (10.3), we need to know how many electrons are exchanged. Notice that Al is multiplied by 2, and its Table 9.1 reaction has 3 electrons. Thus, there are 6 electrons involved. The same can be inferred from the I_2. Thus, n = 6.

$$\Delta G = -n \cdot F \cdot E^o = -6 \cdot 96{,}485 \frac{\text{J}}{\cancel{\text{V}} \cdot \text{mole}} \cdot 2.20 \cancel{\text{V}} = -1.27 \times 10^6 \frac{\text{J}}{\text{mole}}$$

<u>The reaction can do up to 1.27×10^6 J of work</u> for every mole of reactants.

10.10 To determine K, we must first find ΔG. Using Table 9.1, you find that the E^o = 0.126 Volts, and that two electrons are transferred. This means ΔG is:

$$\Delta G = -n \cdot F \cdot E^o = -2 \cdot 96{,}485 \frac{\text{J}}{\cancel{\text{V}} \cdot \text{mole}} \cdot 0.126 \cancel{\text{V}} = -2.43 \times 10^4 \frac{\text{J}}{\text{mole}}$$

Now we can find K using the old standby:

$$\Delta G = -R \cdot T \cdot \ln(K)$$

$$-2.43 \times 10^4 \frac{\text{J}}{\text{mole}} = -8.314 \frac{\text{J}}{\text{mole} \cdot \text{K}} \cdot 298\ \text{K} \cdot \ln(\text{K})$$

$$\ln(\text{K}) = \frac{-2.43 \times 10^4 \frac{\cancel{\text{J}}}{\cancel{\text{mole}}}}{-8.314 \frac{\cancel{\text{J}}}{\cancel{\text{mole}} \cdot \cancel{\text{K}}} \cdot 298\ \cancel{\text{K}}} = 9.81$$

$$\underline{\text{K} = 1.82 \times 10^4}$$

EXPERIMENT 10.1 ANSWERS:

1. Zinc and copper sulfate should have reacted. You should have predicted that this would happen.
2. Zinc and ferric ammonium sulfate should have reacted. You should have predicted that this would happen.
3. Zinc and lime water should not have reacted. You should have predicted that this would happen.

REVIEW QUESTIONS

1. Is the following equation balanced?

$$Zn + Ag^{+} \rightarrow Ag + Zn^{2+}$$

Why or why not?

2. Why must you add half-reactions in such a way as to ensure that the electrons will cancel?

3. An equation is balanced by the half-reaction method by one student, while another student uses the change in oxidation number method. Assuming they both do the work correctly, will their answers be the same or different?

4. Why does it matter whether a redox reaction occurs in an acidic or basic solution?

5. What is the oxidation potential of gold (Au)?

6. What is the reduction potential of Fe^{3+}?

7. If an oxidation half-reaction has a negative potential, does that mean it can never happen?

8. Order the following substances in terms of their increasing power as oxidizing agents:

$$I_2, Pt^{2+}, Fe, Cr, Al^{3+}$$

If a substance cannot be an oxidizing agent, simply leave it out of the list.

9. Order the following substances in terms of their increasing power as reducing agents:

$$Al^{3+}, Cu, Cl^{-}, Ca, Mn$$

If a substance cannot be a reducing agent, simply leave it out of the list.

10. To what must a metal be exposed in order for it to corrode?

PRACTICE PROBLEMS

In problems 1-3, balance the redox equations which happen in acidic solution:

1. $Sn^{2+} + Cu^{2+} \rightarrow Sn^{4+} + Cu^{+}$

2. $C + HNO_3 \rightarrow NO_2 + CO_2$

3. $Zn + NO_3^- \rightarrow Zn^{2+} + N_2$

In problems 4-6, balance the redox reactions which happen in basic solution:

4. $MnO_4^- + S^{2-} \rightarrow MnO_2 + S$

5. $Zn + NO_3^- \rightarrow Zn^{2+} + NH_3$

6. $Cl_2 + OH^- \rightarrow Cl^- + ClO_3^-$

7. What is the ΔG of the following reaction?

$$O_2 + 4H_3O^+ + 2Zn \rightarrow 2Zn^{2+} + 6H_2O$$

8. What is the equilibrium constant for the following reaction?

$$Sn + 4AgCl \rightarrow Sn^{4+} + 4Ag + 4Cl^-$$

9. A chemist wants to reduce Fe^{2+} to make Fe. If he has Cd, Cr, and Mg^{2+}, which substance should he use?

10. A chemist wants to turn Mn^{2+} into MnO_4^-. If she has F_2, I_2, and Cl_2, which gas should she use?

Module #11: Chemical Kinetics

Introduction

In your previous chemistry course, you should have learned that *whether or not* a chemical reaction is spontaneous is not the only factor to consider when studying a chemical reaction. The other thing that you have to consider is the *rate* of the chemical reaction. After all, if a chemical reaction is spontaneous but is painfully slow, then it would not be practical to use that reaction to produce the products of interest. As a result, you must not only consider whether or not a reaction will occur spontaneously, but you must also consider how fast the reaction will occur.

When we determine whether or not a chemical reaction is spontaneous, chemists say that we are looking at the **thermodynamic properties** of the chemical reaction. When we consider the rate at which the chemical reaction occurs, we are looking at the **kinetic properties** of the chemical reaction. All chemical reactions represent a balance between thermodynamics and kinetics. It does little good for the thermodynamic properties of a reaction to indicate that the reaction is spontaneous if the kinetic properties indicate that the reaction proceeds too slowly to be of any value. Over a good fraction of this course so far, I have concentrated on the thermodynamic properties of chemical reactions. In this module, I want to focus on the kinetic properties of chemical reactions.

A Little Bit of Review

In your first-year chemistry course, you should have been introduced to the concept of **reaction rate**. You were probably shown an equation that looked something like this:

$$R = \frac{\Delta[\text{product}]}{\Delta t} = -\frac{\Delta[\text{reactant}]}{\Delta t} \qquad (11.1)$$

Notice that the rate is defined as the change in the concentration of product divided by the change in time. Thus, reaction rate has units of M/s. In addition, notice that you can express the reaction rate in terms of either products or reactants. Since reactants disappear as the chemical reaction proceeds, the change in concentration of the reactants is negative. Reaction rate must always be positive, and that's why there's a negative in the right-hand side of Equation (11.1).

In your first-year course, you should have also been introduced to the **rate equation** for a chemical reaction. For the following reaction:

$$aA + bB \rightarrow cC + dD \qquad (11.2)$$

The rate can be expressed as:

$$R = k \cdot [A]^x \cdot [B]^y \qquad (11.3)$$

Where "R" is the rate of the chemical reaction, k is the **rate constant**, and "x" and "y" are exponents which do not necessarily relate to anything in the chemical equation. The units of the rate constant change with respect to the values of "x" and "y." You will see what I mean in a moment.

If "x" and "y" do not necessarily relate to anything in the chemical equation, how are they determined? Well, in your first-year course, you should have learned that they are determined by experiment. In order to refresh your memory on this point, study the following example.

EXAMPLE 11.1

A chemist studied the instantaneous rate of the following reaction:

$$2NO(g) + H_2(g) \rightarrow N_2O(g) + H_2O(g)$$

The data gathered were as follows:

	[NO] (Molarity)	$[H_2]$ (Molarity)	Instantaneous Rate (M/s)
Trial 1	0.30	0.35	2.835×10^{-3}
Trial 2	0.60	0.35	1.134×10^{-2}
Trial 3	0.60	0.70	2.268×10^{-2}

What is the rate equation for this reaction?

The rate equation for this reaction is as follows:

$$R = k \cdot [NO]^x \cdot [H_2]^y$$

Thus, we need to determine x, y, and k. If we look at how the trials were set up, we can see that from trial 1 to trial 2, the concentration of hydrogen remained the same. The only thing that changed was the concentration of nitrogen monoxide (it was doubled). Thus, any changes in the rate between these two trials must be solely due to the effect of NO. From trial 1 to trial 2, the rate increased by a factor of 4. If the concentration of NO doubled and the result was a four-fold increase in rate, what does that tell you about the value of "x?" The only way a doubling of the concentration of NO can result in a 4-fold increase in rate is if x=2. That way, when the concentration doubles, the rate increases by a factor of 2^2, which is 4.

What about "y?" Remember, in order to do this kind of analysis, we have to find two trials where the only thing that changes is the concentration of the reactant we are working with. Since we need to find the value of "y," we need to find two trials in which the NO concentration stayed the same but the H_2 concentration varied. Between trials 2 and 3, the NO concentration was constant, but the H_2 concentration doubled. Between these two trials, the rate doubled. What does that tell you about "y?" If a doubling of concentration results in a doubling of rate, y = 1. Thus, the rate equation is:

$$R = k \cdot [NO]^2 \cdot [H_2]$$

Now we can figure out the value of "k." After all, in each of the three trials listed above, we have numbers for every variable in the equation except "k." Thus, we can choose *any* of the trials above, plug in the numbers, and calculate "k."

$$R = k \cdot [NO]^2 \cdot [H_2]$$

$$2.835 \times 10^{-3} \frac{M}{s} = k \cdot (0.30\ M)^2 \cdot (0.35\ M)$$

$$k = \frac{2.835 \times 10^{-3} \frac{\not{M}}{s}}{(0.30\ M)^2 \cdot (0.35\ \not{M})} = 0.027 \frac{1}{M^2 \cdot s}$$

Notice the units on "k." They are determined by the values of "x" and "y." Had x been equal to 1, not 2, the units would have been 1/(M·s). Just like the equilibrium constant, then, the units of "k" vary with the details of the reaction. The final rate equation, is:

$$\underline{R = \left(0.027 \frac{1}{M^2 \cdot s}\right) \cdot [NO]^2 \cdot [H_2]}$$

Now that you've dusted the cobwebs out of your mind when it comes to determining the rate equation for a chemical reaction, I want you to actually do such an experiment.

EXPERIMENT 11.1

The Rate of an Iodine Clock Reaction

Supplies:

From the laboratory equipment set:

- Sodium thiosulfate
- Potassium iodide
- Five droppers
- Starch packing peanut
- Two test tubes
- Test tube cap
- Test tube rack

From the supermarket:

- Hydrogen peroxide solution
- Distilled water
- Clear vinegar

From around the house:
- Measuring 1/2 teaspoon
- Measuring cup
- Four small glasses (like juice glasses)
- Stopwatch
- Someone to help you

In this experiment, you will determine the rate equation for the following reaction:

$$3I^- (aq) + H_2O_2 (aq) + 2H_3O^+ (aq) \rightarrow I_3^- (aq) + 4H_2O (l)$$

When trying to do such an experiment, the most important thing to have is some measure of how much reactant is used up within a certain time frame. After all, you need to determine the rate of the chemical reaction in each trial, in order to get data like that which was shown in the example. To do this, then, you need to know the change in concentration of a reactant or product as well as the change in time.

All of the reactions and products in this equation are colorless. How, then, will we know how quickly they are made or used up? Well, in the presence of I_3^-, starch will form a complex that turns a deep blue color. Thus, when I_3^- is made in a solution of starch, a blue color will appear. Now of course, if we just mix I^-, H_2O_2, H^+, and starch together, the blue color will appear right away, because *some* I_3^- will be made right away. How, then, does this help us determine rate?

Well, by itself, it doesn't. However, when we add one more chemical ($S_2O_3^{2-}$), something very interesting happens:

$$I_3^- (aq) + 2S_2O_3^{2-} (aq) \rightarrow 3I^- (aq) + S_4O_6^{2-} (aq)$$

It turns out that this equation is *very* fast. It is so fast that when any I_3^- is formed by the first reaction, it is immediately destroyed by this one. As a result, no I_3^- stays around long enough to form a blue color with the starch. Thus, *as long as there is $S_2O_3^{2-}$ in the solution*, no blue color will form. *When the $S_2O_3^{2-}$ runs out*, however, any I_3^- produced by the first reaction will no longer be destroyed, and the blue color will form.

Let's suppose, then, that you mix I^-, H_2O_2, H^+, starch, and $S_2O_3^{2-}$ together in a test tube. The I^-, H_2O_2, and H^+ will begin to react together and start producing I_3^-. At first, however, no I_3^- will survive, because the $S_2O_3^{2-}$ will immediately convert it back to I^-. Once the $S_2O_3^{2-}$ runs out, however, the I_3^- will be able to form a complex with starch and the blue color will form. The more $S_2O_3^{2-}$ you put in the mixture, the longer it will take for the blue color to form. The less $S_2O_3^{2-}$ you put in the mixture, the less time it will take for the blue color to form. In the end, then, the amount of $S_2O_3^{2-}$ you put in the mixture will determine *how much* product needs to be formed in order to see the blue color. This will allow you to get a handle on the rate of the chemical reaction.

If you didn't understand all of the discussion, don't worry. Once you see the reaction and read my discussion that follows, you should be able to understand it fully. For right now, just do the experiment. To start, you need to rinse everything out thoroughly with distilled water. This includes the droppers. To rinse out the droppers, pull as much water as you can into the dropper and then push it out. Do this several times with each dropper.

Next, add 1/2 of a teaspoon of sodium thiosulfate to one of the juice glasses. Then add a cup of distilled water to the same glass. Use one of the droppers to stir. Stir until all crystals of sodium thiosulfate are dissolved. This will take a while, but it is essential that all crystals are dissolved. After you finish making the sodium thiosulfate solution, take one of the packing peanuts that are in the laboratory equipment set and put it in a juice glass; then add water. Use a dropper to stir the packing peanut in the water. Unlike many packing peanuts, the ones in your laboratory equipment set are made of starch. Thus, the packing peanut will eventually dissolve. It needn't dissolve completely. Just stir it so that a lot of the peanut dissolves. Finally, pour some of the hydrogen peroxide into one of the empty juice glasses, and pour some vinegar into the last empty juice glass. Also, open the potassium iodide solution container.

Now, you need to take each dropper and "wet" them with the proper solution. To do this, take a dropper and assign it to a particular solution (one for hydrogen peroxide, one for the starch solution, one for the vinegar, one for the sodium thiosulfate solution, and one for the potassium iodide solution). Pull a little bit of each solution into the dropper, and then swirl it around in the dropper bulb. Then push it back out. Do this twice for each dropper. Be careful with the dropper assigned to potassium iodide. There isn't a lot of extra for this solution. Thus, use only a little of it each time.

Now you are finally ready to start the experiment. Take one test tube and use the appropriate dropper to add starch solution. Note the dashes on the test tube. Add enough starch solution so that the bottom of the meniscus is at the third dash. Next, add enough vinegar so that the bottom of the meniscus is now touching the fifth dash on the test tube. Finally, add enough potassium iodide solution so that the bottom of the meniscus is at the 7th dash. That prepares your first test tube. The solution should be clear.

To prepare the next test tube, add enough starch solution so that, once again, the bottom of the meniscus is touching the third dash. Then add enough hydrogen peroxide so that the bottom of the meniscus is now at the fifth dash. Finally, add enough sodium thiosulfate so that the bottom of the meniscus is at the seventh dash. If the solution in the test tube is clear, the second test tube is ready.

To run the reaction, have your helper hold the stopwatch. Pour one of the test tubes into the other and, as soon as your helper sees the two solutions come into contact, he or she should start the stopwatch. Next, cap the tube and begin slowly rocking the test tube back and forth so as to keep mixing the solution. For a while, the solution should stay clear. All of the sudden, it should turn blue. At the instant it turns blue, your helper should stop the stopwatch. Record the time it took for the solution to turn blue.

Now realize what happened here. From the moment the two solutions mixed, hydrogen peroxide, hydronium ions, and iodide ions began reacting to form I_3^-. However, as soon as any I_3^- was formed, thiosulfate ions immediately destroyed it. Eventually, however, the thiosulfate ions ran out. At that point, the concentration of I_3^- began to build, which caused a starch complex that turned the solution blue. Thus, the time that your helper measured tells you how long it took the reaction to produce just a little more I_3^- than what the thiosulfate ions could destroy.

You need to perform the reaction three more times, with varying concentrations of reactants. Use the table below to determine how each test tube should be filled. You need not rinse out the droppers each time, because they are dedicated to individual solutions. However, you need to *thoroughly* rinse out the test tubes and the test tube cap between each trial.

TEST TUBE #1

Trial	Add starch to this line	Then add vinegar to this line	Then add potassium iodide to this line
2	First	Fifth	Seventh
3	First	Third	Seventh
4	Third	Fifth	Seventh

TEST TUBE #2

Trial	Add starch to this line	Then add hydrogen peroxide to this line	Then add sodium thiosulfate to this line
2	Third	Fifth	Seventh
3	Third	Fifth	Seventh
4	First	Fifth	Seventh

For each trial, record how many seconds it took for the blue color to appear.

Now you can analyze the data. Notice that in all trials, the same total volume exists in each test tube. Thus, the volume has been held constant through all trials. Also, notice that the amount of sodium thiosulfate in each trial is the same. In each trial, test tube #2 had enough sodium thiosulfate added to the test tube in order raise the solution level a total of two lines. Thus, each trial used the same amount of sodium thiosulfate.

What does this mean? It means that each trial has the same "Δ[product]." Remember, the blue color will appear only after enough I_3^- is made to "eat up" all of the thiosulfate ion. After that, the concentration of I_3^- builds up and the blue color appears. Thus, in calculating the rate that you measured, you would use the same value for Δ[product] in each trial. As a result, the time that you measured is directly reflective of the rate. If the time *decreased* by a factor of two, this tells you that the rate *increased* by a factor of two.

Now let's look at the individual trials. In trial 2, the same amount of iodide ions and hydrogen peroxide molecules were used as were in the first trial. The amount of vinegar used, however, was doubled. This results in a doubling of the hydronium ions. Thus, the only difference between trial 1 and trial 2 was that the concentration of hydronium ions was doubled. What happened to the rate? If things went well, you should have seen that the time it took for the blue color to appear was roughly the same in trial one and trial two. They may be off by as much as 20%, but they are still roughly the same. That tells you that the rate stayed the same in each trial. What does that tell you about how hydronium ions affect the rate? It tells you that rate is *independent* of hydronium ion concentration. Thus, the hydronium ion concentration *doesn't even appear* in the final rate equation!

Now compare trial 3 to trial 1. In trial 3, everything is the same as trial 1 except for the amount of iodide ions used. In trial 3, the amount of iodide ions was doubled. What happened to the rate? You should see that the time it took for the blue color to appear in trial 3 is roughly half that of trial 1. This means the rate *doubled* when the iodide concentration doubled.

Finally, look at trial 4 and compare it to trial 1. In this case, nothing was changed except for the amount of hydrogen peroxide. It was doubled. What happened to the rate? The difference between the times should indicate that the rate doubled as well. Thus, when the concentration of hydrogen peroxide was *doubled*, the rate *doubled.*

Putting the results of the four trials together, we get the following rate equation for the chemical reaction we are studying:

$$R = k \cdot [I^-] \cdot [H_2O_2]$$

Notice that the hydronium ions aren't in the rate equation. That's because the rate is independent of the hydronium ions concentration, as you saw by comparing trial 1 and trial 2. Notice also that you cannot determine the value for "k." That's because you do not have a sensitive enough balance to measure the concentration of the sodium thiosulfate solution, so there is no way to get an actual number for the rate. Finally, notice that even though I^- has a stoichiometric coefficient of "3" in the chemical equation, the rate law has I^- raised to the first power. This demonstrates that "x" and "y" are not necessarily related to the stoichiometric coefficients.

There is one more thing related to the rate equation that you need to remember from your first-year course. Once you have determined the rate equation, the values of "x" and "y" are called the **reaction orders** with respect to each reactant. For example, in the rate equation that you just determined, the reaction is said to be first-order with respect to I^- because the exponent for I^- in the rate equation is equal to 1. Similarly, the reaction is first-order with respect to hydrogen peroxide because its concentration is also raised to the first power. What about H_3O^+, however? What is the order of the reaction with respect to H_3O^+? Well, since the reaction does not depend on the concentration of H_3O^+, it is *zero order* with respect to H_3O^+. The **overall order** of a chemical reaction is determined by summing up all of the individual orders. Thus, this reaction has an overall order of 2.

ON YOUR OWN

11.1 A chemist studied the instantaneous rate of the following reaction:

$$NO_3(g) + O_3(g) \rightarrow NO_2(g) + 2O_2(g)$$

The data gathered were as follows:

	$[NO_3]$ (Molarity)	$[O_3]$ (Molarity)	Instantaneous Rate (M/s)
Trial 1	1.00×10^{-6}	3.00×10^{-6}	6.60×10^{-5}
Trial 2	1.00×10^{-6}	6.00×10^{-6}	1.32×10^{-4}
Trial 3	2.00×10^{-6}	6.00×10^{-6}	2.64×10^{-4}

What is the rate equation for this reaction? What is the overall reaction order as well as the reaction order with respect to each reactant?

Another Way to Look at The Kinetics of a Chemical Reaction

The rate equation is one way of trying to get a handle on how fast a chemical reaction proceeds, but it is not the only way. The rate equation tells you something important. At any time, you can plug in the concentration of reactants and determine, for that instant, how fast the reaction is occurring. Of course, as soon as the reaction proceeds even a little, the concentrations change and, as a result, the rate changes as well. Thus, the rate equation is sort of a snapshot in time. It tells you, for one instant, how fast the reaction is proceeding. As a result, the rate that one calculates from the rate equation is called the **instantaneous rate**, because it is valid only for an instant.

Another way to look at the kinetics of a chemical reaction is to try and develop an equation for how the concentration of a reactant or product changes with time. For example, suppose I was studying the following reaction:

$$2H_2O_2(aq) \rightarrow 2H_2O(l) + O_2(g) \quad (11.4)$$

Let's suppose I had an instrument that allowed me to measure the concentration of hydrogen peroxide at any time during the reaction. What would I see? Well, for this reaction, the data could be graphed as shown in Figure 11.1.

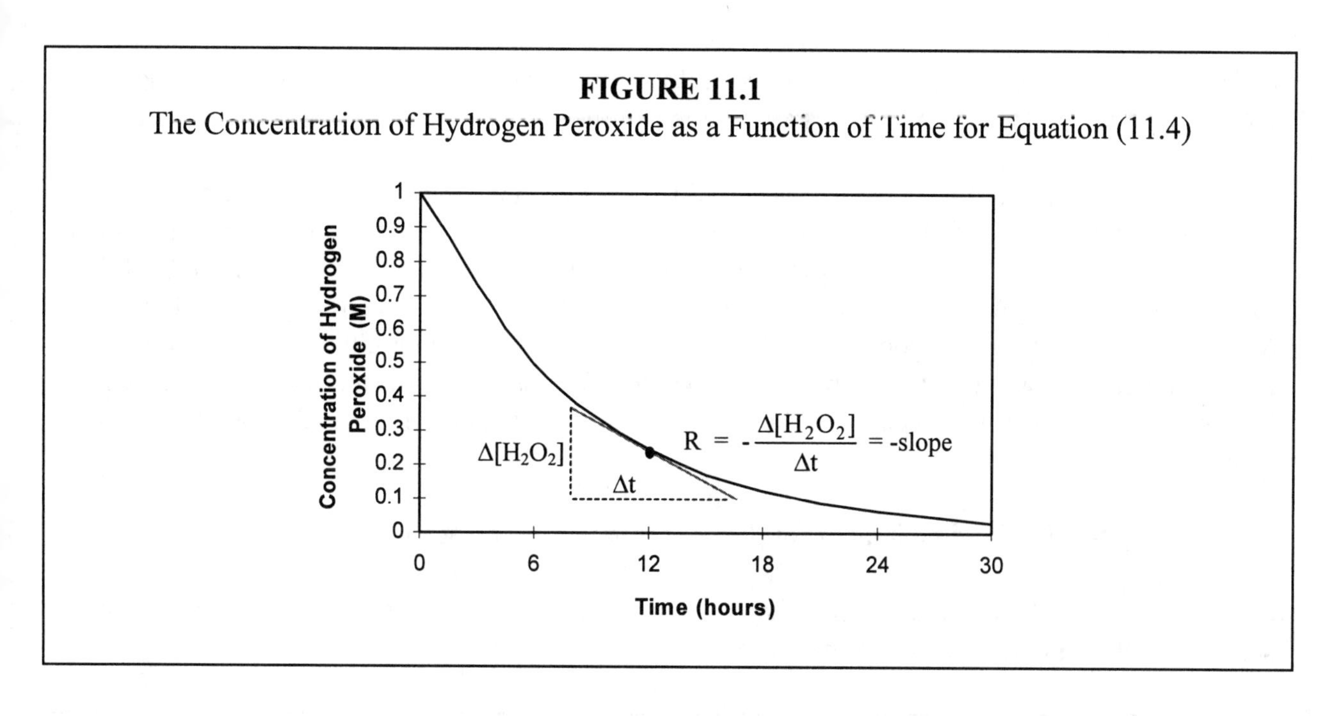

FIGURE 11.1
The Concentration of Hydrogen Peroxide as a Function of Time for Equation (11.4)

Before you concentrate on the details of the figure, first notice the overall trend of the data (represented by the solid line). At time = 0, the concentration of hydrogen peroxide is 1.0 M. That's the initial condition of the reaction. As time goes on, notice that the concentration of hydrogen peroxide gets lower. That should make sense. After all, hydrogen peroxide is a reactant. It will get used up in the reaction, so its concentration lowers.

Notice how quickly the concentration changes. At first, the concentration drops remarkably fast. From 0 - 6 hours, for example, the concentration falls from 1.0 M to 0.5 M, for a change of 0.5 M. Over the next 6 hours, however, the concentration drops from 0.5 M to 0.25M. That's only a change of 0.25 M. Thus, the reaction proceeds quickly at first, but then it slows down. That should make sense to you as well. The rate of a chemical reaction depends on the concentration of the reactants. As the concentration of the only reactant in the reaction decreases, the rate of the reaction will decrease as well.

Now focus on the other details in the figure. Suppose I wanted to know the reaction rate when t = 12 hours. If I had the rate equation, I could look at the graph and see that the concentration of hydrogen peroxide at that time is 0.25 M. I could then put that value in the rate equation and determine how quickly the reaction is proceeding at that time. There is, however, another way I could determine the rate. If I were to draw a line tangent to the curve at t= 12 hours (the gray line in the figure), I could calculate the *slope* of the line. The negative of that slope is the reaction rate.

How does the slope of the tangent line give me the rate of the chemical reaction? Well, a line drawn tangent to any curve tells you how the variable on the y-axis is changing with respect to the variable on the x-axis. Since this curve plots concentration versus time, the line tangent to this curve tells you how concentration changes with time. That's reaction rate! Notice how the slope of the line is calculated in the figure. By using the definition of slope ($\Delta y/\Delta x$), you get the definition of rate. Thus, the rate of the chemical reaction is simply the slope of the line tangent to the curve which plots concentration as a function of time.

I want you to notice one other aspect of the graph. The concentration of hydrogen peroxide starts out at 1.00 M and reduces to 0.500 M in 6 hours. Thus, the concentration was cut in half in 6 hours. What happens in the next 6 hours? The concentration changes from 0.500 M to 0.250 M. Once again, the concentration was cut in half in 6 hours. What happens in the next 6 hours? The concentration goes from 0.250 M to 0.125 M. Once again, the concentration of hydrogen peroxide was cut in half in a 6-hour period. Because the concentration of the reactant in this chemical reaction gets cut in half every 6 hours, we say that 6 hours is the **half-life** of this reaction.

Half-life - The time it takes for the concentration of a reactant to be cut in half

Study the example below and perform the "on your own" problems that follow in order to make sure you understand the concept of half-life.

EXAMPLE 11.2

A decomposition reaction has a half-life of 5.0 minutes. If the initial concentration of the reactant is 2.0 M. What is the concentration of the reactant after 15.0 minutes?

In 15.0 minutes, the reaction will have passed through 3 half-lives. In the first 5.0 minutes, the initial concentration will be cut from 2.0 M to 1.0 M. That accounts for the first half-life. In the next 5.0 minutes, the *remaining* concentration will be cut in half. Thus, at the end of those 5.0 minutes, the concentration will be 0.50 M. In the last 5-minute interval, *that* concentration will be cut in half to 0.25 M. After 15.0 minutes, then, the concentration of the reactant will be 0.25 M.

ON YOUR OWN

11.2 Suppose a chemist studies the decomposition of hydrogen peroxide and starts the reaction with a hydrogen peroxide concentration of 6.0 M. If the half-life is 6 hours, what will the hydrogen peroxide concentration be in a full day (24 hours)?

11.3 The half-life of a chemical reaction is 3.0 seconds. If the concentration of a reactant in that chemical equation goes from 1.000 M to 0.03125 M, how much time has passed?

First-Order Chemical Reactions

In your first-year chemistry course, you should have learned about various factors which affect the rate of a chemical reaction. The rate equation tells you how the rate depends on the concentration of the reactants, but there are other factors which govern how quickly a chemical reaction occurs. The temperature, for example, strongly affects the rate of a chemical equation. In addition, the surface area of the reactants can change the reaction rate. Finally, the presence of a catalyst can either speed up or slow down a chemical reaction, depending on the nature of the catalyst. If Equation (11.3) tells you the rate of a chemical reaction, where are all of these factors in the rate equation?

These factors are all "lumped in" to the rate constant, "k." The rate constant changes with temperature, reactant surface area, and the presence of a catalyst. The rate constant, then, tells us a lot about the rate of a chemical reaction. It contains so much information, in fact, that the concentration of a compound in the reaction can be related to time via the rate constant. The way in which concentration relates to time varies based on the overall order of the chemical reaction, however.

For reactions that are first-order overall, the concentration of any substance in the reaction at a given time can be calculated with the following equation:

$$[A] = [A]_o \cdot e^{-kt} \quad \text{(for first-order reactions)} \qquad (11.5)$$

In this equation, which can be derived using calculus, the concentration of any substance in the reaction is represented by "[A]," while its initial concentration is represented by "$[A]_o$." Finally, "k" stands for the rate constant, and "t" is the time. Using this equation, then, you can actually calculate the concentration of any substance in the chemical equation as long as you know the time, the initial concentration, and the rate constant.

EXAMPLE 11.3

The decomposition of hydrogen peroxide is a first-order reaction. If the initial concentration of hydrogen peroxide is 1.0 M, what is it 15.0 hours later? The rate constant under the conditions at which the reaction was run is 0.1155 1/hr.

We are given the initial concentration (1.0 M), the time (15.0 hrs), and the rate constant. We can therefore use Equation (11.5) to determine the concentration:

$$[A] = [A]_o \cdot e^{-kt}$$

$$[A] = (1.0\text{M}) \cdot e^{-(0.1155\frac{1}{\cancel{hr}})\cdot(15.0\ \cancel{hr})} = \underline{0.18\text{ M}}$$

Although Equation (11.5) is important, there is something even more important that can be derived from it. Suppose I was running a chemical reaction and measured the time it took for exactly half of the reactant to disappear. I would be measuring the half-life, right? What would that look like in terms of Equation (11.5)? Well, at that point, the actual concentration would be half of the initial concentration. Thus, I could replace "[A]" with "$\frac{1}{2}\cdot[A]_o$." Also, I could stick "$t_{1/2}$" in for "t," just to remind myself that the time at which the concentration is half its initial value is defined as the half-life. If I make those substitutions and do a bit of algebra, I get:

$$[A] = [A]_o \cdot e^{-kt}$$

$$\frac{1}{2}\cdot[A]_o = [A]_o \cdot e^{-kt_{1/2}}$$

$$\frac{1}{2} = e^{-kt_{1/2}}$$

$$-0.693 = -k \cdot t_{1/2}$$

Rearranging that final equation gives me a very important relationship between the rate constant and the half-life.

$$t_{1/2} = \frac{0.693}{k} \quad \text{(for first-order reactions)} \qquad (11.6)$$

For first-order chemical reactions, then, the half-life is inversely proportional to the rate constant.

EXAMPLE 11.4

A chemist determines the half-life of a first-order reaction to be 12.1 minutes. What is the rate constant for the reaction?

This is a simple application of Equation (11.6):

$$t_{1/2} = \frac{0.693}{k}$$

$$12.1\ \text{min} = \frac{0.693}{k}$$

$$\underline{k = 0.033\ \frac{1}{\text{min}}}$$

Notice the unit for the rate constant. In a first-order reaction, the rate equation has a concentration raised to the first power. Thus, the unit for "k" is multiplied by molarity raised to the first power. The result is M/min, which is an acceptable unit for rate. Thus, this equation keeps the units correct for the rate constant.

ON YOUR OWN

11.4 The decomposition of cyclobutane is first-order:

$$C_4H_8 \rightarrow 4C + 4H_2$$

At 500 °C, it takes 150.6 seconds for the amount of C_4H_8 to be reduced to one-fourth of its original value. What is the rate constant for this reaction?

11.5 Given your answer to the previous problem, suppose you start with a cyclobutane concentration of 5.5 M. What concentration of cyclobutane would be left after 501 seconds?

Second-Order Reactions

Equations (11.5) and (11.6) are quite useful, but please remember that they apply to first-order reactions only. If a reaction is second-order, things get a lot more complicated. Remember, there are two ways an equation can be second-order overall. It can have only one reactant concentration in the rate equation, but if that is the case, the reactant concentration must be squared. Otherwise, there can be two reactant concentrations in the rate equation, each of which are raised to the first power. Either way, the reaction is considered second-order.

Second-order reactions are complicated to analyze, but if I limit my discussion to certain types of second-order reactions, the math is simple enough to discuss. If a reaction is second-order because its rate equation has only one reactant concentration that is squared, the math is reasonably simple. It turns out that even in the case where the reaction is second-order because it has two concentrations raised to the first power in the rate equation, the math is still reasonably simple as long as the concentration of both reactants starts out the same. For either of those two cases, the relevant equations are:

$$\frac{1}{[A]} - \frac{1}{[A]_o} = k \cdot t \quad \text{(for certain second-order reactions)} \qquad (11.7)$$

$$t_{1/2} = \frac{1}{k \cdot [A]_o} \quad \text{(for certain second-order reactions)} \qquad (11.8)$$

These equations once again come from calculus. Notice that in this case, the half-life of the reaction actually depends on the initial concentration of the reactant or reactants. This is different than the case of first-order reactions, where the half-life depends only on the rate constant.

ON YOUR OWN

11.6 Using Equation (11.7) and the same kind of substitutions that I used in the previous section, derive Equation (11.8).

The Collision Theory of Chemical Kinetics

In order to explain why temperature, surface area, concentration, and the presence of a catalyst affect the rate of a chemical reaction, chemists have come up with the **collision theory** of chemical kinetics. This theory states that chemicals react when the individual molecules or atoms collide with one another. During a collision, it is possible that electrons will be re-arranged, resulting in the products of the chemical reaction.

Now if you think about it, this idea already explains two very important aspects of chemical reactions. From your first-year chemistry course, you should remember that chemical reactions can be "pictured" in terms of a potential energy diagram:

FIGURE 11.2
The Potential Energy Diagram of a Chemical Reaction

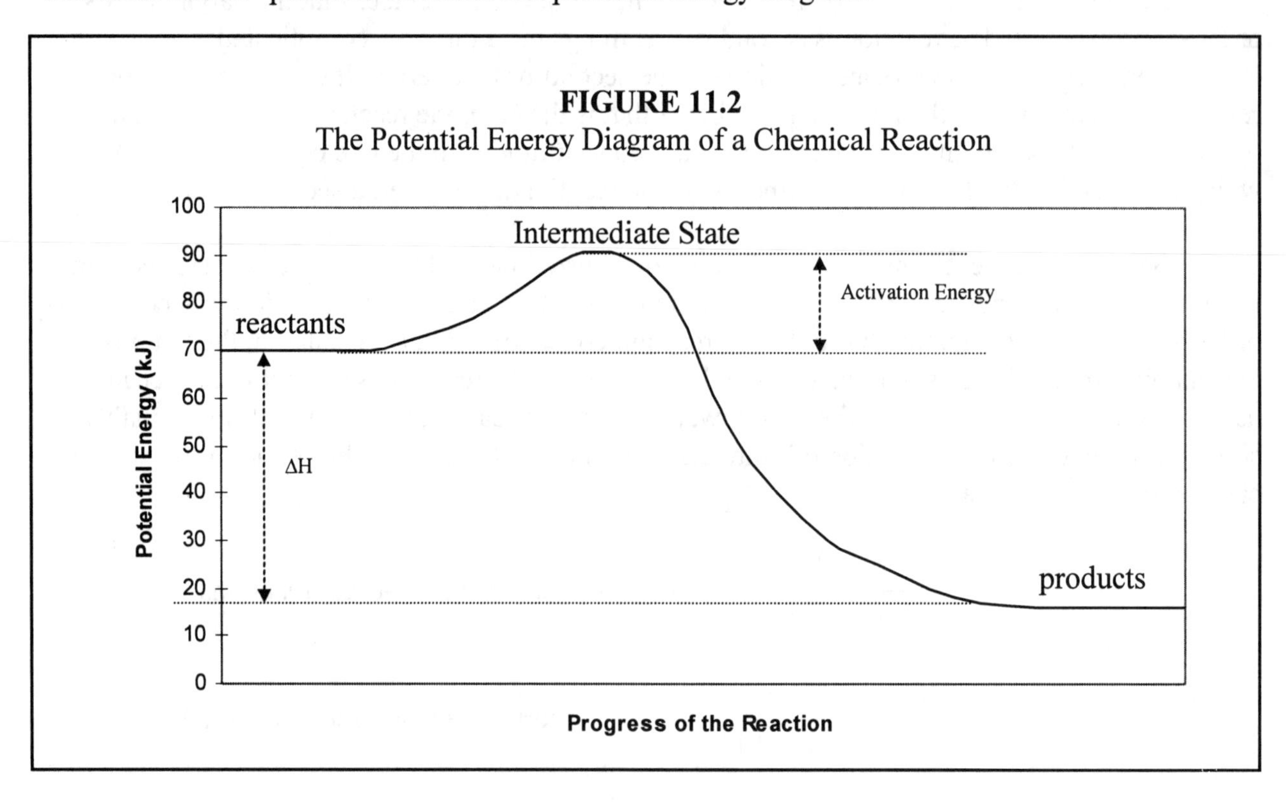

In this "picture" of a chemical reaction, the reactants exist on the left-hand side of the figure. These reactants have a certain amount of potential energy, which determines where the curve in the figure starts out. At the end of the reaction (on the right-hand side of the graph), the reactants have formed the products, which have a different potential energy. The difference in height between the products and the reactants tells you the ΔH of the reaction. This reaction is exothermic, since the products have less energy than the reactants. This tells you that there was energy "left over" at the end of the reaction, which was transferred to the surroundings.

Notice the hump that exists between the reactants and products. The hump tells us the **activation energy** of the reaction. The higher the hump, the larger the activation energy. Despite the fact that this reaction is exothermic, it requires some energy initially in order to make it over that hump. For example, once you strike a match, it burns, releasing a lot of heat. Thus, the combustion reaction for a match is exothermic. Nevertheless, had you not initially given the reaction some energy by striking the match, the reaction would have never gotten started. That's what activation energy is all about. All chemical reactions need some amount of energy in order to get started. The collision theory of chemical kinetics tells us *why* this is the case.

If two chemicals must collide in order to react, they must get very close to one another. Without some energy, that can be a problem. After all, atoms have their positive charges deep at their centers, in the nucleus. What one atom "sees" when it gets close to another atom is just a big cloud of electrons. Those electrons are all negatively-charged. Despite the fact that the atom has no net charge, all of its negative charges are right there at the surface of the atom, while the positive charges which cancel them out are "hidden" deep inside the electron cloud. As a result, when one atom gets close to another atom, the first electrical charge it experiences is the negative electrical charge of the other atom's electrons. These negative charges repel the negative charges of its own electron cloud. The result is that two atoms repel each other when they get close to one another.

This is why every chemical reaction, whether it is exothermic or endothermic, needs activation energy. In order to react, atoms must get close enough to collide with one another. However, because atoms are "covered" in electron clouds, those electron clouds repel each other. As a result, there must be a certain amount of force *pushing* those atoms together. The source of that force is the activation energy.

This also explains why an increase in temperature increases the rate of a chemical reaction. Remember from Module #4 that as a group, atoms or molecules have various speeds, depending on the temperature.

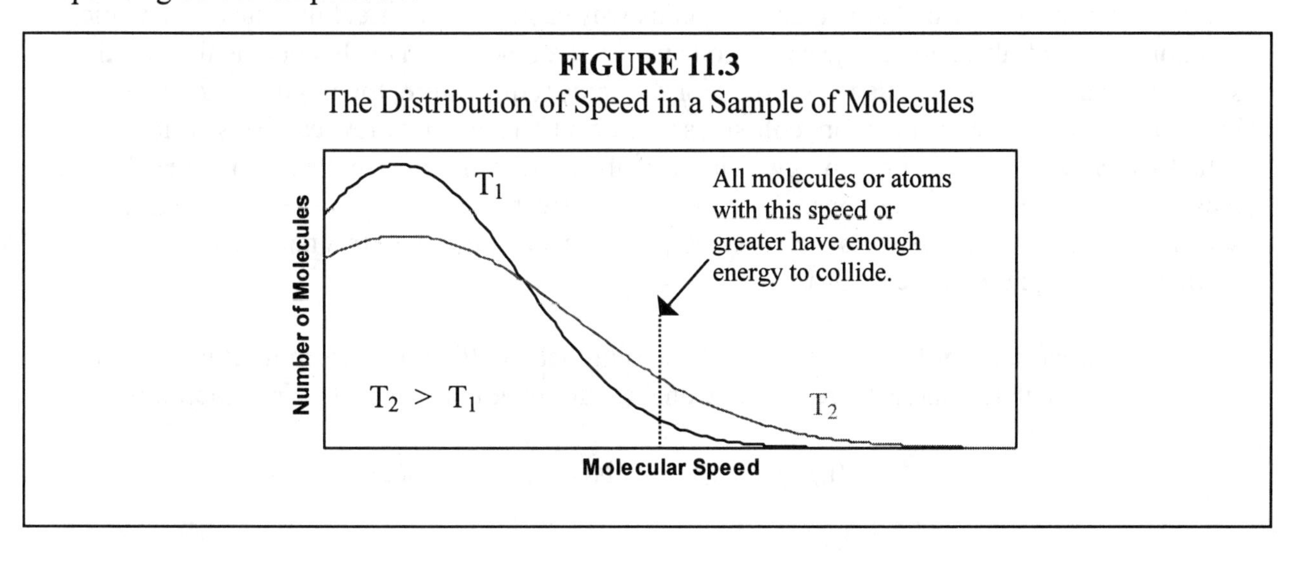

FIGURE 11.3
The Distribution of Speed in a Sample of Molecules

Now if you think about it, a molecule or atom will be able to overcome its repulsion to another molecule or atom as long as it is traveling fast enough. The kinetic energy it has as a result of its motion will be able to overcome the repulsion and a collision will result. Thus, collisions have some sort of "threshold" speed. If an atom or molecule has that speed or greater, then it will be able to collide with another reactant and a chemical reaction might result. In the figure, this threshold speed is drawn with a dashed line.

Notice the difference between the speed distributions for T_1 and T_2. Since T_2 is higher, there is a larger number of atoms or molecules which have the threshold speed or higher. Thus, at the higher temperature, there are *more atoms or molecules which have enough energy to collide with the other reactants*. As a result, there are more collisions. This will lead to more chemical reactions. Thus, an increase in temperature will increase the reaction rate simply because it makes it possible for more reactants to collide with one another.

Just by assuming that molecules or atoms must collide in order to react, then, we have already explained why all reactions have activation energies as well as the fact that an increase in temperature increases the rate of a chemical reaction. It turns out that the collision theory also explains why the other things I have mentioned affect the rate of a chemical reaction as well. It is easy to see why the concentration of the reactants affects the chemical reaction rate. After all, the more concentrated the reactants, the more likely a collision becomes. As a result, more reactions will occur in a given time frame.

Even the effect of surface area can be explained. Suppose, for example, you were reacting a solid with a liquid. If the solid is in a big clump, then many of the solid's molecules are not even exposed to the liquid's molecules. As a result, there is no chance for collision with those molecules. If the solid is crushed and mixed with the liquid, however, then all of the solid's molecules are exposed to liquid molecules. As a result, there will be many more collisions, resulting in more reactions in a given time frame.

What about catalysts? You should have learned about them in your previous chemistry course. Once again, the collision theory explains why catalysts can affect the rate of a chemical reaction. Most catalysts lower the activation energy of a reaction. This lowers the threshold speed needed for two reactants to collide. For the same temperature, then, more reactants will have the threshold speed, and more collisions will occur. There are a few catalysts which actually *increase* the activation energy. These catalysts are sometimes called **inhibitors**, because they slow down chemical reactions. After all, if the activation energy is increased, the threshold speed necessary for a collision is increased as well. As a result, fewer reactants will be able to collide, resulting in fewer reactions.

It is important for me to mention at this point that a collision between reactants is not necessarily enough to generate a reaction. Consider, for example, the following reaction:

$$NO_2\,(g) + CO\,(g) \rightarrow NO\,(g) + CO_2\,(g)$$

In order for this reaction to occur, the reactants must collide in such a way as to make an oxygen atom in NO_2 to leave the nitrogen atom and bond instead to the carbon atom. Well, look at the following diagram and try to think about which collision will result in that process happening.

FIGURE 11.4
Possible Collisions Between NO_2 and CO

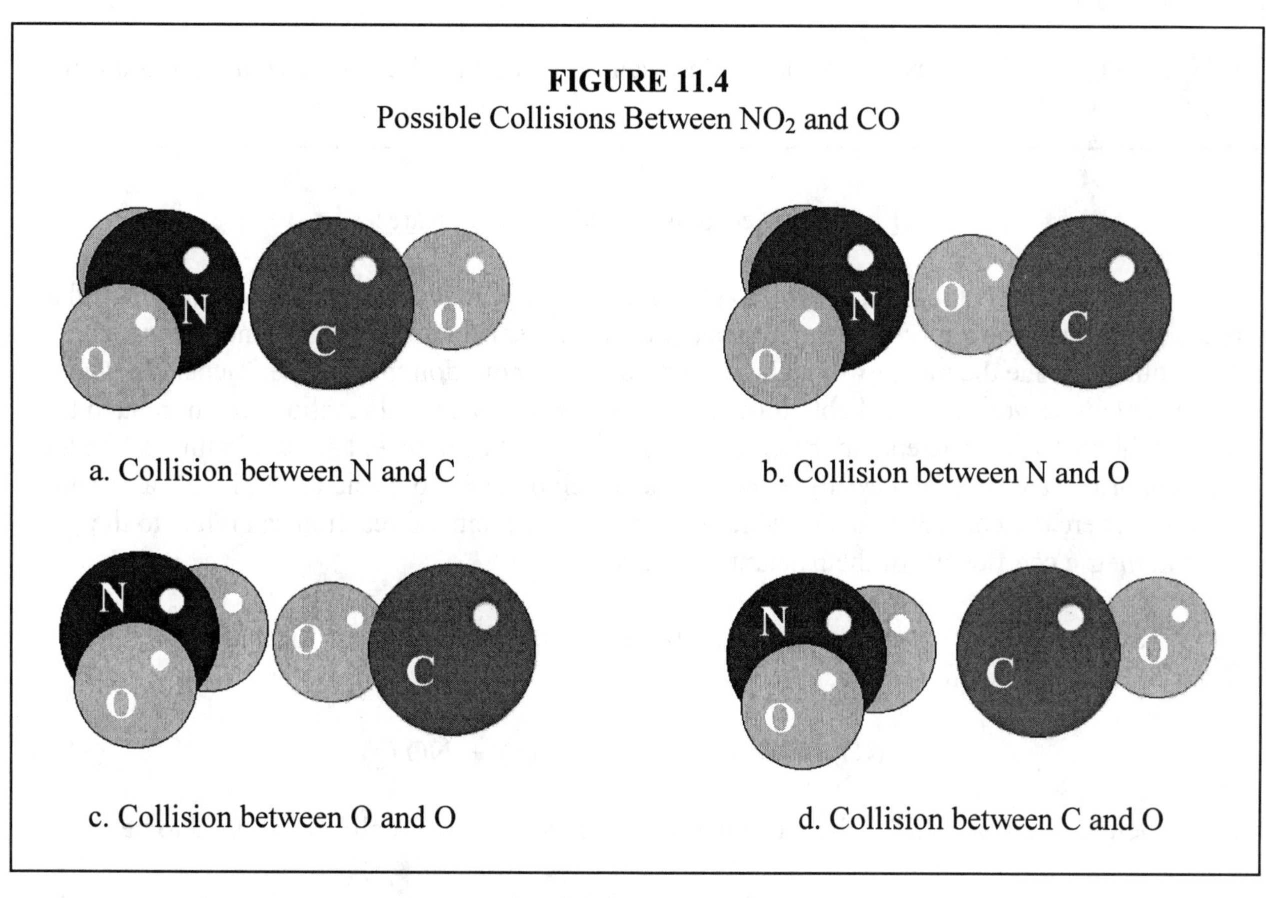

a. Collision between N and C

b. Collision between N and O

c. Collision between O and O

d. Collision between C and O

In collisions (a) through (c), the carbon atom in CO and an oxygen atom in NO_2 don't even touch each other. Thus, there is not a lot of chance for them to re-arrange their electrons so that the oxygen can bond to the carbon. Only in collision (d) does this happen. Thus, for most chemical reactions, having a collision between the reactants is still not enough. You must have *the right* collision between reactants in order for a chemical reaction to occur.

When two reactants collide in just the right way, for a brief instant they are coupled together, as the electrons begin to re-arrange. During that time, we say that the reactants have formed the **intermediate state**, which is sometimes called the **transition state** or the **activated complex**. All of these terms refer to the temporary union that occurs between reactants when the proper collision has taken place. In this reaction, for example, there is a brief time in which an oxygen from NO_2 is sharing electrons with *both* the N in NO_2 and the C in CO. We call that strange union the intermediate state. Of course, the intermediate state does not last long. Once the oxygen, carbon, and nitrogen atoms have finished rearranging their electrons, the products of the reaction are formed. Nevertheless, for a brief moment in time, there is an intermediate state in which the NO_2 and the CO molecules are joined together by the oxygen atom that is being exchanged.

ON YOUR OWN

11.7 Is it possible for the intermediate state to have *less* potential energy than the reactants? Why or why not?

11.8 If you lower the temperature enough, you can stop an otherwise spontaneous reaction from occurring. Why?

Reaction Mechanisms and Reaction Rates

In Module #1, I introduced you to the concept of a **reaction mechanism**. It turns out that reaction mechanisms are *very* relevant to chemical kinetics, because a reaction mechanism can help you determine the rate law of a chemical reaction without doing any experiments. In your first-year course, you were probably told that the orders of a chemical reaction are not related to the stoichiometric coefficients in the chemical reaction. In this course, I have said things like the reaction orders are "not *necessarily* related to the stoichiometric coefficients." There's a reason for this. There *are* conditions under which the orders of a chemical reaction can relate to the stoichiometric coefficients of the reactants in the reaction.

In order for you to see what I mean, I want to discuss the same chemical reaction I discussed in the previous section:

$$NO_2\ (g) + CO\ (g) \rightarrow CO_2\ (g) + NO\ (g) \qquad (11.9)$$

At temperatures *above* 225 °C, the rate equation has been experimentally determined to be:

$$R = k\cdot[NO_2]\cdot[CO] \qquad (11.10)$$

Notice that the stoichiometric coefficients of both reactants are "1," as are the orders of the chemical reaction with respect to each reactant. It turns out that this is not a coincidence. At temperatures greater than 225 °C, this reaction is called an **elementary reaction**.

Elementary Reaction - A reaction that contains only one step: a direct collision between all reactants in the reaction

As I discussed in the previous section, this reaction occurs because a single NO_2 molecule collides in just the right way with a single CO molecule. When the collision is just right, an oxygen will be transferred from the NO_2 molecule to the CO molecule.

When a reaction is an elementary reaction, the values of the exponents in the rate equation are precisely equal to the value of the stoichiometric coefficients in the reaction. This is a very important point.

For *elementary reactions only*, the exponents in the rate equation are equal to the stoichiometric coefficients of the chemical reaction

Thus, if we know a reaction is an elementary reaction, we can determine the rate law *without doing an experiment*! We can simply write the rate law, using the stoichiometric coefficients as the exponents in the rate law.

Now, despite the fact that the reaction between nitrogen dioxide and carbon monoxide is an elementary reaction for temperatures *above* 225 °C, things change when the temperature dips *below 225* °C. The rate law for the *same reaction* at temperatures below 225 °C has been experimentally measured to be:

$$R = k \cdot [NO_2]^2 \qquad (11.11)$$

Now wait a minute. How can the reaction be first-order with respect to each reactant for temperatures above 225 °C but second-order with respect to only one reactant at temperatures below 225 °C? Well, the answer lies in a discussion of activation energy and reaction mechanisms.

Although a direct collision between NO_2 and CO can lead to a reaction between them, the collision requires quite a bit of activation energy. When temperature is high, a fraction of the molecules in a mixture of NO_2 and CO will have enough to make these collisions possible. At temperatures below 225 °C, however, there are simply not enough molecules with that amount of energy. As a result, the reactants cannot collide properly in order to produce a reaction. Thus, a lack of activation energy keeps the elementary reaction from occurring.

These two reactants, however, "really want" to react. Thus, when an elementary reaction between the reactants becomes impossible due to a lack of activation energy, the reactants find another way to react. At temperatures lower than 225 °C, the reactants still react, but they do so according to the following mechanism:

$$NO_2\,(g) + NO_2\,(g) \rightarrow NO_3\,(g) + NO\,(g) \qquad (11.12)$$

$$NO_3\,(g) + CO\,(g) \rightarrow CO_2\,(g) + NO_2\,(g) \qquad (11.13)$$

Now if you add Equations (11.12) and (11.13) together, you will get the original equation (11.9). Thus, at temperatures below 225 °C, two NO_2's first react together to form NO and NO_3. After that, NO_3 reacts with CO to form CO_2 and NO_2.

Now it turns out that Equation (11.12) is very slow, because the activation energy for the proper collisions between two NO_2 molecules is rather high. Equation (11.13) is much faster, however. Both of these reactions are elementary. Thus, the rate equation for Equation (11.12) can be easily determined. The exponents of the rate equation are just the stoichiometric coefficients of the reaction:

$$R = k \cdot [NO_2] \cdot [NO_2] = k \cdot [NO_2]^2 \qquad (11.14)$$

That's the experimentally-determined rate equation for the overall reaction! It turns out that the rate of a chemical reaction is determined by the rate equation of the *slowest step* in the reaction mechanism.

If a reaction is not elementary, its rate is determined by the rate equation of the slowest step in the reaction mechanism.

Since Equation (11.9) occurs via a mechanism for temperatures below 225 °C, its rate is determined by the rate equation of the slowest mechanism step. That step is represented by Equation (11.12). Thus, the rate of the chemical reaction is determined by the rate equation of that particular step. As a result, the slow step in a reaction mechanism is often called the **rate-determining step.**

So, the reaction given by Equation (11.9) is actually very illustrative of how we understand chemical kinetics today. For temperatures above 225 °C, the reaction is elementary, and the resulting rate equation is first-order with respect to both reactants. At temperatures below 225 °C, however, the reaction occurs via a 2-step mechanism. The rate of the reaction is thus determined by the rate law of the slowest step in the reaction mechanism. Since the slowest step has only 2 NO_2 molecules in it, the rate equation is second-order with respect to NO_2 and is not dependent on the concentration of CO. Study the following example to make sure you understand what I'm talking about.

EXAMPLE 11.5

The following reaction:

$$O_3\text{ (g)} + 2NO_2\text{ (g)} \rightarrow O_2\text{ (g)} + N_2O_5\text{ (g)}$$

Occurs according to the following mechanism:

$$O_3\text{ (g)} + NO_2\text{ (g)} \rightarrow NO_3 + O_2 \qquad \textbf{(slow)}$$
$$NO_3\text{ (g)} + NO_2\text{ (g)} \rightarrow N_2O_5\text{ (g)} \qquad \textbf{(fast)}$$

What is the rate law for this reaction?

When a reaction is elementary, the stoichiometric coefficients of the reactants can be used as the exponents in the rate law. The reaction itself is *not* elementary, but the steps of a reaction mechanism are. Thus, the rate law of the overall reaction is determined by the rate law of the slowest step in the mechanism:

$$\underline{R = k\cdot[O_3\text{ (g)}]\cdot[NO_2\text{ (g)}]}$$

That's pretty easy, isn't it? All you do is look at the slow step of the reaction mechanism and write the rate law with the stoichiometric coefficients in that step as the exponents of the rate equation. Well, you should have learned by now that chemistry is never as easy as it first appears. Study the following example to see what I mean.

EXAMPLE 11.6

The following reaction:

$$2NO\ (g) + 2H_2\ (g) \rightarrow N_2\ (g) + 2H_2O\ (g)$$

occurs according to the following mechanism:

$$2NO\ (g) \rightleftarrows N_2O_2\ (g) \qquad \text{(fast)}$$
$$N_2O_2\ (g) + H_2\ (g) \rightarrow N_2O\ (g) + H_2O\ (g) \qquad \text{(slow)}$$
$$N_2O\ (g) + H_2\ (g) \rightarrow N_2\ (g) + H_2O\ (g) \qquad \text{(fast)}$$

What is the rate law for the reaction?

The slow step of the mechanism is the second step. Thus, we get a rate law of:

$$R = k\cdot[N_2O_2]\cdot[H_2]$$

Is this the answer? Well, not quite. Notice that one of the components in the rate equation (N_2O_2) is not a reactant in the *original* equation. N_2O_2 does not exist for very long, because it is made in one step of the reaction and then destroyed in the next. Thus, we say that N_2O_2 is an **intermediate**. A rate equation cannot contain intermediates! It must contain only the reactants in the equation. It need not contain *all* of the reactants in the equation, because sometimes the rate is independent of the concentration of a certain reactant. Nevertheless, a rate equation cannot have any substance which is *not* a reactant in the reaction. Thus, we need to get rid of the N_2O_2 in the rate equation.

How can we do that? Well, look at the first equation. It's an equilibrium. What do we know about equilibria? They have a equilibrium constants:

$$K = \frac{[N_2O_2]}{[NO]^2}$$

Notice that the equilibrium constant relates the concentration of N_2O_2 to the concentration of one of the reactants, NO. If we rearrange the equation a bit, we get an expression for N_2O_2 in terms of NO:

$$[N_2O_2] = K\cdot[NO]^2$$

We can take *that* expression for the concentration of N_2O_2 and stick it in the rate law that we just determined:

$$R = k \cdot [N_2O_2] \cdot [H_2]$$
$$R = k \cdot K \cdot [NO]^2 \cdot [H_2]$$

Now the only substances in the rate law are reactants in the original equation. Thus, we are almost done. All we have to realize is that k·K is still a constant. Since it is a constant, we can simply call it the rate constant, k. Thus,

$$\underline{R = k \cdot [NO]^2 \cdot [H_2]}$$

Now please realize that I did NOT drop "K." The "k" that was originally in the rate law we first wrote down was the rate constant for the slow step of the mechanism. The "k" that is in the answer is not the same. It is the rate constant for the overall reaction. We could have easily called it "c," "K," or even "q." The letter is simply an indication that there is a *constant* there. Since k·K is a constant, we can call it anything we want, so we call it "k."

As this example shows, the determination of a rate equation from the reaction mechanism can become complicated. If, in the end, the rate law you determine has an intermediate in it, you need to relate the concentration of that intermediate to the concentration of one or more of the reactants in the original equation. Typically, you do this by looking for an equilibrium which will allow you to relate one substance to another. Once you have done that, lump all of the constants together into one constant, and call it "k." The result will be the rate law. Study one more example of this to make sure you understand what I'm talking about.

EXAMPLE 11.7

When nitrogen monoxide and oxygen react:

$$2NO\ (g) + O_2\ (g) \rightarrow 2NO_2\ (g)$$

they do so according to the following mechanism:

$$NO\ (g) + O_2\ (g) \rightleftarrows NO_3\ (g) \quad \text{(fast)}$$
$$NO_3\ (g) + NO\ (g) \rightarrow 2NO_2 \quad \text{(slow)}$$

What is the rate law of the reaction between nitrogen monoxide and oxygen?

The rate of the overall reaction is determined by the slow step of the mechanism:

$$R = k \cdot [NO_3] \cdot [NO]$$

This is not the answer, however. Nitrogen trioxide is not a reactant in the overall reaction. It is an intermediate. We must therefore relate it to the concentration of substances which are reactants in the overall equation. The equilibrium step gives us such a relationship:

$$K = \frac{[NO_3]}{[NO]\cdot[O_2]}$$

$$[NO_3] = K\cdot[NO]\cdot[O_2]$$

If we substitute this expression for the concentration of NO_3 into the rate equation we just determined, we get:

$$R = k\cdot[NO_3]\cdot[NO]$$

$$R = k\cdot K\cdot[NO]\cdot[O_2]\cdot[NO]$$

$$\underline{R = k\cdot[NO]^2[O_2]}$$

Now notice that, in the end, the rate law determined by this process has exponents which are equal to the stoichiometric coefficients of the chemical reaction. This leads me to an important point. Just because the rate law has the same exponents as the stoichiometric coefficients, that does not mean the reaction is elementary.

When the exponents in a rate law are equal to the stoichiometric coefficients, the equation is not necessarily elementary. The reaction mechanism might have just worked out that way.

So remember, *when* a reaction is elementary, *you know* that the exponents of the rate equation are the same as the stoichiometric coefficients in the reaction. However, the reverse is not necessarily true! If the exponents of the rate equation happen to be equal to the reaction's stoichiometric coefficients, that does not necessarily mean that the reaction is elementary!

ON YOUR OWN

11.9 The decomposition of NO_2Cl is given by:

$$2NO_2Cl \rightarrow 2NO_2 + Cl_2$$

It is thought to happen according to the following mechanism:

$$NO_2Cl \rightarrow NO_2 + Cl \quad \text{(slow)}$$
$$NO_2Cl + Cl \rightarrow NO_2 + Cl_2 \quad \text{(fast)}$$

What is the rate equation for this reaction?

11.10 The formation of hydrogen monoiodide is given by:

$$H_2 + I_2 \rightarrow HI$$

It is thought to happen according to the following mechanism:

$$I_2 \rightleftarrows 2I \quad \text{(fast)}$$
$$2I + H_2 \rightarrow 2HI \quad \text{(slow)}$$

What is the rate equation for this reaction?

ANSWERS TO THE ON YOUR OWN PROBLEMS

11.1 In order to determine the rate equation, we need to determine the values of "x" and "y." To do that, we compare 2 trials in which one reactant's concentration remained constant and the other reactant's concentration changed. Between trials 1 and 2, for example, the NO_3 concentration stayed constant, but the O_3 concentration doubled. Any change in the rate, then, would be due solely to the change in the O_3 concentration. When the O_3 concentration doubled, the rate doubled as well. This means that the reaction is first-order with respect to O_3. Between trials 2 and 3, the O_3 concentration remained the same, and the NO_3 concentration doubled. The rate between these trials also doubled. This means the reaction is first-order with respect to NO_3. Summing up those orders, we have determined that the overall reaction order is 2.

Now that we know the reaction orders, we know the basic rate equation:

$$R = k \cdot [NO_3] \cdot [O_3]$$

Now that we have the basic rate equation, we can use the data from any trial to determine the value of the rate constant:

$$6.60 \times 10^{-5} \ \frac{M}{s} = k \cdot \left(1.00 \times 10^{-6} \ M\right) \cdot (3.00 \times 10^{-6} \ M)$$

$$k = \frac{6.60 \times 10^{-5} \ \frac{\cancel{M}}{s}}{\left(1.00 \times 10^{-6} \ M\right) \cdot (3.00 \times 10^{-6} \ \cancel{M})} = 2.20 \times 10^{7} \ \frac{1}{M \cdot s}$$

This means that the overall rate equation is:

$$\underline{R = \left(2.20 \times 10^{7} \ \frac{1}{M \cdot s}\right) \cdot [NO_3] \cdot [O_3]}$$

11.2 The half-life tells us that every 6 hours, half of the hydrogen peroxide decays away. If the chemist starts with a concentration of 6.0 M, the concentration will decrease to 3.0 M in 6 hours. In the next 6 hours, half of that would go away, leaving a concentration of 1.5 M. In the next 6 hours, half of that would go away, leaving a concentration of 0.75 M. In the final 6 hours (4 sets of 6 hours is 24 hours), the concentration decreases to 0.375 M. We can only have 2 significant figures, however, so the answer is 0.38 M.

11.3 In one half-life, the concentration will decrease to 0.5000 M. In the next half-life, it decreases to 0.2500 M. In the next half-life, it decreases to 0.1250 M. In the next half-life, it decreases to 0.06250 M. In the next half-life, it decreases to 0.03125 M. This tells us it takes 5 half-lives to get down to a concentration of 0.03125 M. Since each half-life is 3 seconds long, the total time is 15 seconds.

11.4 Since the reaction is first-order, we can relate the rate constant to the half-life using Equation (11.6). We need to determine the half-life first, however. In order to be reduced to one-fourth of its original concentration, the reaction must have passed through 2 half-lives. This means that 150.6 seconds represents 2 half-lives. Thus, the half-life is 75.30 seconds. Now we can use Equation (11.6).

$$t_{1/2} = \frac{0.693}{k}$$

$$75.30 \text{ sec} = \frac{0.693}{k}$$

$$k = \frac{0.693}{75.30 \text{ sec}}$$

$$\underline{k = 0.00920 \frac{1}{\text{sec}}}$$

11.5 Unfortunately, 501 seconds isn't a multiple of the half-life. Thus, we can't use the half-life method to determine the concentration. However, since we know the rate constant and we know that the equation is first-order, we can use Equation (11.5).

$$[A] = [A]_o \cdot e^{-kt}$$

$$[A] = (5.5\text{M}) \cdot e^{-(0.00920 \frac{1}{\cancel{\text{sec}}}) \cdot 501 \cancel{\text{sec}}}$$

$$\underline{[A] = 0.055 \text{ M}}$$

11.6 We start with Equation (11.7):

$$\frac{1}{[A]} - \frac{1}{[A]_o} = k \cdot t$$

After one half-life, the concentration of A will be half its original concentration. Thus:

$$[A] = \frac{1}{2} \cdot [A]_o$$

We can substitute this into the equation to get:

$$\frac{1}{\frac{1}{2}\cdot[A]_o} - \frac{1}{[A]_o} = k\cdot t_{1/2}$$

Now we can do some algebra:

$$\frac{2}{[A]_o} - \frac{1}{[A]_o} = k\cdot t_{1/2}$$

$$\frac{1}{[A]_o} = k\cdot t_{1/2}$$

$$t_{1/2} = \frac{1}{k\cdot[A]_o}$$

That's Equation (11.8).

11.7 It is not possible for the intermediate state to have less potential energy than the reactants. Remember, *all* chemical reactions have an activation energy. The potential energy of the intermediate state is the potential energy of the reactants *plus* the activation energy. See Figure 11.2.

11.8 If you lower the temperature enough, you can have a situation in which essentially no molecules have enough energy to make a collision. Without collisions, there will be no reaction.

11.9 Since this reaction is not elementary, we must look to its mechanism for the rate. The slow step of the mechanism will determine the rate. Since the mechanism is composed of elementary reactions, we can just use the stoichiometric coefficients as the reaction order. Thus, the slow step of the reaction has a rate equation of:

$$R = k\cdot[NO_2Cl]$$

Since this rate equation has no intermediates in it, we are done.

11.10 In this reaction mechanism, the slow step has the following reaction mechanism:

$$R = k\cdot[I]^2\cdot[H_2]$$

This can't be the final rate equation, however, because it contains an intermediate. We need to get rid of that for the real rate equation. We can do this because the previous reaction is an equilibrium. This allows us to relate I to I_2:

$$K = \frac{[I]^2}{[I_2]}$$

$$[I]^2 = K \cdot [I_2]$$

We can use this expression to get rid of the intermediate in the rate equation:

$$R = k \cdot K \cdot [I_2] \cdot [H_2]$$

$$\underline{R = k \cdot [I_2] \cdot [H_2]}$$

This is another example of a reaction whose rate equation has exponents equal to the stoichiometric coefficients even though the reaction is not elementary.

REVIEW QUESTIONS

1. A reaction is third-order overall. What are the units of the rate constant?

2. Which of the following plots is a possible graph of the rate of a chemical reaction as a function of time?

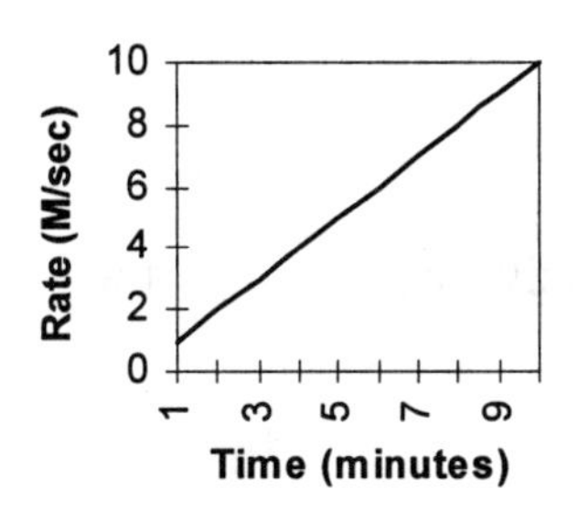

a.

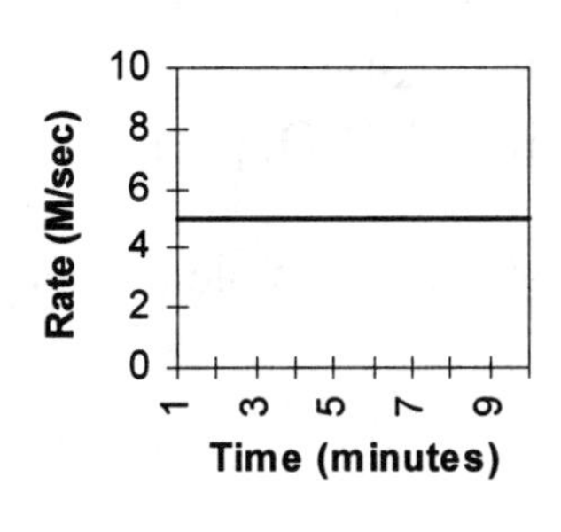

b.

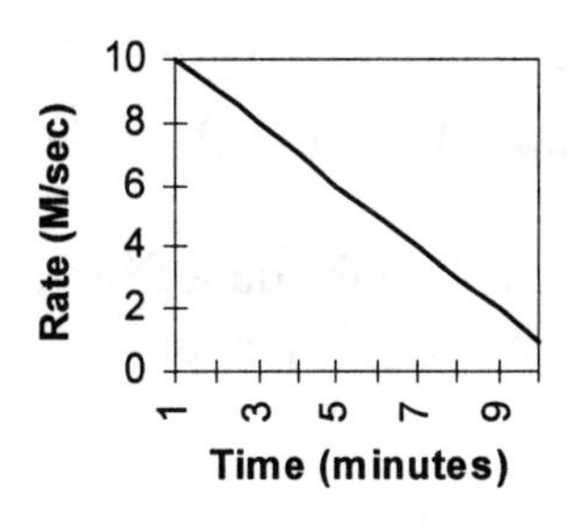

c.

3. If a chemist plots the concentration of a reactant as a function of time, how can the chemist determine the rate of the reaction at any point on the graph?

4. In terms of the collision theory of chemical kinetics, explain why temperature, surface area, activation energy, and concentration affect reaction rate. Where are all of these effects in the rate equation?

5. If a reaction has a large activation energy, can you conclude that the reaction is endothermic? Why or why not?

6. If the activation energy is large, is the rate constant large or small?

7. The following potential energy diagrams are both for the same reaction. One represents the reaction in the presence of a catalyst. One represents the reaction without a catalyst. Which is which?

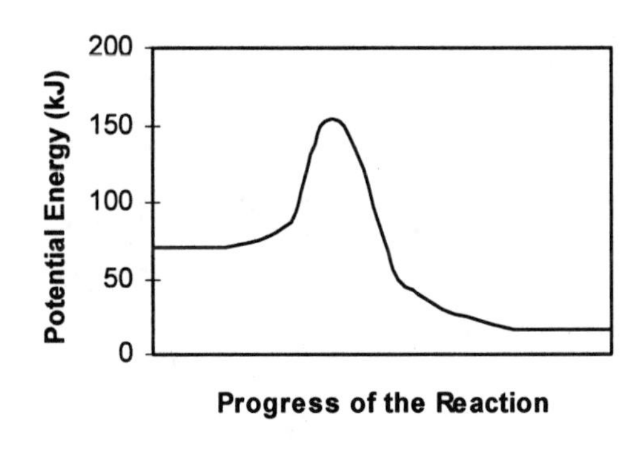

a.

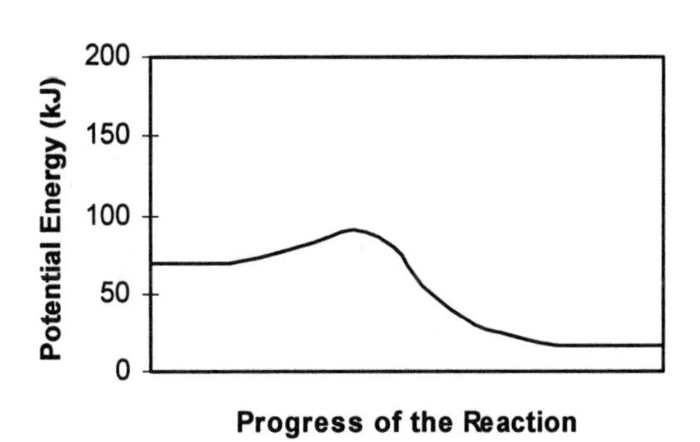

b.

8. Do all collisions between reactants lead to a reaction? Why or why not?

9. For the following reactions, the rate laws are listed. Which reactions are definitely not elementary reactions?

a. $Cl_2 + CO \rightarrow Cl_2CO$ $R = k \cdot [Cl_2]^{3/2} \cdot [CO]$
b. $PCl_3 + Cl_2 \rightarrow PCl_5$ $R = k \cdot [PCl_3] \cdot [Cl_2]$
c. $2NO + 2H_2 \rightarrow N_2 + 2H_2O$ $R = k \cdot [NO] \cdot [H_2]$
d. $2NO + O_2 \rightarrow 2NO_2$ $R = k \cdot [NO]^2 \cdot [O_2]$
e. $NO + O_3 \rightarrow NO_2 + O_2$ $R = k \cdot [NO] \cdot [O_3]$

10. Based on the data from question #9, is it possible to state conclusively that any of the reactions *are* elementary?

PRACTICE PROBLEMS

1. A chemist studied the instantaneous rate of the following reaction:

$$2ClO_2 + 2OH^- \rightarrow ClO_3^- + ClO_2^- + H_2O$$

The data gathered were as follows:

	$[ClO_2]$ (Molarity)	$[OH^-]$ (Molarity)	Instantaneous Rate (M/s)
Trial 1	0.060	0.030	0.0248
Trial 2	0.020	0.030	0.00276
Trial 3	0.020	0.090	0.00828

What is the rate equation for this reaction? What is the overall reaction order as well as the reaction order with respect to each reactant?

2. The decomposition of ethylene oxide, C_2H_4O, has a half-life of 15.0 minutes at 500 K. If a chemist starts with ethylene oxide at 6.00 M, what will the concentration be after 1 hour?

3. A chemist observes a chemical reaction in which the reactant concentration drops from 2.0 M to 0.50 M in 25.3 seconds. What is the half-life of the chemical reaction?

4. In the presence of excess thiocyanate ion (SCN^-), the following reaction is first-order:

$$Cr^{3+} + SCN^- \rightarrow Cr(SCN)^{2+}$$

If the rate constant is 2.0×10^{-6} 1/sec, what is the half-life of the reaction?

5. If the concentration of the thiocyanate ion is equal to the concentration of the chromium ion, the reaction is Problem #4 becomes a second-order reaction. If the rate constant under those conditions is 4.5×10^{-5} 1/[M·sec] and the initial concentration of each reactant is 1.5 M, what is the half-life of the reaction?

6. Given the same conditions as problem number 5, what concentration of each reactant will remain after 1.0 hour has passed?

7. The decomposition of dinitrogen pentaoxide is a first-order chemical reaction. A chemist observes a solution of dinitrogen pentaoxide in which the concentration of the reactant decreases from 3.0 M to 0.50 M in 16 hours. What is the rate constant for this reaction?

8. Nitric oxide (NO) is believed to react with chlorine according to the following mechanism:

$$NO + Cl_2 \rightleftarrows NOCl_2$$
$$NOCl_2 + NO \rightarrow 2NOCl$$

What is the overall reaction? What is the intermediate?

9. Tertiary butyl chloride, $(CH_3)_3CCl$ reacts in basic solution according to the following overall reaction:

$$(CH_3)_3Cl + OH^- \rightarrow (CH_3)_3COH + Cl^-$$

The generally-accepted mechanism for this reaction is:

$$(CH_3)_3CCl \rightarrow (CH_3)_3C^+ + Cl^- \quad \text{(slow)}$$
$$(CH_3)_3C^+ + OH^- \rightarrow (CH_3)_3COH \quad \text{(fast)}$$

What is the rate law for the overall reaction? What intermediates exist in the reaction mechanism?

10. Urea, $(NH_2)_2$, can be prepared by heating ammonium cyanate, NH_4CNO.

$$NH_4CNO \rightarrow (NH_2)_2CO$$

The mechanism for this reaction is:

$$NH_4CNO \rightleftarrows NH_3 + HCNO \quad \text{(fast)}$$
$$NH_3 + HCNO \rightarrow (NH_2)_2CO \quad \text{(slow)}$$

What is the rate law for the overall reaction? What are the intermediates in the mechanism?

Module #12: An Introduction To Organic Chemistry

Introduction

The chemistry you have learned up to this point has mostly ignored one of the more fascinating aspects of chemistry: organic chemistry. Organic chemistry makes up the backbone of biochemistry, which is the study of the chemistry that is involved in life. Obviously, then, organic chemistry is quite interesting. Because life is so complicated, however, organic chemistry is complicated as well. Thus, an entire field of chemistry is devoted solely to organic chemistry, and another entire field of chemistry is devoted to biochemistry.

In this module, I will give you a bare-bones introduction to the field of organic chemistry. If you study chemistry or a chemistry-related subject in college, you will not really see any organic chemistry until your sophomore year. At that point, however, the entire year will be devoted to it. Many prospective doctors end up not making it into medical school specifically because they cannot pass organic chemistry. Thus, organic chemistry is a vast, complex subject. This introduction will not be vast nor will it be complicated. It will, however, give you a flavor for this fascinating field.

The first thing you need to know about organic chemistry is how to distinguish **organic** molecules from **inorganic** molecules. An organic molecule cannot contain a metal and must also have carbon in it.

Organic molecule - A molecule containing carbon atoms and no metals

If a molecule is not organic, then it is inorganic. Organic molecules are placed into several different classes based on the way the atoms are bonded and the type of atoms present. I will discuss four of these classes in subsequent sections of this module and more in the next module.

ON YOUR OWN

12.1 Which of the following molecules are organic?

$CaCO_3$, CO_2, $NaHCO_3$, CH_4, $C_6H_{12}O_6$, H_2, O_2

Saturated Hydrocarbons

The most basic organic molecule is the **saturated hydrocarbon**. Now the "hydrocarbon" part of that term is easy to understand. It refers to a molecule which contains only carbon and hydrogen. What, however, does "saturated" mean? Well, the term "saturated" means that the molecule has as many hydrogen atoms as it possibly can. In other words, the molecule is "saturated" with hydrogens. How do you tell if a molecule is saturated with hydrogens? Well, it has to have a hydrogen everywhere it can. If you think about it, that means there can be no double or triple bonds in the molecule. Thus, a saturated hydrocarbon is a molecule that contains only carbon and hydrogen and has no double bonds.

Saturated hydrocarbons come in two classes: **alkanes** and **cycloalkanes**. Figure 12.1 shows you the Lewis structures of several alkanes and cycloalkanes.

FIGURE 12.1
Alkanes and Cycloalkanes

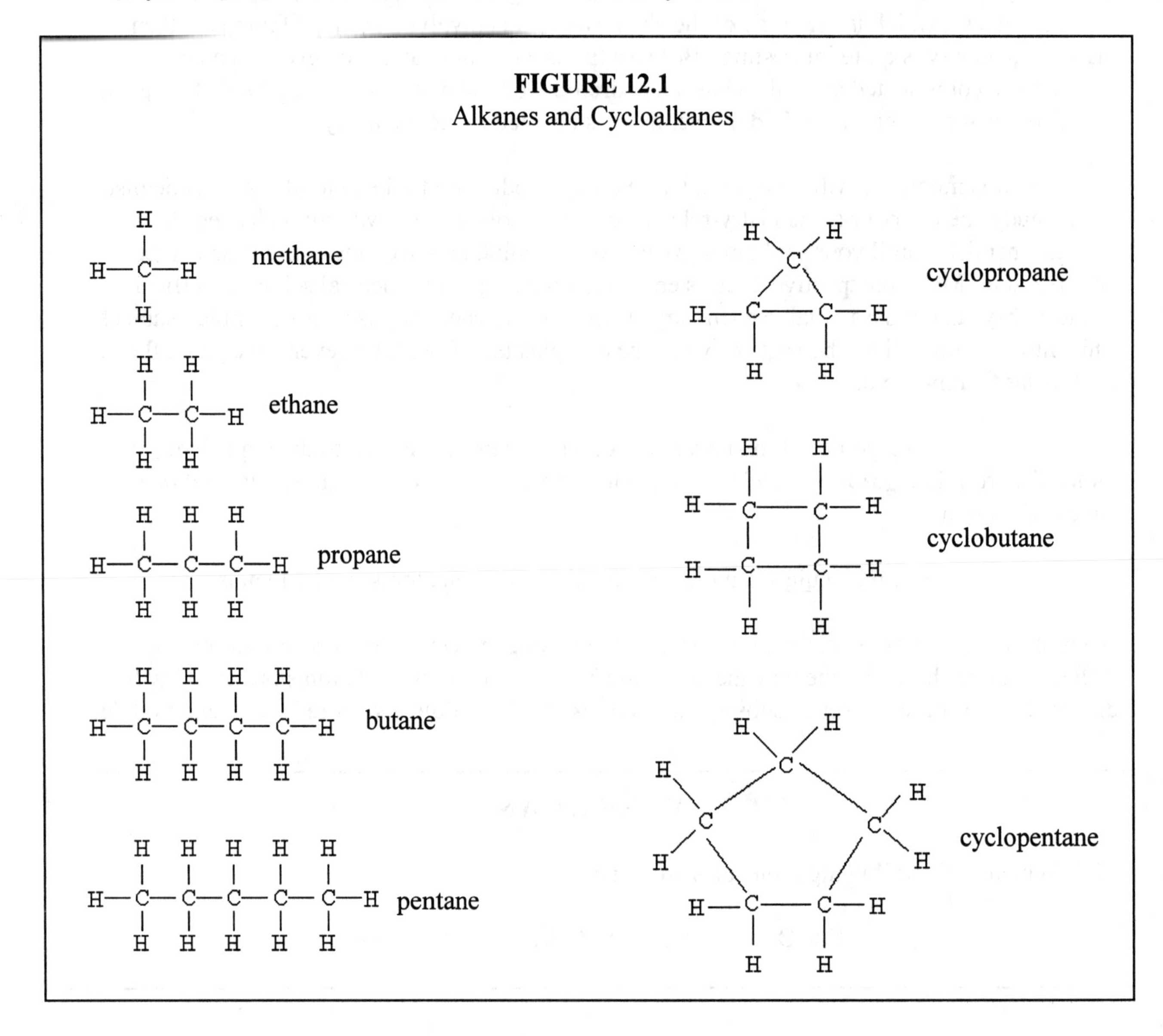

Notice that each molecule shown has only carbons and hydrogens and has no double bonds. That's what makes them saturated hydrocarbons. Notice also that the carbons are all bonded to one another. This is typical of organic compounds. The carbon atoms bond together, making the "carbon backbone" or "carbon chain" of the molecule, and then everything else attaches to the carbon backbone. For those saturated hydrocarbons shown in the figure, only hydrogens are attached to the carbon backbone.

Notice also that all of the names in the figure end in "ane." That tells you they are all alkanes. In organic chemistry, the naming system is completely different than anything you have learned before. You've already learned from your first chemistry course that ionic compounds have one naming system while covalent compounds have another. Well, organic compounds are all covalent compounds (remember, they cannot have metals in them), nevertheless, they have their own naming system. That's part of what you will learn in this module.

Before I tell you how to name saturated hydrocarbons, I want to discuss some of their properties. First of all, saturated hydrocarbons are flammable. The ΔH of a combustion reaction for a saturated hydrocarbon is typically large, and the activation energy is usually low. This tells you that saturated hydrocarbons typically burn hot and burn fast. Some of the alkanes listed in Figure 12.1 should be familiar to you. Methane is the principle component of natural gas, which is used to heat homes, cook food, and heat water in many homes. If you have a gas grill on your back porch or patio, it typically runs on propane. Finally, pocket cigarette lighters typically are fueled by butane. It also turns out that gasoline is comprised mostly of saturated hydrocarbons. I will discuss this a little more in a moment. Thus, we make quite a bit of use out of the combustibility of saturated hydrocarbons.

The properties of saturated hydrocarbons depend principally on how many carbons they have. This is a general rule of thumb in all of organic chemistry. As you continue through this module, you will see that the number of carbons in an organic molecule tells you quite a bit about that molecule. Indeed, the entire system of nomenclature (naming) in organic chemistry starts with counting the number of carbons in the molecule. Notice in the figure, for example, that the alkane which had 3 carbons is called propane. What is the cycloalkane with three carbons called? It's called cyclopropane. The "pro" part of the name tells you there are three carbons in the molecule. You will learn more about this later.

Saturated hydrocarbons are no exception to this general rule of thumb in organic chemistry. For example, when an alkane has 4 carbon atoms or fewer, it is a gas at room temperature and pressure. If the molecule has 5 to 13 carbons, it is a liquid at room temperature and pressure, and when the molecule has more than 13 carbon atoms, it is a solid at room temperature and pressure. The reason for this is quite simple: alkanes are non-polar. As a result, the only intermolecular force at their disposal is the London dispersion force. Do you remember what the London dispersion force depends on? It depends on molecular mass. Thus, as an alkane gets heavier, its London dispersion force increases, and the molecules in a sample can get closer together. The more carbons an alkane has, the heavier it is. Thus, the more carbons it has, the closer its molecules can get to one another.

Now wait a minute. You've probably seen a pocket cigarette lighter whose outer casing was clear. You can actually see the butane in the lighter. It is a liquid. according to our general rule of thumb, it is supposed to be a gas. Well, it is a gas, *at room temperature and pressure*. In a pocket cigarette lighter, the butane is stored under high enough pressure to make it a liquid. When you press down on the button to make the lighter work, it exposes the liquid to air, reducing the pressure. Some of the liquid immediately turns to gas in response to the decreased pressure, and *that's* what the lighter burns. Because the butane is pressurized and liquefied, a cigarette light can store *a lot more* of it and thus the lighter will last a lot longer. Nevertheless, when the lighter actually burns butane, it is burning it in the gaseous state.

It turns out that the chemical formula of a saturated hydrocarbon follows a general formula that is based (not surprisingly) on the number of carbons it has. For alkanes, the generalized formula is C_nH_{2n+2}. In other words, the alkane that has 5 carbons has a chemical formula of C_5H_{12}. For cycloalkanes, the generalized equation is slightly different. In a cycloalkane there is an extra carbon-carbon bond as compared to the alkane with the same number of carbons. As a result, there are two less hydrogen atoms in a cycloalkane as there are in the alkane with the same number of carbons. Thus, the generalized equation for a cycloalkane is C_nH_{2n}. The cycloalkane with 5 carbons, for example, has a chemical formula of C_5H_{10}.

Alkanes have another property that is common among organic molecules. They have **structural isomers** (eye' so murs).

Structural isomers - Two or more molecules with the same chemical formula but different Lewis structures

For example, consider the following alkanes, both of which have the chemical formula C_4H_{10}:

FIGURE 12.2
Structural Isomers of Butane

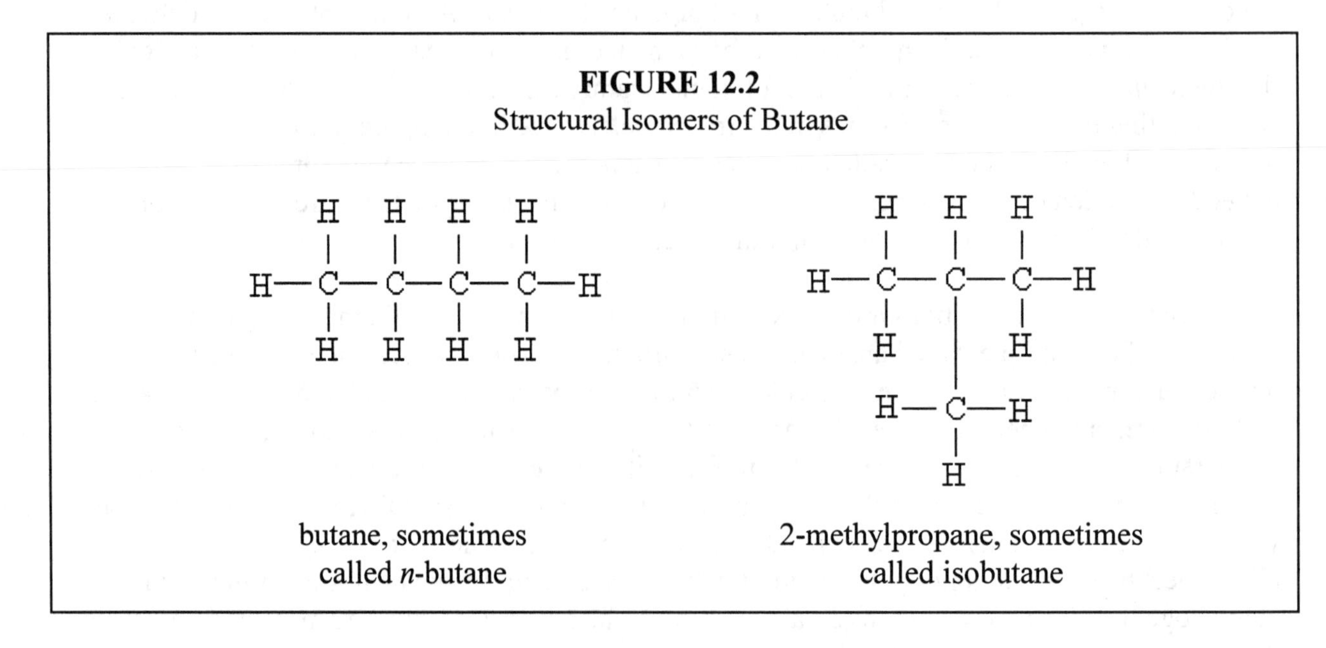

butane, sometimes called *n*-butane

2-methylpropane, sometimes called isobutane

Notice that while the chemical formula of these molecules are identical, the Lewis structures are quite different. They each have 4 carbons, but in butane, the 4 carbons are linked to one another in a straight chain. In 2-methyl propane, the carbons are in a chain of three and then one other carbon attaches to that chain. These different Lewis structures result in slightly different properties, so these molecules do behave differently, even though they have the same chemical formula.

Structural isomers are everywhere in organic chemistry, so you need to get used to them. For example, all alkanes with 4 or more carbons have structural isomers. As you might expect, the more carbons a molecule has, the more structural isomers it will have. Butane, the alkane shown above, has only one structural isomer, which is shown in the figure. The alkane with 5 carbons (pentane), however, has 3.

FIGURE 12.3
Structural Isomers of Pentane

```
 H   H   H   H   H          H   H   H   H                 H
 |   |   |   |   |          |   |   |   |                 |
H—C—C—C—C—C—H            H—C—C—C—C—H              H—C—H
 |   |   |   |   |          |   |   |   |             H   |   H
 H   H   H   H   H          H   H   |   H             |   |   |
                                 H—C—H             H—C—C—C—H
                                   |                  |   |   |
                                   H                  H   |   H
                                                       H—C—H
                                                          |
                                                          H
```

pentane, sometimes called *n*-pentane

2-methylbutane, sometimes called isopentane

2,2-dimethylpropane, sometimes called neopentane

You might think that there are a lot more than 3 structural isomers for pentane. After all, there must be more ways to draw Lewis structures for an alkane with 5 carbons. Well, there are a lot of different ways you can draw the Lewis structure for an alkane with 5 carbons, but in the end, they all end up the same as one of the molecules shown in Figure 12.3. For example, Figure 12.4 shows three Lewis structures which at first glance appear to be different. However, they are all actually identical.

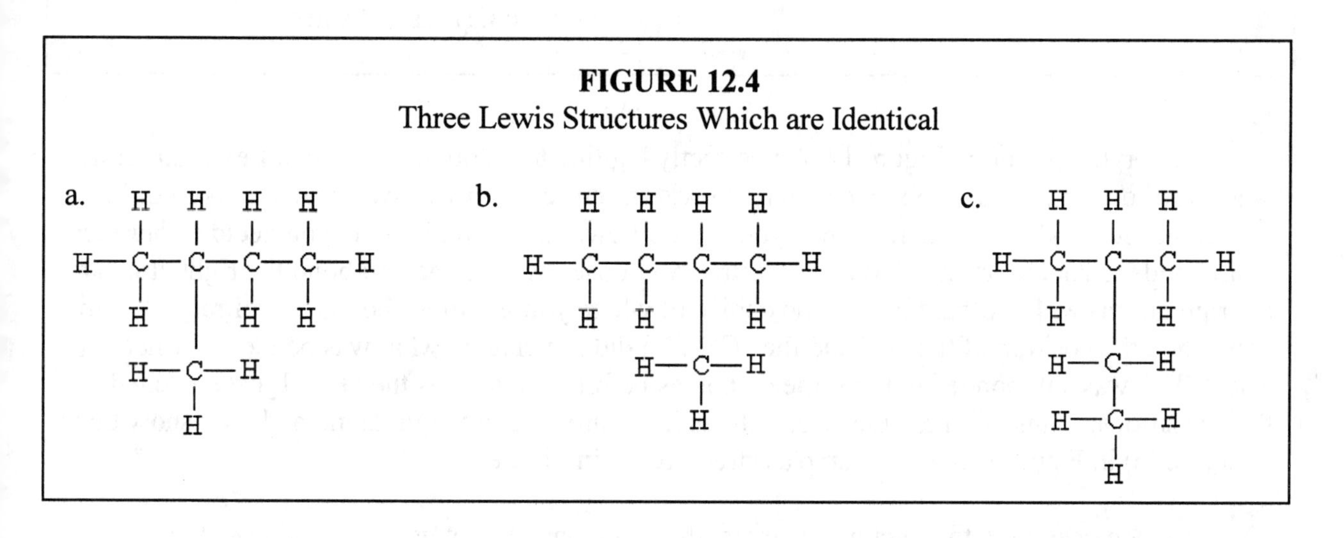

FIGURE 12.4
Three Lewis Structures Which are Identical

Why are all of these Lewis structures identical? Well, suppose you took structure (b) and flipped it over so that the last carbon is now the first carbon. What would you get? You would get structure (a). Thus, structures (a) and (b) are the same, they are just drawn in such a way as to look different. This is one of the first things you need to learn when looking at organic chemicals. You need to learn to look at Lewis structures and turn them in different orientations to see if they are really unique.

What about structure (c)? You might think that there is no way to turn structure (c) around in your mind so as to make structure (a). Well, believe it or not, there is. First of all, you need to realize that the way in which a Lewis structure is drawn is somewhat arbitrary. Any bends, corners, or kinks in a Lewis structure can be straightened out. The only thing that you need to preserve is what atom is bonded to what atom with what kind of bond. Other than that, you can change virtually anything else in a Lewis structure. Here's how you turn structure (c) into structure (a):

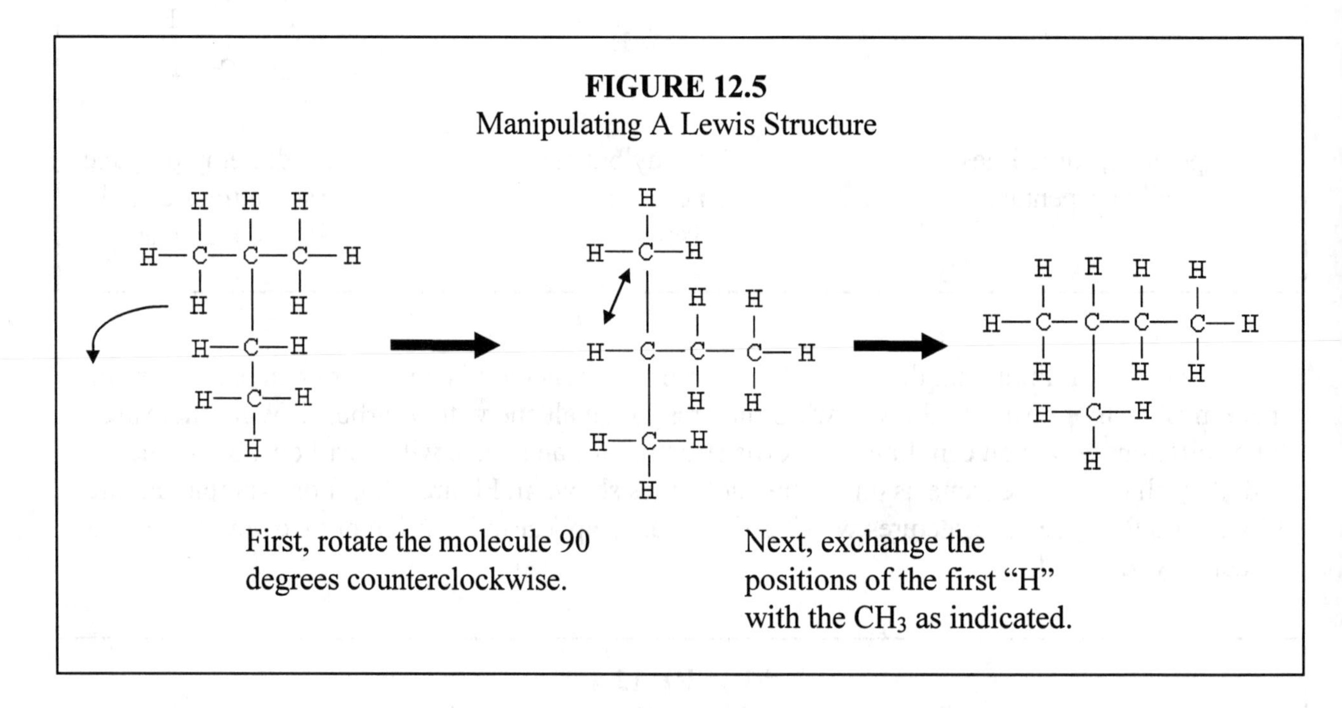

FIGURE 12.5
Manipulating A Lewis Structure

First, rotate the molecule 90 degrees counterclockwise.

Next, exchange the positions of the first "H" with the CH_3 as indicated.

Everything I did in Figure 12.5 is perfectly legitimate. You do not alter a Lewis structure when you rotate it. You do not even alter a Lewis structure when you switch atoms like I did in the second step. Think about it. When you draw a Lewis structure, how do you decide *where* an atom bonds to another atom? You just arbitrarily choose. As long as you bond the right atoms to the right atoms with the right bonds, the choice of where you put those bonds is arbitrary. When I switched the position of the "H" and the "CH_3," I did not change what was bonded to what. The "CH_3" was still bonded to the same carbon as before, and so was the "H." I just switched their positions. Thus, all three Lewis structures in Figure 12.5 are equivalent, and they show that structure (c) in Figure 12.4 is the same as structure (a) in Figure 12.4.

When dealing with structural isomers, then, you have to get used to playing with the Lewis structures in your head. Two Lewis structures are only isomers if they have the same chemical formulas *and* they cannot be manipulated into each other by means such as I employed in Figure 12.5.

ON YOUR OWN

12.2 Will a liquid alkane mix with water to form a homogeneous solution? Why or why not?

12.3 An alkane has 7 carbons. What is its chemical formula? If a cycloalkane had the same number of carbons, what would its formula be?

12.4 Of the following Lewis structures, two are identical. Which two are they? Of the rest, one is a structural isomer of the two identical structures. Which one is it?

```
a.   H   H   H   H   H        b.   H   H   H   H   H        c.   H   H   H          d.  H   H   H   H
     |   |   |   |   |             |   |   |   |   |             |   |   |              |   |   |   |
 H — C — C — C — C — C — H     H — C — C — C — C — C — H     H — C — C — C — H      H — C — C — C — C — H
     |   |   |   |   |             |   |   |   |   |             |   |   |              |   |   |   |
     H   |   H   H   H             H   H   H   H   H             H   |   H              H   |   H   H
         |                                                           |                      |
     H — C — H                                                   H — C — H              H — C — H
         |                                                           |                      |
         H                                                       H — C — H              H — C — H
                                                                     |                      |
                                                                 H — C — H                  H
                                                                     |
                                                                     H
```

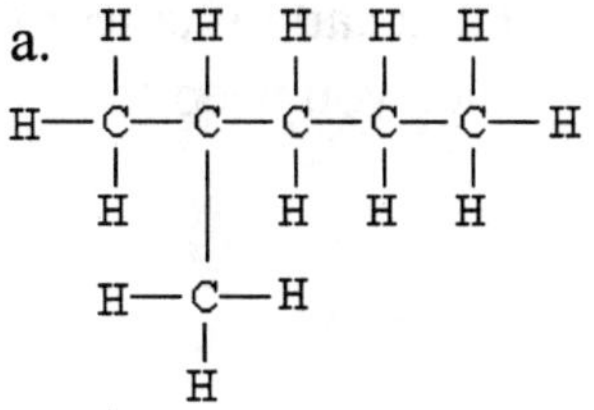

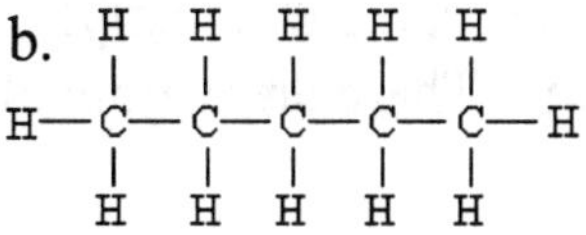

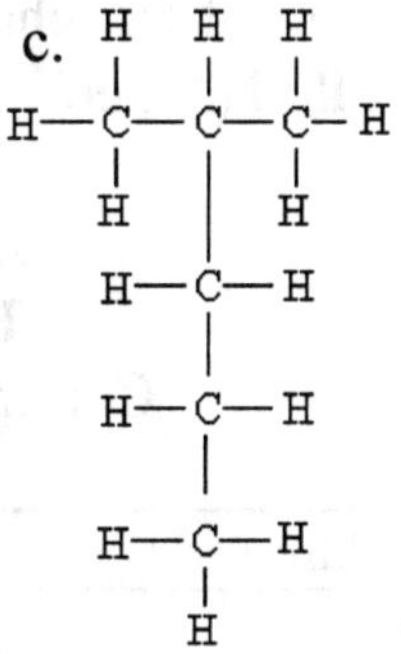

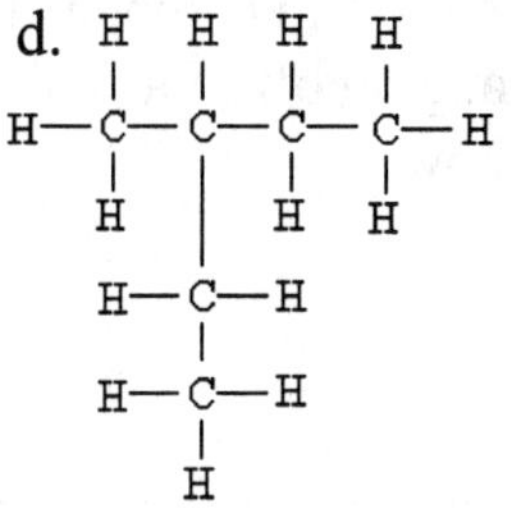

Naming Saturated Hydrocarbons

As I mentioned before, organic chemistry has its own nomenclature system. You will begin to learn it in this section. Regardless of the type of organic molecule you are naming, there are certain general rules. The first is that the basis of the molecule's name is how many carbons it has. There is a prefix you use in the name that tells you the number of carbon atoms the molecule has. Those prefixes are given in the table below. You will need to memorize them.

TABLE 12.1
Prefixes Based on the Number of Carbon Atoms

Number of Carbon Atoms	Prefix Used		Number of Carbon Atoms	Prefix Used
1	meth		6	hex
2	eth		7	hept
3	prop		8	oct
4	but		9	non
5	pent		10	dec

The next general principle is that the end of the name is characteristic of the *type* of organic molecule. All alkanes and cycloalkanes, for example, end in "ane." Thus, if you look back at Figure 12.1, you will see that the first molecule on the left of the figure is called "methane." That's because it has one carbon (thus its prefix is "meth") and it is an alkane (thus it ends in "ane"). Similarly, the 2-carbon alkane is ethane, the 3-carbon alkane is propane, and so on. For cycloalkanes, the only difference is that the prefix "cyclo" is added to the very beginning. Thus, the cycloalkane with 3 carbons is cyclopropane.

Now that's not too bad, is it? As I am sure you are aware by now, however, chemistry is *a lot* more complicated than it first seems. For example, look back to Figure 12.2. The molecule on the right-hand side of the figure is called 2-methylpropane. Why? It has 4 carbons. Shouldn't its name have something to do with butane? No, it shouldn't. The way we determine the base name of an organic carbon is to look at the number of carbons in the *longest chain* of carbons in the molecule. There are 3 carbons strung together in a line. That's the longest chain. The fourth carbon is hanging off of that chain. In organic chemistry, we say that a carbon unit hanging off of a chain is an **alkyl** (al keel') **group**. There are certain alkyl groups you need to know:

TABLE 12.2
Common Alkyl Groups

Alkyl Group Name	Alkyl Group Structure
methyl	CH_3—
ethyl	CH_3CH_2—
n-propyl	$CH_3CH_2CH_2$—
isopropyl	| CH_3CHCH_3
n-butyl	$CH_3CH_2CH_2CH_2$—
sec-butyl	| $CH_3CHCH_2CH_3$
isobutyl	CH_3CHCH_2— | CH_3
t-butyl	| CH_3CCH_3 | CH_3

In this table, the lines that have nothing on the other side indicate the point at which the alkyl group attaches to the carbon chain.

When an alkyl group is attached to a carbon chain, you need to specify which carbon it attaches to. In Figure 12.2, for example, we have 2-methylpropane. We get than name because the longest carbon chain in the molecule is 3 carbons long. That gives us propane. On that three carbon chain, there is a methyl group attached to the middle carbon. If we count carbons in the chain, that would be carbon number 2. Thus, we precede "propane" with "2-methyl," to indicate that there is a methyl alkyl group on the second carbon.

Notice also in the figure that this same molecule is sometimes called "isobutane." That is not the *proper* name of the molecule. It is a nickname that tells you the molecule is an isomer of butane. You should never use nicknames. You should always name a molecule according to its proper name, which is sometimes called its **IUPAC name**. IUPAC stands for the International Union of Pure and Applied Chemistry. It is the international body that agrees to standards which determine how molecules and atoms are named.

In naming a saturated hydrocarbon, then, you take the following steps:

1. **Determine the longest chain. This gives you the base name of the compound. If the carbons are arranged in a circle, the number of carbons in the circle determines the base name, which is then preceded by a "cyclo" prefix.**

2. **Look for any alkyl groups attached to the longest carbon chain or ring. Include their names at the beginning, using numbers to indicate which carbons they are attached to. Always count carbons so that the numbers are as low as possible.**

Study the following examples so that you see how these rules are applied.

EXAMPLE 12.1

Give the IUPAC name of the following molecule:

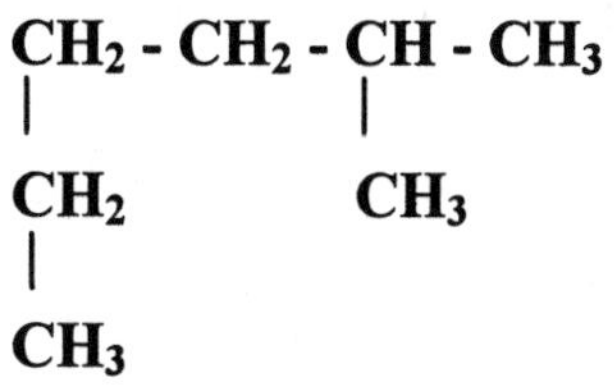

$$
\begin{array}{llll}
CH_2 - & CH_2 - & CH - & CH_3 \\
| & & | & \\
CH_2 & & CH_3 & \\
| & & & \\
CH_3 & & &
\end{array}
$$

This is not a real Lewis structure. Nevertheless, organic chemists often use such notation to abbreviate organic Lewis structures. After all, you know that the hydrogens listed are attached to the carbons with which they are listed, so this "simpler" version of a Lewis structure tells you everything you need to know as far as which atoms are attached to which.

Now you might think that this is a butane with 2 alkyl groups. It is not. Remember, you have to find the *longest* chain of carbons. You do this by dragging a pencil or pen over the carbons, seeing how many you can pass over without lifting your pencil or pen. In this case, here is the longest chain:

```
CH2 - CH2 - CH - CH3
|           |
CH2         CH3
|
CH3
```

There are 6 carbons in this chain, so the molecule is a hexane. On this chain, there is only one alkyl group, and it is a methyl group. Now we just have to figure out a number to indicate which carbon in the chain the methyl group is attached to. If we start counting carbons from the bottom left, the methyl group is on the 5th carbon. The rules tell us to make the number as low as possible, however. If we start counting from the right, then the methyl group is on the second carbon. That keeps the number low. Thus, this is 2-methylhexane.

Name the following molecule with the IUPAC system:

```
                    CH2
                  /     \
CH3 -CH2 - CH             CH - CH2 - CH3
           |              |
           CH2 ---------- CH2
```

This is a cycloalkane. That makes things a little easier because the base name is always given by the number of carbons in the ring when you are dealing with a cycloalkane. Thus, this is a cyclopentane. There are two alkyl groups on this cyclopentane. They are both ethyl groups. What number carbons are they attached to? Well, we need to keep the numbers as low as possible, so we might as well start the numbering at one of the carbons where an ethyl group is attached. Thus, there are two possible ways to number:

```
               5                                            2
                 CH2                                       CH2
           4  /       \  1                             1 /     \ 3
CH3 -CH2 - CH            CH - CH2 - CH3   CH3 -CH2 - CH           CH - CH2 - CH3
           |             |                           |            |
           CH2 -------- CH2                         CH2 -------- CH2
            3            2                          5             4
```

Which is right? We need to keep the numbers low, so the numbering system on the right is the best. Thus, this is 1-ethyl, 3-ethylcyclopentane. When you have more than one of the same alkyl group, you can shorten the notation by adding prefixes of di, tri, etc. Thus, the proper name of this compound is 1,3 diethylcyclopentane.

Alkanes can have groups attached to their carbon chains other than alkyl groups. It is not uncommon, for example, for atoms from group 7A (this group is called the "halides") of the periodic chart to be attached to an alkane's carbon chain. If this happens, the molecule is not technically a saturated hydrocarbon anymore, because it has an atom other than carbon or hydrogen. Nevertheless, the molecule is named using the same method as you use for an alkane. Study the following example to see what I mean.

EXAMPLE 12.2

Give the IUPAC name of the following molecule:

```
CH3 - CH2 - CH - CH2 - CH2 - CH - CH3
             |                |
            CH2               F
             |
            CH - F
             |
            CH3
```

In this molecule, the longest carbon chain is 8 carbons long. Thus, it is an octane. This octane has three substitutions. It has two "F's" attached to it, and there is also an ethyl group. If you start numbering from the bottom left, one "F" is attached to carbon 2, the ethyl group is attached to carbon 4, and the other "F" is on carbon 7. If you number the carbons from the top right, one "F" is on carbon 2, the ethyl group is on carbon 5, and the other "F" is on carbon 7. Thus, the numbers are slightly lower if you count carbons from the bottom right. When you have more than one type of group attached to the carbon chain, you list them in alphabetical order according to the name of the attached group. Thus, the name is 4-ethyl-2,7-difluorooctane. Had there been "Cl's" instead of "F's," we would have used "chloro" instead of "fluoro." Similarly, iodines are signified with "iodo."

ON YOUR OWN

12.5 Name the following molecule with the IUPAC system:

```
CH3 - CBr - CH2 - CH2 - CH2 - CH3
       |
      CH2 - CH2 - CH3
```

12.6 Give the IUPAC name for:

```
              CH2 —— CH2                 CH3
             /          \                 |
CH3 - CH                 CH-CH2 - CH
             \          /                 |
              CH2 —— CH2                 CH3
```

12.7 Give the IUPAC name for the following molecule:

```
        H  CH3
        |  |
CH3 - C - CH - CH2 - CHF - CH3
        |
       CH3
```

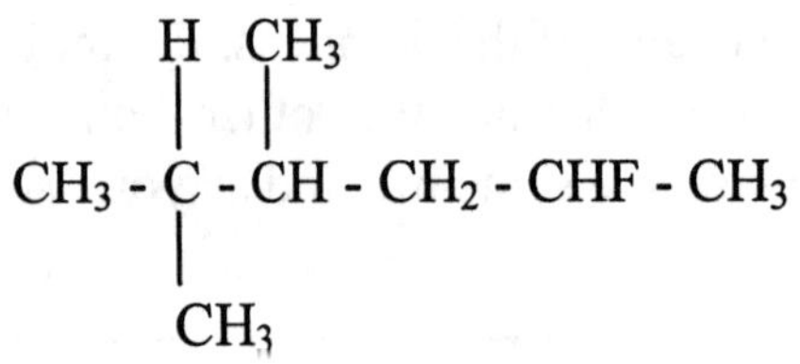

Alkenes and Alkynes

In the previous sections, I have limited my discussions to saturated hydrocarbons. I still want to continue to talk about hydrocarbons, but now I want to discuss **unsaturated hydrocarbons**. As the name implies, these molecules contain only carbon and hydrogen, but they have fewer hydrogens than they could possibly have. What kinds of hydrocarbons have fewer hydrogens than is possible? Well, any hydrocarbon that has a double or triple bond will have fewer hydrogens than is possible.

For example, consider the following molecule:

```
    H        H    H    H
    |        |    |    |
H — C = C — C — C — C — H
        |    |    |    |
        H    H    H    H
```

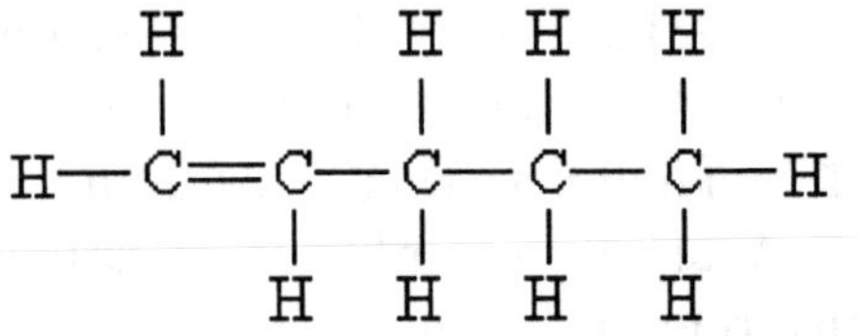

abbreviated as $CH_2 = CH - CH_2 - CH_2 - CH_3$

This hydrocarbon has 5 carbons. If it were a saturated hydrocarbon, its chemical formula would be C_5H_{12}. Instead, it is C_5H_{10}. The double bond takes the place of two hydrogen atoms. Thus, two extra hydrogen atoms could be put on the molecule, if it weren't for that double bond. Thus, we call the double bond a **unit of unsaturation**. In general, hydrocarbons with double bonds in them are called **alkenes**.

Alkene - A hydrocarbon with at least one double bond

As the definition implies, alkenes can have several double bonds. If an alkene has three double bonds, for example, we would say that it has 3 units of unsaturation.

Since an alkene is unsaturated, it is possible to add hydrogen to it. In the presence of an appropriate catalyst (such as platinum), the following reaction can occur:

$$CH_2 = CH - CH_3 + H_2 \xrightarrow{\text{catalyst}} CH_3 - CH_2 - CH_3$$

This kind of reaction is called an **addition reaction**, because hydrogens are being added across the double bond. Notice that the product is a saturated hydrocarbon. Thus, an addition reaction can take an unsaturated hydrocarbon and make it saturated. The word "catalyst" above the arrow in the chemical reaction is merely to remind us that this reaction is so slow that it needs a catalyst to be an effective reaction. Whenever something is necessary for a chemical reaction but is not a reactant or product in the reaction, it is always written above the arrow.

As I mentioned before, if an alkene has more than one double bond, it has more than one unit of unsaturation. There is, however, another way for a hydrocarbon to have more than one unit of unsaturation. It could have a triple bond. Consider, for example, the following molecule:

```
              H    H    H
              |    |    |
H—C≡C—C—C—C—H
              |    |    |
              H    H    H
```

abbreviated as $CH \equiv C - CH_2 - CH_2 - CH_3$

This molecule's chemical formula is C_5H_8. If it were saturated, its formula would be C_5H_{12}. Thus, the triple bond is taking the place of 4 hydrogens. Thus, a triple bond is the same as two units of unsaturation. Hydrocarbons with triple bonds are called **alkynes.**

Alkyne - A hydrocarbon with at least one triple bond

Since alkynes are unsaturated, they also can participate in addition reactions. Because a triple bond is worth two units of unsaturation, however, more hydrogen is needed.

$$CH \equiv C - CH_3 + 2H_2 \xrightarrow{\text{catalyst}} CH_3 - CH_2 - CH_3$$

Alkenes and alkynes are named much like alkanes. You start by finding the longest carbon change to get the prefix of the base name. Instead of ending the base name in "ane," however, you end it in "ene" for alkenes and "yne" for alkynes. Also, you need to indicate which carbon the double or triple bonds are on. You do this with a number. That number determines how you count carbons, because the numbers associated with the double or triple bonds must be the lowest numbers possible. You then keep that numbering system to indicate the groups that are attached to the carbon chain. Study the following examples to see what I mean:

EXAMPLE 12.3

Give the IUPAC name for the following molecule:

```
CH3 - CHI – C= CH2
            |
           CH3
```

This is an alkene, because it has a double bond. Its longest chain is 4 carbons, so it is a butene. We now need to determine how to number the carbon atoms. We need to do this so as to keep the number associated with the double bond as low as possible. Thus, the carbon to the far right will be carbon number 1. This tells us that the base name is 1-butene. The "but" tells us its longest carbon chain is 4, the "ene" tells us there is a double bond, and the "1-" tells us where the double bond is.

Now for the groups attached to the chain. There is a methyl on carbon 2. We must call that carbon 2 because the numbering system is determined by the double bond. There is also an iodine on the third carbon. Since we have more than one type of group, we list them alphabetically. Thus, this molecule is 3-iodo-2-methyl-1-butene.

Name the following molecule with the IUPAC system:

$$\begin{array}{l} CH_3 \\ | \\ CH - C \equiv C - C \equiv CH \\ | \\ CH_3 \end{array}$$

This molecule has two triple bonds, so it is an alkyne. The longest carbon chain is 6 carbons long, so this molecule is a hexyne. There are two triple bonds, so we need two numbers to indicate where those triple bonds are. To keep them as low as possible, we must number from the right. This means the base name is 1,3-hexyne. There is only one group on the chain. It is a methyl group on carbon 5. Thus, this molecule is 5-methyl-1,3-hexyne.

ON YOUR OWN

12.8 An unsaturated hydrocarbon with only 2 carbons is put through an addition reaction. Experimentally, the chemist finds that two moles of hydrogen are used for every one mole of hydrocarbon. Is this an alkene or alkyne?

12.9 Name the following molecules with their IUPAC names.

a. $CH_2 = CH - CH = CH - CH_2Br$

b. $$\begin{array}{l} CH_3 - C \equiv C - CH - CH_3 \\ \qquad\qquad\qquad\;\; | \\ \qquad\qquad\qquad CH_2 - CH_2 - CH_3 \end{array}$$

Aromatic Compounds

Although most hydrocarbons with double bonds fall into the category of alkenes, there is a group of hydrocarbons with double bonds that are so special they must be classified differently. We call these molecules **aromatic** molecules. As their name implies, most aromatic compounds have a strong odor. That is not why they are classified separately from alkenes, however.

Aromatic compounds are special because of the way in which electrons are shared within a molecule.

The most basic aromatic compound is **benzene**, C_6H_6. Two different Lewis structures for benzene are shown below:

FIGURE 12.6
Two Lewis Structures for Benzene

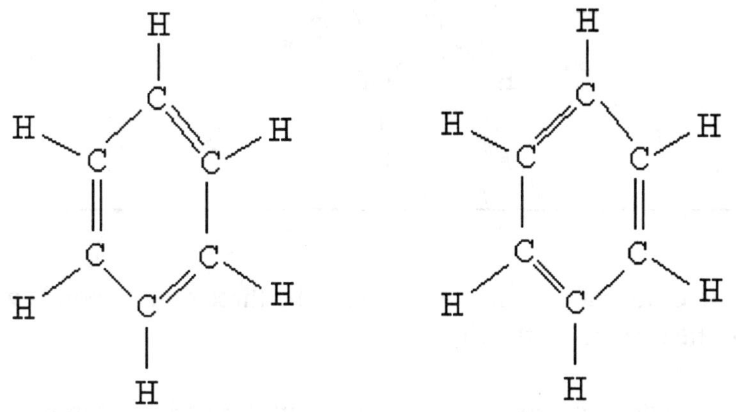

Notice that the only difference between the two Lewis structures in the figure is where the double bonds are placed. Thus, these are not structural isomers. In structural isomers, the actual placement of the atoms in the molecules are different. In these structures, the atoms are all placed exactly the same. The only difference between them is where the double bonds are drawn. As you should have learned in your first-year chemistry course, these kinds of structures are called **resonance structures**.

When a molecule has resonance structures, *both* Lewis structures exist simultaneously. Thus, the double bonds "move around" within the molecule. Sometimes they exist as shown on the right-hand side of the figure. Sometimes, they exist as shown on the left-hand side of the figure. In essence, then, the three double bonds are actually shared by all six carbons in the ring.

You can actually see this fact if you look at the hybridization of the molecular orbitals on each carbon of benzene. Each carbon has sp^2 hybridization. For a given carbon atom, one sp^2 orbital makes the bond with the hydrogen atom. The other two sp^2 orbitals make the single bonds that exist between the carbon of interest and the two carbons close to it. Since the hybridization is sp^2, the carbon has an extra p-orbital. Since each carbon in the molecule is like this, each carbon has an extra p-orbital. This means that each carbon is prepared to have a pi-bond. Thus, the three pi-bonds that make up the three double bonds are shared equally among the 6 carbons. Because of this interesting situation, benzene is often drawn as follows:

FIGURE 12.7
The Way Benzene is Typically Drawn

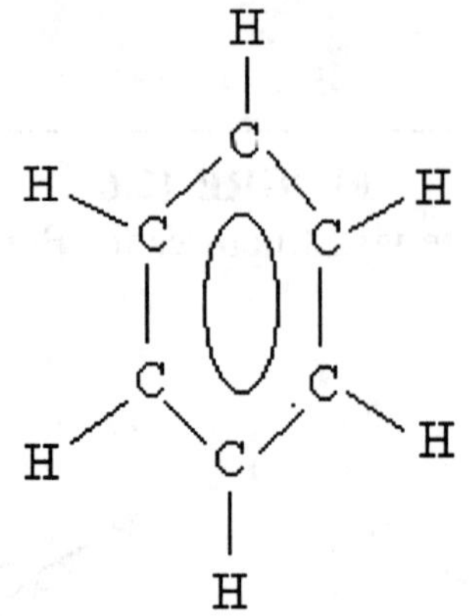

The oval in the middle of the carbon ring symbolizes the three double bonds which are shared equally among all of the carbons in the ring.

Because the three double bonds are shared amongst 6 carbon atoms, chemists say that the double bonds are **delocalized**. This delocalization causes the double bonds to behave differently than "standard" double bonds which are not delocalized. As a result, benzene will not behave like an alkene. It will not, for example, undergo addition reactions. That's why aromatic compounds are given their own classification. Their delocalized electrons cause them to behave so differently that they are a class unto themselves.

Essentially any compound that contains a benzene ring is considered an aromatic compound. Three important aromatics are shown in the figure below:

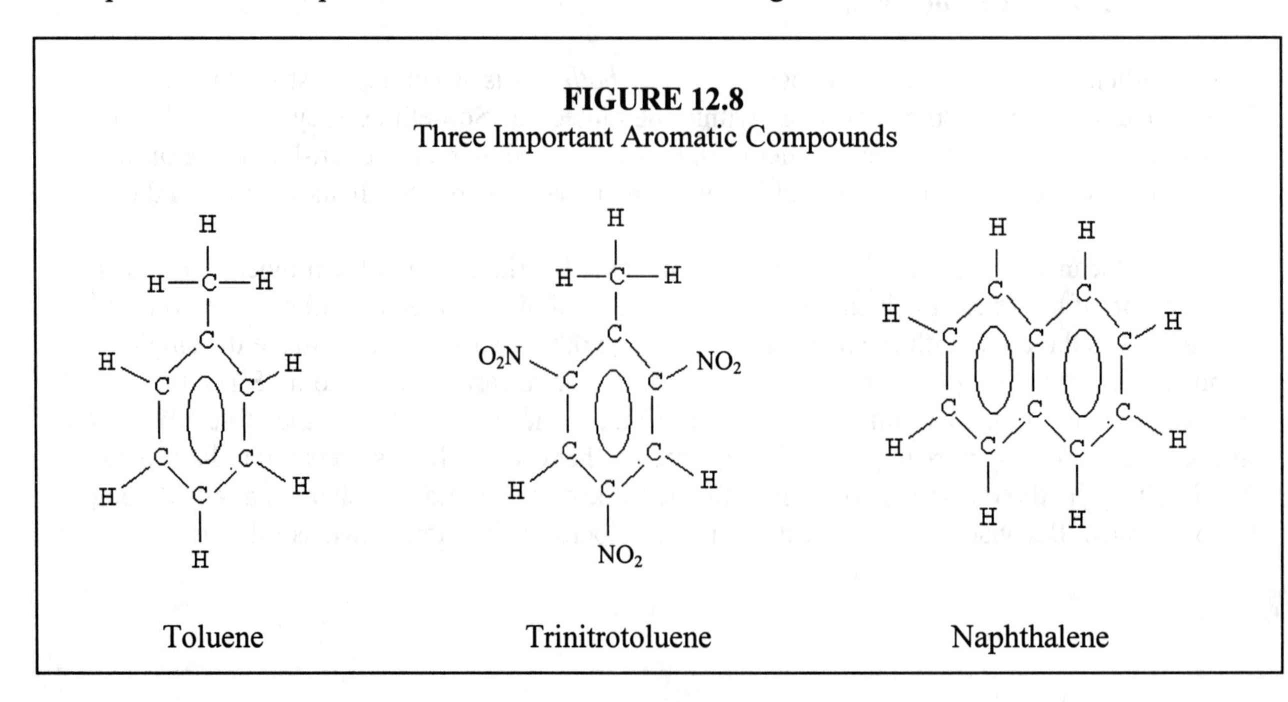

FIGURE 12.8
Three Important Aromatic Compounds

Toluene — Trinitrotoluene — Naphthalene

Toluene is used as a solvent in many industrial processes. A derivative of toluene, trinitrotoluene (TNT) is a powerful explosive that is used in mining processes and war. Finally, naphthalene is the principle component of mothballs. Naphthalene sublimes at room temperature and atmospheric pressure. Thus, when mothballs are put in a closed area, the area fills with naphthalene vapor, which is repellent to most insects.

ON YOUR OWN

12.10 Is the following molecule aromatic? Why or why not?

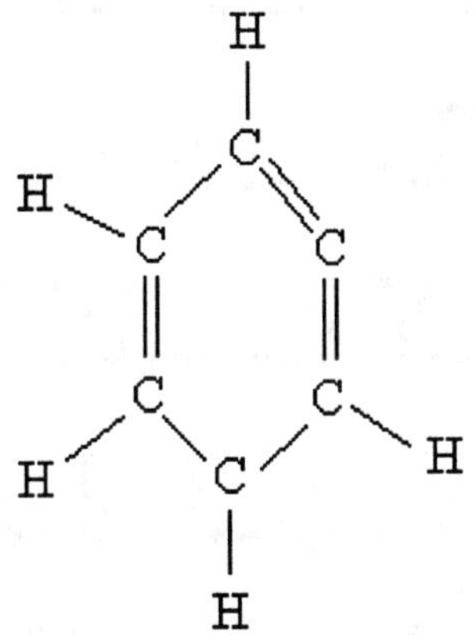

Petroleum

As I mentioned when I was discussing saturated hydrocarbons, the combustibility of alkanes is used heavily in industry. The main source of these alkanes is **crude oil**. Also called petroleum, crude oil is a complex mixture of hydrocarbons. It is *mostly* saturated hydrocarbons, but there are also some alkenes, alkynes, and aromatics in crude oil.

In order to get something useful from crude oil, it must be processed. Typically, processing petroleum involves **fractional distillation**. I have already discussed fractional distillation (Module #5). The fractional distillation of petroleum, however, is worth discussing. The figure below is an idealized schematic of a petroleum distillation facility.

FIGURE 12.9
The Fractional Distillation of Petroleum

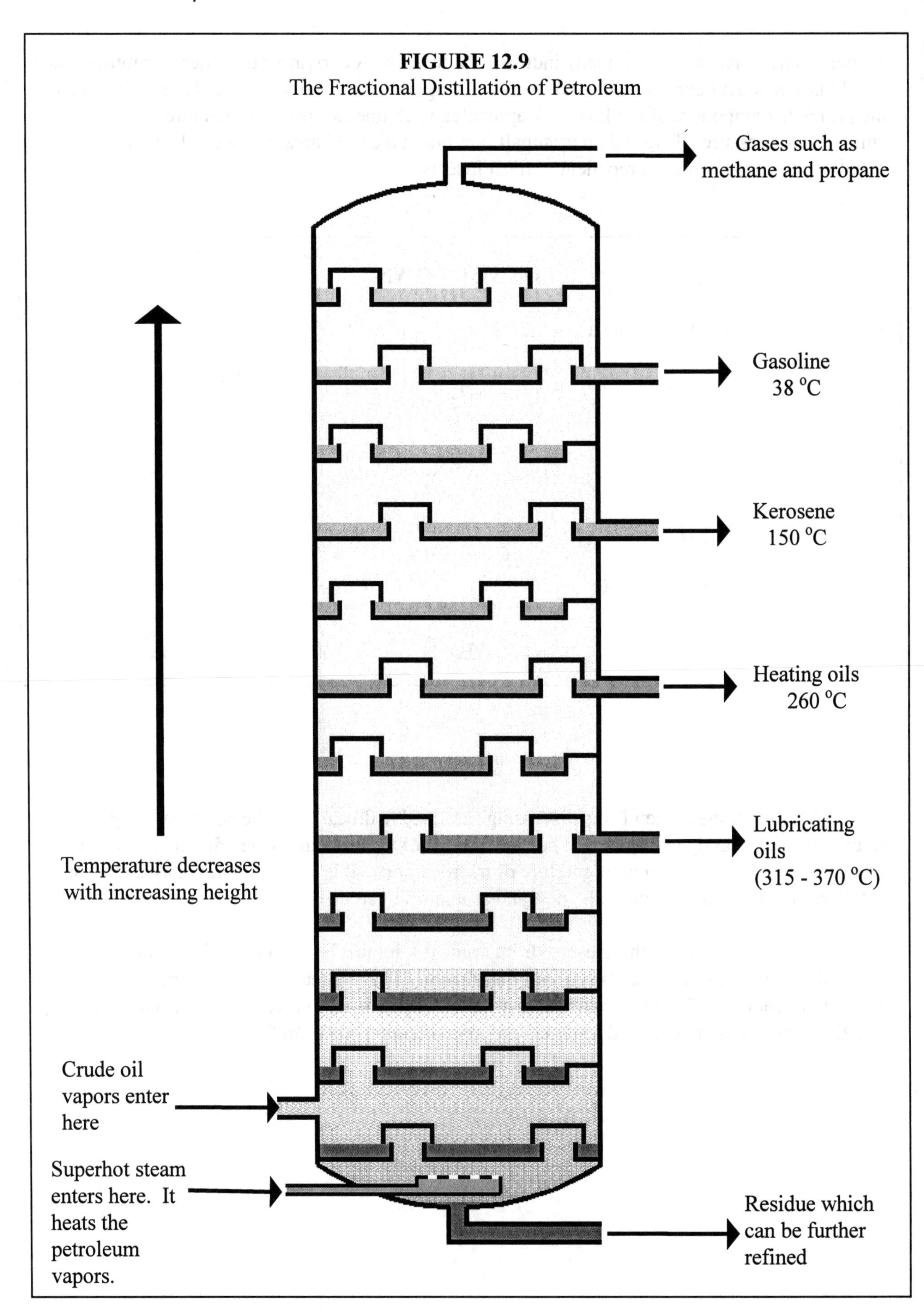

When crude oil is processed, it is first heated and pushed into the fractional distillation column. Solids drop to the bottom of the column and are pulled off as residue. The residue is used to make petroleum jellies such as Vaseline®, paraffin, asphalt, and coke (**not** Coca-Cola®). As the crude oil vapors rise, the temperature decreases. Thus, the vapors will condense according to their mass. The heavier alkanes will condense at higher temperatures, so they can be pulled off at the lower levels of the column. The higher you pull off alkanes, the lower their mass. At the very top, the gaseous alkanes are pulled off.

In the end, this fractional distillation process turns crude oil into many useful products. The lightest liquid alkanes (typically those alkanes with 5-13 carbons) make up gasoline. Slightly heavier alkanes (10 carbons to 16 carbons) make up kerosene and other heating oils, and liquid alkanes with more than 16 carbons make up lubricating oils.

The mixture of alkanes which forms gasoline is rated on a scale known as the **octane rating**. This scale is based on many experiments which indicated that the most efficient gasoline would be one which was 100 % 2,2,4-trimethylpentane (an isomer of octane). Thus, gasoline that has an octane rating of 88 burns with 88% the efficiency of pure 2,2,4-trimethylpentane. It was once thought that nothing could burn more efficiently than 2,2,4-trimethylpentane. Thus, it was assumed that no gasoline could ever have a better octane rating than 100. As time went on, however, chemists were able to synthesize certain compounds that can burn better than 2,2,4-trimethylpentane. As a result, we now have gasolines (usually used by race car drivers) with octane ratings higher than 100.

Polymers

In the 1920's chemists ushered in a brand new age of chemistry when they discovered **polymers**. These chemicals have huge molecular masses and, as a result, unique properties. **Plastics**, such as polyethylene and polystyrene, are polymers which are made from petroleum. Rubber is made up of polymers. Synthetic fibers such as nylon, rayon, and Dacron® are also made of polymers.

A polymer is made up of many smaller molecules called **monomers**. When monomers link up over and over again to form long, long chains, the result is a polymer. For example, under the right conditions, ethylene will react with another ethylene molecule according to the following reaction:

$$CH_2 = CH_2 + CH_2 = CH_2 \rightarrow CH_3 - CH_2 - CH = CH_2$$

If the product of this reaction were to react with another ethylene molecule, the result would be:

$$CH_3 - CH_2 - CH = CH_2 + CH_2 = CH_2 \rightarrow CH_3 - CH_2 - CH_2 - CH_2 - CH = CH_2$$

If this would continue, the resulting molecule would get longer and longer. Eventually, it would become so long that it would be considered a polymer. Since this particular polymer is made by adding ethylenes together, it is called **polyethylene**. In general, these reactions are abbreviated as:

$$n\ CH_2 = CH_2 \rightarrow (CH_2 - CH_2)_n \qquad \text{where n is large} \qquad (12.1)$$

Polyethylene is a popular polymer used in the manufacture of many household items. One such household item is the plastic trash bag. Investigate the properties of a polymer by examining a plastic trash bag.

EXPERIMENT 12.1
Investigating the Properties of Polyethylene

Supplies
- Plastic trash bag (Try to get a white one.)
- Scissors

Cut two 10 cm by 10 cm squares out of the trash bag. Hold one of those squares up to a light, making sure it is stretched tightly. Look through the plastic while the light is behind it. You should see fine parallel lines in the plastic. Those are actually the polymer fibers that make up polyethylene. Orient the square so that the fibers you see run horizontally. Next, grasp both sides of the square, one with each hand, and pull hard. Note how far you can stretch the plastic square before it breaks. Finally, take the other square and orient it so that the fibers run vertically up and down the square. Once again, grasp the square with one hand on each side, and once again pull. Note once again how far you can stretch the plastic before it breaks.

You should have noticed that the polyethylene stretched farther when the polyethylene fibers were oriented vertically as opposed to horizontally. That's because the fibers are already stretched pretty tightly when the polyethylene is made. Thus, when you pull along the fibers, they do not have as far to stretch. Also, the attractive forces *between* polymer molecules are large, because the polymers are so long and lie right next to each other. Thus, the plastic stretches because the polyethylene fibers move in order to stay close to one another.

The length of polymer molecules is what gives them such unique properties. Make your own polymer to see what I mean.

EXPERIMENT 12.2
Making Slime

Supplies:

From the laboratory equipment set:
- Polyvinyl alcohol adhesive
- Borax

- Test tube
- funnel

From around the house:

- Measuring tablespoon
- Measuring ¼ teaspoon
- Two plastic or Styrofoam cups
- Food coloring (optional)
- Spoons for stirring
- Water

Let the tap water run for a few minutes so that it gets really warm. Add 2 tablespoons of that warm tap water to the cup. Add ¼ of a teaspoon of borax to the warm water. Stir until all of the borax is dissolved. That may take a while. Next, use the funnel to pour some of the warm borax solution into the test tube. Pour enough solution so that the test tube is filled to the fourth mark. This is 2.5 mL. If you want, you can add a few drops of food coloring to the test tube to give your slime some color.

Next, pour the entire contents of the polyvinyl alcohol adhesive bottle into the *other* cup (the one that's still empty). Finally, pour the borax solution in the test tube into the polyvinyl alcohol adhesive that is in the cup. Start stirring immediately. The slime will form quickly. Note its properties.

What gave the slime that you made its properties? Well, polyvinyl alcohol adhesive is a polymer, but it is not very long. When you added the borax to the polyvinyl alcohol adhesive, the polyvinyl alcohol adhesive molecules began to start liking to each other, forming a network of molecules that resembles a solid. However, as the network was forming, water became trapped inside it. As a result, the product of this reaction has some of the properties of a solid (from the polymer) as well as some of the properties of water (because water is trapped in the network of the polymer).

For example, slime can hold its shape for a moment or two. Well, solids can hold their shape, but liquids cannot. Since slime can hold its shape for a limited amount of time, it acts as something like a "cross" between a solid and a liquid. Slime tends to ooze around, so in some ways, you can say that it "pours." Well, a liquid pours, but a solid does not. Slime, then, acts like something in between: It pours, but it does not pour quickly.

The process in which polymer molecules link together to make bigger molecules is called **cross-linking**, and it is illustrated in Figure 12.10.

FIGURE 12.10
Cross Linking

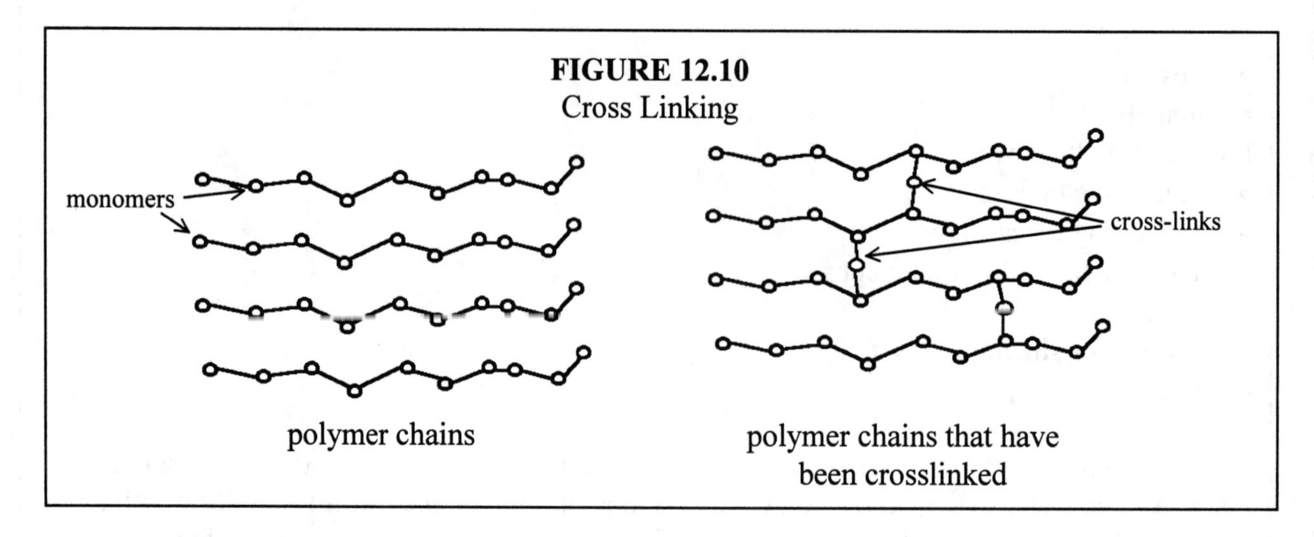

Because cross-linking keeps polymer molecules together, it strengthens the polymer but allows for flexibility. See what I mean by doing your own cross-linking.

EXPERIMENT 12.3
Crosslinking a Polymer

Supplies:
From the laboratory equipment set:
- Borax
- Chemical scoop
- Test tube
- Test tube cap

From around the house:
- White school glue (like Elmers® glue)
- Plastic or Styrofoam cup
- Old spoon

Add four measures of borax to the test tube and fill it halfway with water. Cap and shake the test tube until the borax is dissolved. Next, put about 1/2 of a cup (no need to measure) of glue into the cup and add the borax solution. Use your old spoon to stir. Note the change that takes place. As the polymer you are making takes on a rubbery feel, pull it out of the cup and begin rolling it in your hand. Make a ball with it. See how the ball behaves by bouncing it.

In the experiment, you crossed-linked polymers in glue. Glue is made from the polymer polyvinyl alcohol. When glue "dries," it hardens because the polymers get very close together and bond to one another. When you added Borax to the glue, the Borax catalyzed a reaction in which one atom in a polyvinyl alcohol molecule bonded to another atom in another polyvinyl alcohol molecule. This resulted in a cross-linked polymer, which was strong yet flexible. That's why the glue took on a rubbery quality once you cross-linked it.

Another way to change the properties of a polymer, of course, would be to change its monomers. For example, polyvinyl chloride (PVC) is a hard plastic that is used to make plumbing pipes and building materials. It is made up of 1-chloro ethylene ($CH_2 = CHCl$) monomers. Notice that this is very close to ethylene, which is the monomer of polyethylene, a polymer used to make household trash bags. Note that a small change in the monomer can result in a major change in the nature of the polymer.

The chemical discovery of plastics revolutionized industry. Plastics can be made very inexpensively. They are typically strong and durable. Thus, the use of plastics has skyrocketed since their introduction into society. This has resulted in making most things that we buy less expensive and more durable. Recently, however, there has been a backlash against plastics because they do not biodegrade. Thus, plastics that end up in a landfill will be there until the Lord comes. To some environmentalists, this is a terrible thing. These environmentalists, therefore, try to reduce plastic use by encouraging people to use paper instead of plastic.

Although these initiatives sound good, one must look at the entire picture in order to understand the environmental consequence of any action. For example, one reason polymers are so inexpensive is that they take little energy to produce. Paper, on the other hand, takes an enormous amount of energy to make. Do you know what the single, largest source of pollution is? The production of energy. The production of electricity in particular causes an enormous amount of air pollution. The major source of most air pollutants is the electrical power plant. Since air pollution causes acid rain, electricity is therefore a major contributor to acid rain as well.

Several chemists have demonstrated that paper takes so much energy to produce that it results in as much as three times the pollution that plastic does. Thus, is it more "friendly" to our environment to use paper or plastic? Well, if you use paper, you will reduce the amount of garbage in the landfills, but you will pollute the air three times as much, which will also lead to more acid rain. Which is worse, air pollution (and the acid rain that results) or full landfills? That's the question one must ask in order to determine whether to use paper or plastic, and that's just the question which is **not** asked by most environmentalists.

ANSWERS TO THE ON YOUR OWN PROBLEMS

12.1 Organic compounds must have carbon, but cannot have a metal. Thus, $CaCO_3$ and $NaHCO_3$ are not organic because they have metals in them. H_2 and O_2 are not organic because they have no carbon. As a result, CO_2, CH_4, and $C_6H_{12}O_6$ are organic.

12.2 A liquid alkane will not mix with water homogeneously because alkanes are purely covalent. Water will only dissolve ionic molecules and polar covalent molecules.

12.3 Alkanes have the general formula C_nH_{2n+2}, while cycloalkanes have the generalized formula C_nH_{2n}. Thus, the alkane with 7 carbons is C_7H_{16}, and the cycloalkane with 7 carbons is C_7H_{14}.

12.4 Structures (a) and (c) are identical. Starting with (c), for example, I can make (a):

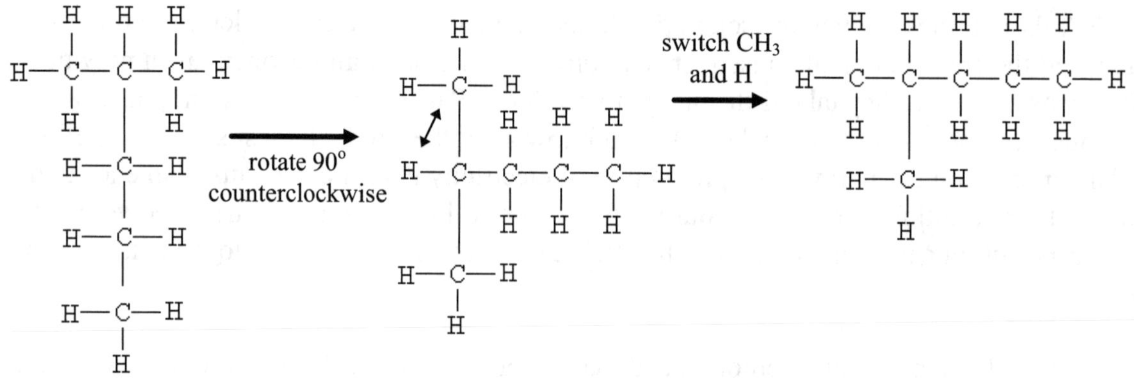

rotate 90° counterclockwise

switch CH_3 and H

This structure is an isomer with (d). Despite the fact that the chemical formulas are the same, the one Lewis structure cannot be turned into the other. Structure (b) is not an isomer because it has a completely different chemical formula as compared to (a) and (d).

12.5 The longest chain that has no branches is 8 carbons long. Thus, this is an octane. The fourth carbon (if you count the other way it is fifth, but you need to count the fewest carbons) has a methyl group on it and a bromine. Thus, this is 4-bromo-4-methyloctane.

12.6 This is a cycloalkane, so we just count the carbons in the circle to determine the base name. There are 6 carbons on the ring, so this is a cyclohexane. One carbon has a methyl group and the other has an isobutyl group on it. Since we list the substitutions in alphabetical order, we might as well give the number 1 to the carbon with the isobutyl group. Thus, the name is 1-isobutyl-4-methylcyclohexane. If you had 1-methyl-4-isobutylcyclohexane, that's fine.

12.7 This molecule has a 6-carbon chain. Thus, this is a hexane. There are three substitutions: two methyls and one fluorine. Since we want the numbers to be as low as possible, we need to give the carbon with the first methyl group the number 2. Thus, this is 2,3 dimethyl-5-fluorohexane.

12.8 The hydrocarbon has only 2 carbons. Thus, there is either a single bond, double bond, or triple bond between them. Since it can be used in an addition reaction, it must have at least one unit of unsaturation. Thus, it must have either a double bond or a triple bond. Since *two* moles of hydrogen can be added to one mole of the hydrocarbon, that means 4 H atoms were added (remember, one mole of hydrogen gas is H_2). Thus, there must have been a triple bond, because for a double bond, only one mole of hydrogen gas could be added per mole of saturated hydrocarbon. Thus, this is an alkyne.

12.9 a. This hydrocarbon has 5 carbons. It also has double bonds. Thus, this is a pentene. We determine the numbers for the carbons by making the numbers on the carbons with the double bonds as small as possible. Thus, this is a 1,3-pentene. Finally, there is a bromine substitution at the end. Thus, this is a 5-bromo-1,3-pentene.

b. This hydrocarbon has a triple bond, and its longest chain without a branch is 7 carbons. Thus, this is a heptyne. There is a methyl group on the fourth carbon. We know it is the fourth carbon because we must count the carbons so that the number of the carbon with the triple bond is as low as possible. Thus, this is 4-methyl-2-heptyne.

12.10 This is not an aromatic, because the double bonds cannot be shared between all carbons.

REVIEW QUESTIONS

1. Which of the following molecules are organic?

$MgCO_3$, CH_2O, F_2, $C_6H_{10}O_6Br_2$, H_2O

2. A chemist studies two hydrocarbons. The first reacts readily with hydrogen in the presence of a catalyst to form a new hydrocarbon. The second does not react with hydrogen at all. Which is the saturated hydrocarbon?

3. For each of the following alkanes, indicate whether it is a solid, liquid, or gas at room temperature and pressure: C_3H_8, C_7H_{16}, $C_{25}H_{52}$.

4. An alkane has 22 hydrogens. What is its chemical formula?

5. A cycloalkane has 22 hydrogens. What is its chemical formula?

6. One alkane has 6 carbons and another has 12. Which has the most structural isomers?

7. An alkene reacts with hydrogen so that 6 moles of H_2 are used up for every one mole of alkene. How many units of unsaturation does the alkene have?

8. Why are aromatic compounds not considered alkenes, despite the fact that they are hydrocarbons with double bonds in them?

9. What is petroleum? How do we get gasoline from it? What other useful products come from it?

10. What has a higher molecular mass, a monomer or a polymer?

PRACTICE PROBLEMS

1. Draw the Lewis structure of 2,3-dimethyl-5-fluoroheptane.

2. Name the following compound with the IUPAC system:

```
     H   H   H   H
     |   |   |   |
 H — C — C — C — C — H
     |   |   |   |
     H   H   |   H
             |
         H — C — Br
             |
         H — C — H
             |
             H
```

3. Name the following compound with the IUPAC system:

```
         H
         |
     H — C — H
 H       |   H   H   H   H   H
 |       |   |   |   |   |   |
 C ————— C — C — C — C — C — C — H
H—C — C — C — C — C — C — C — H
 |   |   |   |   |   |   |
 H   H   |   H   H   H   H
             |   H   H   H
             |   |   |   |
         H — C — C — C — C — H
             |   |   |   |
             |   H   H   H
             |
         H — C — H
             |
             H
```

4. Name the following compound with the IUPAC system:

```
     H   H   H       H   H
     |   |   |       |   |
 H — C — C — C ————— C — C — H
     |   |   |       |   |
     H   H   |       |   H
         H — C ————— C — H
             |       |
             H       H
```

5. There are 5 structural isomers of hexane. Draw them.

6. 2-methyl-1-hexene is reacted with hydrogen in the presence of a catalyst. What is the name of the product?

7. One mole of 5-ethyl-1-heptyne reactant is completely reacted with hydrogen in the presence of a catalyst. How many moles of H_2 will be used?

8. Name the following compound:

```
        CH3
         |
  CH3 - C = C - CH3
             |
             CH2 - CH2 - CH2 - CH3
```

9. Draw the Lewis structure for 3-ethyl-2-nonene.

10. Which of the following could be considered aromatic?

```
a.
            H   Br
            |   |
   H        C — C
   |       /|   |\
H— C — C   H   H  C — H
   |       \\       //
   H         C == C

b.
                 H   Br
                 |   |
      H          C == C
      |         /      \
   H— C — C             C — H
      |      \\        //
      H        C —— C
               |    |
               H    H

c.
                   C ≡ C
      H           /      \
      |          /        \
   H— C — C                C — H
      |      \\          //
      H        C —— C
               |    |
               H    H
```

Module #13: Functional Groups in Organic Chemistry

Introduction

In the previous module, you were given an introduction into the fascinating field of organic chemistry. In this module, I want to continue the introduction by telling you about **functional groups** in organic molecules. As you already know, all organic molecules contain carbon atoms. In fact, the vast majority of organic molecules have carbon chains or rings like those discussed in the previous module. Sometimes these chains or rings are saturated, sometimes they are not saturated. Nevertheless, a carbon chain can be thought of as the "basic" organic molecule.

Most of the other organic molecules can then be thought of as derivatives of this "basic" organic molecule. Consider, for example, a simple ethane molecule:

```
    H   H
    |   |
H—C—C—H
    |   |
    H   H
```

Suppose I replace the last hydrogen atom with an OH:

```
    H   H
    |   |   ..
H—C—C—O—H
    |   |   ..
    H   H
```

The result is a molecule of ethanol, which is the intoxicating ingredient in alcoholic beverages. Now remember, ethane is a gas (all alkanes of 4 or fewer carbons are gases), but ethanol is a liquid. Ethane is poisonous if large quantities are inhaled. Ethanol is not poisonous unless huge quantities are consumed. It does, however, have other negative effects on the body.

The point to this discussion is that the only difference between a molecule of ethane and a molecule of ethanol is that ethanol has an "OH" where ethane has an "H." Ethanol, therefore, can be considered a derivative of ethane. The derivative has enormously different chemistry than the basic molecule, and that is due entirely to the fact that it has an "OH" where the basic molecule has an "H." Thus, the "OH" is called a functional group, because it changes the chemical function of the basic carbon chain.

Functional groups are so important in organic chemistry that chemists often abbreviate the carbon chain so as to highlight just the functional group. For example, the Lewis structure for ethanol shown above can be abbreviated to:

$$R\text{-}CH_2\text{-}OH$$

In this abbreviation, the R represents both carbon atoms and all of the hydrogens attached to them. Organic chemists employ this abbreviation scheme quite frequently, because the carbon chain that composes an organic molecule is not nearly as important (from a chemical point of view) as the functional group that is attached to the chain. This abbreviation scheme emphasizes this fact, by lumping the entire carbon chain into the R, and leaving only the functional group in detail. As you go through this module, you will get more and more comfortable with this abbreviation scheme.

Alcohols

Since I used the "OH" functional group as an example, I might as well start out by describing alcohols. The "OH" functional group is called the **hydroxyl** functional group. You have already seen this functional group before, but not in the context of organic chemistry. If a molecule has both a metal and a hydroxyl group, it is considered a base. For example, NaOH, $Al(OH)_3$, and $Mg(OH)_2$ are all bases. Organic molecules with the hydroxyl group do not act as bases, however. Remember, an organic molecule cannot contain a metal atom. Thus, none of the bases like NaOH are organic molecules, because they contain a metal atom.

When an organic molecule has a hydroxyl group, it is called an **alcohol**. There are probably three alcohols with which you are at least somewhat familiar: **methanol**, **ethanol**, and **2-propanol**. Methanol, which is often called "methyl alcohol" or "wood alcohol," has the formula CH_3OH. It is manufactured commercially from carbon monoxide or carbon dioxide:

$$CO + 2H_2 \rightarrow CH_3OH \tag{13.1}$$

$$CO_2 + 3H_2 \rightarrow CH_3OH + H_2O \tag{13.2}$$

Methanol is a colorless liquid that smells and tastes just like ethanol. It also has the same intoxicating effects as does ethanol. Unfortunately, its effects on the human body are quite severe. If you drink methanol (or even breathe too many fumes), it can cause blindness or even death.

Methanol is used commercially for many things. It is often the starting component for the formation of many other organic compounds. For example, formaldehyde (the smelly stuff that biologists use to preserve specimens) is made from methanol. Methanol is also used directly as a solvent for many hard-to-dissolve compounds. Methanol is also used to **denature** ethanol.

Denature - To poison a product so that it is not consumable

Please realize that this is not the only definition of denature. It is the definition that applies to ethanol and other consumable products, however.

Why poison ethanol? Well, ethanol is used in a lot of commercial applications, as I will discuss in a moment. However, ethanol is also the intoxicating component in liquor. In order to keep people from taking commercially-prepared alcohol and adding it to beverages so as to make them intoxicating, industry adds poisonous compounds to the ethanol they produce. These compounds are specially chosen so that they are hazardous to humans but they do not harm the usefulness of the ethanol. The most popular denaturing agent for ethanol is methanol, because methanol is so chemically similar to ethanol that it will not harm the usefulness of the ethanol in industrial applications. It will, however, harm anyone who tries to drink it!

Why do companies that produce alcohol do this? They are clearly not doing it to protect people. After all, people will go blind or die if they drink denatured ethanol! They do it because the federal government requires them to do it. Why does the government require them to do it? Well, the government makes an enormous amount of money from the taxes that it levies against alcoholic beverages. It does not want to tax industrially-used ethanol, however, because that would increase industrial costs, making U.S. products less competitive in the world market. At the same time, however, it does not want to lose any tax revenue from people taking tax-free ethanol and making their own intoxicating drinks. Thus, in order to be sure that no one gets tax-free alcoholic beverages, the Federal government requires industry to poison the ethanol they produce.

Since I am on the subject of ethanol, CH_3CH_2OH, I might as well discuss it a bit. Ethanol, also called "ethyl alcohol," "grain alcohol," or just plain "alcohol," is the most popular alcohol produced in the world. In the alcoholic beverage industry, it has long been produced by the **fermentation** process. In this process, starch, cellulose, and various sugars are consumed by yeast. The result is alcohol and carbon dioxide. For example, when glucose is exposed to yeast, the following reaction occurs:

$$C_6H_{12}O_6 \xrightarrow{\text{yeast}} 2CH_3CH_2OH + 2CO_2 \qquad (13.3)$$

In wine making, for example, yeast is added to crushed grapes. The yeast begins converting the sugars in the grapes into ethanol. Observe the fermentation process yourself by performing the following experiment.

Experiment 13.1

Yeast and the Fermentation Process

Supplies:

- Packet of active dry yeast (can be purchased at a grocery store)
- Warm water
- Tablespoon
- Measuring cup
- Glass that holds at least 1 cup of water
- Sugar

Mix one packet of yeast with one cup of warm water in the glass. Add a tablespoon of sugar. Stir gently. Let the mixture stand at least five minutes. Active dry yeast contains the living spores of baker's yeast for use in home baking. Adding them to water and sugar causes them to begin growing.

As the mixture stands, you should observe bubbles beginning to form. Most likely, the bubbles formed will be very small. The best way to observe them is to watch the top of the mixture through the side of the glass. A layer of foam caused by the bubbles will appear and grow thicker as time goes on. The bubbles are caused by the carbon dioxide produced in the fermentation process, as shown in Equation (13.3). If you sniff the contents of the glass after a while, you should note a sharp smell that is caused by the alcohol produced.

Interestingly enough, yeast cannot survive high concentrations of alcohol. Thus, as they continue the fermentation process, the increasing amount of alcohol actually ends up killing them. Wild forms of yeast can stand only a 4% level of alcohol before they begin to die off. Wineries and breweries, however, have bred strains of yeast that can survive levels of up to 12%. A mixture of wild yeast and specially-bred yeast are used in the making of most alcoholic beverages. Alcoholic drinks (such as whiskey) that have levels of alcohol much greater than 12% are made by taking a drink with 12% alcohol and distilling it. When a mixture of alcohol and water is distilled, the alcohol tends to boil off first. If you collect those vapors and condense them, the result is a solution with a higher concentration of alcohol.

Although the fermentation process has been used to produce alcohol for thousands of years, it is not the process used by the chemical industry. Industry uses so much ethanol that it must be produced faster and more efficiently than fermentation. Typically, the chemical industry adds water to ethene in order to make large quantities of ethanol.

$$H_2C = CH_2 + H_2O \xrightarrow{\text{acid}} CH_3CH_2OH \tag{13.4}$$

Ethanol is one of industries most popular solvents. It is used heavily in the lacquer industry. Also, it is the principal solvent in most tinctures or extracts. If you buy vanilla extract for baking, for example, the solvent is almost certain to be ethanol. Since extracts must be eaten, the ethanol used in extracts is not denatured. Ethanol is also used as starting material in the manufacture of ether, dyes, and perfumes. Finally, it is even used to a limited extent as a motor fuel.

The last alcohol with which you have at least some experience is **2-propanol**:

$$\begin{array}{c} CH_3CHCH_3 \\ | \\ OH \end{array}$$

Although you have almost certainly used this alcohol, you probably do not recognize its proper name. That's because it is usually called "rubbing alcohol" or "isopropyl alcohol." It is used in the home primarily as a disinfectant for minor cuts and scrapes.

The chemistry of an alcohol depends, to some extent, on where the hydroxyl group is placed on the molecule. If the hydroxyl group is at the very end of the carbon chain, the carbon that has the hydroxyl group bonded to it will be bonded to only one other carbon. In this case, the alcohol is called a **primary alcohol**. If the hydroxyl group is bonded to a carbon that has two other carbons bonded to it, the alcohol is a **secondary alcohol**. Finally, if the hydroxyl group is bonded to a carbon that is bonded to three other carbons, it is called a **tertiary alcohol**. These three types of alcohols are shown in Figure 13.1.

FIGURE 13.1
Primary, Secondary, and Tertiary Alcohols

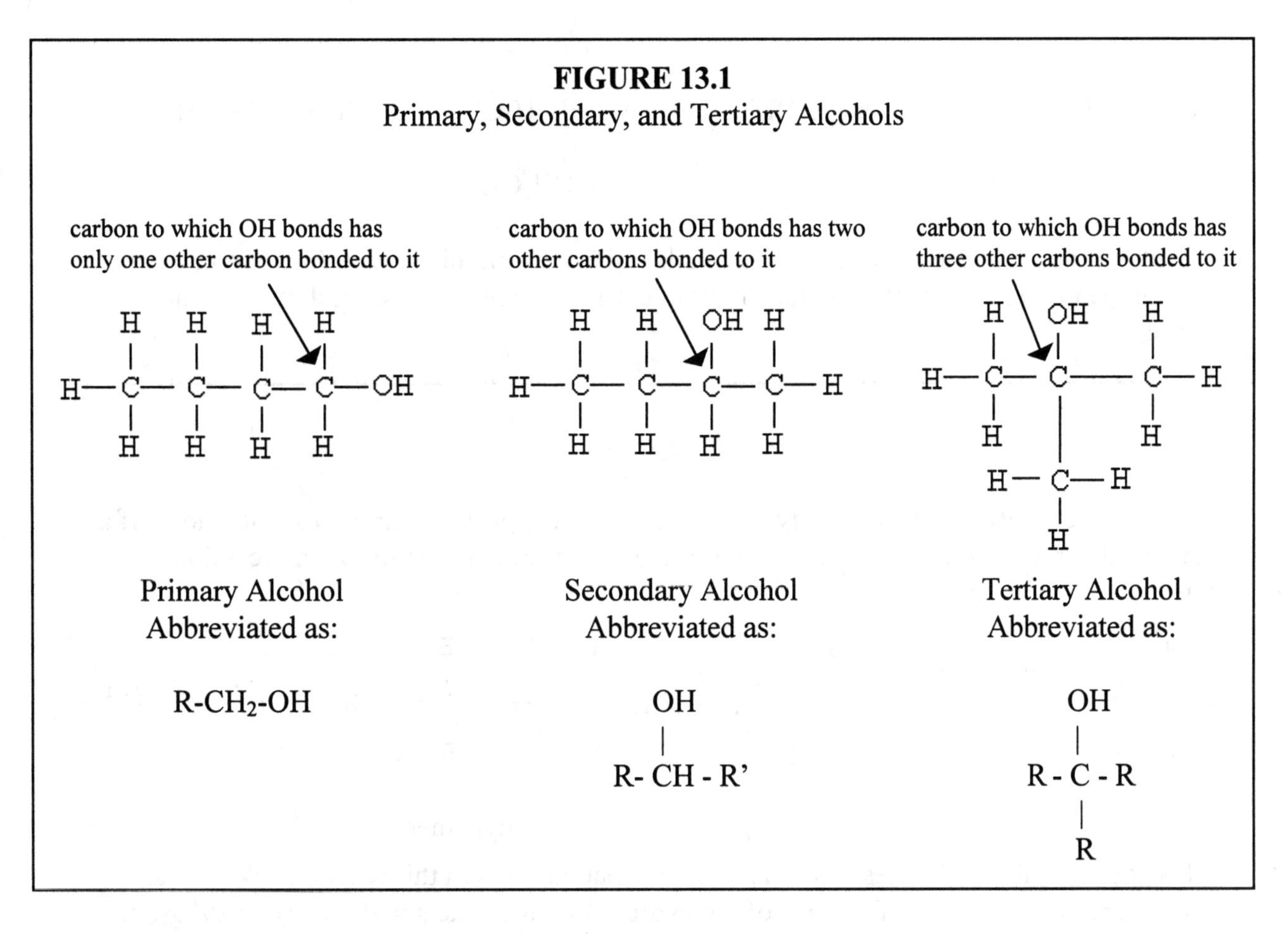

In general, primary alcohols are more reactive than secondary alcohols, which are more reactive than tertiary alcohols.

Notice the abbreviations in the figure. For the primary alcohol, the R represents the entire carbon chain. In the abbreviation for the secondary alcohol, there are two R's. The first is given by R and the second is given by R'. Why did I do that? Well, look at the Lewis structure. The hydroxyl group is on a carbon that is not at the end of the chain. To emphasize the hydroxyl group, then, I must abbreviate the carbon chain. The problem is, I can't use an R to do that. On the left side of the carbon that contains the hydroxyl group, there is a CH_3CH_2 carbon chain. On the right side, there is just a CH_3 attached. Thus, I used the R to represent the CH_3CH_2 and the R' to represent CH_3. In the tertiary alcohol, I just have 3 R's. I can do that because the same thing (CH_3) is attached three times to the carbon that contains the hydroxyl group. Thus, each of those R's represents a CH_3. It is important for you to get comfortable with this abbreviation scheme. As you will see in the next section, it is rather important.

ON YOUR OWN

13.1 Identify each of the following as an alcohol or not an alcohol

```
                                                          OH
                                                          |
a. CH3CH2CH2CH2OH      b. Sr(OH)2      c. CH3CCH2CH3      d. CH3CH2COCH3
                                                          |
                                                          CH2CH3
```

13.2 Identify each of the three specific alcohols I discussed in this section (methanol, ethanol, and 2-propanol), as well as the alcohols in problem 13.1, as primary, secondary, or tertiary alcohols.

Ethers

I want to discuss ethers briefly because ethers are typically formed from alcohols. If a strong acid is mixed with ethanol and the mixture is then heated, the following reaction will occur:

```
   H  H                   H  H                H  H      H  H
   |  |    ..      ..     |  |     H2SO4      |  |  ..  |  |              ..
H—C—C—O—H   +   H—O—C—C—H    →    H—C—C—O—C—C—H  +  H—O—H     (13.5)
   |  |    ..      ..     |  |       Δ        |  |  ..  |  |              ..
   H  H                   H  H                H  H      H  H

                                                  diethyl ether
```

In this equation, the "Δ" represents heat. Notice what happens in this reaction. When the alcohols are heated, the "H" from one of the hydroxyl groups reacts with the hydroxyl group of another alcohol to make water. This leaves an oxygen on one alcohol and a carbon on the other alcohol that each need something else to bond with. They end up bonding with each other, and the result is an ether.

Notice what makes this molecule an ether. The functional group for an ether is an oxygen atom bonded to two carbons. In the typical abbreviation scheme used by organic chemists, then, an ether can be abbreviated as R-O-R'. The R represents any carbon chain attached to one side of the oxygen, while the R' represents any carbon chain attached to the other side of the oxygen atom.

The kind of reaction given by Equation (13.5) is called a **dehydration reaction,** because a water molecule is extracted from the reactants. From a chemical point of view, then, you can say that the reactants are dehydrated. Dehydration reactions are very common in organic chemistry, so you will see several of them in this module.

I want to use the reaction given by Equation (13.5) to introduce one of the most important aspects of organic chemistry reactions: the fact that functional groups govern most of what happens in an organic chemistry reaction. Notice in the equation that the carbon chains which are attached to the hydroxyl groups really don't do anything in the reaction. They are just "along for the ride." It's the hydroxyl groups that do all of the work. The hydroxyl groups react to form the water, and the oxygen left over from that reaction then goes and bonds to the carbon chain left over from the other alcohol.

Because it is really the functional groups which govern how organic chemicals react, equations like Equation (13.5) can be generalized as follows:

$$\text{R-}\boxed{\text{OH + HO}}\text{-R'} \xrightarrow[\Delta]{H_2SO_4} \text{R-O-R'} + H_2O \quad (13.6)$$

What does this generalized reaction tell us? Remember that R's represent any carbon chain. The fact that there is an R on the first alcohol and an R' on the second alcohol really just means that the two alcohols need not be the same. One can have a certain carbon chain, while the other can have a completely different carbon chain. The reaction tells us, then, that any two primary alcohols can react in this way to form any ether. Equation (13.5) is simply a specific case of this equation, where both R and R' are equal to CH_3CH_2. Study the following example to see what all of this means.

EXAMPLE 13.1

Methanol (CH_3OH) and 1-propanol ($CH_3CH_2CH_2OH$) are mixed with sulfuric acid and heated. What product will be formed when these two alcohols react?

We've already learned that when two alcohol molecules are mixed with sulfuric acid and heated, an ether is formed according to the generalized reaction given by Equation (13.6). What we need to do is fit our specific reactants into this generalized equation. How can we do that? Well, our first reactant is methanol. In order for it to fit as the first reactant in Equation (13.6), R must represent CH_3. That way, R-OH will mean CH_3-OH. In the same way, if R' is used to represent $CH_2CH_2CH_3$, then HO-R' will represent HO-$CH_2CH_2CH_3$.

Now remember, the carbon chains represented by R and R' *do not change during the course of the reaction.* They may move around and bond to different things, but if R represented CH_3 before the reaction started, it represents CH_3 after the reaction finishes. Thus, everywhere we see an R in the equation, we can substitute CH_3 and everywhere you see an R', you can substitute $CH_3CH_2CH_2$. Thus, the reaction becomes:

$$CH_3\text{-OH} + \text{HO-}CH_2CH_2CH_3 \rightarrow CH_3\text{-O-}\ CH_2CH_2CH_3 + H_2O$$

The ether produced, then, is <u>CH_3-O- $CH_2CH_2CH_3$</u>.

Notice what we needed to know in order to solve the example problem. First of all, we needed to know the generalized reaction. How do we know it? Basically, we memorize it. Once we know the generalized equation, we simply have to take the specific reactants given in the problem and use them to determine what the R's represent in the generalized equation. From there, we just plug our specifics into the R's, producing the specific reaction about which the problem is asking. Knowing these generalized equations, then, is an important part of organic chemistry and will be an important part of this module.

Now that I have discussed the idea of generalized equations in organic chemistry, it is time to get back to discussing ethers. Typically, ethers have a very low boiling point and are quite flammable. Extreme caution must be used when using ethers in a lab where sparks or flames might be present. For example, I personally witnessed a 20'-long stream of fire produced when a student opened a can of diethyl ether. This flame was produced because a student *on the other side of the room* had an open flame in his experimental setup. The ether evaporated from the open can, and the vapors spread until they encountered the flame. Ether is so flammable that the flame traveled along the trail of the ether vapors in order to catch the entire can of ether on fire!

Ethers are also excellent organic solvents. Diethyl ether, for example, is used as a solvent for gums, waxes, fats, and resins. Since 1846, it has also been used as a general anesthetic for surgery. Because it is so flammable, however, it makes surgery a potentially dangerous place, so modern surgeons use a variety of other compounds for anesthetics.

ON YOUR OWN

13.3 What is the product that results when butanol ($CH_3CH_2CH_2CH_2OH$) is mixed with sulfuric acid and heated?

Aldehydes and Ketones

The next functional group I want to discuss is the **aldehyde** group. When an organic molecule ends with the following group:

$$-\overset{\overset{\displaystyle O}{\|}}{C}-H$$

the molecule is called an aldehyde. When chemists do not want to draw the Lewis structure, an aldehyde group is written as -CHO. Thus, aldehydes have the generalized formula R-CHO, where, as usual, R represents any carbon chain or ring.

Aldehydes typically form as a result of the oxidation of alcohols. When a mixture of a primary alcohol and air are passed through a heated tube which contains a catalyst such as silver or iron, the alcohol is oxidized according to the following generalized reaction:

$$2\,R-\underset{\underset{H}{|}}{\overset{\overset{H}{|}}{C}}-\ddot{\underset{..}{O}}-H \;+\; \ddot{\underset{..}{O}}=\ddot{\underset{..}{O}} \xrightarrow{\text{catalyst}} 2\,R-\overset{\overset{:O:}{||}}{C}-H \;+\; 2\,H-\ddot{\underset{..}{O}}-H$$

Notice what happens in this reaction. The carbon that is bonded to the hydroxyl group loses two hydrogen atoms. Those two hydrogen atoms bond to one of the oxygen atoms in O_2 to make water. As a result, the carbon and oxygen each need another bond, so a double bond forms between them. I can write this equation without Lewis structures as well:

$$2R\text{-}CH_2\text{-}OH + O_2 \xrightarrow{\text{catalyst}} 2R\text{-}CHO + 2H_2O \qquad (13.7)$$

The simplest aldehyde is **formaldehyde**. It is made when methanol is oxidized:

$$2CH_3\text{-}OH + O_2 \xrightarrow{\text{catalyst}} 2HCHO + 2H_2O$$

At room temperature, formaldehyde is a colorless gas. It dissolves readily in water, however. In biology, a solution of formaldehyde and water is used to preserve specimens. The proper name of such a solution is "formalin," but most biology students erroneously call it formaldehyde. **Acetaldehyde** is another important aldehyde. It is a liquid at room temperature, and it has the scent of freshly cut green apples. It is used in the manufacture of rubber and certain dyes.

A very similar class of compounds, called **ketones**, are made from the oxidation of alcohols as well. Rather than being made from primary alcohols (as is the case with aldehydes), ketones are formed from secondary alcohols. The generalized reaction is as follows:

$$\underset{\text{secondary alcohol}}{R\text{-}\overset{\overset{OH}{|}}{CH}\text{-}R'} \xrightarrow[\Delta]{\text{catalyst}} \underset{\text{ketone}}{R\text{-}\overset{\overset{O}{||}}{C}\text{-}R'} + H_2 \qquad (13.8)$$

Notice how similar ketones are to aldehydes. The only real difference is that aldehydes always have one hydrogen bonded to the carbon that is double-bonded to the oxygen. In a ketone, the carbon that is double-bonded to the oxygen has 2 other carbons bonded to it.

The most important ketone is **acetone**. It is formed from 2-propanol ($CH_3CHOHCH_3$):

$$\underset{\text{2-propanol}}{CH_3\text{-}\overset{\overset{OH}{|}}{CH}\text{-}CH_3} \xrightarrow[\Delta]{\text{catalyst}} \underset{\text{acetone}}{CH_3\text{-}\overset{\overset{O}{||}}{C}\text{-}CH_3} + H_2 \qquad (13.9)$$

Acetone is the active component of most nail polish removers. It is also the solvent used for most varnishes. It is even used to make the explosive known as "cordite." Chemists often use acetone to wash their hands free of organic contamination.

ON YOUR OWN

13.4 For each of the following alcohols, indicate whether or not it can be used to make an aldehyde, a ketone, or neither.

a. $CH_3CH_2CH_2CH_2CH_2\text{-}OH$

b. $CH_3\text{-}C(OH)(CH_3)\text{-}CH_3$ (OH above and CH_3 below the central C)

c. $CH_3CH_2CH(OH)CH_2CH_3$ (OH above the central CH)

13.5 What product is formed when ethanol (CH_3CH_2OH) is oxidized in the presence of a catalyst?

Carboxylic Acids

The next functional group I want to discuss is the **carboxyl** group. When a molecule has the following functional group:

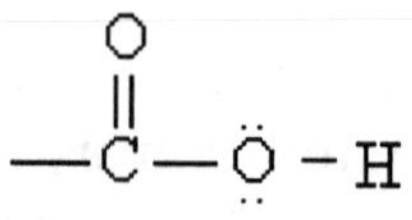

$$-\overset{\overset{\displaystyle O}{||}}{C}-\ddot{\underset{..}{O}}-H$$

the molecule is a **carboxylic acid**. To indicate this functional group without a Lewis structure, chemists use the abbreviation $-CO_2H$ or $-COOH$. Thus, any molecule of the form $R\text{-}CO_2H$ or $R\text{-}COOH$ is a carboxylic acid.

As the name implies, these compounds act like acids. Remember what acids do: they donate H^+ ions. When a carboxylic acid does this, the H that gets donated comes from the "OH" part of the carboxyl group. This leaves a negative sign on the oxygen atom that is part of the "OH." **Acetic acid** (CH_3CO_2H), for example, is a carboxylic acid. It reacts with the base OH^- as follows:

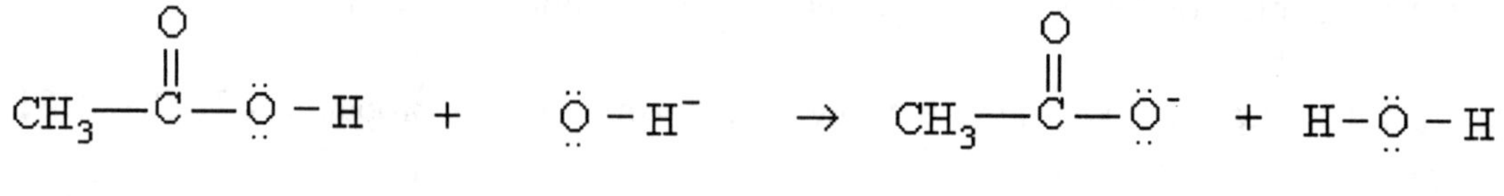

$$CH_3-\overset{\overset{\displaystyle O}{||}}{C}-\ddot{\underset{..}{O}}-H \;+\; \ddot{\underset{..}{O}}-H^- \;\rightarrow\; CH_3-\overset{\overset{\displaystyle O}{||}}{C}-\ddot{\underset{..}{O}}^- \;+\; H-\ddot{\underset{..}{O}}-H$$

In abbreviated form, the reaction is written:

$$CH_3CO_2H + OH^- \rightarrow CH_3CO_2^- + H_2O$$

Carboxylic acids are made from the oxidation of aldehydes according to the following reaction:

$$2\text{R-CHO} + O_2 \xrightarrow[\Delta]{\text{catalyst}} 2\text{R-CO}_2\text{H} \qquad (13.10)$$

Acetic acid, for example, is formed by the oxidation of CH_3CHO:

$$2CH_3CHO + O_2 \xrightarrow[\Delta]{\text{catalyst}} 2CH_3CO_2H$$

As you probably already know, acetic acid is the active component of vinegar. Most vinegars are 3-6% acetic acid and 94% - 97% water.

The first carboxylic acid that was discovered is also the simplest. **Formic acid**, HCO_2H, was isolated by the distillation of red ants. This tells you where it gets its name. In Latin, *formicus* means "ant." One of the reasons that red ant bites sting so much is because the ant uses its bite to inject formic acid into your skin!

Another very useful carboxylic acid is **benzoic acid**:

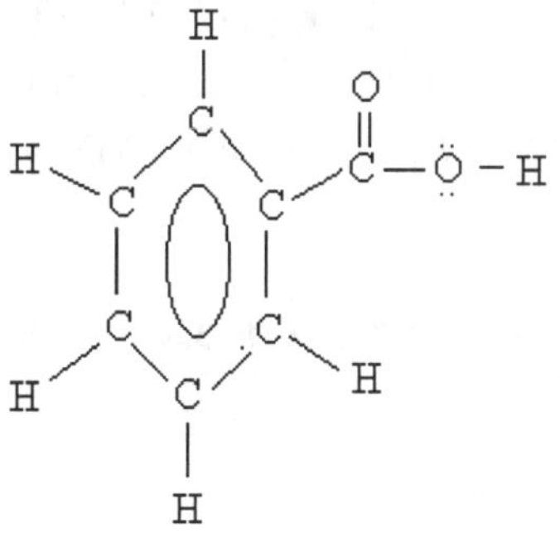

You have probably never heard of this acid before. However, when reacted with the base NaOH, it forms the ionic solid **sodium benzoate**:

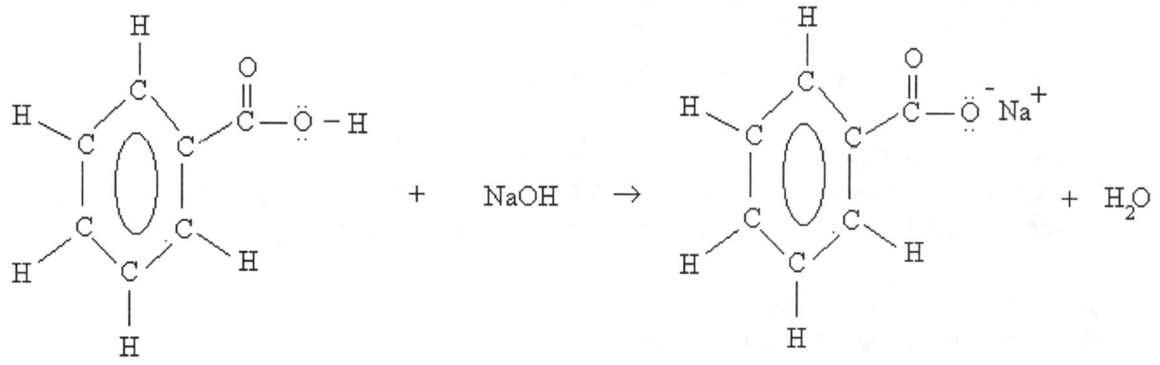

+ NaOH → + H_2O

sodium benzoate

Sodium benzoate is a popular preservative. You will find it in fruit juices and tomato ketchup, for example.

ON YOUR OWN

13.6 When trying to make a carboxylic acid, many chemists do not start with an aldehyde. Instead, they start with a primary alcohol. How can a primary alcohol be used to make a carboxylic acid?

Esters

Just as ketones are very similar to aldehydes, esters are very similar to carboxylic acids. The generalized formula of an ester is as follows:

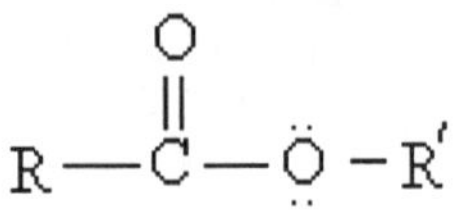

$$R-\overset{\overset{\displaystyle O}{\|}}{C}-\ddot{\underset{..}{O}}-R'$$

Notice that the only difference between a carboxylic acid and an ester is that where the acid has an "H," the ester has a carbon chain. Now since the "H" that is bonded to the oxygen is the "H" that is donated by a carboxylic acid, esters obviously have no H^+ to donate and are therefore not acids at all. Although esters look very similar to carboxylic acids, their chemical properties are quite different.

Esters are typically made via a dehydration reaction between carboxylic acids and alcohols.

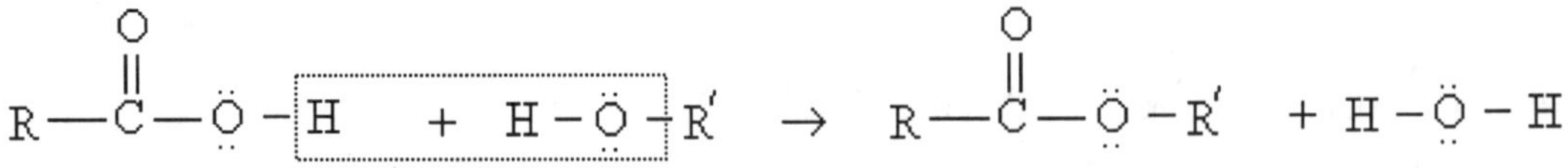

$$R-\overset{\overset{\displaystyle O}{\|}}{C}-\ddot{\underset{..}{O}}-\boxed{H \quad + \quad H-\ddot{\underset{..}{O}}}-R' \rightarrow R-\overset{\overset{\displaystyle O}{\|}}{C}-\ddot{\underset{..}{O}}-R' \quad + H-\ddot{\underset{..}{O}}-H$$

Without the Lewis structures, this equation becomes:

$$R\text{-}CO_2H + HO\text{-}R' \rightarrow R\text{-}CO_2\text{-}R' + H_2O \qquad (13.11)$$

Esters have a distinct, often pleasant aroma. The pleasant scent of flowers is usually due to esters that the flower manufactures in order to attract birds and insects that will help transfer the flower's pollen to other flowers. Many perfumes use esters to create their pleasant scent.

Perhaps the "most feared" class of esters in America are the result of long carboxylic acids reacting with an alcohol called glycerol:

```
    H      O  H H H H H H H                 H    O  H H H H H H H
    |      || | | | | | | |                 |    || | | | | | | |
H—C—OH HO-C -C-C-C-C-C-C-C-H            H—C—O—C -C-C-C-C-C-C-C-H       O
    |         | | | | | | |                 |       | | | | | | |      / \
    |         H H H H H H H                 |       H H H H H H H     H   H
    |
    |      O  H H H H H H H                 |    O  H H H H H H H
    |      || | | | | | | |                 |    || | | | | | | |
H—C—OH HO-C -C-C-C-C-C-C-C-H   ——>      H—C—O—C -C-C-C-C-C-C-C-H       O      (13.12)
    |         | | | | | | |                 |       | | | | | | |      / \
    |         H H H H H H H                 |       H H H H H H H     H   H
    |
    |      O  H H H H H H H                 |    O  H H H H H H H
    |      || | | | | | | |                 |    || | | | | | | |
H—C—OH HO-C -C-C-C-C-C-C-C-H            H—C—O—C -C-C-C-C-C-C-C-H       O
    |         | | | | | | |                 |       | | | | | | |      / \
    H         H H H H H H H                 H       H H H H H H H     H   H
```

Glycerol 3 fatty acid molecules a lipid (fat) molecule 3 water molecules

Glycerol is an alcohol that has three functional groups. As a result, it is called a "triol." When glycerol reacts with the carboxylic acids, three dehydrations occur which attach the three carboxylic acids to the glycerol. The result is a molecule that has 3 ester functional groups. The resulting molecule is what we call a **lipid**, the more general term for **fat**. Because the carboxylic acids in this reaction end up making a fat molecule, they are often called **fatty acid** molecules, although they are really nothing more than carboxylic acids.

Now the main thing to realize about lipid molecules is that there is an enormous variety to them. In principle, *any* carboxylic acid can react with glycerol to make a lipid molecule. In addition, the 3 fatty acid molecules need not even be the same. This leads to large variety in the class of molecules called lipids.

Since there is so much variety in the lipids, we further classify them into one of two groups: **saturated fats** and **unsaturated fats**. Saturated fats are made from fatty acids that are saturated. Remember that an organic molecule is saturated if it has as many hydrogens as it can possibly have. In essence, this means that there can be no double or triple bonds between carbon atoms in the carbon chain. If this is the case, the carboxylic acid is saturated. If there are double or triple bonds in the carbon chain of a carboxylic acid, it is considered unsaturated and will therefore produce unsaturated fats when reacted with glycerol.

There is an observable difference between saturated and unsaturated fats. In general, the Van der Waals forces between saturated fats are much stronger than those between unsaturated fats. As a result, saturated fats tend to be solid at room temperature whereas unsaturated fats tend to be liquids at room temperature. Why is there a difference in the Van der Waals forces? Well, each double bond in the carbon chain of a carboxylic acid causes the carbon chain to bend. Thus, there are "kinks" in the carbon chain of an unsaturated fatty acid. As a result, the fat that it makes is bent and kinked. This makes it hard for the molecules to get close to one another, because all of those bends and kinks get it the way. Saturated fats, however, are more flat. As a result, they can get closer to one another. This in analogous to what happens when you pile wood. If the wood is reasonably flat, you can stack the pieces of wood close to one another, making the wood pile dense. Heavily-branched sticks (with lots of bends and kinks), however, cannot be stacked close to one another, and the resulting wood pile is not very dense.

Not only do the bends and kinks in unsaturated fats cause them to be liquids at room temperature, they also make unsaturated fats more healthy than saturated fats. Notice from the reaction given by Equation (13.12) that fats are very bulky molecules. As fat molecules are carried throughout the body via the twists and turns of the blood vessels, they can get stuck in one of those twists or turns. As the fats pile up there, the blood vessel becomes clogged, leading to a host of maladies. This is much like a bunch of sticks floating down a twisting, turning stream. As the sticks are carried by the water, some will get stuck in the twists and turns of the stream. This will cause other sticks to pile up at the same point, clogging the stream.

Continuing with the stick analogy for a moment, which kind of stick is more likely to get caught in the bends and twists of the stream: a long, straight stick or a bent, kinked stick. Obviously, the long, straight stick will have a rougher time making it through the twists and turns. Thus, the long, straight stick is more likely to get caught up and start clogging the stream. In the same way, saturated fats tend to get caught in your blood vessels more than unsaturated fats. Thus, saturated fats tend to lead to health problems more than unsaturated fats. This is why most dietitians recommend that you eat more unsaturated fats a less saturated fats. For most people, however, saturated fats are more tasty than unsaturated fats!

ON YOUR OWN

13.7 What is the product when acetic acid (CH_3CO_2H) is reacted with ethanol (CH_3CH_2OH)?

13.8 Which of the following carboxylic acids would be considered an unsaturated fatty acid?

a. $CH_3CH_2CH_2CH_2CH_2CO_2H$ b. $CH_3CH_2CH_2CH_2CO_2H$ c. $CH_3CH_2CH_2CH{=}CHCH_2CO_2H$

Amino Acids and Proteins

Now I want to talk about the most important organic compounds in the chemistry of life: **amino acids** and **proteins**. Amino acids are organic molecules with two functional groups: an **amine** group (-NH_2) and a carboxyl group (-CO_2H). When these two functional groups are attached to the same carbon in an organic molecule, the molecule is called an amino acid. The generalized formula of an amino acid, then, is:

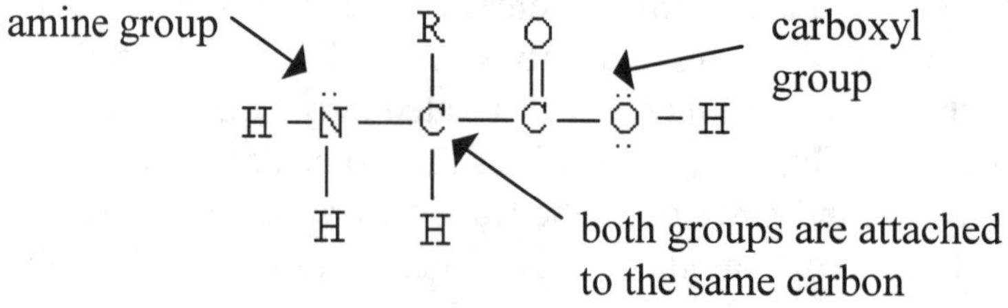

amine group

carboxyl group

H–N–C(R)(H)–C(=O)–O–H, with a second H on the N

both groups are attached to the same carbon

A few of the amino acids used in the chemistry of life are given in Figure 13.2.

FIGURE 13.2
Examples of the Amino Acids Which are Important in the Chemistry of Life

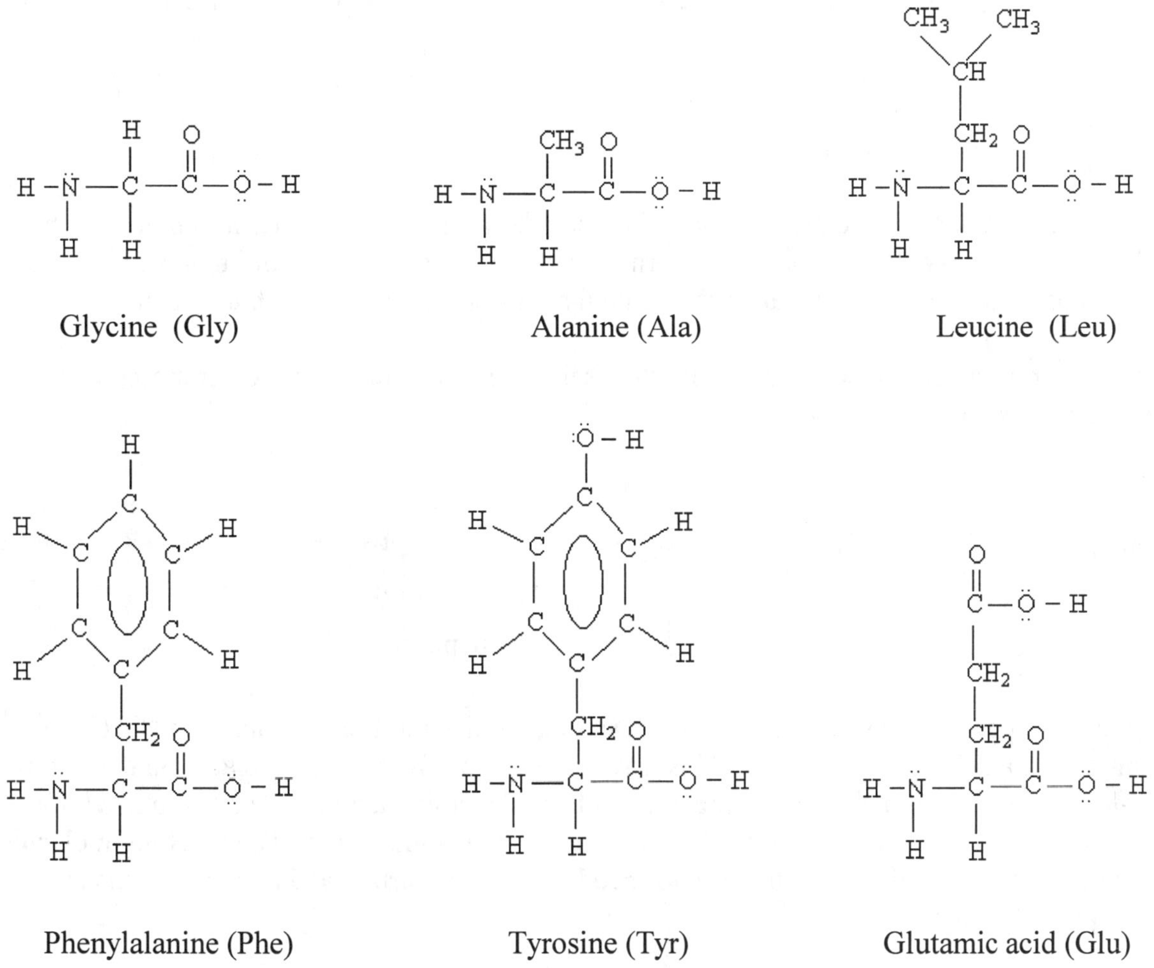

Glycine (Gly)

Alanine (Ala)

Leucine (Leu)

Phenylalanine (Phe)

Tyrosine (Tyr)

Glutamic acid (Glu)

Notice that the only difference between all six of the amino acids in the figure is what R represents. In the case of glycine, R represents a hydrogen atom, whereas in the case of Tyrosine, R represents a much more complex arrangement of carbon atoms. Notice also the three-letter abbreviations that follow each name. You will see the importance of these abbreviations later.

Since amino acids have both an amine group and a carboxyl group, something interesting can happen. As you already know, the carboxyl group acts as an acid because the hydrogen atom bonded to the oxygen can be donated as an H^+ ion. Well, it turns out that the amine group acts as a base, because the lone pair of electrons on the nitrogen atom can *accept* H^+ ions. As a result, *amino acids can react with themselves* by allowing the H^+ to leave the carboxyl group and bond to the amine group:

```
         R   O                          +  H   R   O
   ..    |   ||                            |   |   ||    ..
H -N — C — C — O - H        →          H -N — C — C — O⁻        (13.13)
   |     |       ..                        |   |         ..
   H     H                                 H   H

     standard form                        zwitteron form
```

Although amino acids are usually written in their standard form, it is important to note that they mostly exist in their **zwitteron** form. The term "zwitteron" is a German word that means "hybrid ion." It is used to refer to any covalent molecule that has electrical charges but is neutral overall.

Not only can amino acids react with themselves, they can react with other amino acids via a dehydration reaction:

```
      R   O                  R′  O                        R   O     R′  O
  ..  |   ||     ..          |   ||    ..             ..  |   ||    |   ||   ..           ..
H -N—C—C—[O - H  +  H]-N—C—C—O - H   →   H -N—C—C——N—C—C—O - H  +  H-O-H   (13.14)
   |  |                  |   |                         |  |     ↑ |  |
   H  H                  H   H                         H  H     | H  H
                                                              peptide bond
```

Notice what happens in this reaction. As is the case in all dehydration reactions, an "OH" combines with an "H" to make water. This leaves the carbon from the amino acid on the left in need of a bond. Also, it makes the nitrogen atom from the amino acid on the right in need of a bond. Those two atoms bond to each other, linking the two amino acids. The resulting molecule is called a **peptide**, and the bond that was formed between the carbon and nitrogen atoms is called a **peptide bond**.

Now look at the peptide for a moment. Notice that, just like an amino acid, it has a carboxyl group on the far right side of the molecule and an amine group on the far left. What does this tell you? It tells you that *the peptide can undergo more dehydration reactions*. After all, notice that in order to make this reaction work, one reactant had to have a carboxyl group and the other had to have an amine group. Well, the peptide has both. Thus, it can undergo another dehydration reaction with another amino acid. It turns out that the peptide formed in that reaction *can also* undergo another dehydration reaction. This can go on forever! In the end, then, as long as there are lots of amino acids around, peptides can get longer and longer by undergoing dehydration reactions over and over again.

When peptides are comprised of many, many amino acids, they are eventually classified as **proteins**. The type of amino acids that are linked together as well as the order in which they link up determines the specific properties of the protein that is formed. You have certainly heard the term "protein" before, but you probably have no idea how important proteins are to the chemistry of life. It turns out that virtually every chemical process that occurs in a living organism is governed by proteins. Life simply could not exist without proteins!

Unfortunately, it is impossible to give you a structural formula for a protein. As you can see from Figure 13.2, amino acids are rather complex molecules. Even the simplest protein of life contains 124 amino acids linked together through peptide bonds. Obviously, then, drawing the structural formula of a protein would be quite a job. The situation would be even worse for the "average" protein, which contains several thousand amino acids! Now remember, the type of amino acids that are bonded together and *the order in which they are bonded* determine the properties of a protein. Two proteins can have the same number of amino acids and the same types of amino acids, but if they are linked together in a different order, the two proteins have vastly different chemical properties.

Even though it is impossible to give you a structural formula for a protein, it is possible to give you some idea of what one looks like. Figure 13.3 is a schematic representation of a protein.

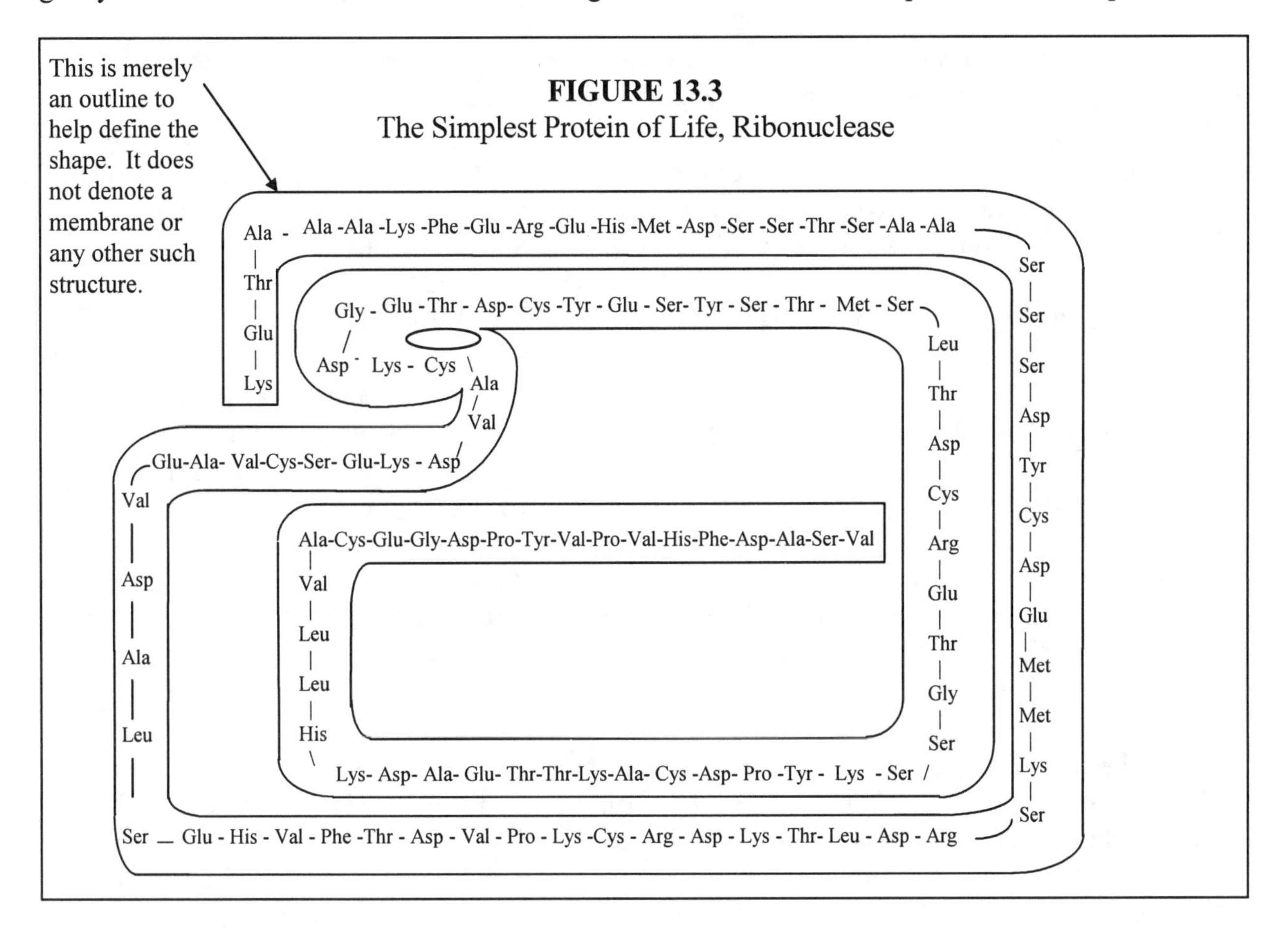

FIGURE 13.3
The Simplest Protein of Life, Ribonuclease

What does this schematic mean? Well, remember the three-letter abbreviations that I gave you for the amino acids in Figure 13.2? The three letter groups you see here are those abbreviations. Since the structural formula of a protein is too complex to draw, the typical way chemists draw a protein is to abbreviate each amino acid with three letters. The letters "Lys," for example, stand for the amino acid called "lysine," while the letters "Glu" stand for glutamic acid. This schematic, then, tells you the type of amino acid and the order in which it appears in the protein ribonuclease (rye' boh new clee' ayse). In addition, the shape that the amino acids are lined up in is roughly equivalent to the shape of ribonuclease. Of course, since the paper is only two-dimensional, it is impossible to draw the entire three-dimensional shape, but what is drawn here is a reasonable approximation.

When biochemists really began to understand the structure of proteins and how integral they are to the chemistry of life, it should have caused the death of the hypothesis of evolution. After all, remember what the hypothesis of evolution says. It says that life is the result of random chemical reactions that occurred over eons of time. If this is the case, then, random chemical reactions would have to make proteins. Without this happening, life could never appear because proteins are such a fundamental component of the chemical reactions that make life possible. Any chemist who understands proteins must conclude that the formation of these complex molecules from random chemical reactions is simply impossible.

Consider, for example, the protein pictured in Figure 13.3. When assembled as shown in the figure, the 124 amino acids form a protein, ribonuclease, that performs a vital task in the chemistry of life. If the amino acids were to link up in the wrong order, or if even one of the amino acids were the wrong type, then ribonuclease would be *completely unable* to perform its job. Thus, if the first amino acid were "Glu" (glutamic acid) instead of "Lys" (lysine), the protein would no longer be ribonuclease and would be completely useless. Think, for a moment, about the probability of such a protein forming through random chemical reactions. Is it possible for such a molecule to form by chance from a mixture of lots of amino acids?

Let's make it easy on ourselves and assume that the only amino acids in the mixture are the 17 types needed to make this particular protein. In fact, there are many types of amino acids that exist naturally on earth, but adding more amino acid types would significantly reduce our chance of forming ribonuclease. So, in order to make the outcome more likely, we will restrict ourselves to using only the 17 different types of amino acids that make up this molecule. Making this assumption, we can say that the possibility of forming a protein that has "Lys" as its first amino acid is 1 in 17. Those aren't bad odds at all. However, the chance of forming a protein with "Lys" as its first amino acid and "Glu" as its second amino acid is 1 in 17 *times* 1 in 17, or 1 in 289. Suddenly the odds are looking less favorable.

The probability of forming a protein whose first three amino acids are "Lys," "Glu," and "Thr" in that order are 1 in 17 *times* 1 in 17 *times* 1 in 17, or 1 in 4,913. If you were to complete this calculation, you would find that the odds for making this protein by chance from a mixture of the proper amino acids is approximately 1 in 10^{152}. In order to illustrate just how ridiculously low these odds are, the probability for forming ribonuclease by chance is roughly equivalent to the probability of a poker player drawing a royal flush *19 times in a row without exchanging cards!* Remember, ribonuclease is a “simple” protein. There are proteins in our bodies that contain more than 10,000 amino acids! Clearly the idea that these proteins could form by chance is absurd! This lays to waste the entire hypothesis of evolution!

ON YOUR OWN

13.9 Suppose one glycine molecule and one alanine molecule were to react to form a simple peptide. What would be the Lewis structure of the peptide? (See Figure 16.2 to get the Lewis structures of glycine and alanine. Assume glycine is the first amino acid and alanine is second.)

Carbohydrates

Carbohydrates are organic molecules that contain only carbon, hydrogen, and oxygen. In addition, they have the same ratio of hydrogen atoms to oxygen atoms as does water. For example, one of the simplest carbohydrates is glucose, $C_6H_{12}O_6$. Notice that it has 12 hydrogen atoms and 6 oxygen atoms. In other words, there are twice as many hydrogen atoms as there are oxygen atoms. This is the same as water (H_2O). In fact, that's where the term "carbohydrate" comes from. "Carbo" stands for the carbon in the molecule, and "hydrate," which means "to add water," stands for the fact that there are twice as many hydrogen atoms as oxygen atoms, just like water. The Lewis structure of glucose is:

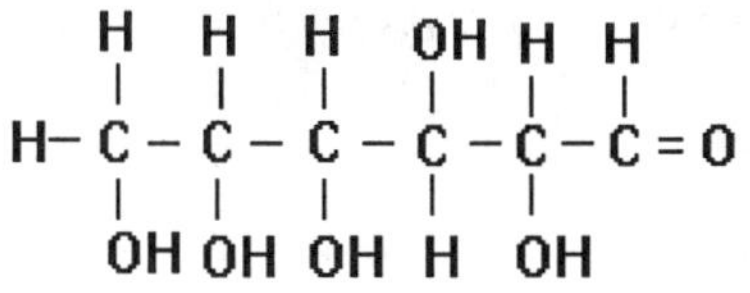

```
   H   H   H   OH  H   H
   |   |   |   |   |   |
H- C - C - C - C - C - C = O
   |   |   |   |   |
   OH  OH  OH  H   OH
```

Looking at the structural formula of glucose a little more carefully, you will see that the molecule has two functional groups, just like amino acid molecules. In the case of glucose, however, the two functional groups are the alcohol group (OH) and the aldehyde group (CHO). This is typical of most organic molecules in the chemistry of life. Carbohydrates are also called "sugars." Glucose, for example, is the sugar that you find in green leaves.

It turns out that glucose has a structural isomer, which is also typical of the organic molecules in life. Although the Lewis structure of glucose as written above is accurate, there is another Lewis structure for glucose:

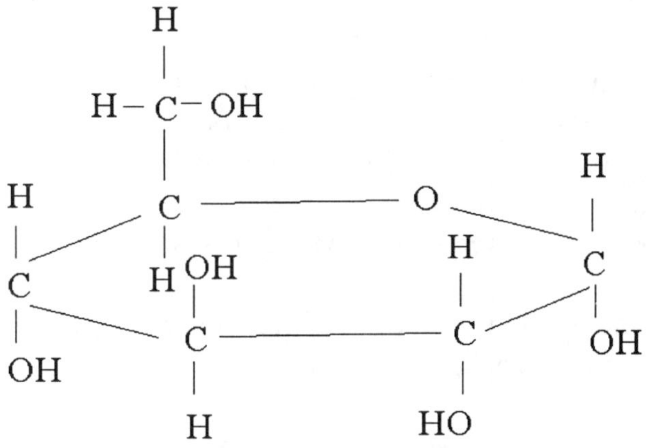

```
              H
              |
          H - C - OH
              |
                                            H
H             C ----------------- O          |
|          /  |                      \      |
C           H  OH             H         C
|  \           |              |      /  |
             C ------------- C          OH
OH           |                |
             H               HO
```

Notice that the chemical formula of this molecule is the same as the other one. If you count the atoms, you will find 6 C's, 12 H's, and 6 O's. The main difference is that the double bond is gone between the oxygen and the carbon and, instead, the atoms have arranged themselves in a ring. Not surprisingly, the first structural formula of glucose is called the **chain structure** while this one is called the **ring structure**. In life, the ring structure is the most prevalent form of glucose.

Not only does glucose have two isomer forms, it turns out that there are a LOT of ways to arrange 6 carbons, 12 hydrogens, and 6 oxygens in a Lewis structure. This leads to several chemicals that each have the chemical formula $C_6H_{12}O_6$ but have significantly different properties. Consider, for example, the carbohydrate known as fructose. It has the same chemical formula as glucose, but its Lewis structure is completely different. As a result, it has different properties.

The ring and chain structural formulas of both glucose and fructose are shown in Figure 13.4.

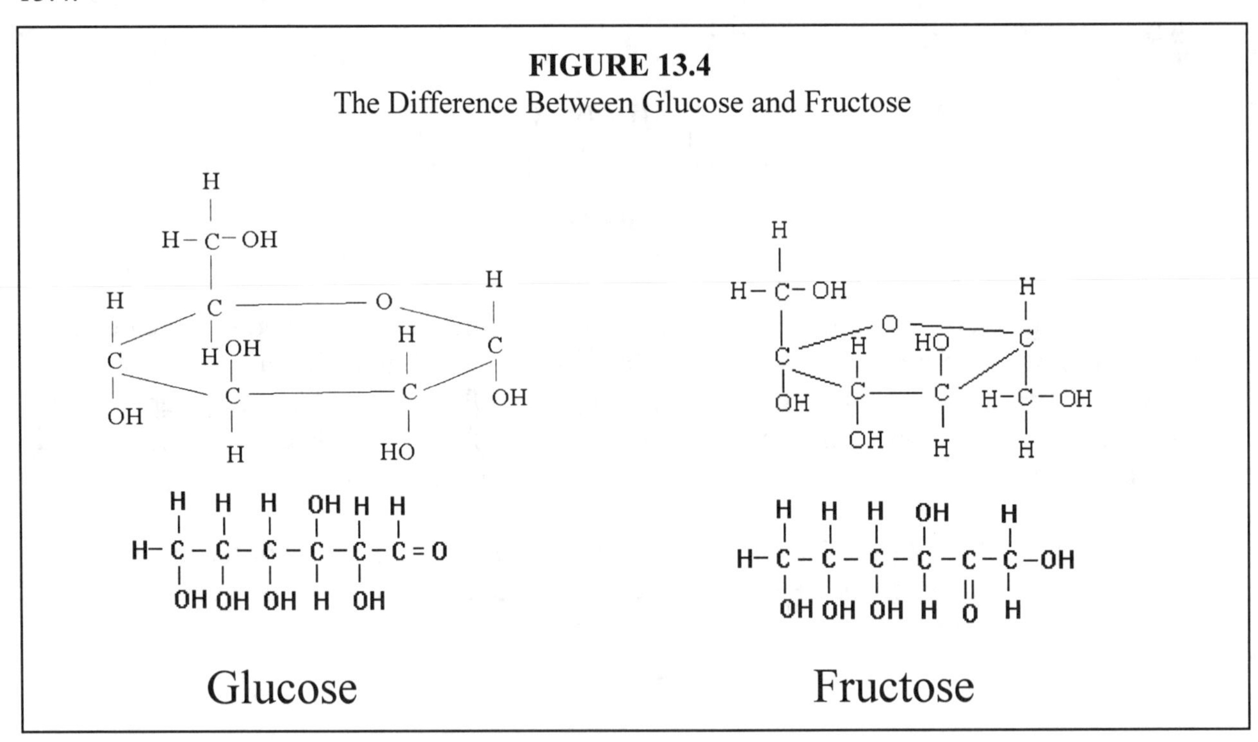

FIGURE 13.4
The Difference Between Glucose and Fructose

The difference in the chemical properties of glucose and fructose is evident in the taste of each compound. Remember, carbohydrates are called sugars. They are sweet. The sweetness of lettuce comes from glucose. Fructose is the sugar than is in most fruits. Thus, the sweetness of fruits comes from fructose. The difference in sweetness between lettuce and fruit is, in part, due to the difference in properties between glucose and fructose.

Glucose and fructose belong to a class of compounds known as **monosaccharides** (mahn uh sak' uh rides), which are also called **simple sugars**.

<u>Monosaccharides</u> - Simple carbohydrates that contain three to ten carbon atoms

The reason that these are called monosaccharides is that they form the basic building blocks of more complex carbohydrates called **disaccharides** (dye sak' uh rides) and **polysaccharides** (pahl ee sak' uh rides).

Disaccharides - Carbohydrates that are made up of two monosaccharides

Polysaccharides - Carbohydrates that are made up of more than two monosaccharides

These more complex carbohydrates form the basis of much of the food that we eat.

For example, table sugar is a disaccharide called sucrose. It is formed when glucose and fructose react in a dehydration reaction, as shown in Figure 13.5:

FIGURE 13.5
Dehydration Reaction That Produces Table Sugar, a Disaccharide

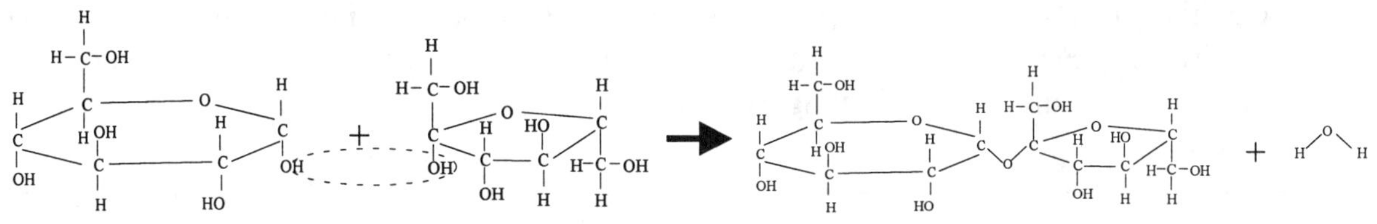

glucose + fructose → sucrose + H_2O

Many other disaccharides make up the sugars that sweeten our foods. Glucose and galactose, for example, combine in a dehydration reaction to make lactose, the sugar that gives milk its sweetness.

When several monosaccharides link together, the result is a polysaccharide. Typically, polysaccharides are not sweet, because the sweetness of the monosaccharides is lost when several combine. Nevertheless, polysaccharides are an important part of our diet. The polysaccharides known as starches, for example, are found in most plants. When a plant has extra monosaccharides, it will store them as polysaccharides by running many dehydration reactions that link the monosaccharides together. Potato starch, corn starch, and starch from wheat, rice, and other grains are a major source of food for people. Humans and animals make their own starch, **glycogen**, when they have excess carbohydrates to store.

As you might have already learned by studying nutrition, carbohydrates are one of the principal sources of food energy for most animals. Interestingly enough, however, most animals can only use monosaccharides for energy. Thus, when an organism eats disaccharides or polysaccharides, it must first break them down into their individual monosaccharide components. How is this done? Well, monosaccharides combine to form disaccharides and polysaccharides by dehydration. To break these complex molecules back down into their monosaccharide components, all you have to do is add water to them. This process is called **hydrolysis**.

Hydrolysis - Breaking down complex molecules by the chemical addition of water

Hydrolysis is essentially the reverse of dehydration, as shown in the figure below.

FIGURE 13.6
An Example of Hydrolysis: Breaking Sucrose Down Into Glucose and Fructose

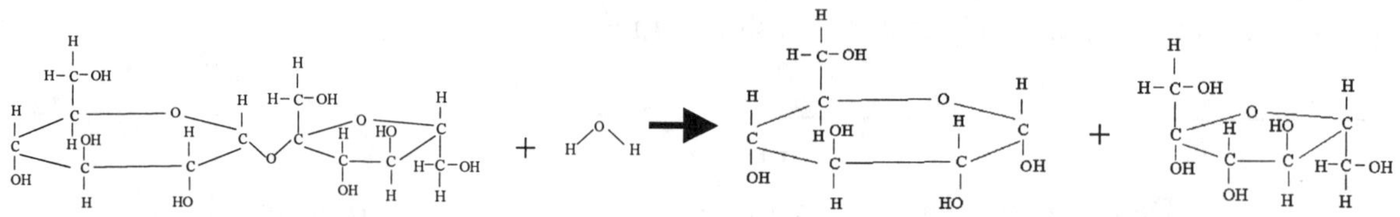

In the chemistry of life, dehydration reactions are typically more thermodynamically favorable than are hydrolysis reactions. As a result, hydrolysis reactions typically have to be forced. See what I mean by performing the following experiment.

EXPERIMENT 13.2
The Hydrolysis of Sucrose

Supplies

From the laboratory equipment set:
- Sodium carbonate
- Calcium hydroxide solution (It's labeled as "Lime Water Solution" in the kit.)
- Copper (II) sulfate
- 2 test tubes
- Test tube rack
- Depression plate (the clear plastic thing with all of the wells in it)
- 4 plastic droppers

From around the house:
- Cream of tartar (a common household spice)
- 2 plastic or Styrofoam cups
- Hot water
- Small plate or bowl
- Bowl (It needs to be big enough for the depression plate to fit inside.)
- Toothpick
- Vinegar
- Teaspoon

The first thing you need to do in this experiment is make the components for **Fehling's solution**. This solution is used to test for monosaccharides. The solution is unstable, however, so you will make the components first and then wait until later to put them together. Start by filling a test tube 1/4 of the way with calcium hydroxide solution. Then, add three measures of sodium carbonate. Stir to dissolve the solid. Put the hottest water from the faucet into one of the cups and let the test tube sit in that hot water for 10 minutes. At the end of the 10 minutes, you will have a solution of sodium hydroxide. **THIS SOLUTION IS CAUSTIC**. Handle it with care. Do not get any of it on you.

While that test tube is soaking, fill the other test tube 3/4 of the way with hot water from the faucet. Then, add 10 measures of cream of tartar to the water. **DO NOT SCOOP DIRECTLY FROM THE CONTAINER**. This will contaminate the rest of the cream of tartar, making it unfit to use with food! Instead, pour some of the spice onto the small plate or bowl and scoop from there. NEVER put the chemical scoop into anything that is meant to be eaten! Mix the cream of tartar in the hot water. It will probably not completely dissolve. You are just making a saturated solution.

Finally, in one of the deep wells of your depression plate, add four measures of copper (II) sulfate. Use one of the droppers to add enough water to nearly fill the well. Mix to make a copper (II) sulfate solution. The three solutions you made are the components to Fehling's solution. Do not mix them yet.

Next, add two measures of table sugar (**once again, don't scoop the sugar out of the sugar bowl!**) to a deep well and use a dropper to fill it mostly full of water. Stir to dissolve the sugar. You have now made a solution of sucrose. Remember, sucrose is a disaccharide, composed of glucose and fructose.

Now you are ready to do a hydrolysis reaction to break down sucrose into its monosaccharides. Fill the other plastic cup halfway with the hottest water you can get from the faucet. Next, add a teaspoon of vinegar and a teaspoon of sugar. Mix thoroughly. The hot water will provide the energy necessary to induce the hydrolysis reaction, and the vinegar will make the solution acidic, which will speed up the reaction. After about a minute of stirring, you should have a solution of vinegar, glucose, and fructose.

Now its time to test and see whether or not the hydrolysis worked. Add the hottest water you can from the faucet to the bowl until there is about an inch of water in the bowl. Float the depression plate in the water, being sure not to spill any of the solutions already there. With a clean dropper, add 15 drops of the cream of tartar solution to a deep well, 10 drops of the sodium hydroxide solution, and 10 drops of the copper (II) sulfate solution. Quickly add 10 drops of the sucrose solution. Repeat this procedure in a different well, this time adding 10 drops of the glucose and fructose solution instead of the sucrose solution.

When you added the cream of tartar solution, the sodium hydroxide solution, and the copper (II) sulfate solution together, you made Fehling's solution. In one of the two wells, then, you have a mixture of Fehling's solution and sucrose. In the other well, if your hydrolysis worked, you should have a mixture of Fehling's solution, glucose, and fructose.

In the presence of monosaccharides, Fehling's solution will react to form a brown color. In the presence of disaccharides and polysaccharides, however, it will do nothing. Thus, as time goes on, you should see nothing happen in the well that has sucrose, while the well that has glucose and fructose should turn brown. This will tell you that you did, in fact, perform a hydrolysis reaction, breaking sucrose down into fructose and glucose.

Now remember, most organisms (including yourself) cannot use disaccharides and polysaccharides for energy. They must first be broken down into monosaccharides. To do this, your body performs hydrolysis reactions, like the one you just performed. Unlike the hydrolysis reaction you performed, however, your body controls these reactions and forces them to proceed by using a special class of proteins called **enzymes**.

Each polysaccharide or disaccharide must have a specific enzyme in order to force its hydrolysis into monosaccharides. Sometimes, a person's body will not be able to produce the proper enzyme and, as a result, the person will not be able to digest a certain disaccharide or polysaccharide. For example, you might know someone who is **lactose intolerant**. These people cannot drink milk or other dairy products because they do not have the enzyme that breaks lactose (a disaccharide in dairy products) down into its monosaccharides. As a result, their bodies can do nothing with lactose, and it continues to build up in their bodies, making them sick.

ON YOUR OWN

13.10 If a hydrolysis reaction is run on lactose, what are the names of the substances produced?

Summing Up Organic and Biochemistry

This module ends your introduction into the fascinating worlds of organic chemistry and biochemistry. If you go on to study chemistry in college, organic chemistry will take your entire sophomore year. Obviously, then, what I have shown you here only scratched the surface. In addition, biochemistry is usually reserved for those students who are concentrating in the chemistry of life, and there are several junior and senior-level courses devoted to the intricacies of the chemistry of life. Once again, then, what you have learned here is a tiny, tiny portion of the chemistry of life. Nevertheless, I hope you have been given some appreciation for how interesting and complex these subjects are!

ANSWERS TO THE ON YOUR OWN PROBLEMS

13.1 Alcohols must be organic compounds with an "OH" group on them. The compound in (b) is not organic (Sr is a metal), and the compound in (d) has no "OH" group. It has O's and H's, but they are not together as an OH group. The molecules in (a) and (c) are alcohols, however, because they are organic molecules with "OH" groups.

13.2 Methanol and ethanol are both primary alcohols because the OH group is attached to a carbon that has only one other carbon attached to it. 2-propanol is a secondary alcohol. If you look at the Lewis structure given, the carbon to which the OH is attached is attached to 2 other carbons. The molecule in 13.1 (a) is a primary alcohol, because once again the carbon to which it is attached is attached to only one other carbon. Finally, the molecule in 13.1 (c) is a tertiary alcohol because the carbon to which the OH is attached has 3 other carbons attached to it.

13.3 When an alcohol is heated and mixed with sulfuric acid, the result is an ether. Equation (13.6) tells us that the general form of this reaction is:

$$\text{R-OH} + \text{HO-R'} \xrightarrow[\Delta]{H_2SO_4} \text{R-O-R'} + H_2O$$

In the case presented by this problem, we are dealing with $CH_3CH_2CH_2CH_2OH$. Thus, in order to make the generalized equation fit this problem, both R and R' must represent $CH_3CH_2CH_2CH_2$. To get this particular equation, then, I just substitute for R and R'.

$$CH_3CH_2CH_2CH_2\text{-OH} + \text{HO-}CH_2CH_2CH_2CH_3 \xrightarrow[\Delta]{H_2SO_4} CH_3CH_2CH_2CH_2\text{-O-}CH_2CH_2CH_2CH_3 + H_2O$$

The product of the reaction, then, is $CH_3CH_2CH_2CH_2\text{-O-}CH_2CH_2CH_2CH_3$.

13.4 Aldehydes are made from primary alcohols, while ketones are made from secondary alcohols. Thus, (a) can be used to make an aldehyde while (c) can be used to make a ketone. The alcohol in (b) can be used for neither. Remember, to make an aldehyde or ketone, there must be a double bond between the oxygen and the carbon. If the carbon and oxygen in this molecule had a double bond, the carbon would have 10 electrons around it, which is not allowed in a Lewis structure unless one of the atoms has access to d-orbitals. Neither carbon nor oxygen have such access, so there is no way to turn (b) into a ketone or aldehyde.

13.5 The oxidation of primary alcohols leads to aldehydes according to Equation (13.7):

$$2\text{R-}CH_2\text{-OH} + O_2 \xrightarrow{\text{catalyst}} 2\text{R-CHO} + 2H_2O$$

For this particular problem, then, R must represent CH_3. Thus, this particular reaction becomes:

$$2CH_3\text{-}CH_2\text{-OH} + O_2 \xrightarrow{\text{catalyst}} 2CH_3\text{-CHO} + 2H_2O$$

The product, then, is CH_3CHO.

13.6 Equation (13.7) tells us that primary alcohols can be oxidized into aldehydes. Thus, the primary alcohol is oxidized to make an aldehyde, which is then oxidized further to make a carboxylic acid.

13.7 Equation (13.11) tells us that carboxylic acids react with alcohols to make esters:

$$R\text{-}CO_2H + HO\text{-}R' \rightarrow R\text{-}CO_2\text{-}R' + H_2O$$

In this problem, then, R represents CH_3 and R' represents CH_2CH_3. The specific reaction for this problem, then, is:

$$CH_3\text{-}CO_2H + HO\text{-}CH_2CH_3 \rightarrow CH_3\text{-}CO_2\text{-}CH_2CH_3 + H_2O$$

The product, then, is $CH_3\text{-}CO_2\text{-}CH_2CH_3$.

13.8 Unsaturated fatty acids are carboxylic acids with at least one double bond in the carbon chain. Thus, (c) is an unsaturated fatty acid.

13.9 Amino acids react via dehydration reactions to form peptides. Taking the amino acids in the order they are listed in the problem, the reaction is:

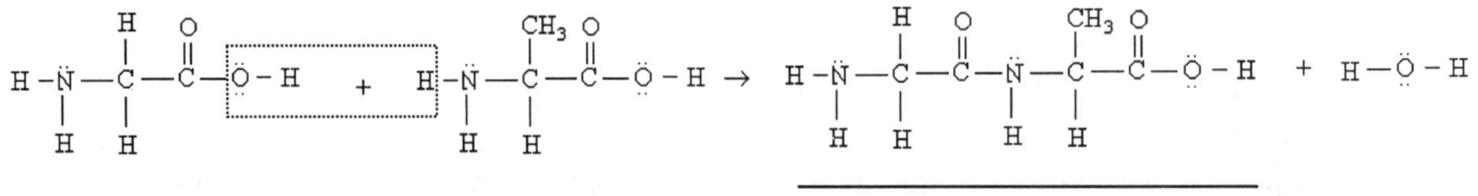

```
        H   O               CH3 O                     H   O       CH3 O
        |   ||              |   ||                    |   ||      |   ||
H -N̈ —C —C —Ö - H   +  H -N̈ —C —C —Ö - H  →  H -N̈ —C —C —N̈ —C —C —Ö - H   +  H —Ö - H
   |   |                  |   |                  |   |       |   |
   H   H                  H   H                  H   H       H   H
```

The Lewis structure of the resulting peptide is underlined above.

13.10 The text tells us that lactose is a disaccharide which is composed of glucose and galactose. Since hydrolysis breaks a disaccharide or polysaccharide into its monosaccharides, a hydrolysis reaction will break lactose down into glucose and galactose. Water will also be produced, of course.

REVIEW QUESTIONS

Consider the following Lewis structures:

a.
```
    H   H   OH  H
    |   |   |   |
H — C — C — C — C — H
    |   |   |   |
    H   H   H   H
```

b.
```
    H   H   O   H   H
    |   |   ||  |   |
H — C — C — C — C — C — H
    |   |       |   |
    H   H       H   H
```

c.
```
    H   H   O
    |   |   ||
H — C — C — C — Ö — H
    |   |
    H   H
```

d.
```
    H   H       H
    |   |       |
H — C — C — Ö — C — H
    |   |       |
    H   H       H
```

e.
```
    H   H
    |   |
H — C — C — OH
    |   |
    H   H
```

f.
```
    H   H   O
    |   |   ||
H — C — C — C — H
    |   |
    H   H
```

g.
```
      CH3   CH3
         \ /
          CH    O
          |     ||
H — N — C — C — Ö — H
    |   |
    H   H
```

h.
```
    H   O       H
    |   ||      |
H — C — C — Ö — C — H
    |           |
    H           H
```

i.
```
         CH3 O          H   O
         |   ||         |   ||
H — N — C — C — N — C — C — Ö — H
    |   |       |   |
    H   H       H   H
```

1. List all structures that correspond to an alcohol.

2. List all structures that correspond to an ether.

3. List all structures that correspond to an aldehyde.

4. List all structures that correspond to a ketone.

5. List all structures that correspond to a carboxylic acid.

6. List all structures that correspond to an ester.

7. List all structures that correspond to an amino acid.

8. List all structures that correspond to a peptide.

9. Which of the structures in represent(s) a secondary alcohol?

10. What kind of reaction is used to join amino acids into proteins as well as monosaccharides into disaccharides? What kind of reaction breaks proteins and disaccharides down?

PRACTICE PROBLEMS

1. Which type of alcohol cannot be oxidized into an aldehyde or a ketone?

2. Which type of alcohol can be turned into a carboxylic acid?

3. What product will be made when $CH_3CH_2CH_2OH$ reacts with CH_3OH in the presence of sulfuric acid and heat?

4. Give the reaction that represents the oxidation of $CH_2CH_2CH_2CH_2CH_2OH$.

5. Give the reaction that represents the oxidation of $CH_3CH_2CHCH_2CH_3$ (with OH on the CH).

6. Give the reaction that will occur when the product in problem #4 is further oxidized.

7. What is the reaction between $CH_3CH_2CO_2H$ and NaOH?

8. What is the product when $CH_3CH_2CH_2CO_2H$ is reacted with CH_3CH_2OH?

9. Consider the following peptide:

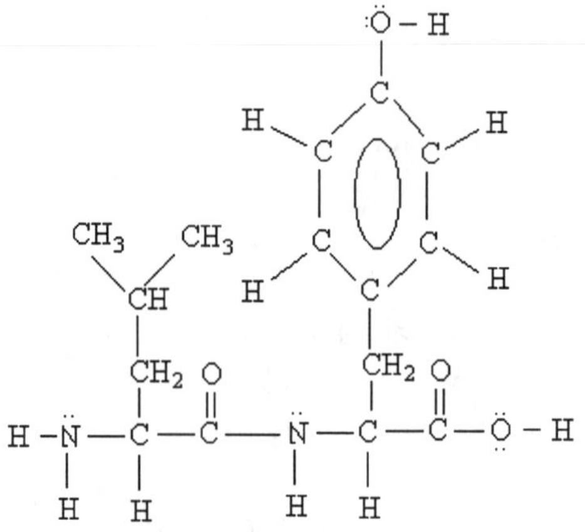

Draw the zwitteron form of each amino acid which makes up this peptide.

10. Consider the following saccharide:

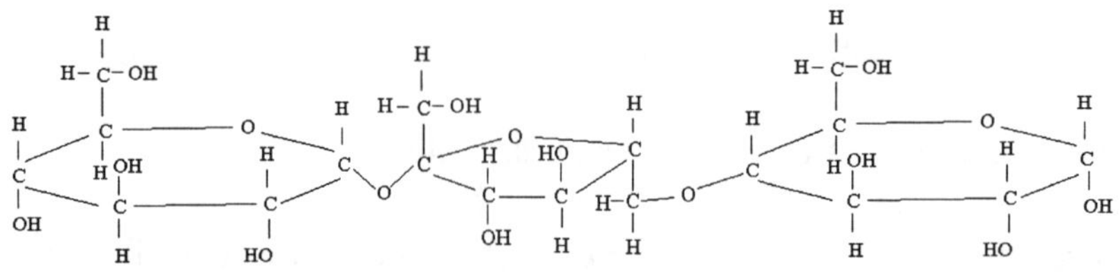

How many monosaccharides make up this molecule?

Module #14: Nuclear Chemistry

Introduction

It is time now to discuss what I consider the most fascinating field of chemistry: nuclear chemistry. In some ways, the term "nuclear chemistry" is contradictory, much like the phrases "jumbo shrimp" and "government assistance." It should already be ingrained in your mind that the electrons in an atom govern virtually all of that atom's chemical properties. Indeed, chemical reactions are simply a matter of atoms either exchanging electrons or re-arranging them. Thus, chemistry is directly linked to electrons. Well, in nuclear chemistry, we ignore the electrons completely, concentrating only on the nucleus. Thus, a nuclear chemist thinks of electrons only as a nuisance, while any other chemist considers electrons the most important part of an atom! That's why I say that the phrase "nuclear chemistry" is somewhat contradictory.

How do nuclear chemists study the nucleus of an atom? After all, an atom is small enough, but a nucleus is even smaller. For example, the average radius of the 1s-orbital of the hydrogen atom is 0.529 angstroms (5.29×10^{-11} m). The radius of the nucleus of that atom (the proton), however, is a mere 1.3×10^{-15} m. That's pretty small. Think about it this way: if a hydrogen atom were expanded until the average radius of its 1s orbital were as big as the walls of a major-league baseball stadium, the nucleus of the atom could be represented by a tiny marble located at the very center of the stadium!

There are three basic ways that nuclear chemists study the properties of the nucleus. First, they study how the mass number of an atom affects its properties. Since the mass number of an atom depends only on its nucleus, the way that the mass number affects the properties of an atom should tell you something about the properties of the nucleus itself. Second, nuclear chemists study radioactivity. This is a process governed almost exclusively by the nucleus, so by studying radioactivity in detail, nuclear chemists can come to a better understanding of the inner-workings of the nucleus. Finally, chemists study what happens when two nuclei collide. If atoms are forced together with enough energy, sometimes the atoms' nuclei will collide and the results can tell us a lot about how the nucleus behaves when it is stressed. In this module, I will discuss each of these means by which nuclear chemists learn about the nucleus.

Binding Energy

When we disregard the electrons in an atom, only the nucleus is left. Since the nucleus contains both protons and neutrons, these particles are generically called **nucleons**.

Nucleon - A term used to refer to both protons and neutrons

Now remember, the nucleus is rather small. Nevertheless, all of the nucleons in an atom are packed tightly together in this small space. Now that should bother you. After all, protons are positively charged and neutrons have no charge. Thus, the nucleus is composed of several positively-charged particles and several neutral particles crammed together in a tight space.

What should those positive charges do to one another? They should repel each other. In fact, the nucleus is so small that the repulsive forces between these positive charges should be enormous. Since there are no negative charges in the nucleus to counteract this repulsive force, you should expect that a nucleus would simply be blown apart because of the repulsion between its protons.

Why doesn't the nucleus explode due to the repulsion between protons? This was one of the great mysteries of science in the early twentieth century. Since nuclei obviously do not explode, nuclear scientists postulated that there was something called the "nuclear force" that was strong enough to hold the nucleus together despite the repulsion between protons. Of course, they had no idea *what* the nuclear force was and *how* it worked. They simply assumed that it must exist, otherwise atoms would not exist.

Scientists began to get a clue as to what holds the nucleus together when nuclear chemists and physicists discovered that the mass of a nucleus is actually *less* than the sum total of the masses of the protons and neutrons which make it up. For example, a ^{4}He atom is composed of 2 protons and 2 neutrons. The mass of an individual neutron is 1.0087 amu, and the mass of an individual proton is 1.0073 amu. Thus, the sum total of the masses of all 4 nucleons that make up a helium-4 nucleus is 4.0331 amu $(2 \times 1.0087 + 2 \times 1.0073)$. Nevertheless, a ^{4}He nucleus (composed of those exact particles) has a mass of only 4.0024 amu. There seems, then, to be a **mass deficit** in this nucleus. The nucleus is 0.0307 amu lighter than the sum of the masses of its individual nucleons. What causes this mass deficit?

The answer to that question can be found in Einstein's **Special Theory of Relativity**, in which he derived the famous equation:

$$E = m \cdot c^2 \qquad (14.1)$$

The Special Theory of Relativity states that energy and mass are really the same thing and are thus interchangeable. Equation (14.1) tells you how much energy it takes to make a certain amount of mass or how much mass can be converted into a certain amount of energy. As a sidelight, it is important to note that Einstein's Special Theory of Relativity is only a special case of his General Theory of Relativity, which tries to explain such esoteric concepts of space and time and how they interact.

How does all of this relate to the nucleus? Well, if the mass of a nucleus is less than the sum total of the mass of its individual nucleons, then the nucleons must "lose" some of their mass when they form a nucleus. This mass is converted to energy, which nuclear scientists call the **binding energy** of the nucleus.

Binding energy - The energy formed from the mass deficit of a nucleus

As long as you know the exact mass of a nucleus, calculating its binding energy is rather easy.

EXAMPLE 14.1

The mass of a ^{7}Li nucleus is 7.0160 amu. What is the binding energy of the nucleus? (The mass of a proton is 1.0073 amu, and the mass of a neutron is 1.0087 amu. The speed of light is 3.00 x 10^{8} m/sec and 1 amu = 1.6605 x 10^{-27} kg.)

Since lithium's atomic number is 3, all lithium atoms have 3 protons. The mass number, which is the sum of the protons and neutrons in a nucleus, therefore indicates that a ^{7}Li nucleus has 4 neutrons. The sum of the masses of 3 protons and 4 neutrons is:

$$3 \times (1.0073 \text{ amu}) + 4 \times (1.0087 \text{ amu}) = 7.0567 \text{ amu}$$

Since the mass of a ^{7}Li nucleus is only 7.0160 amu, there is a mass deficit of 7.0567 amu - 7.0160 amu = 0.0407 amu. This mass deficit is converted to energy according to Equation (14.1). To use this equation, however, we must have consistent units. Since we have the speed of light in m/sec, then the energy will come out in Joules as long as the mass is in kilograms (remember, a Joule is a (kg·m^{2})/sec^{2}). Thus, we must first convert the mass deficit to kg:

$$\frac{0.0407 \cancel{\text{amu}}}{1} \times \frac{1.6605 \times 10^{-27} \text{ kg}}{1 \cancel{\text{amu}}} = 6.76 \times 10^{-29} \text{ kg}$$

Now we can use Equation (14.1):

$$E = m \cdot c^2 = (6.76 \times 10^{-29} \text{ kg}) \cdot (3.00 \times 10^8 \frac{\text{m}}{\text{sec}})^2 = 6.08 \times 10^{-12} \frac{\text{kg} \cdot \text{m}^2}{\text{sec}^2} = \underline{6.08 \times 10^{-12} \text{ J}}$$

Although this doesn't sound like a lot of energy, remember that this is for a *single* atom. In a mole (7.0160 g) of ^{7}Li atoms, the total binding energy is

$$(6.08 \times 10^{-12} \text{ J}) \times (6.02 \times 10^{23}) = 3.66 \times 10^{12} \text{ J}$$

which is quite a bit of energy!

As its name implies, binding energy tells us how tightly bound the nucleons are in the nucleus. The larger the binding energy, the stronger the nucleus holds its nucleons together. If you take the binding energy of a nucleus and divide it by the total number of nucleons in the nucleus, you get the **binding energy per nucleon** for that nucleus. This quantity gives you an idea of how strongly each nucleon is bound within the nucleus. If you think about it, the binding energy per nucleon tells you how stable a nucleus is. After all, if the binding energy per nucleon is high in a nucleus, the nucleus holds tightly to each of its nucleons. If the binding energy per nucleon is low, the nucleus' hold on its nucleons is weak.

If you calculate the binding energy per nucleon for several nuclei, you will find that this important quantity changes from nucleus to nucleus. In other words, some nuclei are more stable than others. Figure 14.1 illustrates a plot of binding energy per nucleon as a function of the mass number of a nucleus.

FIGURE 14.1
The Binding Energy Per Nucleon

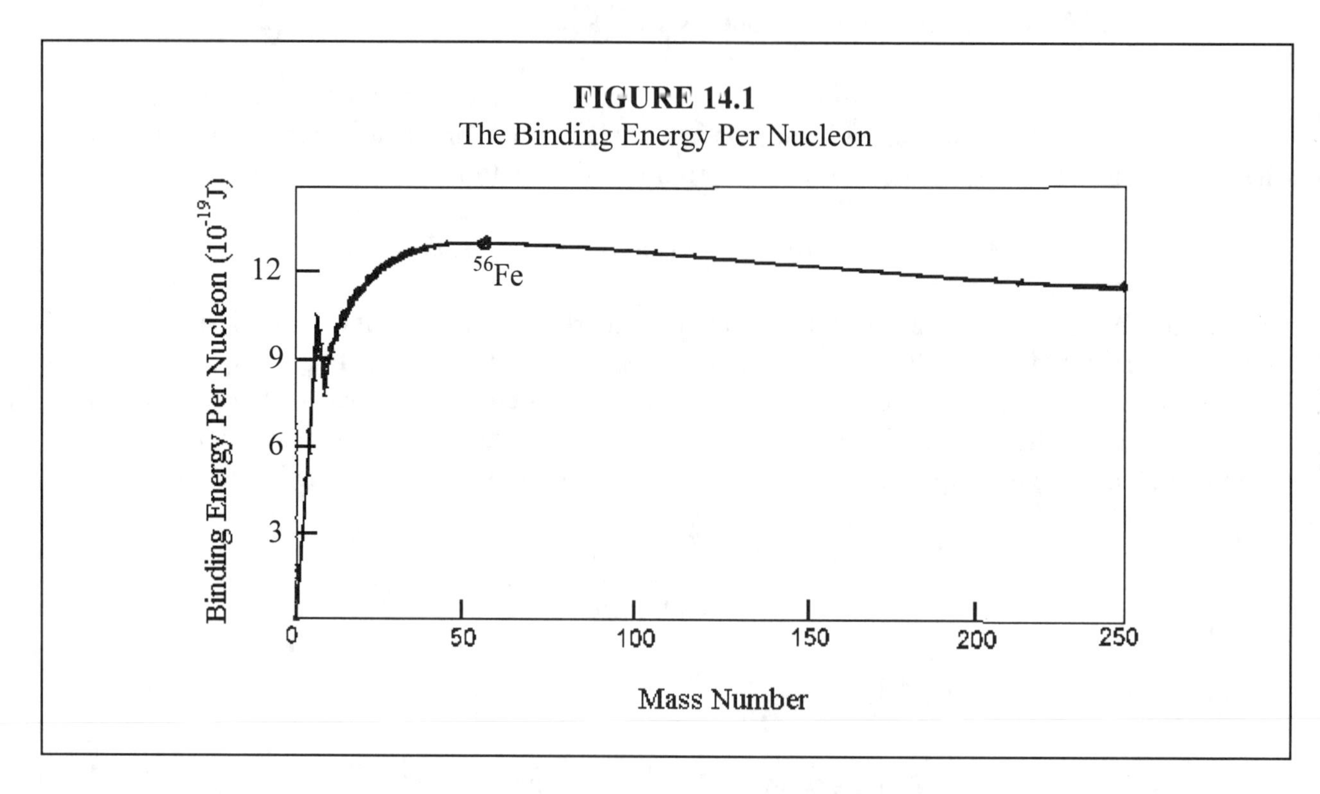

As you can see from the figure, the binding energy per nucleon rises with increasing mass number until the mass number reaches 56, where the maximum binding energy per nucleon exists. For mass numbers higher than 56, the binding energy per nucleon decreases. This tells us that the most stable nuclei are those with mass numbers around 56. In fact, the most stable nucleus in Creation is ^{56}Fe, because it has the maximum binding energy per nucleon.

ON YOUR OWN

14.1 Calculate the binding energy per nucleon for ^{56}Fe. (Use the data given in the example as well as the fact that an ^{56}Fe nucleus has a mass of 55.9349.

14.2 The binding energy of ^{7}Be is 5.739 x 10^{-12} J. What is the mass of ^{7}Be in amu?

The Strong Nuclear Force

So now we know what holds the nucleus together, right? The binding energy of a nucleus binds the nucleons together in the nucleus. That's all well and good, but there is still one important question we must answer: *how does the binding energy do it*? That question was a matter of speculation for quite some time. Some scientists thought that the binding energy formed some sort of "force field" around the nucleus, keeping the nucleons inside. Others thought that the energy somehow acted like glue, "sticking" the nucleons together.

In 1937, a nuclear physicist by the name of Heidiki Yukawa postulated that nucleons stayed together because, at short distances, they exchanged tiny particles called **pions** (pie' ons). Yukawa thought that the binding energy was used to give these pions kinetic energy, allowing them to travel from one nucleon to another. In other words, Yukawa believed that nucleons actually gave up a portion of their mass to form a small particle called a pion. Some of the mass that the nucleons gave up would go towards making the pion, and the rest would be converted to kinetic energy, allowing the pion to travel. Based on the properties of nuclei that were already known, Yukawa actually predicited what the mass of a pion should be.

Yukawa further believed that these pions can only exist for a very short time. As a result, he classified them as **short-lived particles**. Thus, Yukawa believed that a nucleon would form a pion, and the pion would begin to travel away from the nucleon. The pion, however, would not be able to live for very long. Thus, it would quickly encounter another nucleon and be absorbed by that nucleon. Since Yukawa believed that it is beneficial for nucleons to make, exchange, and absorb pions, he believed that nucleons crammed together into the nucleus in order to be able to do those things. Of course, all of this was just an hypothesis until 1947, when nuclear physicists discovered pions and found that they had almost exactly the mass that Yukawa predicted.

As a result of Yukawa's theorizing and the discovery of the pion, nuclear scientists now view the nucleus as a place full of busy activity. Nucleons in the nucleus are continually making, exchanging, absorbing, and re-making pions. The desire for nucleons to do this is so overwhelming that it overcomes the electromagnetic repulsion between protons, allowing protons to live very close to one another. Because pions are short lived, nucleons can only exchange these particles when the nucleons are quite close. Thus, pion exchange exists only in the nucleus.

The binding energy of a nucleus, then, is mostly used to facilitate the exchange of pions. The attraction that nucleons feel as a result of this exchange is called the **strong nuclear force**. The strong nuclear force exists only between nucleons (because only they can exchange pions). It is also a very short-range force, because the pions that are exchanged can only exist for a brief period of time. Thus, a pion must travel from one nucleon to another before its lifetime is up. Finally, for very short distances the nuclear force is incredibly strong, because the desire for nucleons to exchange pions is strong. As a result, for distances on the order of 10^{-15} m, the strong nuclear force is significantly stronger than the electromagnetic force.

The Stability of a Nucleus

Despite the fact that the strong nuclear force is able to hold nucleons together, it is not able to hold just any combination of nucleons together. As a result, a nucleus cannot be made from just any combination of neutrons and protons. Instead, there are certain combinations of neutrons and protons that are stable, and certain combinations that are not.

FIGURE 14.2
A Plot of Neutron Number Versus Proton Number

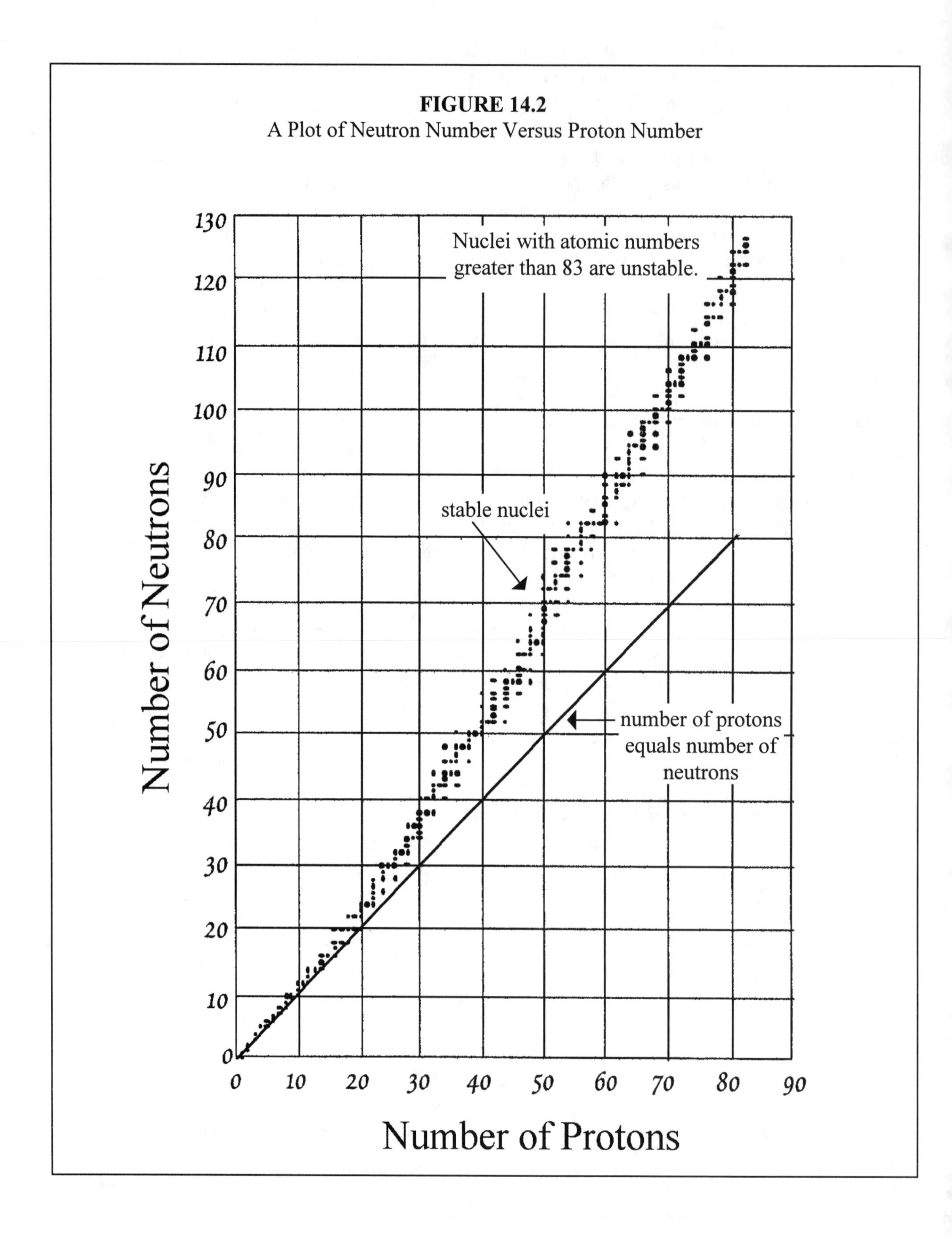

As noted in the figure, the dots represent the known, stable nuclei. Thus, if the number of protons and neutrons in a nucleus places it at one of the dots in the graph, then the nucleus is stable. If not, the nucleus is not stable. The reason that some nuclei are stable and some are not is beyond the scope of this course. In part, this is because nuclear scientists really don't fully understand the intricacies of what makes a nucleus stable and what makes it unstable.

If you look at the figure, you will see that for small nuclei (those that have only a few nucleons), the dots lie right along a line that corresponds to the number of protons equal to the number of neutrons. This means that for small nuclei, an equal number of neutrons and protons leads to a stable nucleus. Indeed, nuclei like ^{4}He, ^{16}O, and ^{40}Ca are all stable. Notice, however, that as the number of nucleons in the nucleus increases, the dots begin to rise farther and farther above the line that indicates an equal number of protons and neutrons. This tells us that as a nucleus gets larger, it can only be stable if it has more neutrons than protons.

Does this mean that the only nuclei in Creation are the ones represented by the dots? No, of course not. It means that the only *stable* nuclei in Creation are the ones represented by the dots. There are plenty of *unstable* nuclei in Creation as well. This might surprise you a bit. After all, if a nucleus is unstable, how can it exist? The answer is quite simple: It can exist, but only for a certain amount of time. When a nucleus is unstable, we call it **radioactive**. A radioactive nucleus, also called a **radioactive isotope**, will eventually change into another nucleus. As you will soon see, however, that can often take a *significant* amount of time.

ON YOUR OWN

14.3 Using Figure 14.2, determine whether or not each of the following nuclei is stable.

a. ^{237}U b. ^{57}Fe c. ^{100}La d. ^{12}C

Radioactivity

The thick curve formed by the dots in Figure 14.2 is often called the **valley of stable nuclei**. If a nucleus is not in the valley of stable nuclei, then in order to become stable, it must get there. How can it do that? A nucleus can move to the valley of stable nuclei by changing its number of protons and/or neutrons. It accomplishes this through a process called **radioactive decay**.

Consider, for example, the nucleus ^{241}Am. According to its symbol, this nucleus has 95 protons and 146 neutrons. As stated in Figure 14.2, all elements with atomic number greater than 83 are unstable. To try and reach stability, then, the ^{241}Am nucleus emits a ^{4}He nucleus. Remember, a ^{4}He nucleus has 2 protons and 2 neutrons. Thus, the ^{241}Am nucleus actually loses 2 protons and 2 neutrons, which it ends up "spitting out" in the form of a ^{4}He nucleus. So, ^{241}Am has 95 protons and 146 neutrons. If it loses 2 protons and two neutrons, the result is a nucleus with 93 protons and 144 neutrons. This nucleus, ^{237}Np, is still unstable, so *it* will emit an alpha particle, continuing the decay. This will happen over and over again until the resulting nucleus is stable.

This process can be represented in the form of a reaction. Please understand that this is not a *chemical* reaction, it is a *nuclear* reaction:

$$\underset{\text{Radioactive isotope}}{^{241}_{95}Am} \rightarrow \underset{\text{Daughter product}}{^{237}_{93}Np} + \underset{\text{Alpha particle}}{^{4}_{2}He} \tag{14.2}$$

There are three important things I need to tell you about this equation. First notice that there are subscripts before the atomic symbols as well as superscripts. As always, those superscripts represent the mass number of the nucleus. The subscripts, on the other hand, represent the atomic number of the nucleus. Thus, the "95" subscript tells you that the atomic number of Am is 95. Now, of course, those subscripts are NOT necessary. After all, the symbol of the nucleus tells you the atomic number. All Am nuclei have 95 protons, so they all have an atomic number of 95. Even though the subscripts are not necessary, they will make things convenient a little later on, so I will keep using them throughout the module.

The next thing I want you to notice about the equation is that the subscripts on one side of the equation add up to the sum of the subscripts on the other side of the equation. This should make sense to you. After all, the subscripts represent the number of protons in each nucleus. On the reactant side of the equation, then, there are 95 protons. Since protons are not destroyed in this process, there should be 95 protons on the products side of the equation as well. In the same way, the mass numbers (the superscripts) represent the total nucleons in each nucleus. Since nucleons are not being destroyed in the process, the sum of all nucleons in the reactants must be equal to the sum of the nucleons in the products. When these two conditions are met, the nuclear equation is said to be **balanced**.

Finally, notice how I have labeled the participants in the equation. The only reactant is called the **radioactive isotope**. Some nuclear chemists also call it the **parent isotope**. The nucleus that results from the process is called the **daughter product**. Finally, the ^{4}He that is emitted in the reaction is called an **alpha particle**. Why is it called an alpha particle rather than a helium nucleus? The reason is historical. Scientists had determined that radioactivity exists long before they understood it. All they knew was that there were three distinctly different radioactive processes which resulted in three distinctly different particles being emitted. Since scientists at that time did not know what those particles were, they simply called them "alpha particles," "beta particles," and "gamma particles" (alpha, beta, and gamma are the first three letters in the Greek alphabet). Despite the fact that we now know what all of these particles are, we still cling to the old terminology. Thus, the radioactive process which produces alpha particles is typically called **alpha decay**.

Now that you know what alpha decay is, I will continue by discussing **beta decay**. When a nucleus undergoes beta decay, an amazing thing happens: *a neutron turns into a proton*! How does this happen? Well, look at the mass of a neutron (1.0087 amu) as compared to a proton (1.0073 amu). Also, consider the fact that a neutron is neutral, whereas a proton is positive. Suppose a neutron were able to emit an electron. What would happen? Its mass would decrease, and it would turn positive. After all, if a neutral particle releases a negative charge, the particle left over must be positive. Thus, *a neutron turns into a proton by emitting an electron.* This process can be written as the following reaction:

$$^{1}_{0}n \rightarrow ^{1}_{1}H + ^{0}_{-1}e \tag{14.3}$$

Now what in the world do those symbols mean? Well, the symbol on the reactant side represents a neutron. After all, how many protons are in a neutron? None, of course! Thus, the subscript for a neutron in a nuclear equation must be zero. What is the mass number of a neutron? Well, the mass number is the sum of all neutrons and protons. Thus, for a single neutron, the mass number is 1. That's why we use the symbol $^{1}_{0}n$ to represent a neutron. In the same way, what is a proton? A proton is simply a hydrogen-1 nucleus. Thus, the symbol $^{1}_{1}H$ is used in a nuclear reaction to represent a proton. Finally, the symbol $^{0}_{-1}e$ is used to symbolize an electron in a nuclear equation. After all, an electron has no protons and no neutrons in it, so its mass number is zero. Since it has the opposite charge of a proton, you can think of the atomic number of an electron as -1.

Notice that using these symbols, Equation (14.3) balances just like Equation (14.2). The total number of protons on the reactant side of the equation is 0. The sum total of subscripts on the products side of the equation is also zero. The total number of nucleons on the reactant side of the equation is 1, and the sum of all superscripts on the products side of the reaction is also 1. Because these kinds of symbols for protons, neutrons, and electrons allow us to balance nuclear equations, I will continue to use them throughout this module.

It is important for you to know that Equation (14.3) represents a spontaneous reaction that will occur with *all* neutrons that are not a part of the nucleus. The rate of this reaction is a bit slow (it has a half-life of 10 minutes); nevertheless, it will eventually happen to any free neutron. Interestingly enough, this reaction is usually not spontaneous for a neutron that exists in a stable nucleus. It is theorized that the pion exchange which goes on in a nucleus stabilizes the neutrons so that they do not decay into protons and electrons. Without that pion exchange, the neutron is not stable, and it will eventually decay into a proton and an electron.

Even if a neutron is in a nucleus, it can still decay if the nucleus itself is not stable. For example, the nucleus ^{90}Sr lies just to the left of the valley of stable nuclei in Figure 14.2. In order to move to the valley of stable nuclei, one if its neutrons will release an electron to make a proton.

$$\underset{\text{Radioactive isotope}}{^{90}_{38}Sr} \rightarrow \underset{\text{Daughter product}}{^{90}_{39}Y} + \underset{\text{Beta particle}}{^{0}_{-1}e} \tag{14.4}$$

Notice that this equation is balanced, because the superscripts on one side add up to the superscripts on the other, and the subscripts on one side add up to the subscripts on the other. Also, even though the electron produced in this process is just an electron, we still call it a **beta particle**. As a result, this process is called beta decay.

The last type of radioactive decay that I want to discuss is **gamma decay**. It turns out that a gamma particle (also called a **gamma ray**) is really just a photon. Remember, photons are light "particles," so a gamma particle is really just a "piece" of light. The light has a very large energy (thus a short wavelength), but it is still just light. Since light has no protons and no neutrons in it, a gamma particle is symbolized as ${}^{0}_{0}\gamma$, where the symbol γ is the lower-case Greek letter gamma. Since a gamma particle has no neutrons or protons, the emission of a gamma particle does not affect the identity of the nucleus. However, it does remove energy from the nucleus. Thus, if a nucleus is stable but has too much energy, it will rid itself of the extra energy by emitting a gamma particle. For example, if a ${}^{90}Y$ nucleus has too much energy, it will emit a gamma particle.

$$ {}^{90}_{39}Y \rightarrow {}^{90}_{39}Y + {}^{0}_{0}\gamma \quad (14.5) $$

This process, not surprisingly called **gamma decay**, simply rids the ${}^{90}Y$ nucleus of its excess energy. For historical reasons, gamma rays are also called **X-rays**. When you get an X-ray in order for a doctor to diagnose a condition, gamma rays are being shot at you.

Now that you have been introduced to the three forms of natural radioactivity, study the examples and solve the "on your own" problems that follow to be sure you understand how to deal with nuclear equations.

EXAMPLE 14.2

${}^{14}C$ is a radioactive isotope that goes through beta decay. What is the daughter product of this decay? Write a balanced equation for the decay process.

According to the periodic chart, carbon has an atomic number of 6. This tells us that a ${}^{14}C$ atom has 6 protons and 8 neutrons in its nucleus. When a radioactive isotope undergoes beta decay, one of its neutrons turns into a proton. Thus, it will end up with one more proton and one less neutron. The daughter product (the nucleus that results from the beta decay), then, will have 7 protons and 7 neutrons. According to the chart, all atoms with 7 protons are symbolized with an "N." The mass number of this nitrogen atom will be 7+7=14. Thus, the daughter product is ${}^{14}N$.

Now that we know the daughter product, the balanced equation is rather simple. Using the notation for a beta particle that was discussed above, the equation is:

$$ {}^{14}_{6}C \rightarrow {}^{14}_{7}N + {}^{0}_{-1}e $$

Note that this equation is balanced, because the subscripts on one side add up to the subscripts on the other, as do the superscripts.

^{232}Th is a radioactive isotope that goes through alpha decay. What is the resulting daughter product?

According to the chart, thorium (Th) atoms have 90 protons. Thus, this particular atom has 90 protons and 142 neutrons in it. When it goes through alpha decay, it actually spits out 2 protons and two neutrons in the form of a helium-4 nucleus. The result will be only 88 protons and 140 neutrons in the daughter product. The chart tells us that Ra is the symbol for all atoms with 88 protons. The mass number of the resulting nucleus will be 88+140 = 228. Thus, the daughter product is $\underline{^{228}\text{Ra}}$.

Now that we know the daughter product, the balanced equation is rather simple. Using the notation for an alpha particle that was discussed above, the equation is:

$$\underline{{}^{232}_{90}\text{Th} \rightarrow {}^{228}_{88}\text{Ra} + {}^{4}_{2}\text{He}}$$

Note that this equation is balanced, because the subscripts on one side add up to the subscripts on the other, as do the superscripts.

Write a balanced reaction for the gamma decay of ^{22}Na.

Gamma decay simply takes energy away from the nucleus in the form of light. It does not change the identity of the nucleus. Thus, the daughter product is still ^{22}Na. The equation, then, is particularly easy to produce:

$$\underline{{}^{22}_{11}\text{Na} \rightarrow {}^{22}_{11}\text{Na} + {}^{0}_{0}\gamma}$$

A radioactive decay process starts with a ^{234}Th nucleus and produces a ^{234}Pa nucleus. What kind of radioactive decay is this?

In this case, we are asked to figure out the radioactive decay by examining the radioactive isotope (^{234}Th) and the daughter product (^{234}Pa). In ^{224}Th, there are 90 protons and 144 neutrons. In ^{234}Pa, there are 91 protons and 143 neutrons. Thus, this must be <u>beta decay</u>, because the daughter product has one more proton than the radioactive isotope and one less neutron. This can only happen if a neturon turns into a proton.

ON YOUR OWN

14.4 Write a balanced nuclear equation for the beta decay of ^{87}Rb.

14.5 The daughter product of an alpha decay process is ^{220}Rn. What was the radioactive isotope that went through alpha decay?

14.6 A nucleus goes through radioactive decay but does not change its number of neutrons or protons. What kind of decay process did the nucleus undergo?

Before I leave this section, I want to point out something. Radioactive decay reactions as discussed in this section of the module are highly exothermic. If you add up the masses of the products produced in radioactive decay, you will find that the sum is less than the mass of the radioactive isotope. The "missing mass" is converted into energy according to Equation (14.1). Most of that energy is released as heat. In fact, radioactive decay processes are so exothermic that geophysicists speculate they are partially responsible for keeping the earth's interior hot!

Artificial Radioactivity

The three kinds of radioactivity that I discussed so far make up the phenomenon known as "natural radioactivity." The reason for this term is simple. These three types of radioactive decay are the only ones that occur naturally here on earth. As technology has improved, however, scientists have become able to synthesize their own nuclei in a nuclear chemistry/nuclear physics lab. As a result, scientists have artificially produced nuclei that decay via other mechanisms. The two "artificial" forms of radioactive decay are **electron capture** and **positron emission**.

In electron capture, a proton in a nucleus captures an electron (typically from the electrons that surround the nucleus). What will be produced when a proton captures an electron? Well, when a neutron releases an electron, the result is a proton. Electron capture is the reverse of this process. Thus, when a proton captures an electron, the result is a neutron. The following is an example of an electron capture reaction:

$$^{14}_{8}O + ^{0}_{-1}e \rightarrow ^{14}_{7}N$$

In this reaction, ^{14}O has too many protons and not enough neutrons. To fix this problem, one of the protons captures an electron from the electron orbitals and the result is a stable ^{14}N nucleus.

In positron emission, a proton emits a positron to become a neutron. What is a positron? Well, it is a form of **antimatter**. A positron is, in fact, an anti-electron. Although this sounds a bit like Star-Trek, it is reality. A positron has positive charge and behaves just the opposite of an electron. Not surprisingly, then, a positron is symbolized in with a $^{0}_{+1}e$ in nuclear chemistry. Interestingly enough, when a positron and an electron encounter one another, they destroy each other, leaving nothing behind but a gamma ray (high energy light).

$$^{0}_{+1}e + ^{0}_{-1}e \rightarrow ^{0}_{0}\gamma$$

This process, called **annihilation**, is what makes a positron antimatter. Matter and antimatter destroy each other. Thus, a positron is an anti-electron because, when it encounters an electron, the two particles destroy each other, leaving only energy (no mass) behind.

In the radioactive process known as positron emission, a proton emits a positron to turn into a neutron:

$$^{8}_{5}\text{B} \rightarrow \, ^{8}_{4}\text{Be} + \, ^{0}_{+1}\text{e}$$

Notice that this process has the same effect as electron capture, because it transforms a proton into a neutron.

The Rate of Radioactive Decay

As I said before, the reason that unstable nuclei exist in Creation is that although all unstable nuclei *eventually* decay into stable nuclei, this can often take quite some time. Remember from your first year chemistry course that there are always two things you must consider when studying a reaction. You first have to ask whether or not the reaction is spontaneous. That is a thermodynamics issue. You then have to ask how quickly the reaction proceeds. That is a kinetics issue. The kinetics of radioactive decay are worth studying.

All known radioactive decay processes are governed by first-order kinetics. In other words, the rate of the reaction is directly proportional to the amount of radioactive isotope that exists. If you recall from Module #11, the concentration of a first-order *chemical* reaction can be related to time according to the following equations:

$$[A] = [A]_o \cdot e^{-kt} \qquad (11.5)$$

$$t_{1/2} = \frac{0.693}{k} \qquad (11.6)$$

In these equations, [A] represents the concentration of the reactant at any time, $[A]_o$ represents the original concentration of the reactant, k is a constant that changes from reaction to reaction, and $t_{1/2}$ is the half-life of the reaction. In radioactive decay, concentration is not an issue. However, the *number* of radioactive isotopes is what's important. Thus, Equation (11.5) changes just a bit for radioactive decay:

$$N = N_o \cdot e^{-kt} \qquad (14.6)$$

In this equation, N represents either the *number* or *mass* of radioactive isotopes at any given time, N_o is the initial number or mass of radioactive isotopes, and k is a constant that changes from radioactive isotope to radioactive isotope. The half-life of radioactive decay reactions is related to k through Equation (11.6), just as chemical reactions are.

Like most chemical reactions, some radioactive decay reactions proceed quickly. For example, the alpha decay of ^{214}Po into ^{210}Pb has a half-life of 0.00016 seconds. That's a pretty fast reaction. On the other hand, some radioactive processes are incredibly slow. For example, ^{238}U alpha decays via a reaction whose half-life is 4.41×10^9 *years*! Now *that's* a slow reaction! Thus, unstable nuclei abound in Creation because many of them take a long, long time to decay.

Since we went through first-order rate kinetics in Module #11, I will not spend any time reviewing it here. Nevertheless, I do want you to be able to apply Equations (14.6) and (11.6) to radioactive decay. Therefore, please answer the following "on your own" problem to make sure you remember how to use these equations.

ON YOUR OWN

14.7 The alpha decay of ^{210}Bi proceeds with a half-life of 5.00 days. A chemist makes 100.0 grams of the isotope.

a. How many grams will be left in 15.00 days?
b. How many grams will be left in 18.00 days?

The Dangers of Radioactivity

Now that you know a little bit about radioactivity, you might be interested in knowing why everyone is so afraid of it. Well, part of the fear is based totally on ignorance, and part of the fear is based on fact. Radioactivity *can be* dangerous, but it is *not always* dangerous. That's a good thing, too, because we are *constantly* being exposed to radioactivity. If you have brick or mortar in the walls of your home, they are radioactive. By standing near them, you are exposed to beta particles. You are exposed to gamma rays when you are outside in the sun. If you have a smoke detector in your house, you are exposed to alpha particles, because the main detection component of a smoke detector undergoes alpha decay. In fact, you are exposed to beta particles each time you get close to someone, because people themselves are radioactive! It's a good thing, then, that radioactivity is not always dangerous.

The first thing you have to understand is why radioactivity can be dangerous. Radioactivity does not act like a poison. A poison is dangerous because it chemically reacts with your body, causing chemical processes to occur in your body which should not occur. This upsets your body's chemistry, causing sickness or even death. Some poisons actually build up in your body. As you take them in small doses, they do not cause you any problems. However, as they continue to build up in your body, they eventually start causing chemical reactions that shouldn't happen in your body, and that's when you are in trouble.

Unlike poisons, radioactivity is not dangerous because it can upset your body's chemistry. It also cannot build up in your body. Instead, radioactivity affects your body much like a tiny machine gun. You see, the danger in radiation comes from the particles that are emitted during the radioactive decay. Depending on the isotope involved, radioactive decay involves a nucleus "spitting out" something. In alpha decay, the nucleus spits out an alpha particle. In beta decay, it spits out a beta particle. In gamma decay, the nucleus spits out high energy light. There is nothing chemically poisonous about these things. They are dangerous, however, because they have a lot of energy.

When alpha, beta, or gamma particles collide with atoms or molecules in their way, the energy of the collision can ionize the atom or molecule with which the particle collides. Thus, alpha, beta, and gamma particles are referred to as **ionizing radiation**, because they ionize matter as they pass through it. If you happen to be unfortunate enough to be in the way of the emitted particle, it might collide with one of your cells. The vast majority of the time, when an alpha, beta, or gamma particle collides with a cell, it results in the cell's death, because it ionizes chemicals in the cells that should not be ionized. Every now and again, however, the cell will not die. If the particle hits the cell just right, the resulting ionization might mutate the cell's DNA rather than kill the cell.

Do you see why I say that radioactivity acts like a tiny machine gun? When you have a sample of radioactive material, each atom in that sample can "shoot" one "bullet" (an alpha particle, a beta particle, or a gamma ray). Since there are trillions and trillions of atoms in even a small sample of matter, that means that a sample of radioactive isotopes can shoot off trillions and trillions of these 'bullets." If you happen to be in the path of these bullets, each bullet that hits you will most likely kill an individual cell. Thus, a radioactive sample is like a tiny machine gun that kills you one cell at a time. Every now and again, however, rather than killing a cell, the particle will cause a mutation in the cell's DNA.

Sounds dangerous, doesn't it? Well, it *can* be dangerous, but *not necessarily*. You see, your body *expects* cells to die. God therefore designed your body to reproduce cells. This helps you grow and mature, and it also replaces cells that die. When you scratch an itch, for example, you actually kill as many as several hundred cells. This is no problem, as your body quickly replaces them. Thus, as long as your cells do not die faster than they can be replaced by your body, there is no real problem.

When your cells are being destroyed by the little "bullets" that are being "shot" from a sample of radioactive isotopes, then, there is no problem as long as the "bullets" are not killing your cells faster than your body can replace them. If you are exposed to too much radiation too quickly, then your cells will be killed faster than your body can replace them, leading to radiation burns, organ damage, and the like.

What about the chance for mutating a cell's DNA? Isn't that bad? Well, yes, but once again, it depends on the amount of mutation that is going on. Everyone's body has a few mutant cells. Most of them simply die off. The bad thing about mutation is that a mutant cell can result in cancer or some other sickness. This happens only rarely, however, so once again, a few mutant cells in your body is not a bad thing. Everyone has them. The problem only occurs when you have too many mutant cells. Thus, as long as you are not exposed to too much radiation, the danger is minimal.

In the end, then, the important thing to remember about the dangers of radioactivity is that it depends on the level of radioactivity to which you are exposed. A small amount of radioactivity is reasonably safe, a large amount is not. How much radioactivity exposure is too much? Well, nuclear scientists have come up with ways of measuring how much ionizing radiation people are exposed to. They refer to this as the **dose** of radiation to which a person is exposed.

There are two units nuclear scientists use to measure radiation dosage. They are the **rad** (radiation absorbed dose) and the **rem** (roentgen equivalent in man). The rad is the amount of radiation that will deposit 100 Joules of energy into a kilogram of living tissue. This is a fine measure of radiation exposure, but it neglects the fact that certain types of radiation are more damaging to biological systems than others. Alpha particles, for example, do more damage to tissue per Joule they deposit because of the details of how alpha particles ionize matter. As a result, alpha particles are considered "more effective" at destroying living tissue.

To take this into account, nuclear scientists have come up with the **RBE** (relative biological effectiveness) factor. This factor is different for each type of ionizing radiation. Alpha particles, for example have an RBE factor of 4 while gamma and beta particles have an RBE factor of 1. When the number of rads are multiplied by the RBE factor, the result is the dosage in rems.

$$\text{rems} = (\text{RBE}) \times \text{rads} \tag{14.7}$$

Thus, if you are exposed to 0.010 rads of beta particles, your radiation dose is 1 x 0.010 = 0.010 rems. If you are exposed to 0.010 rads of alpha particles, your radiation dose 4 x 0.010 = 0.040 rems.

Remember when I said that brick and mortar are radioactive, as well as smoke detectors and other people? If you add up all of the radiation you are exposed to from such sources, your average radiation dose each year would be about 0.2 rems. Since studies by radiation biologists indicate that a lethal dose of radiation is about 470 rems, the dose of radiation you get as a result of everyday activity is simply too minimal to be worried about.

Even if you are in a position in which you are exposed to large amounts of radioactivity, there are ways you can protect yourself. For example, it is possible to stop the little "bullets" before they ever reach your body. For example, alpha particles are extremely weak in terms of how much matter they can travel through. If you put a piece of paper between you and the radioactive source emitting the alpha particles, the vast majority of those alpha particles will stop in the paper. As a result, they will never hit you. Beta particles can travel through obstacles a bit better. It typically takes a thin sheet of metal to stop most of the beta particles coming from a radioactive isotope that emits them. Finally, gamma rays are the strongest type of radiation, requiring several inches of lead to stop them.

Thus, one way you can protect yourself is to block the radiation before it hits you. This method is called "shielding." The other way you can protect yourself from an intensely radioactive source is to simply move away from it. The farther you move away, the fewer "bullets" can hit you. Of course, most people will never be exposed to a large amount of radiation in their lifetime, so they will never be faced with such a situation.

ON YOUR OWN

14.8 People who regularly work with large samples of radioactive isotopes sometimes wear special suits that are lined with a thin layer of lead or other heavy material. What kinds of radiation are these people protected from when wearing such a suit?

Radioactive Dating

The fact that radioactive isotopes decay at a measurable rate allows scientists to use radioactive decay as a means of dating objects whose age we do not know. This is known as **radioactive dating**. Although radioactive dating can be accurate under certain circumstances, it is important to note that it has some serious weaknesses as well. As a result, radioactive dating techniques must be viewed rather critically. Despite the fact that some scientists will try to convince you that radioactive dating is an accurate means of determining the age of an object, the scientific facts tell quite a different story.

The best way of examining the strengths and weaknesses of radioactive dating is to examine one of the radioactive dating methods in detail. Since ^{14}C is probably the best known radioactive dating technique, I might as well discuss that one. As I have already mentioned, ^{14}C decays by beta decay with a half-life of 5,730 years. It turns out that all living organisms contain a certain amount of ^{14}C, which is part of the reason that all living organisms are radioactive.

Interestingly enough, living organisms continually exchange ^{14}C with their surroundings. Human beings, for example, exhale carbon dioxide, some of which contains ^{14}C. In addition, human beings eat other organisms (plants and animals), which contain ^{14}C as well. Finally, part of the air that we inhale is made up of carbon dioxide, some of which contains ^{14}C. Thus, organisms are continually exchanging ^{14}C with their environment. The practical result of all of this exchange is that, at any time when an organism is alive, it contains the same amount of ^{14}C as does the atmosphere around the organism.

This changes when the organism dies, however. At that point, the ^{14}C exchange ceases. Thus, the organism cannot replenish its supply of ^{14}C, and the amount of ^{14}C in the organism begins to decrease because of the beta decay of ^{14}C. The half-life of this process is 5730 years, so the decay happens slowly. Nevertheless, it is a measurable effect. In general, then, organisms that have been dead a long time tend to have less ^{14}C in them as compared to those that have been dead for only a short time.

Now if you think about it, this fact can be used to measure the length of time that an organism has been dead. After all, if we know how much ^{14}C was in an organism when it died, and if we measure the amount of ^{14}C in it now, the difference will be the amount of ^{14}C that has decayed away. With that information, Equation (14.6) will tell us how long the organism has been dead. Pretty simple, right?

Well, it *would* be simple, *if* we knew how much ^{14}C was in the organism when it died. The problem is, how do we figure that out? After all, no one was around to measure the amount of ^{14}C in the organism when it died; thus, we must make an *assumption* about how much ^{14}C would have been measured if someone had been there to measure it. Now there is nothing wrong with making assumptions in science. The trick is that you have to know your assumptions are accurate.

In the case of ^{14}C dating, scientists assume that, on average, the amount of ^{14}C in the atmosphere has never really changed that much. They assume that the amount of ^{14}C in the atmosphere today is essentially the same as it was 100 years ago, 1,000 years ago, etc. Thus, when the age of a dead organism is being measured with ^{14}C dating, we assume that the amount of ^{14}C it had when it died was the same as the amount of ^{14}C that is in the atmosphere now. That gives us a value for how much ^{14}C was initially in the dead organism. We can measure the amount of ^{14}C that is in the organism now and then determine how long the organism has been dead.

Notice, however, that the age we get from this process is completely dependent on the assumption that we made about how much ^{14}C was in the organism when it died. If that assumption is good, the age we calculate will be accurate. If that assumption is bad, the age we calculate will not be accurate. So the question becomes, "Is the assumption accurate?" In short, the answer is, "No."

Through a process involving tree rings, there is a way we can measure the amount of ^{14}C in the atmosphere in years past. When you cut down a tree, you can count the rings in the tree's trunk to determine how old it is. Each ring represents a year in the life of the tree. We know which ring corresponds to which year by simply counting the rings from the outside of the trunk to the inside. Well, it turns out that through a rather complicated process, you can actually measure the amount of ^{14}C in a tree ring and use it to determine how much ^{14}C was in the atmosphere during the year in which the tree ring was grown. As a result, scientists have determined the amount of ^{14}C in the atmosphere throughout a portion of the earth's past.

It turns out that scientists have studied the ^{14}C content in tree rings that are as many as 3,000 years old. From these measurements, scientists have determined the amount of ^{14}C in the atmosphere over the past 3,000 years. What they have seen is that the amount of ^{14}C has varied by as much as 70% over that time period. The variation is correlated to certain events that occur on the surface of the sun. As a result, *we know* that the amount of ^{14}C in the atmosphere has not stayed constant. Instead, it has varied greatly. Thus, *we know* that the initial assumption of ^{14}C dating is wrong. Thus, one must take most ^{14}C dates with a grain of salt. After all, we know that the assumption used in making those dates is wrong. Consequently, we cannot put too much trust in the results!

Notice that I said we must take "most" ^{14}C dates with a grain of salt. Why "most?" Why not "all?" Well, it turns out that since we can determine the amount of ^{14}C in the atmosphere during the past using tree rings, we can actually use that data to help us make our initial assumption. As a result, the assumption becomes much more accurate. The problem is, however, that we don't have ^{14}C measurements for tree rings that are older than 3,000 years. Thus, we can only make an accurate assumptions for organisms that have died within the last 3,000 years. As long as the organism died in that time range, we can use tree ring data to help us make an accurate assumption of how much ^{14}C was in the organism when it died. For organisms that have died longer than 3,000 years ago, we have no tree ring data, so we have no way to make an accurate assumption. As a result, we cannot really believe the ^{14}C date.

In the end, then, the ^{14}C dating method can be believed for organisms that have been dead for 3,000 years or less. Thus, it is a great tool for archaeology. If an archaeologist finds a manuscript or a piece of cloth (both cloth and paper are made from dead plants), the archaeologist can use ^{14}C dating to determine its age, provided all of the experimental techniques of ^{14}C dating have been followed accurately. As long as the result is about 3,000 years or younger, the date can be believed. If the date turns out to be older than 3,000 years, it is most likely wrong.

So you should see that radioactive dating involves a pretty important assumption. If the assumption is good, the date you get from radioactive dating is good. If the assumption is bad, the result you get from radioactive dating will be bad. Now there are a lot of other radioactive dating techniques besides ^{14}C dating. Unfortunately, they all suffer from a similar malady. In every radioactive dating technique, you must make assumptions about how much of a certain substance was in the object originally. Such assumptions are quite hard to make accurately.

The difficulty of making these assumptions can be seen in the fact that radioactive dates have been demonstrated to be wrong in many, many instances. John Woodmorappe, in his book *Studies in Flood Geology*, has compiled more than 350 radioactive dates that conflict with one another or with other generally accepted dates. These erroneous dates demonstrate that the assumptions used in radioactive dating cannot be trusted. As a result, the dates that one gets from radioactive dating cannot be trusted, either.

Unfortunately, many in the scientific community are unwilling to admit to the inadequacies of radioactive dating, because many scientists like its *results*. Because certain radioactive decay schemes have long, long half-lives, the dates that one calculates from these methods can be breathtakingly large. For example, there are rocks on the planet that radioactive dating techniques indicate are more than 4 *billion* years old. It turns out that many scientists *want* the earth to be that old because they believe in the discredited hypothesis of evolution. This hypothesis *requires* a very old earth, and radioactive dating techniques provide dates that indicate the earth is very old. As a result, they turn a blind eye to the inadequacies of radioactive dating, because it gives them an answer that they want! Hopefully, as time goes on, this unfortunate situation will change!

Other uses of Radioactivity and Ionizing Radiation

Radioactivity has far more reliable applications than the tenuous process of radioactive dating. For example, radiation has revolutionized medicine. When a doctor wants to look at your bones, the doctor gives you an X-ray. This is accomplished by placing the portion of your body that needs to be examined between a sheet of film and a high-intensity gamma ray source. As you are exposed to the gamma rays, some pass through your body and hit the film, while others collide with cells in your body and stop. Gamma rays collide more with the dense portions of your body (the bones) than with the fleshy parts of your body. As a result, the film gets hit by gamma rays more frequently when bone is not between the gamma ray source and the film. When the film is developed, this will result in the parts of the film behind your bones being much whiter than those parts of the film behind the rest of your body. As a result, the gamma rays form an image of your bones on the film.

Now when you get an X-ray, the gamma rays used to make the X-ray work are killing your cells and mutating some DNA. Nevertheless, as I have discussed before, that is not a problem as long as you do not get too many X-rays. Any risk caused by your exposure to gamma rays is far outweighed by the medical benefit of being able to see your bones without surgery. Of course, the person *giving* you the X-ray would be exposed to gamma rays all day if he or she were not shielded from them. That's why the person giving you the X-ray stands behind thick shielding during the X-ray process.

Radioactive isotopes are also used to track things like blood flow inside the body. If a gamma emitting isotope is injected into your bloodstream, doctors can analyze how the blood flows to different parts of your body by detecting where the gamma rays are coming from inside your body. Once again, although this exposes you to gamma rays, the risk is low as long as the amount of exposure is minimized. The diagnostic benefit to such a procedure outweighs the risk of the gamma ray exposure.

Ionizing radiation is even used to kill cancerous cells in tumors and the like. People with thyroid cancer often are given radioactive iodine (^{131}I) to drink. Since iodine collects in your thyroid, drinking radioactive iodine (usually called "the cocktail") will concentrate radiation in your thyroid, killing cancerous cells. The healthy cells will die as well, but your body is more likely to replace the healthy cells and not the cancerous cells, so this is a very popular treatment for thyroid cancer.

Finally, ionizing radiation is even used to keep you safe from fire. Most homes have a smoke detector. In a smoke detector, there are two metal plates hooked to a wire. One plate is hooked to the positive side of the battery and is thus positively-charged. The other is hooked to the negative end of the battery and is thus negatively-charged. An ^{241}Am source is then placed under the plates, and it emits alpha particles through a small hole in the bottom plate. As the alpha particles collide with the molecules in the air between the plates, the molecules are ionized. The positive ions travel to the negative plate and the negative ions travel to the positive plate.

When there is no smoke in the air, the ^{241}Am source shoots a steady stream of alpha particles, resulting in a constant rate of ion production. The electronics in the smoke detector detect those ions when they hit their respective plates. When smoke gets between the plates, however, the smoke traps the ions, not allowing them to hit the plates. This causes a drop in the number of ions detected by the electronics, and that causes the alarm to go off.

As with all forms of ionizing radiation, there is a small amount of inherent risk in having a radioactive isotope (the ^{241}Am in the smoke detector) in your house. Nevertheless, the chance of you dying or being hurt in a fire is *millions of times* greater than any chance of your being hurt by one of the alpha particles in the smoke detector. As a result, they are used in homes despite the fact that they are radioactive.

Nuclear Reactions

Although radioactive decay is, by far, the most common nuclear process that occurs on earth, there are other types of nuclear reactions that occur over and over again in outer space as well as in nuclear power plants and nuclear research facilities. These nuclear reactions can usually be classified as one of two types:

Nuclear fusion - The process by which two or more small nuclei fuse to make a bigger nucleus

Nuclear fission - The process by which a large nucleus is split into two smaller nuclei

You have probably heard of both of these processes before, but I want to make sure that you understand them in a thorough way.

I will begin with nuclear fission, which is the basis for nuclear power plants and most nuclear bombs. Usually, nuclear fission begins with a neutron colliding with a large nucleus. For example, if a neutron were to collide with a ^{235}U nucleus, the nucleus would become very unstable. It would be so unstable that it would, in fact, break apart into two smaller nuclei. One possible reaction would be as follows:

$$^{1}_{0}n + ^{235}_{92}U \rightarrow ^{112}_{45}Rh + ^{121}_{47}Ag + 3^{1}_{0}n$$

Notice what this equation says happened in the reaction. A neutron collided with a ^{235}U nucleus. The result is a ^{112}Rh nucleus, a ^{120}Ag nucleus, and 3 neutrons. Don't be alarmed about the coefficient of 3 next to the neutron on the products side of the equation. Like chemical equations, nuclear equations can also have coefficients. The meaning of these coefficients is the same in both cases. Thus, the 3 simply tells you that 3 neutrons are produced.

Notice also that the equation is balanced. On the reactants side, there are a total of 92 (92 + 0) protons (subscripts). On the products side, there are also 92 (45 + 47 + 3x0) protons. On the reactants side, the mass numbers total to 236 (235 + 1). On the products side, the mass numbers also total to 236 (112 + 121 + 3). Although this is the first nuclear equation that you have seen with a coefficient, it should not disturb you. You use it just as you would if it were in a chemical equation. Since there is a 3 in front of the neutron, you multiply the numbers associated with the neutron by 3.

Now it is very important for you to realize that although the equation above represents a valid reaction that occurs when a neutron collides with a ^{235}U nucleus, it is not the only possible reaction that occurs under those conditions. One of the very interesting aspects of nuclear reactions is that, unlike chemical reactions, the same reactants will not always produce the same products. When a neutron collides with a ^{235}U nucleus, sometimes the nucleus will split apart to give the products listed above. Many times, however, it will split so as to produce other products. For example, the following reaction is even slightly more likely to occur than the one listed above:

$$^{1}_{0}n + ^{235}_{92}U \rightarrow 2\,^{117}_{46}Pd + 2\,^{1}_{0}n$$

Notice that this reaction is balanced. In this case, remember that you have to multiply the subscripts and superscripts for Pd by 2 because there is a coefficient in from of Pd as well. It turns out that there are a host of possible products for the neutron-induced fission of ^{235}U. The two equations I have shown you are just a couple of the possible reactions that will occur when a neutron collides with a ^{235}U nucleus. When a bunch of neutrons collide with a bunch of ^{235}U nuclei, many different kinds of products are produced.

Notice something else about both of the reactions I have listed. In each case, the reaction produces more than one neutron. Since a neutron is one of the reactants in the fission reaction, this sets up a very interesting situation. Imagine that you have a large sample of ^{235}U. Suppose one neutron collides with one nucleus in this sample and a fission reaction results. What will happen next? Well, the neutrons produced in this reaction can go out *and start more fissions reactions, each of which will produce even more neutrons which can go out and start even more reactions*. Thus, a single neutron can start a series of events that will result in more and more fission reactions occurring.

What do we call this? We call it a **chain reaction**. Nuclear fission can result in chain reactions because the very process of nuclear fission forms one of the reactants. Thus, as long as there is enough ^{235}U around, the number of fission reactions occurring each second can grow and grow and grow.

What's the practical upshot of all of this? Well, if you sum up the masses of the products in a fission reaction, you will find that the sum is less than that of the reactants. This means that fission reactions are exothermic, because the mass that is "missing" in the product gets converted directly to energy. Well, if the number of fission reactions grows each second, then the amount of energy being released grows each second. If the amount of energy released gets large enough, an explosion will occur.

This is, of course, the idea behind a nuclear bomb. In a nuclear bomb, the chain reaction goes out of control, producing an enormous amount of energy in a short period of time. This results in a devastating explosion. Remember, however, that you can only produce enough energy to make an explosion if there is *enough* ^{235}U. The amount of ^{235}U necessary to allow a fission chain reaction to sustain the process of fission indefinitely is called the **critical mass** of ^{235}U.

Critical mass - The amount of fissioning nucleus necessary for the chain reaction to be self-sustaining

If you have a critical mass of ^{235}U just "lying around" in the right geometry, a self-sustained reaction is inevitable. That's because the earth is constantly being bombarded by neutrons from the sun. Thus, eventually a single neutron will start a single fission reaction, and the chain reaction will keep the reaction going indefinitely. However, that in itself might not lead to an explosion. In order for an explosion to occur, the chain reaction must spin out of control. This will happen only if the critical mass of ^{235}U is highly concentrated.

Luckily, of course, the two nuclei that are used in nuclear bombs (^{235}U and ^{238}Pu) are very rare. The isotope ^{235}U, in fact, makes up only 0.7% of all uranium on the earth. In order to make a bomb, then, the amount of ^{235}U in a sample of uranium must be increased. This process, called **isotopic enrichment**, separates out the other isotopes of uranium, trying to leave behind only ^{235}U. This allows nuclear chemists to concentrate ^{235}U so as to get a critical mass of the isotope. The problem is, since ^{235}U is nearly identical to the other isotopes of uranium, this process is *very* difficult. In fact, the only thing that is secret about making a nuclear bomb is the means by which ^{235}U (or ^{238}Pu) is enriched from natural supplies. Anyone with even the most basic nuclear training can make a nuclear bomb. Making the *fuel* for the bomb is impossible, however, unless you know the secret technological steps necessary to enrich the ^{235}U (or ^{238}Pu) content in order to achieve a concentrated critical mass.

The other kind of nuclear reaction that I want to discuss is nuclear fusion. You can view nuclear fusion as the opposite of nuclear fission, because it takes small nuclei and makes them bigger. For example, ^{2}H nuclei can collide with each other and stick together, forming ^{3}He and a neutron:

$$2\,{}^{1}_{1}\text{H} \rightarrow {}^{3}_{2}\text{He} + {}^{1}_{0}\text{n}$$

This kind of reaction is also exothermic, because there is less mass in the products than there is in the reactants.

You should notice a contrast between nuclear fusion and nuclear fission. Nuclear fusion has small nuclei as the reactants and nuclear fission has large nuclei as reactants. If you think about this in terms of nuclear binding energy, this should make some sense to you. Go back and take a look at Figure 14.1. According to this figure, the most stable nucleus in Creation is ^{56}Fe, because it has the most binding energy per nucleon. Now think, for a moment, about what that means. Remember that in chemistry, all atoms try to attain the electron configuration of noble gases. Why? Because noble gases are the most stable of all atoms. In the same way, all nuclei would "like" to be ^{56}Fe, because it is the most stable of nuclei.

What does this mean? Well, it means that as long as nuclei are smaller than ^{56}Fe, they are "willing" to fuse with other nuclei so as to become more like ^{56}Fe. ^{56}Fe nuclei (and those heavier than ^{56}Fe), however, have no desire to fuse, because they would "like" to lose nucleons so as to become more like ^{56}Fe. Thus, nuclear fusion reactions between nuclei lighter than ^{56}Fe are spontaneous, whereas nuclear fusion reactions between ^{56}Fe nuclei and those that are heavier are not spontaneous. In the same way, nuclear fission reactions are spontaneous for nuclei heavier than ^{56}Fe, but not for ^{56}Fe and those nuclei that are lighter.

Wait a minute. If fusion reactions between nuclei lighter than ^{56}Fe are spontaneous, why don't all light nuclei fuse until they become ^{56}Fe. In the same way, why don't all heavy nuclei fission until they because ^{56}Fe? Remember, there are two things you must consider for every reaction: spontaneity *and* kinetics. Although fusion reactions are spontaneous for light nuclei, they proceed so slowly as to be non-existent unless the nuclei can be forced close to one another. This is tough because, since nuclei are positively charged, they repel each other. Thus, unless there is enough activation energy to push the nuclei very close to one another, the nuclei will never fuse at any kind of appreciable rate.

Is there any place that such activation energy exists? Well, it exists in nuclear research labs where huge instruments called **particle accelerators** accelerate nuclei to such high speeds that they have enough energy to get close to and fuse with other nuclei. It also exists in stars, where the gravitational force is so strong and the temperature is so high that nuclei have enough energy to get close enough to fuse. In fact, most (if not all) of the sun's energy comes from the fusion of light nuclei into heavier nuclei.

Study the following examples so that you are sure you understand how to deal with fission and fusion reactions.

EXAMPLE 14.3

Fill in the blank for the nuclear fusion reaction below:

$$2\,^{3}_{1}\mathrm{H} \rightarrow \,^{4}_{2}\mathrm{He} + \_\_\,^{1}_{0}\mathrm{n}$$

In order for this to be a valid reaction, it must balance. This means that the sum of the superscripts on both sides of the equation must equal each other as well as the sum of the subscripts. This is already the case for the subscripts. However, in order for the superscripts on the products side to equal 6, there must be 2 neutrons. Thus, the answer is:

$$\underline{2\,^{3}_{1}\mathrm{H} \rightarrow \,^{4}_{2}\mathrm{He} + 2\,^{1}_{0}\mathrm{n}}$$

What is the missing reactant in the following equation? Is this fusion or fission?

$$^{1}_{0}\mathbf{n} + \underline{\quad} \rightarrow\ ^{112}_{45}\mathbf{Rh} +\ ^{123}_{49}\mathbf{In} + 3^{1}_{0}\mathbf{n}$$

In order to get the subscripts to balance, the subscript in the blank must be 94. The chart tells us, then, that the symbol is Pu. To get the superscripts to balance, the superscript in the blank must be 237. That way the sum of the mass numbers on the reactants side (1 + 237) equals the sum of the mass numbers on the products side (112 + 123 + 3x1). Thus, the missing reactant is $\underline{^{237}_{94}\text{Pu}}$. This is fission, because a large nucleus is splitting into smaller nuclei.

ON YOUR OWN

14.9 What is the missing reactant in the following nuclear equation?

$$^{27}_{13}\text{Al} +\ ^{4}_{2}\text{He} \rightarrow \underline{\quad} +\ ^{1}_{0}\text{n}$$

14.10 Is the reaction above a fusion or fission reaction?

Using Nuclear Reactions to Make Energy

As I have already noted, nuclear fission is used in today's nuclear power plants in order to make electricity. If you took physics, you will know that power plants turn the mechanical motion of magnets and loops of wire into electricity. To produce the motion, they use steam that comes from boiling water. The fuel in an electrical power plant, then, is simply used to boil water. Coal-burning power plants use the heat of combustion of coal to boil water. Nuclear power plants simply use the heat of a nuclear fission reaction to boil the water.

The wonderful thing about using nuclear fission to make electricity is that the fuel for nuclear fission is reasonably cheap and will last a long, long time. The downside is that nuclear fission can be quite dangerous. Now it is important to realize that the danger of nuclear fission is *not* that a nuclear power plant can create a nuclear explosion. That's physically impossible! In order to make a nuclear explosion, you must have a critical mass of the fissioning isotope and it must be highly concentrated. Since nuclear power plants *do not* have a concentrated critical mass of the fissioning isotope, they *cannot* explode.

Even though nuclear power plants cannot explode, other nasty things can happen. In a normally operating nuclear power plant, the rate at which the fission processes occur is heavily controlled. If the control operations fail, then the chain reaction starts producing too much energy. This will not lead to an explosion, but it can produce so much heat that everything in the vicinity, including the reactor itself, will begin to melt. When this happens, it is called a **meltdown**, and the results can be devastating.

This is what happened at the Chernobyl nuclear power plant in the Soviet Union. This particular nuclear power plant did not have many safety protocols and, when the primary cooling system which helps control the rate of the nuclear reaction failed, there was nothing that could keep the reaction from running out of control. As a result, the reactor began to melt. This caused widespread fire throughout the plant and resulted in the release of an enormous amount of radioactive isotopes. More than 30 people were killed as a result of the fires and structural damage in the power plant itself, and thousands were exposed to high levels of radiation. To this day, no one can live near where the plant was, because the radioactive contamination is so high.

Nuclear power in the form of nuclear fission, then, can be quite dangerous. You have to understand, however, that *all* forms of power production are dangerous. Since 1900, for example, more than 100,000 people have been killed in American coal mines due to mining accidents and black lung, a malady that is caused by exposure to too much coal dust. Coal is used primarily for the production of energy. Studies indicate that nuclear power is responsible for less death and fewer health maladies than any other form of power production that we have today.

Nuclear power in the form of fission also has another serious drawback: the by-products are radioactive. We have no safe way of disposing this radioactive waste. This can eventually lead to serious environmental problems. Of course, other forms of energy production also lead to serious environmental problems. Coal-burning power plants, for example, dump pollution into the air. The *amount* of pollution they dump into the atmosphere has been reduced considerably (remember the discussion in Module #2). Nevertheless, they still emit pollutants. They are, in fact, the principal contributors to the acid rain problem.

Although nuclear power in the form of nuclear fission can be dangerous and polluting, it is not clear that it is any more dangerous and polluting than other forms of energy production. There are those who think it is, in fact, one of the safest and cleanest forms of energy production. In France, for example, the scientific community is so convinced that nuclear power is (overall) the safest form of power production that more than 90% of the country runs on electricity produced by nuclear power plants.

In order to make energy production safer, better for the environment, and longer-lasting, scientists are trying to use nuclear fusion instead of nuclear fission to produce electricity. Nuclear fusion has no harmful by-products. Remember, when nuclear fusion occurs between two hydrogen atoms, the products are helium and a free neutron. Helium is not radioactive, and has no toxic chemical properties either. Thus, using nuclear fusion to produce electricity would completely eliminate the radioactivity problem caused by nuclear fission. It is also much safer than nuclear fission. Experiments indicate that nuclear fusion is much easier to halt, allowing for the nuclear fusion process to be stopped quickly. This would avert any meltdown possibilities. Finally, the fuel for nuclear fusion (^{2}H) is virtually unlimited and very inexpensive. Nuclear power from nuclear fusion, then, would be safe, cheap, and almost limitless.

Why don't we use nuclear fusion to make electricity, then? The answer is that from a *technological* viewpoint, we have not mastered the process yet. We *know* that nuclear fusion can be used to make energy. After all, it powers the sun. However, nuclear fusion can happen in the sun because of the intense heat and pressure in the sun's core. In order to get nuclear fusion to work, we have to essentially re-create that environment here on earth. That's a tough job! Right now, nuclear physicists can, indeed, cause nuclear fusion to occur in a variety of different ways. However, in each way used so far, there is an enormous amount of energy wasted in order to create the conditions necessary for the nuclear fusion. As a result, the total energy produced is rather small. In other words, right now we have to put an enormous amount of energy into a nuclear fusion reaction, and we don't get much more than that amount of energy back. As a result, nuclear fusion is not an economically viable process for the large-scale production of energy.

In the end, then, we know that there are some drawbacks to nuclear fission. Some consider those drawbacks to be quite serious, others consider them to be about the same or even a little less than what other forms of power production have. If scientists are ever able to overcome the technological problems associated with nuclear fusion, the result would be a much safer, cleaner, and cheaper form of power production. Whether that will ever happen, however, remains to be seen.

ANSWERS TO THE ON YOUR OWN PROBLEMS

14.1 Since iron's atomic number is 26, all Fe atoms have 26 protons. The mass number therefore indicates that a ^{56}Fe nucleus has 30 neutrons. The sum of the masses of 26 protons and 30 neutrons is:

$$26 \text{ x } (1.0073 \text{ amu}) + 30 \text{ x } (1.0087 \text{ amu}) = 56.4508 \text{ amu}$$

Since the mass of a ^{56}Fe nucleus is only 55.9349 amu, there is a mass deficit of 0.5159 amu. This mass deficit is converted to energy according to Equation (14.1). To use this equation, however, we must have consistent units. Since we have the speed of light in m/sec, then the energy will come out in Joules as long as the mass is in kilograms (remember, a Joule is a $(\text{kg}\cdot\text{m}^2)/\text{sec}^2$). Thus, we must first convert the mass deficit to kg:

$$\frac{0.5159 \, \cancel{\text{amu}}}{1} \times \frac{1.6605 \times 10^{-27} \text{ kg}}{1 \, \cancel{\text{amu}}} = 8.567 \times 10^{-28} \text{ kg}$$

Now we can use Equation (14.1):

$$E = m \cdot c^2 = (8.567 \times 10^{-28} \text{ kg}) \cdot (3.00 \times 10^8 \frac{\text{m}}{\text{sec}})^2 = 7.71 \times 10^{-11} \text{ J}$$

This is not the answer. The question asks for the binding energy per nucleon, so we must divide this by the total number of nucleons in the nucleus, which is 56.

$$\frac{7.71 \times 10^{-11} \text{ J}}{56 \text{ nucleons}} = \underline{1.38 \times 10^{-12} \frac{\text{J}}{\text{nucleon}}}$$

14.2 Since we know that there are 4 protons and 3 neutrons in a ^{7}Be nucleus, we can figure out the mass of the nucleons:

$$4 \text{ x } (1.0073 \text{ amu}) + 3 \text{ x } (1.0087 \text{ amu}) = 7.0553 \text{ amu}$$

This isn't the mass of the nucleus, however, because the nucleons always lose some mass when they form a nucleus. How much mass do these nucleons lose? We can calculate it from the binding energy:

$$E = m \cdot c^2$$

$$5.739 \times 10^{-12} \text{ J} = m \cdot (3.00 \times 10^8)^2$$

$$m = 6.38 \times 10^{-29} \text{ kg}$$

We can convert that to amu:

$$\frac{6.38 \times 10^{-29} \text{ kg}}{1} \times \frac{1 \text{ amu}}{1.6605 \times 10^{-27} \text{ kg}} = 0.0384 \text{ amu}$$

Now remember, this is the mass that the nucleons *lose* when they make a nucleus. Thus, the mass of the nucleus is:

$$7.0553 \text{ amu} - 0.0384 \text{ amu} = \underline{7.0169 \text{ amu}}$$

14.3 a. This is unstable, because the chart tells us that uranium has 92 protons, and the figure tells us that all nuclei with more than 83 protons are unstable.

b. This is stable. There are 26 protons (look at the periodic chart) in an Fe. Thus, there are 30 neutrons in ^{56}Fe. If you find the spot on the graph where atomic number = 26 and neutron number equals 30, you are right in a group of dots.

c. This is not stable. The nucleus has an atomic number of 57, meaning that it has 43 neutrons. If you find that spot on the figure, you are far to the right of the band of dots.

d. This is stable. A nucleus with 6 protons and 6 neutrons is right on the line that indicates the number of protons equals number of neutrons, which for small nuclei is right in the band of dots.

14.4 A ^{87}Rb nucleus has 37 protons and 50 neutrons. In beta decay, a neutron turns into a proton. Thus, the daughter product will have 38 protons and 49 neutrons. This is ^{87}Sr. Thus, the reaction is:

$$\underline{{}^{87}_{37}\text{Rb} \rightarrow {}^{87}_{38}\text{Sr} + {}^{0}_{-1}\text{e}}$$

14.5 A ^{220}Rn nucleus has 86 protons and 134 neutrons. This nucleus is the result of the nucleus in question *losing* 2 protons and 2 neutrons. Thus, the original nucleus must have 88 protons and 136 neutrons, which is $\underline{^{224}Ra}$.

14.6 The only radioactive decay that does not change the type of nucleus is gamma decay.

14.7 a. This part is not so bad. The amount of time elapsed is an integral multiple of the half-life. Thus, after 5.00 days, there are only 50.00 grams left, after the next 5.00 days there are only 25.00 grams left, and after 15.00 days, there are only 12.50 grams.

b. This one is not so easy, because the elapsed time is not an integral multiple of the half-life. Thus, we need to use Equation (14.6). To use that equation, however, we need to know k. This comes from Equation (11.6):

$$5.00 \text{ days} = \frac{0.693}{k}$$

$$k = \frac{0.693}{5.00 \text{ days}} = 0.139 \frac{1}{\text{days}}$$

Now we can use Equation (14.6):

$$N = N_o \cdot e^{-kt}$$

$$N = (100.0 \text{ grams}) \cdot e^{-(0.139 \frac{1}{\text{days}}) \cdot (18.00 \text{ days})} = 8.19 \text{ g}$$

14.8 They will be protected from alpha and beta particles, but there is not enough matter to shield against gamma particles.

14.9 To get the sum of the subscripts to equal each other on each side of the equation, the missing product must have 15 protons. To get the sum of the superscripts equal to each other on both sides of the equation, it must have a mass number of 30 as well. Thus, the missing nucleus is ^{30}P.

14.10 This is two smaller nuclei forming a larger one. Thus, it is fusion.

REVIEW QUESTIONS

1. What is binding energy? What is it used for?

2. What is the most stable nucleus in Creation?

3. What causes the strong nuclear force? Why does it act only over a short distance?

4. Using Figure 14.2, note which of the following nuclei are stable:

$$^{14}N,\ ^{88}Ru,\ ^{118}Sn,\ ^{50}Ca$$

5. What is an alpha particle? What about a gamma ray? What about a beta particle? Which can pass through the most matter? Which can pass through the least?

6. What are the two forms of artificial radioactivity?

7. What happens when a positron collides with an electron? What is the process called?

8. If ionizing radiation is dangerous, why do you get X-rays, and why do we use smoke detectors in our homes?

9. What is the difference between nuclear fission and nuclear fusion? Which nuclei tend to undergo fission? Which tend to undergo fusion?

10. Why don't we use fusion as a way of producing electricity?

PRACTICE PROBLEMS

(The mass of a proton is 1.0073 amu, and the mass of a neutron is 1.0087 amu. The speed of light is 3.00×10^{8} m/sec, and 1 amu = 1.6605×10^{-27} kg.)

1. The mass of a ^{19}F nucleus is 18.9984 amu. What is the binding energy per nucleon of the nucleus?

2. The binding energy of ^{59}Co is 8.326×10^{-11} J. What is its mass in amu?

3. The nucleus ^{131}I is radioactive and decays by beta emission. This is the nucleus most commonly used in the thyroid "cocktail" which is used to treat thyroid disease. Write a nuclear equation for the beta decay of this isotope.

4. The radioactive isotope ^{222}Rn decays into ^{218}Po. What kind of radioactive decay is this?

5. A ^{14}N nucleus is stable but has too much energy. What can it do to release the energy?

6. ^{218}At decays by alpha emission with a half-life of 2 seconds. If you have a 1.00×10^{3} g sample of ^{218}At, how many grams will be left in half of a minute?

7. ^{206}Tl decays by beta emission with a half-life of 4.20 minutes. If a sample of this isotope has 1.2×10^{23} nuclei, how many nuclei will be left in 10.0 minutes?

8. A nuclear chemist studies an unknown radioactive isotope. The sample of isotope has a mass of 14.0 grams when the nuclear chemist begins the study. In 22.2 minutes, the mass is 13.6 grams. What is the half-life of the radioactive isotope?

9. Some nuclear power plants and most nuclear bombs use ^{239}Pu as their fuel. Write an equation for the neutron-induced fission of ^{239}Pu. Assume that 4 neutrons are produced and that the rest of the nucleus splits exactly in half.

10. ^{27}Al and ^{3}H fuse to make ^{27}Mg and one other product. What is that other product?

MODULE #15: Review - Part 1

Introduction

The last two modules of this course will be different than any of the other ones you have studied so far. That's because these modules will be devoted entirely to review. If you think about it, you are probably in desperate need of such a review. After all, you have taken chemistry for almost two years. Although much of what you have learned in this course helped remind you of things you learned in your first-year course, there are still a *lot* of things to remember about the broad subject known as general chemistry. Hopefully, these next two modules will help remind you of everything that you have learned.

Review Questions and Problems

There are many ways that I could review all of the information that you have learned over the course of two years of chemistry study. In my opinion, the most effective way to review is to answer questions and solve problems. That way, you will be forced not only to remember the things you have learned, but you will have to apply them as well. Thus, the rest of this module will simply be a series of questions and problems that you need to solve. Once you have tried to solve the problem or answer the question, look at the solutions in the back of the module to check your work.

Please do not become concerned if you cannot solve many of these problems or answer many of the questions. Remember, these questions and problems have been designed to make you think about what you have learned. Some of the material needed to solve these problems you studied in your first-year chemistry course. Thus, you should not be concerned if you don't remember all of it. You should only be concerned if, once you have looked at my answers, you still do not understand how to answer the question or solve the problem.

1. What principle leads to the conclusion that each atomic orbital can hold only 2 electrons?

2. What is Hund's rule?

3. Which atom has a lower ionization energy: Na or Cs? What about Mg or P?

4. Complete the sentence: Light has the characteristics of both a _______ and a ______.

5. What is the main difference between the Bohr model and the quantum-mechanical model of the atom?

Questions 6-12 refer to the following phase diagram for Wileium:

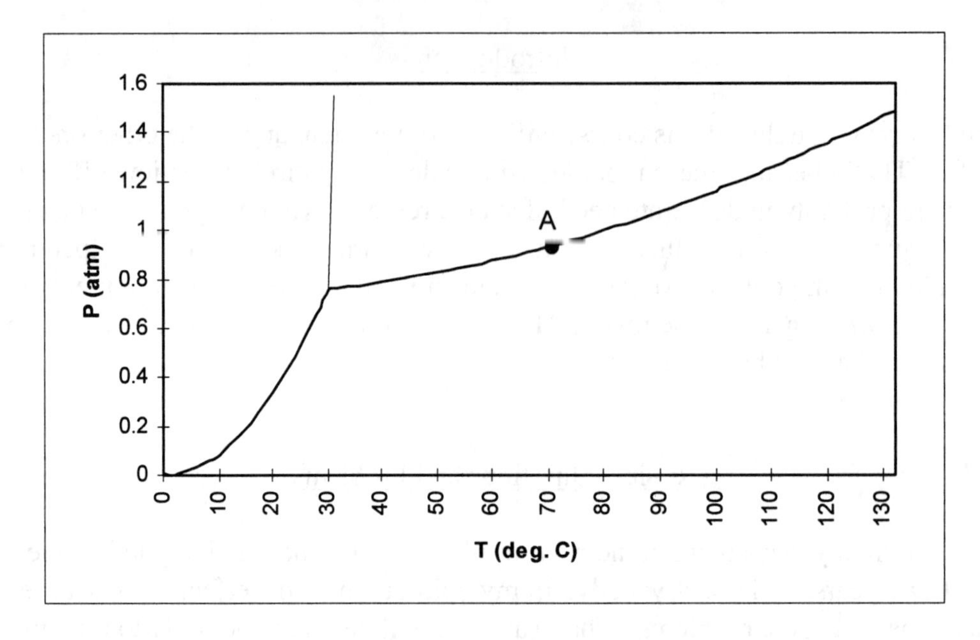

6. A sample of Wileium is at 10 °C and a pressure of 0.5 atm. What phase is it in?

7. If the sample in #6 is heated to 40 °C and the pressure remains fixed, what will happen to the substance? What do we call this process?

8. Once the sample reaches a temperature of 40 °C and a pressure of 0.5 atms, the pressure is then increased to 1.5 atms. This increases the temperature to 50 °C. What phase is the sample in now? What do we call this process?

9. After the sample reaches a steady temperature of 50 °C and pressure of 1.5 atms, the sample is cooled while the pressure remains fixed. At approximately what temperature will the sample freeze?

10. At point "A" noted on the graph, what phases of the sample are present? Are they in equilibrium?

11. What are the temperature and pressure of the triple point? What phases exist there?

12. Although most phase diagrams have the same general shape as the one printed here, there is something unique about water's phase diagram. What is it?

Questions 13 - 16 refer to the following molecules:

N_2, H_2, O_2

13. Which molecule has the largest bond energy?

14. For each molecule, state how many sigma bonds and how many pi bonds exist.

15. Classify each molecule as ionic, purely covalent, or polar covalent.

16. All three of these molecules belong to a generalized class. What is that class called?

17. For each of the following quantities, indicate whether it will always, sometimes, or never be *lower* for a 1.0 M aqueous solution of any solute than for distilled water:

a. Absorption of visible light
b. Boiling point
c. Freezing point
d. Vapor pressure
e. pH
f. Density

18. Which of the following would be an isotope of ^{208}Pb?

$$^{208}Bi,\ ^{210}Pb,\ ^{32}P$$

19. The combustion of butane is used to provide the flame for a pocket lighter. Write a balanced chemical equation for this process.

20. A certain molecule's central atom has sp^3d^2 hybridization. What is the base geometry in which the electron pairs around the central atom find themselves?

21. A 25.0 mL sample of vinegar was titrated with 0.600 M NaOH. It took 32.1 mL of NaOH to reach the endpoint. What is the concentration of acetic acid (CH_3COOH) in the vinegar?

22. Note whether the following situations will cause the rate of a chemical reaction to be relatively slow or relatively fast:

a. High activation energy
b. High reactant concentration
c. High temperature
d. High reactant bond energy

23. For the following reaction:

$$2H_2O + 4MnO_4^- + 3ClO_2^- \rightarrow 4MnO_2 + 3ClO_4^- + 4OH^-$$

Identify the reducing agent and oxidizing agent.

24. In which of the following compounds is the chromium to oxygen mass ratio closest to 1.60?

$$CrO_3,\ CrO_2,\ Cr_2O_3$$

25. Balance the following equation:

$$Ag^+ + AsH_3 + OH^- \rightarrow Ag + H_3AsO_3 + H_2O$$

26. Indicate whether each of the following statements is true or false:

a. Alpha particles have a mass number of 4 and a charge of +2.
b. Alpha particles are more penetrating than beta particles.
c. Alpha particles are helium nuclei.

27. In the following equilibrium:

$$HCO_3^- + H_2O \rightleftarrows H_3O^+ + CO_3^{2-}$$

Name all substances that act as a base.

Questions 28-30 refer to the following reaction mechanism:

$$Ce^{4+}(aq) + Mn^{2+}(aq) \rightarrow Ce^{3+}(aq) + Mn^{3+}(aq)$$

$$Ce^{4+}(aq) + Mn^{3+}(aq) \rightarrow Ce^{3+}(aq) + Mn^{4+}(aq)$$

$$Mn^{4+}(aq) + Tl^{+}(aq) \rightarrow Tl^{3+}(aq) + Mn^{2+}(aq)$$

28. What is the overall reaction?

29. There is a catalyst in this mechanism. What is the catalyst?

30. Is the catalyst in the mechanism heterogeneous or homogeneous?

31. A sample of 0.0100 moles of nitrogen gas is confined to a fixed volume at 37 °C and 0.216 atm. What is the pressure of the gas when the temperature is lowered to 15 °C?

32. Given the following information:

$$H_2(g) + \frac{1}{2}O_2(g) \rightarrow H_2O(l) \qquad \Delta H = -286 \text{ kJ}$$

$$2Na(s) + \frac{1}{2}O_2(g) \rightarrow Na_2O(s) \qquad \Delta H = -414 \text{ kJ}$$

$$Na(s) + \frac{1}{2}O_2(g) + \frac{1}{2}H_2(g) \rightarrow NaOH(s) \qquad \Delta H = -425 \text{ kJ}$$

Calculate the ΔH for the following reaction:

$$Na_2O(s) + H_2O(l) \rightarrow 2NaOH(s)$$

33. The boiling point of a pure liquid in an open container is measured. For each of the following actions, indicate whether the action would lower, increase, or not affect the boiling point:

a. The liquid is moved to a smaller, open container.
b. More liquid is added to the container.
c. The container is moved to a higher altitude.
d. A solute is dissolved in the liquid.

34. What are the quantum numbers for the highest-energy electron in the ground state of a gallium (Ga) atom?

35. A glucose ($C_6H_{12}O_6$) solution is 5.0 % glucose by mass. If the density of the solution is 1.10 g/mL, what is the molarity of the glucose?

Questions 36-40 refer to the diagram below:

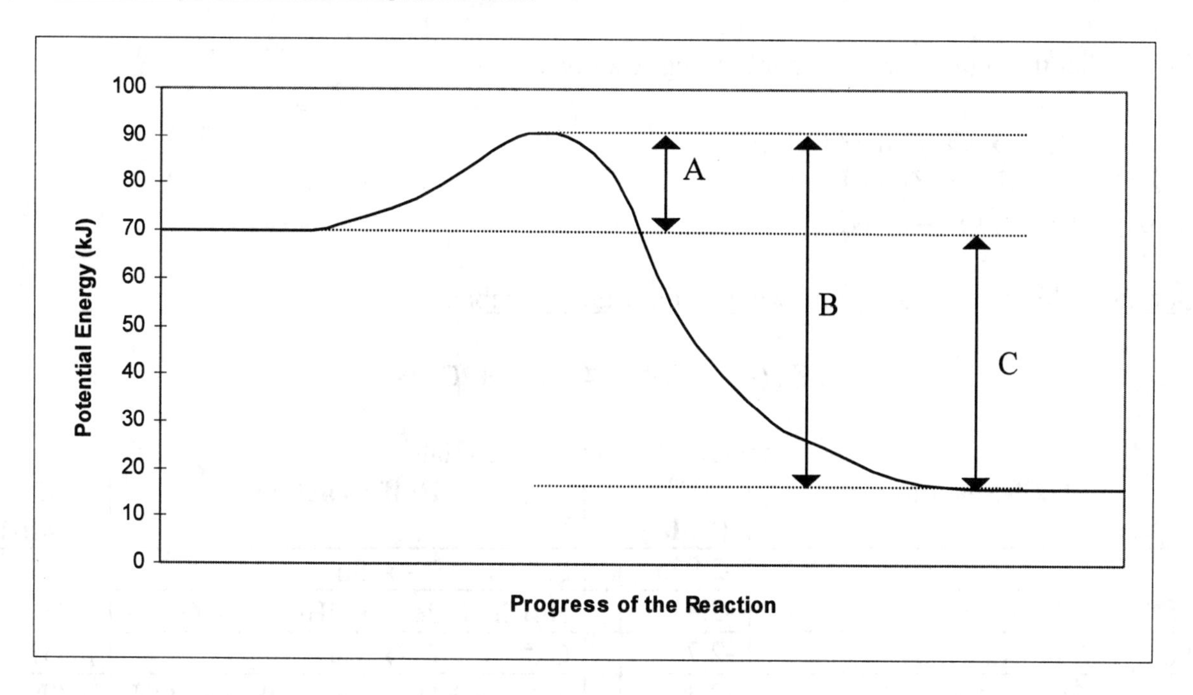

36. Is the reaction exothermic, endothermic, or neither?

37. Of the quantities illustrated in the figure (A, B, or C), which affects the rate of the chemical reaction?

38. Which of the quantities illustrated in the figure (A, B, or C) represents the enthalpy of the reaction?

39. Is it possible for the reaction in the diagram to have a negative ΔS and still be spontaneous?

40. If a catalyst is added to the reaction, which of the quantities illustrated (A, B, or C) will change?

41. Oxalic acid ($H_2C_2O_4$) is a diprotic acid with $K_1 = 5.36 \times 10^{-2}$ and $K_2 = 5.3 \times 10^{-5}$. What is the equilibrium constant of the following reaction?

$$H_2C_2O_4 + 2H_2O \rightleftarrows 2H_3O^+ + C_2O_4^{2-}$$

42. Which substance has the highest boiling point: CH_3CH_2OH or CH_3OCH_3?

43. A gas with an empirical formula of CH_2 has a density of 1.88 g/L at 0 °C and 1.00 atm. What is the substance's chemical formula?

44. Order the following compounds in terms of increasing solubility in water:

a. CH_3-CH_2-CH_2-CH_3 b. CH_3-CH_2-CH_2-OH c. HO-CH_2-CH_2-CH_2-OH

45. Predict the sign of ΔS for the following reactions:

a. $CaCO_3$ (s) → CaO (s) + CO_2 (g)
b. Zn^{2+} + $2OH^-$ → $Zn(OH)_2$ (s)
c. $3H_2$ (g) + N_2 (g) → $2NH_3$ (g)

Questions 46-50 refer to the following equilibrium and table:

$$Zn\,(s) + Cu^{2+} \rightleftarrows Zn^{2+} + Cu\,(s)$$

Standard Reduction Potentials

Half-Reaction	**E° (Volts)**	**Half-Reaction**	**E° (Volts)**
$K^+ + e^- \rightarrow K$	-2.925	$Sn^{4+} + 4e^- \rightarrow Sn$	0.15
$Ca^{2+} + 2e^- \rightarrow Ca$	-2.87	$Hg_2Cl_2 + 2e^- \rightarrow 2Hg + 2Cl^-$	0.27
$Na^+ + e^- \rightarrow Na$	-2.714	$Cu^{2+} + 2e^- \rightarrow Cu$	0.337
$Mg^{2+} + 2e^- \rightarrow Mg$	-2.37	$NiO_2 + 2H_2O + 2e^- \rightarrow Ni(OH)_2 + 2OH^-$	0.49
$Al^{3+} + 3e^- \rightarrow Al$	-1.66	$I_2 + 2e^- \rightarrow 2I^-$	0.5355
$Mn^{2+} + 2e^- \rightarrow Mn$	-1.18	$Fe^{3+} + 3e^- \rightarrow Fe$	0.771
$Zn^{2+} + 2e^- \rightarrow Zn$	-0.763	$Ag^+ + e^- \rightarrow Ag$	0.7991
$Cr^{3+} + 3e^- \rightarrow Cr$	-0.74	$Br_2\,(l) + 2e^- \rightarrow 2Br^-\,(aq)$	1.0652
$Cd^{2+} + 2e^- \rightarrow Cd$	-0.430	$O_2 + 4H_3O^+ + 4e^- \rightarrow 6H_2O$	1.23
$PbSO_4 + 2e^- \rightarrow Pb + SO_4^{2-}$	-0.356	$Cl_2 + 2e^- \rightarrow 2Cl^-$	1.3595
$Ni^{2+} + 2e^- \rightarrow Ni$	-0.250	$MnO_4^- + 8H_3O^+ + 5e^- \rightarrow Mn^{2+} + 12H_2O$	1.51
$Pb^{2+} + 2e^- \rightarrow Pb$	-0.126	$F_2 + 2e^- \rightarrow 2F^-$	2.87
$2H_3O^+ + 2e^- \rightarrow H_2 + 2H_2O$	**0.000**		

46. What is the standard electron potential for this reaction?

47. Is the reaction spontaneous?

48. Would this reaction be the basis of a Galvanic cell or an electrolytic cell?

49. Draw a schematic of the cell for which this reaction is the basis. Label the anode and cathode, as well as the flow of electrons.

50. Is there any situation under which the measured potential of this cell would NOT equal its standard electron potential?

51. To determine the molar mass of a molecule, 3.30 grams of the molecule in its liquid phase is vaporized at a temperature of 150.0 °C. At 1.25 atms, the resulting gas filled a 2.00-liter vessel. What is the molar mass of the gas?

52. Samples of fluorine and xenon gas are mixed in a container whose volume cannot change. Initially, there are 8.0 atms of fluorine gas and only 1.7 atmospheres of xenon gas. A reaction ensues in which only one product is formed. All of the xenon gas is used up, but 4.6 atms of fluorine gas are left over. The temperature remained constant throughout the reaction. What is the chemical formula of the product?

53. The substances in the diagram below are at equilibrium:

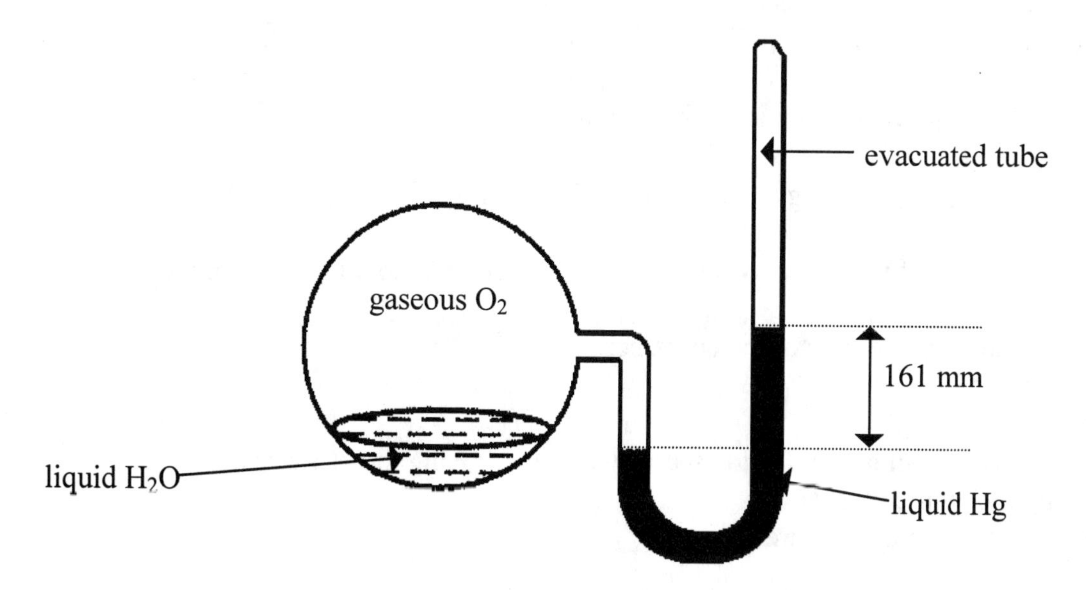

If the vapor pressure of water at this temperature is 28 mm Hg, what is the partial pressure of oxygen gas in the container?

54. The following data were gathered to measure the density of an unknown substance:

Mass of empty beaker:	3.0 g
Mass of beaker with substance:	25.0 g
Volume of substance:	11.0 mL

What is the density of the substance?

55. Two of the following molecules are isomers. Identify them:

a. $CH_3\text{-}CH_2\text{-}CH_2\text{-}CH_3$ b. $CH_3\text{-}CH{=}CH\text{-}CH_3$ c. $CH_3\text{-}O\text{-}CH_2\text{-}CH_3$ d. $CH_3\text{-}CH\text{-}CH_3$ (with CH_3 attached to the central CH)

56. Which of the following aqueous solutions has the lowest freezing point?

a. 0.50 m CH_3OH b. 0.50 m $MgSO_4$ c. 0.50 m $ZnCl_2$ d. 0.50 m $NaNO_3$

57. An ideal gas is in a sealed container of fixed volume. For each of the properties listed below, indicate whether the property will increase, decrease, or stay the same as the gas is cooled.

a. The average speed of the gas molecules
b. The average molar mass of the gas molecules
c. The average distance between the gas molecules
d. The average size of the gas molecules

58. The following equilibrium is established in a laboratory container:

$$PCl_3\ (g) + Cl_2\ (g) \rightleftarrows PCl_5\ (g) \qquad \Delta H < 0$$

Indicate how each of the following changes will affect the amount of PCl_3 in the container.

a. A decrease in the volume of the container
b. Addition of PCl_5 (g)
c. Addition of Cl_2 (g)
d. Selectively allowing PCl_5 (g) to escape, without allowing the other gases to escape
e. Raising the temperature
f. Decreasing the total pressure

59. A chemical reaction has a half-life of 20 minutes. How long will it take for the concentration of a reactant to fall from 1.00 M to 0.25 M?

60. Which of the following substances can be oxidized to a stronger acid?

HCl, HNO_3, H_2SO_3, H_2CO_3

61. Equal concentrations of HCl and O_2 are placed in a container. The following equilibrium is achieved:

$$4HCl\ (g)\ +\ O_2\ (g) \rightleftarrows 2Cl_2\ (g)\ +\ 2H_2O\ (g)$$

At equilibrium, will there be more O_2 (g) or more HCl (g) present?

62. A 30.0-g sample of Na_2CO_3 is dissolved in enough water to make 500.0 mL. What is the concentration of each ion in the solution?

63. What volume of 0.50 M HCl is required to neutralize 50.0 mL of 0.10 M $Ba(OH)_2$?

64. Which of the following statements is not true for the halogens (group 7A)?

a. The ionization potentials decrease with increasing atomic number.
b. Fluorine is the most electronegative of all halogens.
c. They are excellent oxidizing agents.
d. They exist in both ionic and covalent molecules.
e. Each halogen is larger than the atom that lies to its immediate left on the periodic chart.

65. In order to remove SO_2 gas from an industry's exhaust, the exhaust fumes from the industry are bubbled through a vat containing 0.25 M KOH. The reaction between SO_2 gas and aqueous KOH makes aqueous K_2SO_3. What is the maximum mass of SO_2 gas that can be cleaned from the industry's exhaust using 1,000.0 L of the KOH solution?

66. Which of the following molecules have shapes that are planar?

BCl_3, NH_3, Cl_2O, CHF_3

67. A reaction is spontaneous for all temperatures below 300 K. It is not spontaneous for all temperatures above 300 K. What is the sign of ΔG below 300 K, above 300 K, and at 300 K?

68. For the reaction in problem 67, what are the signs of ΔH and ΔS?

69. A 1.00 g sample of a rock was dissolved in acid, liberating 0.38 g of CO_2. If the only source of carbonate in the rock comes from $CaCO_3$, what is the percentage of $CaCO_3$ in the rock (by mass)?

70. Given the following data:

Bond	Bond Energy (kJ/mole)
I-I	149
Cl-Cl	239
I-Cl	208

calculate the ΔH of the following reaction:

$$I_2 + 3Cl_2 \rightarrow 2ICl_3$$

71. Which of the following solutions has the lowest pH?

1 M NaCl, 1M $NaHSO_4$, Na_2SO_4, $NaNO_3$

72. Draw the Lewis structures for each of the following molecules:

H_2S, PH_3, HCN, H_2CO

73. A chemist plates copper onto baby shoes. In the procedure, aqueous $CuCl_2$ is electrolyzed with a 3.00 amp power supply. If the chemist plates for 16.0 hours, what mass of copper will be produced? (F = 96,500 Coulombs)

74. In which of the following 0.010 M solutions is barium sulfate *least* soluble?

$Al_2(SO_4)_3$, $(NH_4)_2SO_4$, NH_3, $BaCl_2$, HCl, Na_2SO_4

75. What is the pH of a 0.0100 M solution of HCN? ($K_a = 4.0 \times 10^{-10}$)

MODULE #16: Review - Part 2

Introduction

In this module, I will complete your review of general chemistry. I must once again remind you that you should not get too discouraged if you have problems with the review. Remember, you have covered a lot of chemistry in your two courses, so you are bound to forget some of it. The main goal of this review is not to test you, but to get you to remember. So, let's tackle a few more problems and questions, shall we?

More Review Questions and Problems

1. Of the molecules below, note those that have a dipole moment of zero?

NH_3, C_3H_6, NO, H_2S, $SiCl_4$, Cl_2

2. What is the daughter product when ^{90}Sr decays by beta decay?

3. The following reaction occurs between sulfur dioxide gas and oxygen gas:

$$2SO_2\ (g) + O_2\ (g) \rightleftarrows 2SO_3\ (g)$$

When 0.40 moles of SO_2 and 0.60 moles of O_2 are placed in a 1.00 L container and allowed to reach equilibrium, there are 0.30 moles of SO_3. What is the equilibrium constant of this reaction?

4. Using the following table:

Standard Reduction Potentials

Half-Reaction	**E° (Volts)**	**Half-Reaction**	**E° (Volts)**
$K^+ + e^- \rightarrow K$	-2.925	$Sn^{4+} + 4e^- \rightarrow Sn$	0.15
$Ca^{2+} + 2e^- \rightarrow Ca$	-2.87	$Hg_2Cl_2 + 2e^- \rightarrow 2Hg + 2Cl^-$	0.27
$Na^+ + e^- \rightarrow Na$	-2.714	$Cu^{2+} + 2e^- \rightarrow Cu$	0.337
$Zn^{2+} + 2e^- \rightarrow Zn$	-0.760	$I_2 + 2e^- \rightarrow 2I^-$	0.5355
$PbSO_4 + 2e^- \rightarrow Pb + SO_4^{2-}$	-0.356	$Ag^+ + e^- \rightarrow Ag$	0.80
$Ni^{2+} + 2e^- \rightarrow Ni$	-0.250	$Cl_2 + 2e^- \rightarrow 2Cl^-$	1.3595
$Pb^{2+} + 2e^- \rightarrow Pb$	-0.126	$F_2 + 2e^- \rightarrow 2F^-$	2.87
$\mathbf{2H_3O^+ + 2e^- \rightarrow H_2 + 2H_2O}$	**0.000**		

determine what will happen when positive and negative electrodes are hooked to a 9-Volt battery and immersed in an acidic solution of KI. Specifically, note what substance will be formed at each electrode.

5. In a saturated solution of MgF_2, the concentration of Mg^{2+} is 1.21×10^{-3} M. Write the expression for the solubility product of MgF_2 and give its value.

6. A 0.100 mole sample of solid NaF is added to 1.00 L of a saturated solution of MgF_2. What will the concentration of Mg^{2+} be when the solution reaches equilibrium? Assume that the volume does not change significantly when the solid is added.

7. Will a precipitate of MgF_2 form if 100.0 mL of 0.00300 M $Mg(NO_3)_2$ is mixed with 200.0 mL of 0.00200 M NaF?

8. At a higher temperature, the concentration of Mg^{2+} in a saturated solution of MgF_2 is 1.17×10^{-3} M. Does MgF_2 dissolve exothermically or endothermically?

Questions 9-13 refer to the following information:

The reaction:

$$2NO\ (g) + 2H_2\ (g) \rightarrow N_2\ (g) + 2H_2O\ (g)$$

is studied in a kinetics experiment. The following data is collected:

Trial	[NO] (initial)	$[H_2]$ (initial)	Initial Rate (M/s)
1	0.0060	0.0010	0.00018
2	0.0060	0.0020	0.00036
3	0.0010	0.0060	0.000030
4	0.0020	0.0060	0.00012

9. What is the order of the reaction with respect to each reactant? What is the overall order?

10. Write the rate equation.

11. Determine the rate constant, including the units.

12. In trial number 1, how much NO is left when exactly 1/2 of the H_2 has reacted?

13. The following mechanism was suggested for this reaction:

$$NO + NO \rightleftarrows N_2O_2$$
$$N_2O_2 + H_2 \rightleftarrows H_2O + N_2O$$
$$N_2O + H_2 \rightarrow N_2 + H_2O$$

a. Show that this mechanism is consistent with the stoichiometry of the reaction.
b. What must the rate determining step of this mechanism be?
c. Identify the catalysts and intermediates in this mechanism.

Questions 14 - 17 refer to the following data:

A student collects hydrogen gas over water. The volume of the gas when it is collected over water is 90.0 mL. The temperature is 25 °C and the atmospheric pressure is 745 mm Hg. The vapor pressure of water at this temperature is 23.8 mmHg. $R = 62.4 \frac{L \cdot mm\,Hg}{mole \cdot K}$. Avagadro's number is 6.02×10^{23}.

14. How many moles of hydrogen gas were collected?

15. How many molecules of hydrogen gas were in the sample?

16. Calculate the ratio of the average speed of hydrogen molecules to the average speed of water molecules in the gas sample.

17. Which of the two gases (H_2 or H_2O) deviates more from ideal behavior? Why?

In questions 18-22, please feel free to refer to any information given in previous problems.

18. A beaker of water (beaker A) and a beaker of a 10% sugar solution (beaker B) are placed in a closed container. Initially, the volume of liquid in beaker A is the same as the volume of liquid in beaker B. As the entire system comes to equilibrium, what will happen to the volume of liquid in each beaker? At equilibrium, will the volume of the liquid in beaker A still be equal to that in beaker B? Compare the concentration of sugar in the sugar solution before and after equilibrium.

19. A beaker of water is placed in a closed container which is connected to a vacuum pump. What happens to the water as the pump evacuates the container?

20. A strip of silver is immersed in a solution of $CuSO_4$. What, if anything, will be observed on the surface of the silver after a long time?

21. A strip of zinc is immersed in a solution of $CuSO_4$. What, if anything, will be observed on the surface of the zinc after a long time?

22. A solution of aqueous I_2 is mixed with the nonpolar solvent TTE in a test tube. Explain what will happen as the mixture settles.

Questions 23-27 refer to the following information:

At 298 K, the ΔH for the reaction below is -145 kJ.

$$2H_2S\ (g) + SO_2\ (g) \rightleftarrows 3S\ (s) + 2H_2O\ (g)$$

23. Predict the sign of ΔS for the reaction.

24. At 298 K, the reaction is spontaneous. What will happen to the value of ΔG as the temperature increases.

25. What happens to the value of the equilibrium constant as the temperature is increased?

26. Suppose the value for the ΔS of this equation is given as "X." Give an equation for the temperature (in terms of X) at which the reaction ceases to be spontaneous.

27. The temperature calculated from the equation you determined in Problem #26 is only an approximation. Why?

Questions 28 - 33 refer to the following situation:

100.0 mL of 0.200 M HCl is added dropwise to 100.0 mL of 0.100 M Na_3PO_4.

28. Write the net ionic equation for the reaction that occurs as the first few drops of HCl are added.

29. As time goes on, the reaction in Problem 28 will eventually stop and a new reaction will start to occur. What is the net ionic equation for the new reaction?

30. How many mL of HCl must be added before the reaction in Problem 29 begins?

31. A chemist measures the pH of the resulting solution as HCl is added drop-by-drop. The chemist notices that after a few drops of HCl are added, the pH ceases to change much with the addition of more HCl. Why?

32. Eventually, as more and more HCl is added, the chemist sees the pH change dramatically. At approximately what volume of HCl will this dramatic change occur?

33. Will the chemist see any more dramatic changes as even more HCl is added?

34. Why does spreading sodium chloride on ice help melt the ice? Why is calcium chloride even more effective?

35. Why is NH_3 a gas at room temperature while H_2O is a liquid, despite the fact that they have nearly the same mass?

36. The substance we call graphite is just solid carbon. So is the substance we call diamond. Why can graphite be used as a lubricant while diamond is an abrasive?

37. The radius of a Ca atom is 0.197 nanometers. The radius of a Ca^{2+} ion is 0.099 nanometers. Why is the ionic radius so much smaller than the atomic radius?

38. Which has the largest lattice energy: CaO or K_2O?

39. The first ionization energy of K is 410 kJ/mole while the first ionization energy for Ca is 590 kJ/mole. Why?

40. Which is smaller: the second ionization energy of Ca or the second ionization energy of K?

41. The first ionization energy of Mg is *greater* than the first ionization energy of Al. Why?

42. What is the temperature of a mixture of ice and water when 90% of the mass is ice and only 10% of the mass is water? What about when the mixture is 10% ice and 90% water?

43. The concentration of a KCl solution is 1.12 M. If the density of the solution is 1.05 g/mL, what is the solution's molality?

44. Indicate the hybridization of each carbon atom in the molecule below:

```
      H
     a|  b    c    d
  H — C — C ═ C — C ≡ N:
      |   |    |
      H   H    H
```

45. What is the geometry of the electrons around each of the carbons in the above molecule?

46. The figure below is a graph of the distribution of molecular speeds in a gas. Sketch what the graph would look like at a higher temperature.

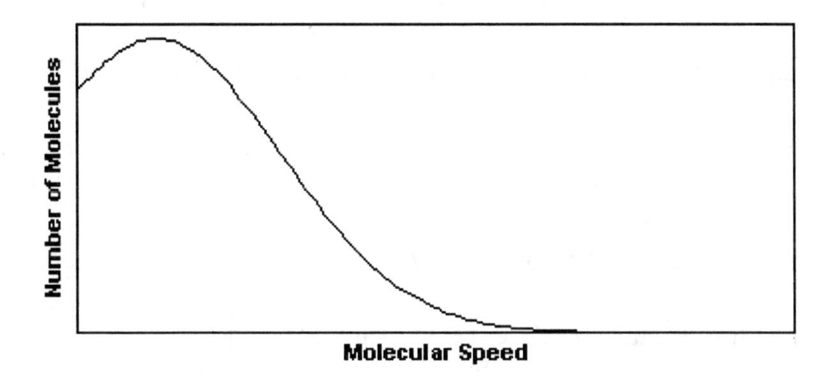

47. What are the three basic Van der Waals forces and what are their relative strengths?

48. Which of the following compounds is most likely to be insoluble in water?

$$NaF, PH_3, CH_3CO_2H, SiCl_4$$

49. When a gas is dissolved in water, will its solubility increase or decrease with increasing temperature?

50. Explain how distillation separates liquids that are mixed together in a solution.

51. 25.0 grams of a solute are mixed with 100.0 g of water (K_f = -1.86 C/m) and the freezing point of the solution is measured to be -2.58 °C. If the solute is a polar covalent molecule, what is its molecular mass?

52. During the course of a chemical reaction, does the reaction rate typically increase, decrease, or stay relatively constant?

53. Aqueous hydrogen peroxide decays slowly over time to water and oxygen. If solid platinum is added, the same decay occurs, but much faster. Is platinum a heterogeneous or homogeneous catalyst?

Questions 54 - 56 refer to the following information:

An ester is made according to the following equilibrium:

$$CH_3CO_2H + C_2H_5OH \rightleftarrows CH_3CO_2C_2H_5 + H_2O$$

The equilibrium constant is 4.0, and water is NOT the solvent.

54. A chemist dumps *all four* chemicals together in a container and, a few minutes later, analyzes it to find the following concentrations:

$$[CH_3CO_2H] = 0.15\ M,\ [C_2H_5OH] = 0.15\ M,\ [CH_3CO_2C_2H_5] = 0.40\ M,\ [H_2O] = 0.40\ M$$

Is the reaction at equilibrium? If not, which way must it shift to reach equilibrium?

55. If the chemist starts again, this time with 1.0 M concentrations of each reactant, what will be the equilibrium concentrations for all reactants?

56. What is the pH of a 0.10 M solution of NaOH?

57. What is an amphiprotic substance? Give an example of one.

58. What is the definition of a Lewis acid? A Lewis base?

59. In the following reaction, would F^- be best described as a Lewis acid, a Lewis base, a Bronstead-Lowry acid, or a Bronstead-Lowry base?

$$F^- + BF_3 \rightarrow BF_4^-$$

Questions 60 - 63 refer to the following graph of reaction rate versus time for both the forward and backward reactions of an equilibrium:

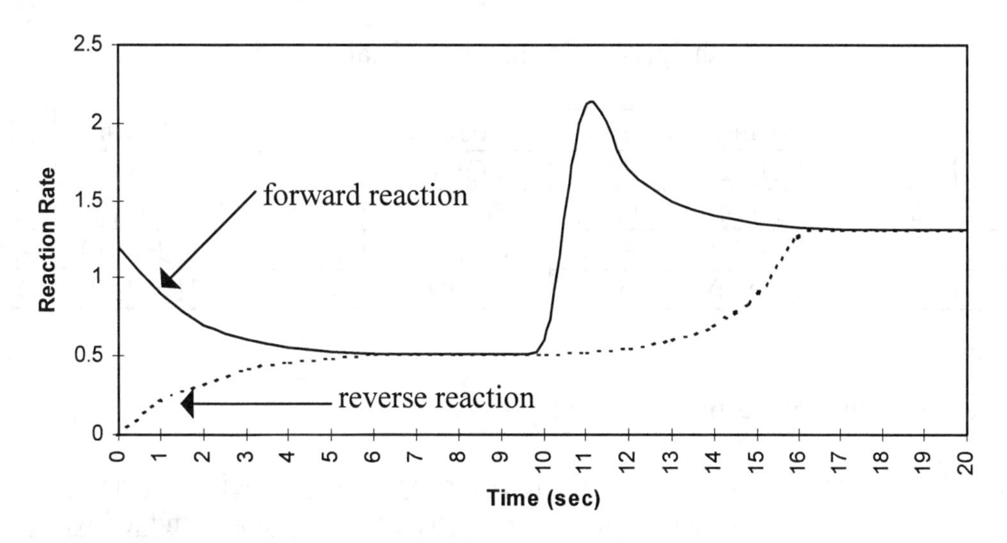

60. At what time is equilibrium first achieved?

61. When is the equilibrium stressed?

62. If the equilibrium was stressed by adding more of one substance, was the substance a reactant or a product?

63. If the equilibrium was stressed by increasing the temperature, is the forward reaction exothermic or endothermic?

64. Given the following table:

Bond Strengths

Chemical Bond	Bond Strength (kcals/mole)		Chemical Bond	Bond Strength (kcals/mole)
C-H	99		H-H	104
C-Cl	79		H-O	111
C-C	83		H-Cl	103
C-N	70		O=O	118
C-O	84		N≡N	226
C=C	147		N-H	66
C=O	170		Cl-Cl	58

What is the ΔH of the following reaction?

$$N_2 (g) + 3H_2 (g) \rightarrow 2NH_3 (g)$$

65. Why is the answer in problem 64 only an approximation?

66. Given the following reaction.

Enthalpies of Formation (in kcals/mole)

Substance	**ΔH_f**		**Substance**	**ΔH_f**		**Substance**	**ΔH_f**		**Substance**	**ΔH_f**
H_2O (g)	-58		NO_2 (g)	8		$Ca(OH)_2$ (s)	-236		C_2H_6O (l)	-66
H_2O (l)	-68		NH_3 (g)	11		$CaCO_3$ (s)	-288		C_2H_4 (g)	13
SO_3 (g)	-94		CO_2 (g)	-94		C_2H_6 (g)	-20		C_4H_{10} (g)	-30
NO (g)	19		CaO (s)	-152		CH_3OH (l)	-57		H_2SO_4 (l)	-194

What is the ΔH for the combustion of butane gas?

For each of the reactions in Problems 67-69, determine whether or not it is a redox reaction. If it is a redox reaction, identify the atoms whose oxidation numbers changed and tell whether they were oxidized or reduced.

67. $Ca (s) + Cl_2 (g) \rightarrow CaCl_2 (s)$

68. $Ag^+ (aq) + OH^- (aq) \rightarrow AgOH (s)$

69. $3IF_5 (aq) + 2Fe (s) \rightarrow 2FeF_3 (aq) + 3IF_3 (aq)$

70. 125 mL of nitric acid (HNO_3) with unknown concentration is titrated against $Mg(OH)_2$ with a concentration of 2.3 M. If 35.4 mL of base are required in order to reach the endpoint, what was the concentration of the acid?

71. Which of the following electron configurations represents an atom that is *not* in its ground state?

a. $1s^2 2s^2 2p^6 3s^2 3p^6 4s^2 3d^{10} 4p^3$ b. $1s^2 2s^2 2p^6 3s^2 3p^6 4s^2 3d^{10} 4p^6 5s^2 4d^6$ c. $1s^2 2s^2 2p^6 3s^1 3p^1$

72. Assign quantum numbers to the valence electrons in a bromine atom.

73. What is wrong with the following quantum number assignment? $n=3, \ell=2, m=3, s=\frac{1}{2}$

74. What is the hybridization of the central Cl atom and the resulting geometry of the ClF_4^+ polyatomic ion?

75. Why is the Kelvin temperature scale called the "absolute" temperature scale?

Summing It All Up

Well, you've reached the end of your second year of high school chemistry. Now that you are done, you have completed what is often called an "advanced placement" (or "AP") chemistry course. What this means is that you have been exposed to all of the chemistry that you would have been exposed to in an above-average university's first-year general chemistry course. In other words, you have really completed a college-level course!

If you like, you can take the AP chemistry test, which is a national, standardized test that determines how well you learned the subject. If you get a good score on the test, most universities will give you credit for their first year chemistry course without you actually taking the course. In general, I do not recommend that you do this. The first year of college is a very stressful time. If you go ahead and take the first year chemistry course in college, it will be a rather easy "A" for you. Thus, you can impress your professors and concentrate more on some of the other stresses of college life.

If you do decide to take the test, please realize that most students who get a "B" in their college chemistry course would still not pass the AP exam. In other words, it's a tough exam. Thus, you need to study for it. The best way to do that is to get an AP chemistry review book that has been specially written to allow you to review and practice for the AP test.

Whether you take the AP test or not, you should take a few moments to reflect on all that you have learned. Think of the enormous amount of information contained in your 2 years of high school chemistry, and then realize that you have really only scratched the surface of this fascinating field! Each of the subjects that you have learned about in this course can be studied at a *much greater* depth. And to top it all off, there are still vast areas of chemistry that we don't understand very well! The science of chemistry is a vast field with unlimited potential. I hope that this course has given you an appreciation of that!

GLOSSARY

The numbers in parentheses refer to the page number on which the definition is presented in the text.

Alkene - A hydrocarbon with at least one double bond (380)

Alkyne - A hydrocarbon with at least one triple bond (381)

Binding energy - The energy formed from the mass deficit of a nucleus (426)

Boiling point - The temperature at which a liquid's vapor pressure equals the current atmospheric pressure (112)

Buffer solution - A solution made up of a weak acid and its conjugate base. These solutions are resistant to changes in pH (230)

Common ion effect - A change in a solute's solubility when put in the presence of a common ion (173)

Critical mass - The amount of fissioning nucleus necessary for the chain reaction to be self-sustaining (447)

Degenerate orbitals - Orbitals that contain electrons of the same energy (59)

Denature - To poison a product so that it is not consumable (398)

Disaccharides - Carbohydrates that are made up of two monosaccharides (417)

Elementary Reaction - A reaction that contains only one step: a direct collision between all reactants in the reaction (354)

Faraday's Law of Electrolysis - The number of moles of products in an electrolytic cell is directly proportional to the current supplied and the time over which it is supplied. (291)

Half-life - The time it takes for the concentration of a reactant to be cut in half (346)

Hess's Law - Energy is a state function and is therefore independent of path. (7)

Hund's rule - For a given value of "ℓ," all orbitals must be singly filled with equivalent values of "s" (usually $+\frac{1}{2}$) before the orbitals are filled with a second electron. (55)

Hybrid orbital - An electron orbital that forms when the atomic orbitals of an atom mix together to form a new kind of orbital (68)

Ideal solution - A solution formed with no accompanying energy change (134)

Lattice energy - The energy required to separate the ions in an ionic solid (120)

Molecular orbital - The energy and region in space in which an electron can be found orbiting the nuclei that share it (68)

Monosaccharides - Simple carbohydrates that contain three to ten carbon atoms (416)

Normal boiling point - The temperature at which a liquid's vapor pressure equals normal atmospheric pressure (760 torr) (112)

Nuclear fission - The process by which a large nucleus is split into two smaller nuclei (445)

Nuclear fusion - The process by which two or more small nuclei fuse to make a bigger nucleus (445)

Nucleon - A term used to refer to both protons and neutrons (425)

Organic molecule - A molecule containing carbon atoms and no metals (369)

Oxidation number - The charge that an atom in a molecule would develop if the most electronegative atoms in the molecule took the shared electrons from the less electronegative atoms (267)

Pauli exclusion principle - No two electrons in the same atom can have the same 4 quantum numbers. (54)

Percent yield - The actual amount of product you make in a chemical reaction, divided by the amount stoichiometry indicates you should have made, times 100 (19)

Pi-bond - A bond in which the electron density is not concentrated along the internuclear axis (75)

Polysaccharides - Carbohydrates that are made up of more than two monosaccharides (417)

Sigma-bond - A bond in which the electron density is concentrated along the internuclear axis (75)

Structural isomers - Two or more molecules with the same chemical formula but different Lewis structures (372)

Sublimation - The process by which a solid turns into a gas without passing through a liquid phase (101)

Triple Point - The temperature and pressure at which all three phases of matter exist together in a given substance (113)

APPENDIX

Difference in Energy Levels for the Bohr Model of Hydrogen

$$\Delta E = (2.18 \times 10^{-18}\,J) \cdot Z^2 \cdot \left[\left(\frac{1}{n_{final}} \right)^2 - \left(\frac{1}{n_{initial}} \right)^2 \right]$$

Gibbs Free Energy

$$\Delta G = \Delta H - T \cdot \Delta S$$

Equilibrium Constant

For the equilibrium: $aA + bB \rightleftarrows cC + dD$

$$K_{eq} = \frac{[C]^c [D]^d}{[A]^a [B]^b}$$

Gibbs Free Energy and the Equilibrium Constant

$$\Delta G = -R \cdot T \cdot \ln(K)$$

The pH Scale

$$pH = -\log([H_3O^+])$$

Henderson-Hasselbalch equation (for buffer solutions)

$$pH = pKa + \log\left(\frac{[A^-]}{[HA]}\right)$$

where $pK_a = -\log(K_a)$

Nernst Equation

$$E_{cell} = E^o_{cell} - \frac{0.05916}{n} \cdot \log(Q)$$

Gibbs Free Energy and Electrochemical Potential

$$\Delta G^o = -n \cdot F \cdot E^o$$

$$\text{where } F = 96{,}485 \frac{J}{V \cdot mole}$$

Reaction Rate

$$R = \frac{\Delta[\text{product}]}{\Delta t} = -\frac{\Delta[\text{reactant}]}{\Delta t}$$

Rate Equations

For the reaction: $aA + bB \rightarrow cC + dD$

The rate can be expressed as: $R = k \cdot [A]^x \cdot [B]^y$

For first-order reactions: $[A] = [A]_o \cdot e^{-kt}$ and $t_{1/2} = \frac{0.693}{k}$

For second-order reactions: $\frac{1}{[A]} - \frac{1}{[A]_o} = k \cdot t \quad t_{1/2} = \frac{1}{k \cdot [A]_o}$

The Visible Spectrum (Wavelengths are in nanometers)

Red	**O**range	**Y**ellow	**G**reen	**B**lue	**I**ndigo	**V**iolet
λ=700-655	λ=655-615	λ=615-570	λ=570-505	λ=505-460	λ=460-420	λ=420-390

Quantum Numbers

Principal Quantum Number: $n = 1, 2, 3, 4...$
Azimuthal Quantum Number: $\ell = 0, 1, ... (n-1)$
Magnetic Quantum Number: m ranges from $-\ell$ to $+\ell$ in integer steps
Spin Quantum Number: $+\frac{1}{2}$ or $-\frac{1}{2}$

Hybridizations and Molecular Geometries

Orbital Hybridization	Base Geometry (no non-bonding electron pairs)	Geometry with one non-bonding electron pair	Geometry with two non-bonding electron pairs	Geometry with three non-bonding electron pairs
sp	linear	linear	**not possible**	**not possible**
sp^2	trigonal	bent	linear	**not possible**
sp^3	tetrahedral	pyramidal	bent	linear
sp^3d	trigonal bipyramidal	see-saw	T-shaped	linear
sp^3d^2	octahedral	square pyramid	square plane	**not considered**

Important Unit Cell Configurations in Creation

Cubic
All three axes are of equal length and are 90° from each other.

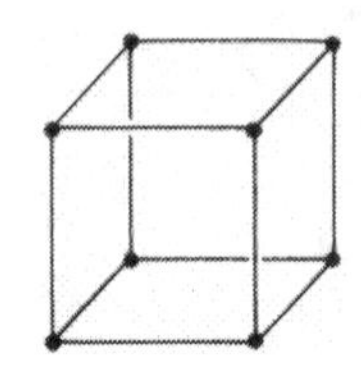

Tetragonal
One axis is different in length from the others, but all are 90° from each other.

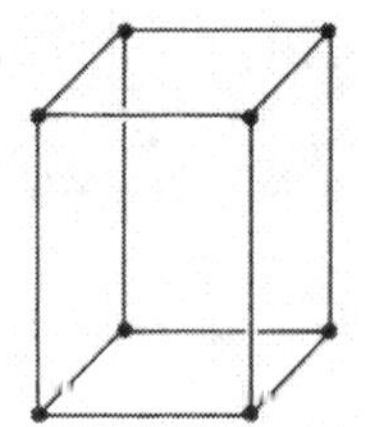

Orthorhombic
All three axes are of different length and are 90° from each other.

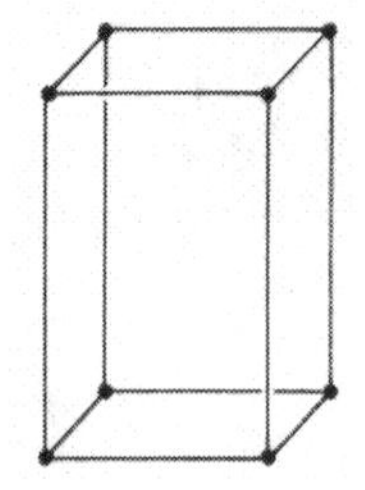

Hexagonal
One axis is different in length from the other. The two that are equal in length are 60^0 from each other (making 6 sides rather than 4), and the other is 90^0 from those two.

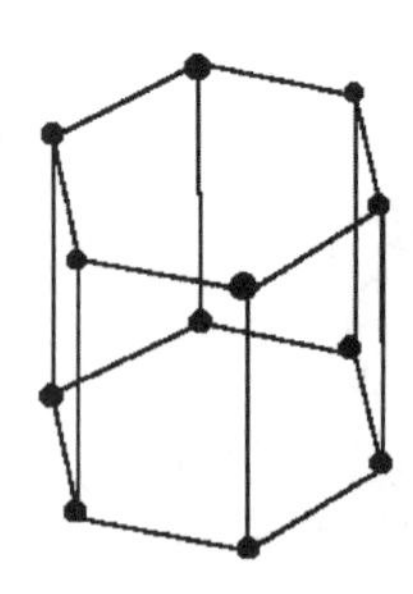

Monoclinic
All three axes are of different length. Two of them are 90° from each other but the third is at an angle other than 90^0

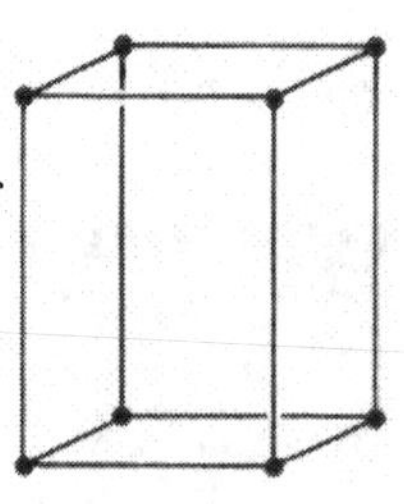

Triclinic
All three axes are of different length. All of them are at angles other than 90^0

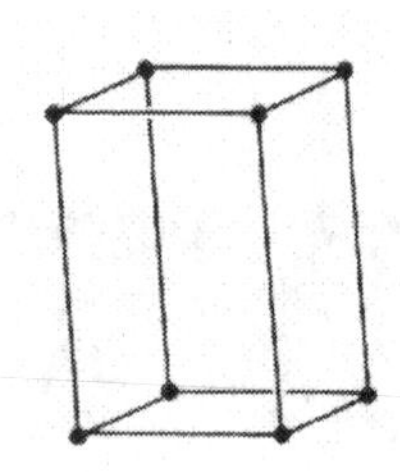

The Major Cubic Unit Cells

UNIT CELL:

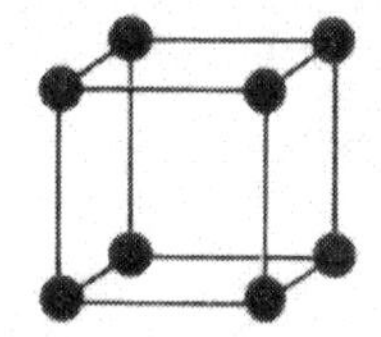

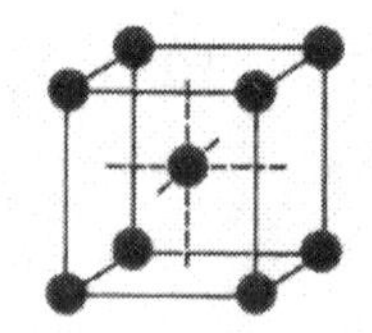

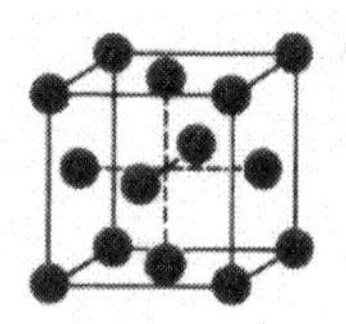

REALISTIC VERSION:

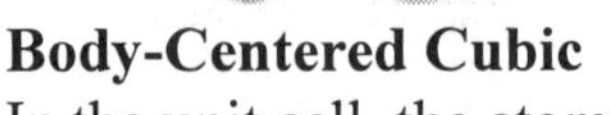

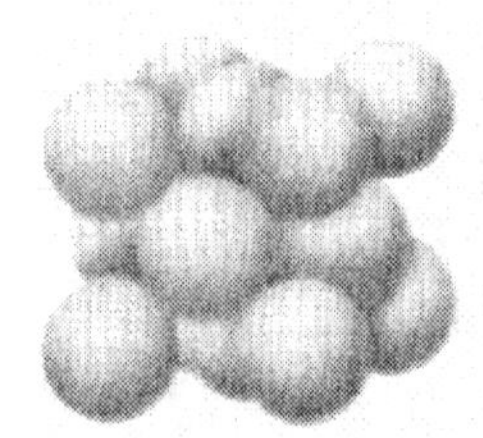

Simple Cubic
In the unit cell, the atoms are arranged on the corners of the cube.

Body-Centered Cubic
In the unit cell, the atoms are arranged on the corners of the cube, and there is one in the very center as well.

Face-Centered Cubic
In the unit cell, the atoms are arranged on the corners of the cube, and there is one at the center of each face of the cube.

The First Law of Thermodynamics

Energy cannot be created or destroyed. It can only change forms.

$$\Delta E_{system} + \Delta E_{surroundings} = 0$$

The Second Law of Thermodynamics

The entropy of the universe always increases or stays the same. It never decreases

$$\Delta S_{systsem} + \Delta S_{surroundings} \geq 0$$

Rules of Solubility in Water

1. Most ionic solids made with Group 1A metals are soluble.
2. All ionic solids made with the ammonium ion are soluble.
3. Most nitrates and acetates are soluble. (Exceptions: $AgC_2H_3O_2$, $HgC_2H_3O_2$, $Cr_2C_2H_3O_2$)
4. Most chlorides are soluble. (Exceptions: HgCl, AgCl, $PbCl_2$)
5. Most sulfates are soluble. (Exceptions: $SrSO_4$, $BaSO_4$, and $PbSO_4$)
6. Most hydroxides (except those made with Group 1A metals and the ammonium ion) are mostly insoluble.
7. Most carbonates, phosphates, and sulfides (except those made with Group 1A metals and the ammonium ion) are mostly insoluble.

Strong Acids

Formula	Name		Formula	Name
HCl	hydrochloric acid		HNO_3	nitric acid
HBr	hydrobromic acid		$HClO_4$	perchloric acid
HI	hydrogen iodide		$H_2SO_4^*$	sulfuric acid

Oxidation Number Rules

1. When a substance has only one type of atom in it (F_2, O_3, or Mg^{2+} for example) the oxidation number for that atom is equal to the charge of the substance divided by the number of atoms present.

2. Group 1A metals (Na, K, Rb, Cs, and Fr) always have oxidation numbers of +1 in molecules that contain more than one type of atom.

3. Group 2A metals (Be, Mg, Ca, Sr, Ba and Ra) always have oxidation numbers of +2 in molecules that contain more than one type of atom.

4. Fluorine always has a -1 oxidation number in molecules that contain more than one type of atom.

5. When it groups with just one other atom that happens to be a metal, H has an oxidation number of -1. In *all other cases* in which it is grouped with other atoms, H has an oxidation number of +1.

6. Oxygen has an oxidation number of -2 in molecules that contain more than one type of atom. H_2O_2 is an exception. Oxygen has an oxidation number of -1 in H_2O_2.

7. If all else fails, assume that the atom's oxidation number is the same as what it would take on in an ionic compound. The atoms that are most likely to follow this rule are in groups 3A, 6A, and 7A.

Prefixes Based for Organic Molecules

Number of Carbon Atoms	Prefix Used	Number of Carbon Atoms	Prefix Used
1	meth	6	hex
2	eth	7	hept
3	prop	8	oct
4	but	9	non
5	pent	10	dec

Standard Reduction Potentials

Half-Reaction	E° (Volts)	Half-Reaction	E° (Volts)
$K^+ + e^- \rightarrow K$	-2.925	$Sn^{4+} + 4e^- \rightarrow Sn$	0.15
$Ba^{2+} + 2e^- \rightarrow Ba$	-2.90	$AgCl + e^- \rightarrow Ag + Cl^-$	0.222
$Ca^{2+} + 2e^- \rightarrow Ca$	-2.87	$Hg_2Cl_2 + 2e^- \rightarrow 2Hg + 2Cl^-$	0.27
$Na^+ + e^- \rightarrow Na$	-2.714	$Cu^{2+} + 2e^- \rightarrow Cu$	0.337
$Mg^{2+} + 2e^- \rightarrow Mg$	-2.37	$NiO_2+2H_2O+2e^- \rightarrow Ni(OH)_2 + 2OH^-$	0.49
$Al^{3+} + 3e^- \rightarrow Al$	-1.66	$I_2 + 2e^- \rightarrow 2I^-$	0.5355
$Zn(OH)_2 + 2e^- \rightarrow Zn + 2OH^-$	-1.245	$MnO_4^-+2H_2O+3e^- \rightarrow MnO_2 + 4OH^-$	0.588
$Mn^{2+} + 2e^- \rightarrow Mn$	-1.18	$Fe^{3+} + 3e^- \rightarrow Fe$	0.771
$Fe(OH)_2 + 2e^- \rightarrow Fe + 2OH^-$	-0.877	$Hg_2^{2+} + 2e^- \rightarrow 2Hg$	0.789
$Zn^{2+} + 2e^- \rightarrow Zn$	-0.763	$Ag^+ + e^- \rightarrow Ag$	0.7991
$Cr^{3+} + 3e^- \rightarrow Cr$	-0.74	$Br_2 (l) + 2e^- \rightarrow 2Br^- (aq)$	1.0652
$Fe^{2+} + 2e^- \rightarrow Fe$	-0.440	$Pt^{2+} + 2e^- \rightarrow Pt$	1.20
$Cd^{2+} + 2e^- \rightarrow Cd$	-0.430	$O_2 + 4H_3O^+ + 4e^- \rightarrow 6H_2O$	1.23
$PbSO_4 + 2e^- \rightarrow Pb + SO_4^{2-}$	-0.356	$Cl_2 + 2e^- \rightarrow 2Cl^-$	1.3595
$Co^{2+} + 2e^- \rightarrow Co$	-0.277	$Au^{3+} + 3e^- \rightarrow Au$	1.50
$Ni^{2+} + 2e^- \rightarrow Ni$	-0.250	$MnO_4^-+8H_3O^++5e^- \rightarrow Mn^{2+}+12H_2O$	1.51
$Sn^{2+} + 2e^- \rightarrow Sn$	-0.136	$PbO_2+SO_4^{2-}+4H_3O^++2e^- \rightarrow PbSO_4+6H_2O$	1.685
$Pb^{2+} + 2e^- \rightarrow Pb$	-0.126	$F_2 + 2e^- \rightarrow 2F^-$	2.87
$2H_3O^+ + 2e^- \rightarrow H_2 + 2H_2O$	**0.000**		

Common Alkyl Groups

Alkyl Group Name	Alkyl Group Structure
methyl	CH_3—
ethyl	CH_3CH_2—
n-propyl	$CH_3CH_2CH_2$—
isopropyl	| CH_3CHCH_3
n-butyl	$CH_3CH_2CH_2CH_2$—
sec-butyl	| $CH_3CHCH_2CH_3$
isobutyl	CH_3CHCH_2— | CH_3
t-butyl	| CH_3CCH_3 | CH_3

THE PERIODIC CHART OF ELEMENTS

1A	2A	1B	2B	3B	4B	5B	6B	7B	8B	9B	10B	3A	4A	5A	6A	7A	8A
1 **H** 1.01																	2 **He** 4.0
3 **Li** 6.94	4 **Be** 9.00											5 **B** 10.8	6 **C** 12.0	7 **N** 14.0	8 **O** 16.0	9 **F** 19.0	10 **Ne** 20.2
11 **Na** 22.0	12 **Mg** 24.3											13 **Al** 27.0	14 **Si** 28.1	15 **P** 31.0	16 **S** 32.1	17 **Cl** 35.5	18 **Ar** 39.9
19 **K** 39.1	20 **Ca** 40.1	21 **Sc** 45.0	22 **Ti** 47.9	23 **V** 50.9	24 **Cr** 52.0	25 **Mn** 54.9	26 **Fe** 55.8	27 **Co** 58.9	28 **Ni** 58.7	29 **Cu** 63.5	30 **Zn** 65.4	31 **Ga** 69.7	32 **Ge** 72.6	33 **As** 74.9	34 **Se** 79.0	35 **Br** 79.9	36 **Kr** 83.8
37 **Rb** 85.5	38 **Sr** 87.6	39 **Y** 88.9	40 **Zr** 91.2	41 **Nb** 92.9	42 **Mo** 95.9	43 **Tc** 97.0	44 **Ru** 101.1	45 **Rh** 102.9	46 **Pd** 106.4	47 **Ag** 107.9	48 **Cd** 112.4	49 **In** 114.8	50 **Sn** 118.7	51 **Sb** 121.8	52 **Te** 127.6	53 **I** 126.9	54 **Xe** 131.3
55 **Cs** 132.9	56 **Ba** 137.3	57 **La** 138.9	72 **Hf** 178.5	73 **Ta** 180.9	74 **W** 183.9	75 **Re** 186.2	76 **Os** 190.2	77 **Ir** 192.2	78 **Pt** 195.1	79 **Au** 197.0	80 **Hg** 200.6	81 **Tl** 204.4	82 **Pb** 207.2	83 **Bi** 209.0	84 **Po** (209)	85 **At** (210)	86 **Rn** (222)
87 **Fr** 223.0	88 **Ra** 226.0	89 **Ac** 227.0	104 (261)	105 (262)	106 (263)	107 (264)	108 (265)	109 (266)									

58 **Ce** 140.1	59 **Pr** 140.9	60 **Nd** 144.2	61 **Pm** 145.0	62 **Sm** 150.4	63 **Eu** 152.0	64 **Gd** 157.3	65 **Tb** 158.9	66 **Dy** 162.5	67 **Ho** 164.9	68 **Er** 167.3	69 **Tm** 168.9	70 **Yb** 173.0	71 **Lu** 175.0
90 **Th** 232.0	91 **Pa** 234.0	92 **U** 238.0	93 **Np** 237.0	94 **Pu** (244)	95 **Am** (243)	96 **Cm** (247)	97 **Bk** (247)	98 **Cf** (251)	99 **Es** (252)	100 **Fm** (257)	101 **Md** (258)	102 **No** (259)	103 **Lr** (260)

INDEX

—A—

—B—

—C—

—D—

—E—